Europe After an American Withdrawal
Economic and Military Issues

sipri
Stockholm International Peace Research Institute

SIPRI is an independent international institute for research into problems of peace and conflict, especially those of arms control and disarmament. It was established in 1966 to commemorate Sweden's 150 years of unbroken peace.

The Institute is financed mainly by the Swedish Parliament. The staff, the Governing Board and the Scientific Council are international.

The Governing Board and the Scientific Council are not responsible for the views expressed in the publications of the Institute.

Governing Board

Ambassador Dr Inga Thorsson, Chairman (Sweden)
Professor Egon Bahr (Federal Republic of Germany)
Professor Francesco Calogero (Italy)
Dr Max Jakobson (Finland)
Professor Dr Karlheinz Lohs (German Democratic Republic)
Professor Emma Rothschild (United Kingdom)
Sir Brian Urquhart (United Kingdom)
The Director

Director

Dr Walther Stützle (Federal Republic of Germany)

sipri
Stockholm International Peace Research Institute
Pipers väg 28, S-171 73 Solna, Sweden
Cable: SIPRI STOCKHOLM
Telephone: 46 8/55 97 00   Telefax: 46 8/55 97 33

Europe After an American Withdrawal
Economic and Military Issues

Edited by
Jane M. O. Sharp

sipri
Stockholm International Peace Research Institute

OXFORD UNIVERSITY PRESS
1990

Oxford University Press, Walton Street, Oxford OX2 6DP
Oxford New York Toronto
Delhi Bombay Calcutta Madras Karachi
Petaling Jaya Singapore Hong Kong Tokyo
Nairobi Dar es Salaam Cape Town
Melbourne Auckland
and associated companies in
Berlin Ibadan

Oxford is a trade mark of Oxford University Press

Published in the United States
by Oxford University Press, New York

© SIPRI 1990

All rights reserved. No part of this publication may be reproduced, stored in a retrieval system, or transmitted, in any form or by any means, electronic, mechanical, photocopying, recording, or otherwise, without the prior permission of Oxford University Press

British Library Cataloguing in Publication Data
Data available
ISBN 0–19–827836–5

Library of Congress Cataloging in Publication Data
Data available
ISBN 0–19–827836–5

Typeset and originated by Stockholm International Peace Research Institute
Printed and bound in
Great Britain by Biddles Ltd.,
Guildford and King's Lynn

Contents

Foreword	xiv
Preface	xvi
Acronyms, abbreviations and conventions	xx

Part I. Introduction and summary

1. Why discuss US withdrawal?
Jane M. O. Sharp

		3
I.	Introduction	3
II.	The US commitment to the defence of Europe	4
III.	Incentives and disincentives for West European autonomy	12
IV.	Perceptions of a reduced WTO threat	18
V.	What role for the USA in Europe after the cold war?	20
VI.	Summary	27
	Notes and references	28
Figure 1.1.	US military forces in NATO Europe, 1945–90	6

2. Summary and conclusions
Jane M. O. Sharp

		34
I.	Economic issues	34
II.	Assessing the military mission gaps	42
III.	Conclusion	48
	Notes and references	50

Part II: Economic implications of a US military withdrawal from Europe

3. US perspectives on the economic costs and benefits of a withdrawal of US troops and facilities from Europe
Alice C. Maroni

		53
I.	Introduction	53
II.	Economic implications of withdrawal of US European-deployed troops	56
III.	Concluding observations	70
	Notes and references	71
Table 3.1.	Incremental cost of operating US Army and Air Force units in Europe, as of 1986	58

Table 3.2.	Estimate of costs/savings of withdrawing all US Army and Air Force units from Europe	61
Table 3.3.	Direct cost of forces deployed in Europe, FY 1989	64
Table 3.4.	Direct US defence expenditures abroad for foreign goods and services in NATO Europe, 1987	65
Table 3.5.	Percentage cost of US European-deployed forces, FYs 1974–82	66
Table 3.6.	US–European NATO defence trade balance summary, 1987	67
Table 3.7.	US–European NATO balance of trade summary, 1984–86	68

4. The impact of a withdrawal on US investments in Europe 74
Joseph R. Higdon and Paul J. Friedrich

	I. Introduction	74
	II. US investment in Europe	75
	III. Gorbachev's challenge	76
	IV. 'Europe 1992' demands US business presence	77
	V. Why US troops in Europe?	78
Table 4.1.	Majority-owned European affiliates with greater than 50 per cent ownership by a US parent, 1985–86	75
Table 4.2.	European affiliates with greater than 10 per cent ownership by a US parent, 1985–86	76
Table 4.3.	Majority-owned European affiliates with greater than 50 per cent ownership by a US parent, 1986	77

5. The economics of US bases and facilities in Western Europe 79
David Greenwood

	I. Introduction	79
	II. The US presence in Europe	81
	III. The economic impact of European-deployed forces	89
	IV. The economics of withdrawal	91
	V. The economics of compensation	93
	Notes and references	95
Table 5.1.	US front-line units in Europe, 1987–88	82
Table 5.2.	US active duty military personnel in Europe, 31 December 1987	84
Table 5.3.	US bases and facilities in Europe: population statistics 1987–88 (estimates)	85
Table 5.4.	US bases and facilities in Europe: financial statistics, 1987-88 (estimates)	87
Table 5.5.	DOD direct defence expenditures abroad for goods and services, 1983–87	88

6. The economic impact of the stationing of US forces in the Federal Republic of Germany 97
Hartmut Bebermeyer and Christian Thimann

	I. Introduction	97
	II. The US force presence as an economic factor	99

III.	Costs to the Federal Republic of Germany	107
IV.	Benefits to the West German economy	109
V.	Cost–benefit analysis	112
VI.	Economic implications of a withdrawal	114
	Notes and references	117
Table 6.1.	US active forces in the Federal Republic of Germany, 1986–87	98
Table 6.2.	US facilities in the Federal Republic of Germany, 1974	99
Table 6.3.	Army and Air Force Exchange Service (AAFES) direct sales in the FRG, FY 1987	102
Table 6.4.	Army and Air Force Exchange Service (AAFES) payments to West German suppliers, 1988	103
Table 6.5.	Average 1987 income of USAREUR military personnel, by rank	104
Table 6.6.	Average 1987 income of USAFE military personnel, by rank	105
Table 6.7.	Per capita incomes and D-mark expenditures as shares of total income of US military and civil personnel in the FRG, 1986–88	106
Table 6.8.	Cost–benefit balance of the US force presence in the Federal Republic of Germany, 1986	113

7. Economic consequences of a withdrawal of US forces from the Benelux states 119
Sami Faltas

I.	The political setting	119
II.	The US military presence	122
III.	The impact of withdrawal	122
IV.	Conclusion	130
	Notes and References	130
Table 7.1.	US defence personnel in the Benelux countries (incl. NATO staff) before (1988) and after withdrawal	124
Table 7.2.	Personal incomes of US defence personnel in the Benelux countries, 1988	127
Table 7.3.	Economic costs and benefits to the Benelux countries of a withdrawal of US personnel and facilities (options A and B)	128

8. Costs and benefits to the United Kingdom of the US military presence 132
Keith Hartley and Nicholas Hooper

I.	Introduction	132
II.	US forces and British defence	132
III.	Assumptions	137
IV.	Evidence	140
V.	The economic impact of British military expenditure	144
VI.	European co-operation	148
	Notes and references	152
Table 8.1.	United Kingdom defence expenditures and manpower, 1987	134
Table 8.2.	Major US military installations in the United Kingdom, 1987–88	135
Table 8.3.	Costs and benefits to the UK of the US force presence, 1987–88	136

Table 8.4.	UK local (base area) and regional unemployment rates, December 1987	142
Table 8.5.	Defence and the British economy	145
Table 8.6.	The economic impact of British defence expenditure	147
Table 8.7.	Immediate effects of a US withdrawal	151

9. The economic impact of a US military withdrawal from NATO's Northern Flank   154
Simon Duke

	I.	Introduction	154
	II.	The base agreements	156
	III.	The economic impact of a US withdrawal	165
	IV.	Summary	174
		Notes and references	175
Table 9.1.		US and Danish areas of financial responsibility for bilateral and joint NATO defence arrangements in Iceland	158
Table 9.2.		US and Icelandic areas of financial responsibility for defence arrangements in Iceland	161
Table 9.3.		US and Norwegian areas of financial responsibility for bilateral and joint defence arrangements in Norway	164
Table 9.4.		Personnel at US military installations in Greenland, 1986	168
Table 9.5.		IDF personnel and dependants, 1984–87	169
Table 9.6.		NATO and US cost shares for constructions at Keflavík AB, 1984	170

10. Economic aspects of the US military presence in Spain   179
José Molero

	I.	Introduction	179
	II.	Historical evolution of the military agreements and treaties signed by Spain and the United States, 1953–82	180
	III.	The 1982 Agreement	187
	IV	Approach to an economic evaluation of the US military bases in Spain	191
	V.	Conclusion: towards a new phase of US–Spanish military relations	198
		Notes and references	201
Table 10.1.		US economic aid to Spain based on 1953 pacts, FYs 1953–58	182
Table 10.2.		US aid to Spain, 1953-63	183
Table 10.3.		Significance of US aid to Spanish foreign trade, 1953–58	184
Table 10.4.		US military and economic aid to Spain, 1970–74	187
Table 10.5.		US loans and grants obligations and loan authorizations to Spain, FY 1946–85	190
Table 10.6.		Number of personnel at US bases in Spain, 1987	192
Table 10.7.		US running costs for installations in Spain, 1985–86	192
Table 10.8.		Committed Spanish Ministry of Defence purchases of US military equipment, 1983-87.	193

Table 10.9.	Main Spanish arms procurement programmes involving US compensation, 31 December 1986	194
Table 10.10.	US compensation to Spanish industry, 31 December 1986	195
Table 10.11.	Distribution of US compensation to Spanish industry by sector, 31 December 1986	196
Table 10.12.	Spanish arms industry technology transfer contracts with foreign firms, 1974–84	197

11. Costs and benefits of the US military presence in Portugal   204
Jane M. O. Sharp

I.	Portugal's security needs	204
II.	US interests in Portugal	204
III.	Portuguese commitments to NATO	209
IV.	Modernizing the defence-industrial base	211
V.	Conclusion	214
	Notes and references	215
Table 11.1.	US public aid to Portugal, 1975–1986	207
Table 11.2.	Activities of Fundação Luso-Americano para o Desenvolvimento (FLAD), 1985-86	208

12. Costs and benefits of the US military bases in Greece   218
Athanassios G. Platias

I.	Introduction	218
II.	Political aspects of Greek–US relations	219
III.	Economic aspects of Greek–US relations	226
IV.	Implications for Greece of a US withdrawal	238
	Notes and references	239
Table 12.1.	Military construction appropriations for US facilities in Greece,	226
Table 12.2.	Hellenikon AB expenditure in the economy of Athens	227
Table 12.3.	Value of real estate utilized by major US bases in Greece	227
Table 12.4.	Total NATO defence spending by country as percentage of GDP, 1985–86	228
Table 12.5.	Greece's defence expenditure as percentage of GDP, 1975–85	229
Table 12.6.	Total NATO active duty military and civilian manpower as percentage of total population, by country, 1985–86	230
Table 12.7.	NATO military manpower as percentage of population, by country, 1985–86	231
Table 12.8.	Total military and civilian defence personnel of the NATO countries as percentage of the labour force, by country, 1980–85	232
Table 12.9.	Changes in the distribution of countries supplying arms to Greece, 1965–85	233
Table 12.10.	US aid (grants and loans) to Greece, 1974–88	236
Table 12.11.	Percentage of new FMS loans indirectly devoted to the repayment of old FMS loans, 1981–85	237
Table 12.12.	Greek obligations for FMS repayment, 1987–96	238

Table 12.13.	Impact for Greece of a US military withdrawal	239
Figure 12.1.	Planned US military assistance to Greece under the 1988 FMS Foreign Assistance Program	234

13. The economic costs and benefits of US–Turkish military relations 243
Saadet Deger

I.	Introduction and background	243
II.	Three questions about Turkish security	244
III.	The tangled web: Turkish–US relations	254
IV.	Turkey with and without the US forces: economic costs and benefits	259
V.	Implications of a US withdrawal	266
VI.	Concluding remarks	271
	Notes and references	272
Table 13.1.	Land area and exposed borders of European NATO allies	246
Table 13.2.	Soviet economic aid to less developed countries (LDCs), 1954–79	247
Table 13.3.	Estimated defence burden of a Turkish professional armed force, 1985–87	250
Table 13.4.	Growth rate of real military expenditure for the United States, European NATO and Turkey, 1980–87	251
Table 13.5.	NATO and WTO force comparison, 1987	252
Table 13.6.	Turkey's debt repayment schedule under US Foreign Military Sales, FYs 1985–90	258
Table 13.7.	Number of personnel at US military installations in Turkey, 1987	260
Table 13.8.	US gross annual expenditure for outlays associated with defence arrangements with Turkey, 1986–87 average	261
Table 13.9.	Annual financial costs and benefits to Turkey relating to the US military presence, 1986–87 average	263
Table 13.10.	Annual financial benefit to the United States of payments on military sales and loans to Turkey, 1986–87 average	264
Table 13.11.	Incremental or additional running costs per year to Turkey of maintaining bases after a US withdrawal (option A)	265
Table 13.12.	Cost comparison of a US withdrawal under options A and B	267

14. Economic costs and benefits of US military bases and facilities in Italy 274
Jennifer Sims

I.	Introduction	274
II.	The US military presence: status and economic impact	275
III.	Implications of a US withdrawal for Italy's defence costs	283
IV.	Conclusion	296
	Notes and references	297
Table 14.1.	USA–Italy defence balance of trade summary, 30 September 1986	280
Table 14.2.	Value of ROICC-controlled construction in Italy, 1982–86	281
Table 14.3.	NATO and WTO land forces in north-eastern Italy, 1988	283

Table 14.4.	NATO and WTO air forces in the southern region, 1988	285
Table 14.5.	The Italian defence budget, 1982–84	291
Table 14.6.	The Italian defence budget as percentage of GDP and total state budget, 1975–84	292
Table 14.7.	Italian defence expenditure per capita: selected years, 1970–86	293

Part III: Military implications of a US military withdrawal from Europe

15. US withdrawal and NATO's Northern Flank: impact and implications 303
Steven E. Miller

I.	Introduction	303
II.	The USA and the Northern Flank: bases and other arrangements	304
III.	Defending the Northern Flank without US involvement	313
IV.	The United States and the defence of the Baltic Approaches	326
V.	Conclusion	332
	Notes and references	334
Figure 15.1.	The Nordic Flank	306
Figure 15.2.	Northern Norway	315
Figure 15.3.	The Baltic region	328

16. Mission gaps in Central Europe 349
Hilmar Linnenkamp

I.	Introduction	349
II.	The structure of the analysis	350
III.	Political choices: gaps in deterrence	351
IV.	Military choices: gaps in forward defence	355
V.	Europe without US troops: the zero-based assessment	365
VI.	Postscript	366
	Notes and references	367

17. The military implications for NATO of a US withdrawal from the southern region 369
George E. Thibault

I.	Purpose and approach	369
II.	Commander's strategic estimate	370
III.	The traditional military mission gaps	384
IV.	The functional military gaps	390
V.	Conclusion	397
	Notes and references	398
Figure 17.1.	Major US military installations on the Azores	372
Figure 17.2.	Major US military installations in Spain	374

Figure 17.3. Major US military installations in Italy   376
Figure 17.4. Major US military installations in Greece   379
Figure 17.5. Major US military installations in Turkey   381

18. The contribution of US C³I to the defence of NATO   400
Paul Stares and John Pike

    I.   Introduction   400
   II.   Intelligence assets   402
  III.   Communications assets   418
  IV.   Impact of a US withdrawal   421
   V.   Conclusion   424
       Notes and references   424

Part IV. Coping with a US withdrawal

19. Could NATO cope without US forces?   431
Martin Farndale

    I.   Introduction   431
   II.   The threat   433
  III.   Meeting the threat   436
  IV.   Filling the gaps   440
   V.   Changes in NATO strategy and operational concepts   450
  VI.   Command and control   453
 VII.   Conclusion   454
       Notes and references   455

20. The United States coping without bases in Europe   456
Robert E. Harkavy

    I.   Introduction   456
   II.   Bases for strategic nuclear forces   459
  III.   Conventional out-of-area problems after a US withdrawal   473
  IV.   Horizontal escalation involving crises arising outside of Europe or the Persian Gulf   476
   V.   Ways in which the United States can compensate for loss of access to Europe: general approaches   477
  VI.   Summary   478
       Notes and references   481

Table 20.1. US satellite control and receiver sites in Europe, 1987   465
Table 20.2. US signals intelligence collection sites in Europe, by country, 1985   469

About the contributors   487

Index   *491*

Europe After American Withdrawal Advisory Committee

Dr Christoph Bertram (Federal Republic of Germany)
Professor Karl Birnbaum (Sweden)
Professor Catherine Kelleher (United States of America)
Professor Dominique Moisï (France)
John Roper (United Kingdom)
Professor Max Schmidt (German Democratic Republic)
Lt-General Peter Tandecki (Rtd) (Federal Republic of Germany)
Professor Angel Viñas (Spain)
Dr Walther Stützle (Chairman)

This book is one of two volumes that result from a three-year project at SIPRI. Both are meant to further the discussion about the future American–West-European relationship and to provide, through information and analysis, a basis for the emerging debate.

In the first volume, *United States Military Forces and Installations in Europe*, Simon Duke documents the nature and extent of the United States military presence in Europe.

This volume, edited by Jane M. O. Sharp, discusses whether Western Europe can assume its security in the absence of present US forces and whether the United States can remain a world power without being, at the same time, a military power in Europe.

Eight non-SIPRI colleagues accepted the invitation and time-consuming task of serving as members of an Advisory Committee to the project. To all of them we are greatly indebted. I want to mention in particular Professor Catherine Kelleher from the University of Maryland, who took the trouble of frequently crossing the Atlantic. The project has greatly profited from her invaluable experience, friendship and seemingly endless energy.

The project would not have been possible without the generous support from the John D. and Catherine T. MacArthur Foundation.

Dr Walther Stützle
Director, SIPRI
1 July 1990

Foreword

Europe in 1990 is a completely different place from what it was in 1987, when this SIPRI research project commenced. What then was a purely hypothetical assumption—what if the US forces were to withdraw from Europe—has meanwhile developed into a policy, agreed and accepted on either side of the Atlantic, aiming at a considerable reduction of the US military presence in Europe.

The Warsaw Treaty Organization (WTO) has changed from an alliance of Communist Party politburos into a grouping of sovereign states, most of which are now striving for a democratic parliamentary system. Soviet troops stationed in Central and Eastern Europe have started to withdraw from Hungary and Czecho-Slovakia, and Moscow has agreed to reduce its troops in both Poland and the German Democratic Republic to 195 000, in return for the reduction of US forces to 225 000 in the Atlantic-to-the-Urals zone.

More importantly the Soviet Union, together with the United States, the United Kingdom and France, has accepted German unification. As the four victorious powers, they have accepted to relinquish their jointly held original rights in and responsibility for Germany as a whole as well as for Berlin. Consequently a completely new German state will be formed—one which never before existed in European history, size-wise and with the proven democratic record of the larger part with which the smaller part will be united. Europe is challenged to manage a political process for which neither the Germans nor their neighbours have any experience to resort to.

The post-war presence of the USA in Europe is a long and complicated story, rich in achievements and disappointments alike. US forces have successfully supported the political objective of reconstructing Western Europe and of containing Soviet military preponderance. However, as the danger of a military threat became less imminent, clichés crept into the mutual perception on both sides of the Atlantic, such as 'the Europeans are soft on defence' and 'the United States should leave Europe to the Europeans'. However, in whatever structure the future Atlantic relationship is going to take shape, two features stand out from the past more than anything else: if the United States is to continue to exercise global responsibility, no other alliance can be more cost-effective than the one with Europe; and as long as Western Europe chooses to seek ultimate nuclear protection from one superpower, the cost-effectiveness of the Alliance with Washington remains second to none. It is unlikely that anyone could have designed a more productive and mutually beneficial alliance if asked to do so way back in 1949. As history moves on, the lesson to be drawn from this experience is this: new challenges require new political answers. Continuity requires adjustment, if it is not to turn into stalemate and paralysis.

It is in this light and against this background that the central question for a newly emerging security structure in Europe has to be seen and dealt with, that is, how to combine the future security status of the united Germany with the process of establishing a co-operative security system in which the United States is explicitly asked by the Soviet Union and the West and East Europeans to continue active participation, bolstered by military presence in Europe, and in which the Soviet Union and its former allies play a co-operative role. It is in the logic of this profound change that the Soviet Union, under President Gorbachev's leadership, has removed all the principal obstacles that have blocked the Conventional Armed Forces in Europe Negotiation and has principally agreed to search for a European security structure, based on identical interpretations of a defensive doctrine and strategy.

Not the least against the background of an increasingly difficult domestic situation in the Soviet Union, the East and West are confronted with a tall and complex agenda, unparalleled in European post-war history. This suggests that we should not expect an accident-free development, that we should sharpen our knowledge about the situation in Europe and that we should make the reduction of American forces in Europe part of a concerted development, rather than a result of unilateral budget considerations in Congress. Now that the 'what if' scenario has become reality, domestic considerations should not become the overriding determinant. It is my hope that this study will contribute to a discussion that seeks to strike the right balance between security policy needs and domestic requirements.

Dr Walther Stützle
Director, SIPRI
1 July 1990

Preface

This book stems from a multi-year research project at SIPRI designed to assess the impact of a hypothetical US withdrawal from Europe. It addresses two questions: Could Western Europe defend itself without US forces, and could the United States remain a global power without its European bases?

As the project began, some critics feared that such a study could become a self-fulfilling prophecy. Others feared the opposite, namely that the project was to maintain the status quo by raising the straw man of withdrawal only to knock it down. The purpose, however, is analysis rather than advocacy. The objective is not to provide a blueprint to fill all the gaps left by a withdrawal, but is rather to assess in detail the economic and military implications of such a withdrawal and thereby provide data and analysis to help policy makers on both sides of the Atlantic, and in both halves of Europe, to respond to a reduced US commitment (either total or partial withdrawal) should that occur in the short term, and to explore alternative security systems that should be necessary in the long term.

States join alliances to pool their military capabilities and financial resources in order to meet a common threat.[1] Each thereby saves resources and gains protection compared to the situation it would face alone. As this book demonstrates, however, national interests rather than altruism determine one state's commitment to another. The US presence in Europe is designed not only to reassure the allies of US support, but also to elicit reciprocal commitments. Thus the United States expects its European allies to provide bases for US forces and facilities, to share the risks associated with the basing of nuclear weapons and to increase their defence budgets by a certain percentage each year.[2] In addition the United States expects support for out-of-area activity, either through the provision of allied troops in distant conflicts, the granting of overflights or the expression of political support.

This book explodes the myth that all members of the Atlantic Alliance enjoy equal status. Manifestly they do not. Each bilateral transatlantic bargain embraces a unique set of rights and obligations since US interests as well as allied needs and capabilities vary widely, not only from region to region but also across time.

Europeans should not need to be reminded that alliance commitments are dynamic rather than static. A particular set of military, political and economic conditions gave birth to the alliance in the late 1940s. President Harry Truman's post-World War II commitment of US troops to Europe stemmed not from the North Atlantic Treaty itself but from the sense of crisis surrounding the onset of

[1] For a discussion, see Walt, S. M., *The Origins of Alliances* (Cornell University Press: Ithaca, N.Y., 1987).
[2] From May 1977 to May 1990 this was set at 3 per cent.

the Korean War and the fear that the Soviet Union might move into Western Europe. It came at a time when the West European democracies needed both economic support and military protection, and the United States could provide both. The commitment of extra troops in the early 1950s was also necessary to overcome French objections to the rearmament of the Federal Republic of Germany.

The US military presence played an important role in creating four decades of stability in Europe, but that presence was never intended to be permanent. The United States anticipated that sooner or later Western Europe would unite politically and be prosperous enough to provide for its own defence. However, prosperity did not bring with it political unity or independence from the United States. On the contrary, as the European Community became an economic competitor to the United States, the NATO allies became more rather than less dependent on the US militarily.

Commitments made in time of crisis are harder to justify in times of stability. At the beginning of the 1990s, West Europeans were no longer as apprehensive about a Soviet military threat, but Europe was hardly stable politically. Turbulence in the Soviet republics, the fall of communist regimes in Eastern Europe and the prospect of German unification made for a confusing political situation in Europe. If fear of instability made some NATO leaders cautious about force reductions, harsh economic realities encouraged both the Soviet and the US governments to look seriously at reallocating resources from the defence to the civilian economy.

In preparing this study the authors did not believe that a total withdrawal of US forces was imminent, as the advocates of such a step were few and far between. Nevertheless, as US trade and budget deficits grew, Congress was less willing to allocate as much of the US gross national product to defence as it had in the past. President Ronald Reagan's fiscal year 1989 defence budget mandated the closure of 86 military bases in the continental US, and in January 1990 President George Bush proposed the closure of additional bases in the USA and nine facilities in Western Europe, as well as the withdrawal of 80 000 US servicemen from Europe in the context of the Conventional Armed Forces (CFE) Negotiation in Vienna. The US commitment to Western Europe was thus being assessed on both sides of the Atlantic to take account of a less threatening adversary, a less prosperous security guarantor and an increasingly uncertain security environment.

As a contribution to the transatlantic debate on the future of the US commitment to Europe, this study took a 'what if' approach to a US withdrawal. Analysts from both sides of the Atlantic were asked to assess the economic and military impact of two hypothetical withdrawal options. Option A assumes that all US forces and facilities have been withdrawn from Europe. The United States remains a member of the alliance, but the troops withdrawn are demobilized and are not earmarked for redeployment to Europe (or anywhere else) in a crisis; essentially, they are gone for good. Option B assumes only a skeleton peacetime presence in Europe, with the bulk of forces and all dependents redeployed back in the continental United States. This case presupposes a clear commitment to return to Europe in a crisis. Under both withdrawal options the Supreme Allied Commander Europe (SACEUR) would be a European rather than, as hitherto, a US officer. NATO would continue to rely on a military strategy of forward defence and flexible response.

The authors were asked not to speculate about the political conditions under which either withdrawal option might have been taken, but to focus instead on the economic and military impact of each case. How would the West Europeans cope without the US presence, and how would the United States cope without its European bases? In particular, how would the two withdrawal options affect national defence budgets and defence industries? Would withdrawal encourage or inhibit transatlantic defence co-operation and co-operation among the different West European countries?

Most of the chapters were first presented at a workshop in April 1988 and revised in the light of the workshop discussions, although economic data presented remains largely for fiscal years 1986–87. The book is in four parts. In Part I an introductory chapter sets the US commitment and the alliance burden-sharing debate in context. A second chapter summarizes the principal findings of the project. Part II addresses the economic costs and benefits of the US presence in Europe, and the likely consequences of withdrawal from the perspective of the United States and of each of the host nations in Europe. Part III identifies the military mission gaps that would be created by US withdrawal from the northern, central and southern regions of Western Europe, as well as the gaps that would arise in NATO's command, control, communication and intelligence (C^3I) capabilities if US assets were no longer available. Part IV addresses the problems that the United States would face in maintaining its position as a world power without access to European bases, and that the European NATO countries would face in seeking to defend themselves under the two hypothetical withdrawal options.

In a companion volume, Simon Duke provides a comprehensive history and analysis of the nature and extent of US forces in Europe, including the different bilateral legal agreements that govern US use of European facilities, and the fluctuations of US forces levels over time.[3]

Acknowledgements

SIPRI wishes to thank all those who contributed to this project: the author's who gamely accepted the eccentric assumptions of the study and risked the raised eyebrows of their collegues; the participants at the April 1988 workshop whose rigorous criticism helped to shape the final product; members of the advisory group who gave generously of their time and advice not only at the

---

[3] Duke, S., *United States Military Forces and Installations in Europe* (Oxford University Press: Oxford, 1989)

workshop but at a number of other meetings. John Roper, David Greenwood, Catherine Kelleher and Peter Tandecki deserve special thanks for chairing the workshop sessions. Gabrielle Bartholomew, Fatima Asp-Barreto, Miyoko Suzuki and Ricardo Vargas-Fuentes typed and retyped many of the manuscripts. Last, but by no means least, Paul Claesson of SIPRI's editorial staff polished our prose, double-checked much of our data, set the book in final camera-ready format, and contibuted many insights to this study from his own work on island basing.

Financial support for the three-year project was generously provided by the John D. and Catherine T. MacArthur Foundation.

Jane M. O. Sharp
SIPRI
1 July 1990

Acronyms, abbreviations and conventions

AAFCE	Allied Air Forces Central Europe
AAFES	Army and Air Force Exchange Service
AB	Air Base
ACE	Allied Command Europe
ACE HIGH	ACE troposcatter communications system
ACLANT	Allied Command Atlantic
AFCENT	Allied Forces Central Europe
AFNORTH	Allied Forces Northern Europe
AFSATCOM	US Air Force Satellite Communications
AFSOUTH	Allied Forces Southern Europe
AIRSOUTH	Allied Air Forces Southern Europe
AMF	Allied Mobile Force
ASAS	All-Source Analysis System
ASAT	Anti-satellite activities
ASW	Anti-submarine warfare
ATACMS	Army Tactical Missile System (US Army)
ATAF	Allied Tactical Air Force
ATTU	Atlantic-to-the-Urals
AUTODIN	Automatic Data and Information Network
AUTOSEVOCOM	Automatic Secure Voice Communications Network
AUTOVON	Automatic Voice Network
AWACS	Airborne Warning and Control System
BALTAP	Commander Allied Forces Baltic Approaches
BETA	Battlefield Exploitation and Target Acquisition (system)
BICES	Battlefield Information and Collection System
BMEWS	Ballistic Missile Early Warning System
C^3	Command, control and communications
C^3I	Command, control, communications and intelligence
CAST	Canadian Air/Sea Transportable (brigade)
CDAA	Circularly Disposed Antenna Array
CDU	Christian Democratic Party (FRG)
CEGE	Combat Equipment Group Europe
CENTAG	Central NATO Army Group
CEWI	Combat Electronic Warfare Intelligence
CFE	Conventional Armed Forces in Europe (Vienna Negotiation)
CHOD Norway	Chief of Defence Norway
CIA	Central Intelligence Agency
CINCEUR	Commander-in-Chief Europe
CINCSOUTH	Commander-in-Chief Allied Forces Southern Europe
CINCUSNAVEUR	Commander-in-Chief US Naval Forces in Europe
COB	Collocated operating base
COCOM	Coordinating Committee on Export Controls
COD	Carrier on-board delivery

COIC	Combat Operations Intelligence Center
COMAFNORTH	Commander Air Forces North
COMAIRBALTAP	Commander Air Baltic Approaches
COMAIRSOUTH	Commander Allied Air Forces Southern Europe
COMFAIRMED	Command Fleet Air Mediterranean
Comint	Communications intelligence
COMLANDSOUTH	Commander Allied Land Forces Southern Europe
COMNAVSOUTH	Commander Allied Naval Forces Southern Europe
CONUS	Continental USA
COS	Chief of Staff
CSCE	Conference on Security and Co-operation in Europe
DARPA	Defence Advanced Research Projects Agency
DCS	Defense Communication System
DEBS	Digital European Backbone System
DECA	Defense and Economic Cooperation Agreement
DEW Line	Distant Early Warning Line
DF	Direction finding
DMSP	Defense Meteorological Satellite Program
DOD	Department of Defense
DPC	Defence Planning Committee (NATO)
DSCS	Defense Satellite Communication Systems
DSP	Defense Support Program
DSSCS	Defense Special Security Communications System
EC	European Community
EFA	European Fighter Aircraft
EHF	Extremely high frequency
ELF	Extremely low frequency
Elint	Electronic intelligence
ENSCE	Enemy Situation Correlation Element
ESF	Economic Support Fund
ETS	European Telephone System
EUCOM	European Command
EUREKA	European Scientific Co-operation Project
FEBA	Forward edge of battle area
FLOT	Forward line of own troops
FLTSATCOM	USN Fleet Satellite Communications System
FMS	Foreign Military Sales
FOFA	Follow-on Forces Attack
FORACS	Forces Range Accuracy Control Station
FOSIC	Fleet Ocean Surveillance Information Centres
GCHQ	Government Communications Headquarters (UK)
GEODSS	Ground-based Electro-Optical Deep Space Surveillance
GIUK	Greenland–Iceland–United Kingdom
GLCM	Ground-launched cruise missile
HAS	Hardened aircraft shelter
HF	High frequency
Humint	Human intelligence

I&W	Intelligence and warning
IBERLANT	Iberian Atlantic Command
ICBM	Intercontinental ballistic missile
IDF	Icelandic Defense Force
IEPG	Independent European Programme Group
IMET	International Military Education and Training
Imint	Imagery intelligence
INF	Intermediate-range nuclear forces
IRBM	Intermediate-range ballistic missile
JSIPS	Joint Service Imagery Processing System
JSTARS	Joint Surveillance and Target Attack Radar System
JUSMAG	Joint US Military Advisory Group
LANDJUT	Allied Forces Schleswig–Holstein & Jutland
LANDSOUTH	Allied Land Forces Southern Europe
LANDSOUTHEAST	Allied Land Forces South East
LF	Low frequency
LOA	Letter of Offer and Acceptance
LOCE	Limited Operational Capability Europe
Loran	Long-Range Aid to Navigation (system)
MAB	Marine Amphibious Brigade
MAC	Military Airlift Command
MAP	Military Assistance Program
MEB	Marine Expeditionary Brigade
MGT	Mobile ground terminal
MIRV	Multiple independently targetable re-entry vehicle
MoD	Ministry of Defence
MOU	Memorandum of Understanding
MTMC	Military Traffic Management Command (US Army)
MW	Microwave (communications)
NADGE	NATO Air Defence Ground Environment
NAMPA	NATO AWACS Management Program Activity
NARS	The North Atlantic Relay System
NAVOCFORMED	Naval On-Call Force Mediterranean
NAVSOUTH	Allied Naval Forces Southern Europe
NAVSPASUR	Naval Space Surveillance System
NDS	Nuclear Detection System
NICS	NATO Integrated Communication Systems
NIP	NATO Infrastructure Program
NOAA	National Oceanic and Atmospheric Agency
NORAD	North American Air Defense
NORTHAG	Northern Army Group
NOSIC	Naval Ocean Surveillance Information Center
NPG	Nuclear Planning Group (NATO)
NRO	National Reconnaissance Office
NSA	National Security Agency
NWS	North Warning System

ACRONYMS, ABBREVIATIONS AND CONVENTIONS

O&M	Operations and maintenance
OASIS	Operational Application and Special Intelligence System
OBS	OSIS Baseline Subsystem
OSIS	Ocean Surveillance Information System (US Navy)
PAS	Primary Alert System (USAFE)
PASOK	Pan-Hellenic Socialist Movement
Photint	Photographic intelligence
POL	Petroleum–Oil–Lubricants
POMCUS	Prepositioned Organizational Material Configured to Unit Sets
PX	Post Exchange
R&D	Research and development
RDT&E	Research, development, test and evaluation
RAF	Royal Air Force (UK)
Reforger	Return of Forces to Germany
ROICC	Resident Officer in Charge of Construction
RPV	Remotely piloted vehicle
SAC	Strategic Air Command
SACEUR	Supreme Allied Commander Europe
SACLANT	Supreme Allied Commander Atlantic
SAS	Special Ammunition Storage Site
SATCOM	Satellite communications
SETAF	Southern European Task Force
SHAPE	Supreme Headquarters Allied Powers Europe
SHF	Super high frequency
Sigint	Signals intelligence
SIXATAF	6th Allied Tactical Air Force
SLOC	Sea lines of communication
SOFA	Status of Forces Agreement
SOON	Solar Optical Observing Network
SOSUS	Sound Surveillance System
SPADATS	Space Tracking and Detection System
SPASUR	Space Surveillance System (US Navy)
SPOT	Système Probatoire d'Observation de la Terre (France)
SPS	Simplified Processing Station
SSBN	Nuclear-powered ballistic missile submarine
SSN	Nuclear-powered submarine
START	Strategic Arms Reduction Talks
STRIKEFORSOUTH	Naval Striking and Support Forces Southern Europe
SURTASS	Surveillance Towed-Array Surveillance System
SVS	Secure Voice System
SWHQ	Static War Headquarters
TACAMO	Take Charge and Move Out (US emergency communications aircraft)
TAN	Tactical Alert Net (US Army)
TAOC	Tactical Operations Center (USMC)
TARE	Telegraph Automatic Relay Equipment (part of NATO's Integrated Communication Systems, NICES)
TENCAP	Tactical Exploitation of National Capabilities

TFS	Tactical Fighter Squadron
TFW	Tactical Fighter Wing
TNW	Theatre nuclear weapons
TTCE	Transportation Terminal Command Europe
TUSLOG	US Logistics Group Turkey
UHF	Ultra high frequency
UKADGE	UK Air Defence Ground Environment
USAF	US Air Force
USAFE	US Air Forces in Europe
USAREUR	US Army Europe
USEUCOM	US European Command
USMC	US Marine Corps
USN	US Navy
USNAVCOMSTA	US Naval Communications Station
USNAVEUR	US Naval Forces Europe
VHF	Very high frequency
VLF	Very low frequency
WESTLANT	Western Atlantic area
WEU	Western European Union
WHNS	Wartime Host Nation Support
WTO	Warsaw Treaty Organization
WWMCCS	World-Wide Military Command and Control System

Conventions used in tables

Bde	Brigade
Comm	Communications
USAr	US Army
2d, 3d, etc.	2nd, 3rd, etc. in US military designations
. .	Data not available or not applicable
–	Nil or a negligible figure
m.	million
b.	billion
$	US $

Part I
Introduction and summary

1. Why discuss US withdrawal?

Jane M. O. Sharp
SIPRI

I. Introduction

SIPRI embarked on this project in the second half of the 1980s for a number of reasons. In the first place, despite a firm commitment by the US executive branch to the present security arrangements in NATO, the US Congress and the media continued to express interest in the reduction or elimination of direct military involvement in the defence of Western Europe.[1] Pressure to bring the troops home was motivated primarily by resentment that the USA carries a disproportionate share of financial burdens and military risks. Americans particularly resent subsidizing the cost of European defence when US trade and budget deficits were high (as they were throughout the 1980s), and when Europeans oppose, or at least do not whole-heartedly support, US policy outside the NATO area. US policy makers were also increasingly reluctant to accept the risks inherent in extending nuclear deterrence over NATO countries, and resented the fact that Europeans allies were less willing to share the risks of nuclear basing.

A second reason to look carefully at the role of US forces in Europe was the growing West European interest in acquiring more autonomy from the USA in defence matters. A third, related to the second, was the steady reduction in Western perceptions of a Soviet military threat since Mikhail Gorbachev came to power in the USSR in 1985. These two reasons lead to a fourth, namely that as the cold war confrontation drew to a close, NATO and the Warsaw Treaty Organization (WTO) no longer seemed the most appropriate security systems for Europe. Both have already lasted longer than most military alliances, but there is no historical reason to suggest they should endure indefinitely. The USSR and Eastern Europe probably changed more radically during the three years this book was being researched and written than in any period since 1945. By early 1990 several Soviet republics had expressed the desire for greater independence from Moscow, parliamentary democracies and market economies were beginning to emerge in the non-Soviet WTO countries, and German unification was on the agenda. While many policy makers in Western capitals entered the 1990s hoping that existing institutions could be adapted to serve new missions, others believed that new conditions required pan-European institutions and a new collective security system. Whether and to what extent US forces would be deployed in Europe under such schemes remains a matter for speculation.

II. The US commitment to the defence of Europe

States join alliances to pool resources with others facing a common threat. Emerging from World War II, the common threat to the Western democracies was twofold: a resurgent Germany and an expansionist USSR. Initially the German threat loomed larger, as reflected in the language of the 1947 Anglo-French Treaty of Dunkirk and the 1948 Brussels Treaty, to which also Belgium, Luxembourg and the Netherlands were parties. Although the Truman Doctrine in March 1947 rationalized US aid to Greece and Turkey as a necessary counter to communist pressure, in June 1947 general economic aid was offered through the Marshall Plan to all of Europe, including the USSR. None of the communist-controlled countries accepted the offer, however, and after the communist *coup* in Czechoslovakia in February 1948, and especially after the Soviet blockade of Berlin in June 1948, the Soviet threat began to dominate West European perceptions, and statesmen on both sides of the Atlantic looked for a structure that would commit the USA to the defence of Western Europe. In April 1949 the USA, together with Canada, Denmark, Iceland, Italy, Norway and Portugal, joined the Brussels Treaty powers in signing the North Atlantic Treaty, designed to convey a clear political message to the USSR that any move westwards would be met by the combined military might of North America and Western Europe. Greece and Turkey joined the Treaty powers in 1952.

Inside the US State Department, George Kennan, then Director of the Policy Planning Staff, believed that Soviet expansionism would best be checked by rejuvenation of the West European economies, not by a transatlantic military alliance. He thoroughly approved of the Marshall Plan for the economic recovery of war-torn Europe, but disapproved of NATO. He argued for a European military alliance backed by a US security guarantee rather than a US-dominated alliance which he feared would result in debilitating European dependencies. He felt that the United States should have been willing to offer security guarantees to a West European alliance without becoming integrated in the alliance itself; a situation that corresponds to the option B withdrawal posited by this study.[2]

Neither as a result of the *coup* in Prague, the Berlin blockade or the signing of the North Atlantic Treaty were any extra US troops committed to Europe, beyond the two divisions already there on occupation duty. It was only after the outbreak of the Korean War in June 1950 that Western leaders felt the need to boost the US military presence in Europe and to rearm the Federal Republic of Germany. These two needs were inextricably linked. While it was the Korean War that provided the impetus for President Harry Truman to send more troops to Europe, the decision was controversial in Washington, where many believed Europe should now be rearming to defend itself. It was the need to make West German re-armament acceptable to France that overcame doubts about committing more US troops and that prompted the Truman Administration to add four divisions to the two already there on occupation duty.[3]

US force levels in Europe and the North Atlantic have fluctuated considerably since then (see figure 1.1). The level peaked in 1955 at approximately 431 000, decreased to 365 000 by 1960, rose again during the Berlin crisis of 1961 to approximately 420 000 in 1962, fell steadily through the 1960s and early 1970s, as forces were redeployed to the Far East, to reach a low of 299 000 in 1973. Once the Viet Nam War was over, US forces returned to Europe, reaching approximately 356 000 in the early 1980s.[4] In 1982 Congress legislated a ceiling of 315 000 on US ground-based troops in NATO countries. This was later raised to 326 414 to accommodate extra personnel deployed with Cruise and Pershing II missiles.

Burden-sharing and offset payments

The issue of burden-sharing is a recurrent bone of contention in any military alliance, although less so in a bipolar than in a multipolar world.[5] Despite the rhetoric about equality, the allies vary enormously in the contributions they make to, and the benefits they draw from, their NATO partners.[6]

The issue of who pays for the US troops in Europe has often been a matter of heated debate, especially in the FRG. Until 1955, when it gained sovereignty and accession to NATO, the FRG paid the costs of all four occupation powers. The USA began to pay stationing costs after 1955, not least to provide the Europeans with dollars to spend on US goods. By the end of the decade, however, US balance of payments deficits with the FRG had almost doubled; from $6.5 billion in 1951–57 to $11.2 billion in 1958–60. In November 1960, Robert Anderson, US Secretary of the Treasury, and C. Douglas Dillon, Deputy Secretary of State for Economic Affairs, went to Bonn to make the first of a series of requests for offset payments. The FRG Foreign Minister, Heinrich von Brentano, refused direct payment for US troops 'for political and psychological reasons', but did agree to cover part of the infrastructure costs and to accelerate repayment of post-war debts to the USA.[7]

In late 1961 the Kennedy Administration signed a new agreement with Bonn that provided for FRG purchases of US equipment to offset the D-mark costs of stationing US forces in the FRG.[8] At the time this arrangement seemed to satisfy both parties but disputes about offset payments would continue for the next two decades.[9] In April 1970, for example, US insistence on offset purchases by the FRG was a serious irritant in Bonn. West German Defence Minister Helmut Schmidt caused quite a stir when he said that no further large purchases of US equipment would be required, as the Bundeswehr had ended its build-up phase. In the event, however, Bonn signed a new offset agreement in December 1971 that increased FRG payments to the USA over those of the previous two years.[10] These offset agreements were allowed to expire in 1975, and the 1983 Defence White Paper of the FRG stated that the extensive contributions made by the Government to the common defence no longer justified additional burdens on the federal budget.[11]

Figure 1.1. US military forces in NATO Europe, 1945–90

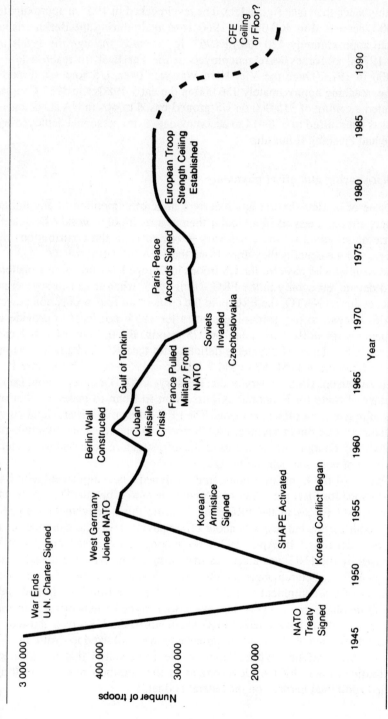

Source: Based on US General Accounting Office, *Military Presence: US Personnel in NATO Europe*, Report to the Chairmen and Ranking Minority Members, Senate and House Committees on Armed Srvices, report no. GAO/NSIAD-90-04 (US GAO: Washington, DC, Oct. 1989), fig. 1, p. 3.

Congressional pressure to bring US troops home

The Viet Nam War highlighted the gap between US overseas capabilities and commitments, and encouraged legislators to seek cuts in US forces in Europe, especially as West European economies continued to prosper and tensions between East and West eased after the Cuban missile crisis in 1962 and the signing of the Partial Test Ban Treaty.

US Senator Mike Mansfield first complained about the over-commitment of US troops abroad in 1960 and urged force reductions talks from 1962. By 1966 his tactics changed from personally pressuring the President to a series of Senate resolutions urging unilateral reductions. The first of these resolutions garnered 44 votes in August 1966, and Senator Mansfield pressed his case every year until he retired from the Senate in the mid-1970s.

President Lyndon Johnson responded to this congressional pressure in a number of ways. He pressed the Bonn Government, even harder than his predecessors had, on financial offset payments.[12] He also withdrew 35 000 men to bases in the USA, where they remained earmarked for duty in Europe; in their place, duplicate sets of equipment were stored—the initial Prepositioned Organizational Material Configured to Unit Sets (POMCUS). This move was of economic as well as political value, since it saved foreign exchange and eased the balance of payments.[13] Finally, as a way of stemming pressure for unilateral withdrawals, not only by the USA but also by Canada and the European allies, the Johnson Administration endorsed a NATO invitation to the WTO to begin negotiations on mutual and balanced force reductions (MBFR) in Europe.[14]

In 1969 President Richard Nixon inherited the domestic turbulence and the growing balance of payments deficits associated with the Viet Nam War and the decreasing US share of world trade. The Nixon Administration resented the fact that the West European allies supported neither the US war effort in Viet Nam nor US policy in the Middle East, and that they objected to not being consulted when US nuclear forces were put on alert in 1973.[15] Irritation with the Europeans increased support for Senator Mansfield. In January 1970 he complained that, 'Europeans have been soft and comfortable in a status quo defended by US troops for twenty years. There is no reason why 250 million Europeans can not organize an effective defence against 200 million Russians who are at the same time trying to contend with 700 million Chinese'.[16] In the early 1970s Senator Manfield moved from introducing recommendations to cut troop strengths abroad to legislation that would, if passed, mandate cuts.

Increasingly, however, the Senate's focus moved away from unilateral cuts to a more sophisticated burden-shifting tactic. In October 1973 Senators Sam Nunn and Henry Jackson co-sponsored an amendment requiring the allies to offset fiscal year (FY) 1974 US–NATO defence balance of payments expenditures in Europe. Failure to do so would require the USA to withdraw forces from Europe by the same percentage as the offset was not met. In the event, the US balance of payments eased considerably in 1975 and in June President

Gerald Ford reported that the US deficit had been more than offset by the European allies. The troop reduction provisions of the Jackson–Nunn amendment were never invoked.

During President Ronald Reagan's first term, East–West relations as well as intra-alliance relations were at a low point. In the aftermath of the Iran hostage crisis, the Soviet invasion of Afghanistan and the tensions over the decision to deploy new intermediate-range nuclear forces (INF) in Europe, US conservatives grew increasingly irritated by lack of support from the European allies for US activities outside the NATO area. In particular, they found the European allies unresponsive to President Carter's call to boycott the 1980 Olympic Games in Moscow after the Soviet invasion of Afghanistan, too mild in their reactions to the imposition of martial law in Poland, too willing to assist in the construction of the Siberian gas pipeline, too hostile to US policy in Central America and too dismissive of President Reagan's proposal for a space-based strategic defense.[17]

These differences brought renewed calls for US troop withdrawals from Europe not only to cut costs but as punitive measures for lack of allied support for US grand strategy. In 1984 Former Secretary of State Henry Kissinger suggested that if the European allies continued to make only token improvements the USA should consider withdrawing half its forces.[18] In 1986, after the European outcry against the US bombing of Libya, Kissinger proposed the withdrawal of US troops from Europe to form a strategic reserve in the USA available for contingencies anywhere on the globe.[19] Former National Security Advisor Zbigniew Brzezinski made a similar proposal in 1987, involving the redeployment of 100 000 US troops back to the continental USA.[20] Melvyn Krauss, an economist at New York University, proposed that the USA should withdraw from Europe altogether.[21]

As they had done while countering Senator Mansfield in the early 1970s, US secretaries of defence continued to acknowledge that the European allies contributed their fair share, and to emphasize that US bases in Europe were the most cost-effective way to project US power globally.[22] Nevertheless, given the growing US trade and budget deficits, US legislators found it increasingly difficult during the 1980s to justify spending so much on US military forces in European countries whose citizens no longer felt threatened by the USSR, and who enjoyed a higher standard of living than most US citizens.

In 1982 Senator Ted Stevens was particularly annoyed to learn that US troops in Europe had increased by approximately 60 000 men since the mid-1970s, while some West European force levels had declined, although European forces in NATO had been modernized during this period. Senator Stevens succeeded in imposing a European Troop Strength (ETS) ceiling of 315 600 for ground-based US troops in NATO-Europe, effective in FY 1983.[23] Senator Nunn and Senator William Roth adopted a carrot-and-stick approach, by threatening to withdraw troops if the allies could not do more, but promising substantial research and development funds for co-operative alliance projects if they could.²⁴ In June 1984 an amendment proposed by Senator William Cohen raised the previously imposed EST ceiling to 326 414, to accommodate extra US troops deployed with the new Cruise and Pershing II missiles.²⁵

According to the US General Accounting Office, there were 319 000 US troops in Europe as of 30 September 1987, including Army, Navy and Air Force personnel.²⁶ In 1988, in the wake of the Treaty between the USA and the USSR on the elimination of their intermediate-range and shorter-range missiles (the INF Treaty), Representatives Patricia Schroeder and Andrew Ireland of the House Armed Services Committee, pointing to the growing US trade and budget deficits and the reduction of the Soviet threat, argued that the USA could not afford to shoulder as much of the NATO defence burden as it had hitherto done. If the European allies would not or could not do more, then the USA should do less, specifically bring home and demobilize the 15 000 troops that had served with the Cruise and Pershing II missile units.²⁷ In the event, approximately 14 000 troops were withdrawn along with these units, leaving 305 000 as the baseline for US reductions at the Conventional Armed Forces in Europe (CFE) Negotiation that began in Vienna in 1989. Further reductions appear inevitable in the 1990s as a result of both negotiated and unilateral measures.

In February 1990 the Bush Administration proposed that an initial CFE agreement set limits of 225 000 US troops in Europe, entailing a cut of 80 000 men. With respect to unilateral measures, in December 1988, during the last days of the Reagan Administration, a bipartisan US Government commission recommended the closure of 86 military bases in the continental USA, and the 're-alignment' of 54 other bases.²⁸ These recommendations were accepted by the incoming Bush Administration and approved by the Armed Services Committees of the House and Senate. In early 1990 Defence Secretary Richard Cheney announced the closures of an additional 35 bases in the US, 3 bases in Korea and 9 in Western Europe.²⁹ The latter include 7 US Air Force bases: 3 in the UK (Greenham Common, Wethersfield and Fairford), 1 in Italy (Comiso), 1 in the FRG (Zweibrücken), one in Turkey (Erhac) and 1 in Greece (Hellenikon). A naval base at Nea Makri in Greece and a munitions storage site at Eskisehir in Turkey were also scheduled for closure.

US Representatives from congressional districts about to suffer unemployment and other disruptions as a result of base closures, urged the Administration to make further cutbacks in Europe.³⁰ Former US Secretary of Defense James Schlesinger suggested that a residual force of 50 000 troops should be kept in Europe.³¹ In April 1990 the US Army Chief of Staff, Carl Vuono, and the Secretary of the Army, Michael P. W. Stone, proposed cutting Reserve and National Guard forces from 776 000 to 645 000 and active-duty forces from 764 000 to 580 000, of which some 150 000 should be retained in Europe, by 1997.³² In the Senate, Senator Nunn, as Chairman of the Armed Services Committee and the Democratic Majority Leader, proposed that between 75 000 and 100 000 US troops be kept as a residual force in Europe, that is less than half the number that President Bush proposed as the floor of US troops to be codified in an initial CFE agreement.[33]

The US nuclear guarantee

If the anticipated benefits of shared resources produce acrimonious disputes about economic burden-sharing, the supposed benefits of shared protection often give rise to disputes about the sharing of risks, especially those associated with NATO's policy of flexible response, embodying a US threat to initiate the use of nuclear weapons on behalf of its allies. Allied governments in Europe tend to oscillate between fear of abandonment by the USA in a crisis and fear of entrapment in a conflict not of their own choosing, at worst a superpower nuclear exchange confined to European territory. These fears create a management problem for the USA, since the steps it takes to relieve allied fears of abandonment tend to generate fears of entrapment, and vice versa. The difficulties are compounded by NATO's geography which creates different perceptions of vulnerability on either side of the Atlantic. The USA shares European interest in preventing war, but if war breaks out then US planners prefer one confined to European territory, while the European allies prefer that any nuclear exchanges occur over their heads. The allies are therefore more reassured by long-range missiles based in the USA than by US short-range systems based in Europe. However, US planners, trying to convince legislators that a nuclear war-fighting capability is what makes deterrence credible to the adversary, like to show that the allies share the risk associated with nuclear basing in Europe.

NATO has on several occasions since 1957 tried to establish guide-lines for nuclear use. In the mid-1960s Defense Secretary McNamara established the Nuclear Planning Group (NPG) to give the European allies a stronger voice in US planning. However, NATO command post exercises regularly demonstrate that US nuclear use is incompatible with European security interests.[34] The problem for NATO is to devise a nuclear policy credible enough to deter a potential aggressor, yet stable enough to reassure the allies. The pattern has been for nuclear deployment decisions to trigger fears of entrapment among the allies, and for Soviet–US arms control diplomacy to trigger fears of abandonment and of the establishment of a superpower condominium at Europe's expense.

European fears of entrapment were manifest in the first 'ban the bomb' movements that followed decisions to deploy US nuclear weapons on the continent in the 1950s, in the debate over whether to deploy a multilateral nuclear force (MLF) in the 1960s and the controversy over the neutron bomb in the 1970s, in the double-track decision in the late 1970s that led to the deployment of Cruise and Pershing II missiles in the early 1980s, and in the debate about whether to modernize short-range nuclear forces in the late 1980s and early 1990s.[35]

Fears of abandonment, on the other hand, were triggered by Soviet–US efforts to limit nuclear weapons, since superpower co-operation inevitably undermines alliance guarantees. West German leaders have perhaps been the most troubled, having renounced independent nuclear forces as a condition of joining NATO in 1954. In the 1960s the nuclear Non-Proliferation Treaty (NPT) was seen as especially discriminatory in Bonn, coming so soon after cancellation of the MLF project. The 1972 US–Soviet Anti-Ballistic Missile (ABM) Treaty was reassuring to the allies because limits on Soviet defences against ballistic missiles enhanced the credibility of the US guarantee to NATO as well as the credibility of British and French nuclear deterrent threats against the USSR, while President Carter's handling of SALT II, especially his insensitivity to the threats that Soviet INF posed to Western Europe, was especially distressing to the FRG, and led directly to the double-track decision to both deploy and negotiate limits on new US INF.

In October 1986 the allies felt abandoned by President Reagan's apparent willingness, at the Reykjavík summit with President Gorbachev, to negotiate away all US ballistic missiles, on which many believe the NATO guarantee depends.[36]

The INF Treaty signed in 1987 was widely applauded by general publics in both halves of Europe, but denounced by conservatives in France and the FRG. French Defence Minister André Giraud went so far as to call the treaty a 'nuclear Munich'.[37] The INF Treaty certainly complicated NATO plans to modernize its stockpile in Europe, since it removed the intermediate-range systems that coupled Western Europe to the USA and left in place the short-range systems that were most worrying for Europeans. As Volker Rühe of the West German Christian Democratic Union put it, 'the shorter the range the deader the Germans'. The Thatcher and the Reagan administrations wanted to solve this dilemma by modernizing and extending the range of the Lance missile to just less than the 500-kilometre limit set by the INF Treaty. By contrast, FRG Chancellor Helmut Kohl, facing elections in late 1990, did not want to be confronted with another nuclear crisis. He opposed an early decision on Lance modernization and urged early negotiations to limit short-range nuclear forces (SNF).

The issues of nuclear risk-sharing and US troops in Europe were joined in the late 1980s when several senior members of the Reagan Administration responded to anti-nuclear protests in Western Europe with a veiled 'no nukes no troops' threat, suggesting that European willingness to accept the risks of nuclear weapons was an essential quid pro quo for continued basing of US troops on the continent.[38] Prime Minister Margaret Thatcher issued a similar threat when she suggested that Britain might no longer feel able to deploy the British Army on the Rhine (BAOR) if the FRG would not accept a follow-on to the Lance missile (FOTL).[39] In April 1989 West German irritation with the Anglo-US position on nuclear modernization was reflected at the highest level in an unusually forceful speech by President Richard von Weizsäcker in

Copenhagen that referred to differences in SNF policy between continental and other members of NATO.[40]

The 40th anniversary NATO summit meeting in late May 1989 demonstrated the difference between the Bush and Reagan administrations on NATO in general, and on nuclear issues in particular. After consultation with President François Mitterrand of France, President Bush resolved the Anglo-German differences with a plan to delay SNF modernization and to initiate talks to limit SNF once progress had been demonstrated at the CFE Negotiation.[41] In the twelve months that followed, the Berlin wall was breached, democracy was introduced in Eastern Europe and the idea of modernizing short-range nuclear weapons looked even less appropriate than it had in 1989.

Thus Congress baulked at the $112 million requested for a FOTL, and in the FRG opposition continued to mount against the deployment of any new nuclear weapons. Accordingly, in a set of speeches in May 1990 the US Administration announced cancellation of the FOTL and of the nuclear artillery modernization programmes as well as its willingness to begin talks aimed at the withdrawal of all land-based SNF from Europe.[42] By then even Prime Minister Thatcher had stopped insisting on a FOTL, although she and President Bush agreed that NATO should develop a new tactical air-to-surface missile (TASM) to be deployed in Europe. President Mitterrand and Prime Minister Thatcher met in Britain immediately after the NATO Foreign Ministers meeting in early May to discuss co-operation on a number of defence issues, including a new TASM.[43] Discussion of a new NATO missile that could strike deep into Soviet territory was not welcome in Bonn, however. Precisely at the time when the first formal 'two plus four' talks were taking place, West German Foreign Minister Hans-Dietrich Genscher wanted to reassure Moscow that NATO would not disturb the military balance in Europe or undermine Soviet security.[44]

III. Incentives and disincentives for West European autonomy

The second reason to look carefully at the role of US forces in Europe was West European interest in acquiring autonomy from the USA on defence policy issues. Incentives for West European defence co-operation stem from at least three sources: anxiety—usually but not exclusively on the part of France—about German power, recurrent US pressure on Europe to carry a greater share of the alliance burden and the desire to speak with one voice where US and West European interests differ.

A truly independent West European defence identity is obviously a long way off, however. Three factors tend to work against co-operation. First, strong nationalist tendencies led both France and Britain to acquire independent nuclear forces that set them apart from each other and from their non-nuclear allies and, in 1966, prompted President Charles de Gaulle to take France out of the integrated NATO military command and push NATO installations out of France. Secondly, strong Atlanticist tendencies in Britain and the FRG have often led policy makers in London and Bonn to set a higher priority on relations with Washington than with their European allies. Third, in the FRG in particular, there is fear that too much West–West co-operation on defence might preclude closer East–West political co-operation in Europe.

Anxiety about Germany

The original NATO bargain was a deal struck between the USA and France about how best to contain Germany after World War II. Britain was an active participant—at times a spoiler, at times a facilitator—but the FRG could only lobby for its interests on the sidelines.[45]

Prior to the formation of NATO in April 1949 the other West European allies had already signed two defence treaties against the threat of a revanchist Germany: the Anglo-French Treaty of Dunkirk in 1947, and the Brussels Treaty (the Dunkirk Treaty powers plus the Benelux countries) in 1948. The crucial factor in persuading the USA to commit troops to Europe was the perception that it would make West German rearmament palatable to France. In Paris, however, President Truman's decision was seen as a necessary but not a sufficient condition—French leaders still wanted direct control over the pace of German rearmament. Accordingly, in October 1950 the French National Assembly approved a plan to create a European Defence Community (EDC) that envisaged an integrated European army within which German units would be absorbed. This army would not be formed, however, until a European parliament could allocate funds and appoint a European minister of defence. The Truman Administration saw this as a tactic to delay German rearmament indefinitely, and proposed a compromise under which France accepted the principle of German rearmament under 'strong provisional controls' and the USA endorsed a long-term plan for an integrated European defence force, appointed a US Supreme Allied Commander, and promised to begin deploying troops to Europe without waiting for West German troops to materialize.[46]

Negotiations toward an EDC began in 1951. French Prime Minister René Pleven envisaged the EDC as a military counterpart to the European Coal and Steel Community (ECSC) then being formed. Precursor to the European Economic Community, the aim of the ECSC was in part to contain German industrial potential.[47] France, the FRG, Italy and the Benelux countries signed the EDC Treaty in May 1952, but the French National Assembly voted against ratification. In an effort to salvage the situation and create a climate conducive to FRG rearmament, the British Foreign Secretary, Anthony Eden, convened a nine-power conference in London (including the six EDC members, the USA, the UK and Canada), at which Britain committed four divisions and a tactical air force to permanent deployment in the FRG.[48] This finally convinced France to accept West German rearmament as well as FRG membership in both NATO and the expanded Western European Union (WEU). These decisions, as well as Italian accession to NATO and the WEU and an end to the occupation regime in the FRG, were formalized in a series of agreements signed in Paris in October 1954.[49]

Since then, NATO's role in binding the FRG firmly to the Western democracies has been as important as its role in deterring and containing the Soviet threat. This West–West stabilising function of NATO is often forgotten in the USA, but rarely so in Western Europe and least of all in the FRG.[50]

If the EDC was the first failure of West European co-operation, its first success was the ECSC, which led to the European Economic Community in 1957. The original six signatories of the Treaty of Rome (France, the FRG, Italy and the Benelux countries) were joined by Britain, Ireland and Denmark in 1973, by Greece in 1981 and by Spain and Portugal in 1986. Advocates of European unity have found it frustrating that EC members have not moved more rapidly towards political union, but in October 1970 the foreign ministers of the original six EC members initiated the process of European Political Co-operation (EPC). The fruits of the EPC were first manifest at the Conference for Security and Co-operation in Europe (CSCE) during the 1970s, when the EPC rather than NATO became the co-ordinating group for West European policy. In general, EPC members have consciously stopped short of co-ordinating their defence policies, an area reserved for NATO consultations. The Tindemans Report[51], published in 1976, recommended that 'security' should be an integral part of the European Union's common foreign policy, that there should be a common defence policy and that the European parliament should take over the functions of the Assembly of the WEU. In the Single European Act, signed in 1986 and ratified in 1987, the parties agree to 'co-ordinate more closely their views on political and economic aspects of security.'[52] Thus EC consultations now regularly include issues that also appear on the NATO agenda.[53] In practice, however, the major West European powers have not yet been willing to give up sovereignty over defence decision-making to a supra-national body.

As the FRG pursued a more ambitious *Ostpolitik* in the late 1980s, several French political and business leaders, fearing a general drift eastwards on the part of the FRG, recommended giving up sufficient sovereignty to form a Franco-German federal union in order to anchor the FRG more firmly to the West.[54] French anxiety grew even more pronounced after the Berlin wall was breached in November 1989.[55] In February 1990, as Chancellor Kohl urged an accelerated pace for German unification, former West German Chancellor Helmut Schmidt and former French President Valéry Giscard d'Estaing emphasized, in an article published simultaneously in Paris, Bonn and London, that German unification must be part of the process leading towards federal union of the EC states.[56] In April 1990, sensitive to growing nervousness about German unification on the part of his European allies, Chancellor Kohl persuaded President Mitterrand to join in a Franco-German initiative to accelerate the timetable for EC political union, as a way to demonstrate Germany's Western credentials and to reassure the allies that a united Germany would remain as anchored in the EC as the FRG had been.[57] Nevertheless, when the

EC foreign ministers met in Ireland in May they clearly preferred modest rather than radical proposals for political union.[58]

Responding to US pressure on burden-sharing

Just as the USA in 1947 made Marshall Plan aid contingent on West European co-operation in the allocation of aid, so in 1950 the deployment of US troops to Europe was contingent on West European efforts to provide for their own defence. When President Truman on 9 September 1950 announced his decision to send four more US divisions to Europe he said: 'A basic element in the implementation of the decision is the degree to which our friends [in Western Europe] match our action in this regard. Firm programs for the development of their forces will be expected to keep full step with the dispatch of additional forces to Europe. Our plans are based on the sincere expectation that our efforts will be met with similar actions on their part'.[59]

Twenty years later US Senator George Aiken was not impressed with the progress towards European integration, noting in early 1971: 'I presume the Common Market aspires to be more than an association of grocers. I assume it is a step towards Western Europe resuming responsibility for its own defence and for its own collective and distinct contribution to a better and safer world.'[60]

During the Kennedy and Johnson years, analysts at the US Department of Defence published estimates that suggested that the NATO–WTO balance came close to parity, implying that a conventional defence of Western Europe was feasible with a little more effort by the Europeans themselves.[61] This contrasted with the received wisdom in Europe that the balance was so overwhelmingly in favour of the WTO that conventional defence was hopeless and that nuclear deterrence was the only hope of deterring or defending against a WTO attack.

Partly in response to Defense Secretary McNamara and to the Mansfield amendments in Congress, 10 West European defence ministers formed the Eurogroup in the late 1960s to better co-ordinate, and advertise, their defence efforts within NATO.[62] The prime mover was British Defence Minister Denis Healey, who believed that there was an urgent need for a common West European position to balance the USA in NATO councils, 'to be able to talk back to teacher' as he put it.[63] Negotiations on the Non-Proliferation Treaty in the 1960s had generated anxiety in Western Europe, especially in the FRG, about the creation of a Soviet–US condominium at the expense of European interest. In anticipation of the scheduled bilateral talks on strategic arms and the multilateral talks on MBFR, West European leaders were anxious for a forum in which their interests could be addressed and properly taken into account.

The Eurogroup decided to enhance NATO defences before beginning formal talks on MBFR with the WTO. In December 1971 the group launched its European Defence Improvement Programme (EDIP). The focus of EDIP was to accelerate work on NATO infrastructure, particularly on communication facilities and shelters for aircraft, and to organize military aid from the wealthier to the poorer European allies, beginning with transfers from the FRG to Turkey. These European efforts were applauded by the Nixon Administration as it lobbied successfully against all the Mansfield amendments through the early 1970s.[64]

In May 1978, responding to a proposal by President Carter at a NATO summit meeting in May 1977, the European allies agreed to a ten-point Long Term Defence Programme (LTDP). More ambitious than the EDIP of 1971, the LTDP was designed primarily to improve European conventional forces and to commit the allies to a 3 per cent increase in their defence budgets. Not all the goals set were achieved, but the result was nevertheless a substantial increase in the European contribution to the overall NATO posture. Although the burden was not shared as equitably as US legislators would like, it was shared far more equitably than in the 1950s and 1960s.[65] The squabbling about who paid for what did not disappear, however, because of growing trade and budget deficits in the USA, and because of fundamentally different attitudes to welfare payments on either side of the Atlantic.[66]

The Western European Union

Of the various institutional foundations on which collective West European security might be built, the Western European Union seemed to offer the most pragmatic possibilities in the late 1980s.[67] The Eurogroup excludes France, while the mandate of the EPC mechanism of the EC only stretches to include the political and economic aspects of security, not military issues. France belongs to the Independent European Programme Group (IEPG), which includes 13 NATO members but excludes Canada, the USA and Iceland and which is confined to narrow discussion of armaments.[68] While Franco-German defence co-operation intensified during the 1980s, problems with the Franco-German brigade suggest that this may not be the foundation for a greater West European army.

The WEU dates back to the anti-German Brussels Treaty of 1948. The FRG and Italy joined later when they joined NATO, and the WEU served as the institution that controlled the rearmament of FRG and Italy, and was therefore unpopular in Bonn and Rome. Until Britain joined the EC in 1973 the WEU served as an important link between London and the EC capitals. After 1973 this link was unnecessary, and the WEU ceased to meet at full ministerial level, remaining virtually moribund until the early 1980s.

France led the effort to revive the West European Union in the early 1980s. The other members at the time—Britain, the FRG, Italy and the Benelux countries—welcomed the French initiative insofar as it might bring France closer to NATO. The French initiative, which also removed the last restrictions on the production of conventional weapons in the FRG, seems to have been part of a campaign to bind the FRG closer to the other West European countries at a time when the USA seemed to be re-assessing its European commitments.

Ministerial meetings began again in 1984 with both defence and foreign ministers in attendance.[69]

British officials, always nervous about weakening the transatlantic link, were initially sceptical of the value of reviving the WEU, but later conceded that it served as a useful device to co-ordinate West European views on the US Strategic Defence Initiative (SDI) and on President Reagan's behaviour at the 1986 Reykjavík summit meeting. The European allies' response to the Reykjavík meeting found expression in the so-called Platform Document, signed by the seven WEU foreign ministers in October 1987. While less than the 'security charter' that French Prime Minister Jacques Chirac had called for at the WEU Assembly in December 1986, the Platform Document was nevertheless considered a landmark in the history of the WEU. It called for a 'more cohesive European defence identity' and a 'strategy based on an adequate mix of appropriate nuclear and conventional forces', and acknowledged the 'contribution to overall deterrence and security' of independent British and French nuclear forces.[70]

Two other signs of WEU revival in the late 1980s were the accession of Spain and Portugal, and the co-ordination of West European naval activity to protect shipping caught in the cross fire of the Iraq–Iran War. Spain in particular appeared to see the WEU as the ideal forum in which to pursue its security interest, as it offered more autonomy from the USA than was possible in the NATO context. The WEU also appeared to act as a caucus for developing West European positions on arms control. As for naval co-operation, Britain had been operating its Armilla Patrol in the Persian Gulf since 1980. In 1987 it was joined by naval units from France, Italy, Belgium and the Netherlands, while the FRG deployed ships to the Mediterranean to cover for those redeployed to the Gulf.[71]

Prospects for a separate West European defence entity?

In theory, West European governments could absorb German power and contain the USSR, but this seems unlikely to happen. West European governments have made a number of bilateral and multilateral efforts to co-ordinate their defence policies both inside and outside the NATO framework. Among the stronger European powers, the most advertised bilateral co-operation is that between France and the FRG, the most quietly effective that between the FRG and Britain, and the least developed that between France and Britain, although Franco-British relations have improved concomitantly with the German unification process. France, Germany and Britain would have to form the core of any West European defence entity that was to balance Soviet power on the continent, but none of the three is likely to subordinate itself to any of the others and have needed, thus far, the USA as a unifying factor.

Other proposals for European military co-operation emerged in the early 1990. In April 1990, at a meeting of WEU foreign and defence ministers hosted by Belgian Defence Minister Gaston Eyskens, Willem van Eekelen, Secretary

General of the Western European Union, proposed an embryonic multilateral European Army as part of the European pillar of NATO.[72] Paddy Ashdown, leader of the British Liberal Democrats, suggested a pan-European Territorial Security Force.[73] Oscar Lafontaine, West German Social Democratic Party Vice-Chairman and Chancellor candidate, proposed that a joint Polish–West German brigade be established as a confidence-building measure.[74] Two West German scholars proposed a Franco-German brigade to serve in the German Democratic Republic as a way of linking German unification to European unification rather than to NATO integration.[75] However, none of these schemes are likely to put NATO out of business in the short term. West Europeans have little interest in a defence entity dominated by Germany. Thus, until such time as an alternative pan-European structure can perform these missions, NATO will still serve to balance Soviet and absorb German power.

IV. Perception of a reduced WTO threat

Perhaps more than any other single factor, President Gorbachev's new foreign and defence policy agenda for the USSR changed perceptions of the threat to Western Europe and generated reassessments of the need to maintain the same level of US forces in Europe. For four decades after World War II, Western analysts argued about the size of WTO forces and the intentions of the Soviet leadership.[76] In the pre-Gorbachev era, the USSR was very secretive about data and reluctant to allow any kind of on-site inspection, but the general consensus was that Soviet military doctrine was offensive and that WTO conventional forces were quantitatively, though not qualitatively, superior to those of NATO. By 1989, however, President Gorbachev's policy of *glasnost* had eliminated much of the ambiguity about Soviet capabilities, with the publication of data on the WTO force posture and the Soviet defence budget.[77] Although not as revealing as that available on US forces, the WTO data showed categories in which the USSR was numerically superior, primarily in tanks and other land forces and equipment, and categories where the West was superior, primarily air and naval assets.

Fired in large part by the need to free up resources for the domestic economy, President Gorbachev launched an extensive programme of arms control and disarmament early in 1986. To justify these cuts, the Soviet leader adopted a more conciliatory view of international relations than his predecessors, and reshaped the Soviet defence and foreign policy agenda to de-emphasize national security and independence in favour of international security and interdependence. In particular, President Gorbachev specifically condemned the WTO invasion of Czechoslovakia in 1968, and rejected the so-called Brezhnev Doctrine sanctioning Soviet military intervention in Eastern Europe to bolster local communist regimes.[78]

A recurrent theme in his speeches on arms control was the recognition that forces between East and West were asymmetric and that to reach parity, the side that was ahead must reduce. Gorbachev was not, however, content to wait for a multilateral agreement to reduce the Soviet military burden, but initiated unilateral cuts even before the CFE Negotiation convened.

At the United Nations on 7 December 1988 President Gorbachev announced that 500 000 men would be cut from the Soviet armed forces by 1991, whereof 50 000 men would be taken, with their equipment (including 5000 tanks), from the groups of Soviet forces stationed in Eastern Europe. These would include 6 tank divisions as well as some independent tank regiments. President Gorbachev promised that the Soviet divisions remaining in Eastern Europe would be restructured to make them strictly defensive. Another 5000 tanks were to be removed from the European part of the USSR, and total reductions from this region of the country and the territory of the WTO allies would amount to 10 000 tanks, 8500 artillery pieces and 800 combat aircraft.[79] After the December 1988 announcement, all WTO countries except Romania followed with unilateral cuts of their own.[80]

In Washington assessment of the threat from the WTO changed radically in the second half of 1989. In August Defense Secretary Richard Cheney told Congress that it was a mistake to think that the Soviet threat was abating. By November, however, after a more sanguine threat assessment by the Central Intelligence Agency and the Defense Intelligence Agency, Cheney judged that the threat was lower than at any time since 1945.

More significant than the cuts in men and equipment was the political collapse of the WTO. The new democratic governments elected in early 1990 all pledged loyalty to the WTO in the short term. However, it was obvious they had no desire to remain in a military security relationship with the USSR via the WTO, or in a economic relationship via the Council for Mutual Economic Assistance (CMEA). While recognizing that it would take time to extricate themselves from military, economic and trade commitments, the non-Soviet WTO countries all wanted to join Western or pan-European institutions as soon as possible. Thus, by 1990, in most of the so-called East–West forums, the USSR stood virtually alone. At the 35-state CSCE, on most issues Moscow was outnumbered by 34:1; at the CFE Negotiation by 22:1; and at the 'two plus four' talks by 5:1, as the GDR usually supported Western positions.

Soviet resistance to the Western position at the 'two plus four' talks, namely that a united Germany should be a fully integrated member of NATO, slowed down diplomatic activity at the CFE Negotiation in Vienna and suspended the scheduled withdrawal of Soviet troops from the GDR.[81] Western leaders were sensitive to the dangers of Soviet isolation, and NATO governments were at pains not to further humiliate the Gorbachev leadership.[82] In May 1990, for example, NATO foreign ministers approved a package of incentives, designed to make membership of a united Germany in the alliance more palatable to the USSR, that include institutionalization of the 35-state CSCE, unilateral cuts in NATO SNF, accelerated talks to limit SNF, and economic support not only to cover the cost of maintaining Soviet troops in the GDR but also to house troops withdrawn from Eastern Europe to the USSR.[83]

V. What role for the USA in Europe after the cold war?

Alliances work better when they are hierarchical and when the common external threat looms large than when the allies yearn for equality and autonomy and the threat appears diffuse. NATO certainly seemed to work more smoothly in the 1950s when the European allies needed protection but could not afford to protect themselves. Under these circumstances the USA could afford to be magnanimous and to expect less reciprocation from its European allies. For the most part, the latter gratefully accepted US aid and protection and willingly accepted US leadership and hegemony.

As the Soviet threat abated under President Gorbachev, and as the EC became a tough economic competitor to the USA, especially as it moved toward 1992 and the single market, intra-NATO relationships were often strained. The USA appeared less willing to provide security guarantees and the European allies less willing to accept US hegemony. That said, the USA clearly maintained a vital interest in keeping Western Europe within its own sphere of influence, and the West Europeans did not establish alternative security arrangements. As the cold war drew to a close the prospect of German unification, in conjunction with Soviet disintegration, suggested a radically different distribution of power from the bipolar stalemate that prevailed from 1945 to 1990. It was not clear what role, if any, NATO would play in the transition to a new security system, nor how the USA and the USSR would fit into the post-cold war Europe. Soviet, US and European attitudes towards a US military presence in Europe vary according to time and circumstance.

Soviet perspectives on the US presence in Europe

Despite the fact that WTO communiqués since 1955 have repeatedly called for the abolition of both NATO and the WTO and the establishment of an all-European collective security system, President Gorbachev's concept of a common European home, discussed with President Mitterrand in October 1985 on a visit to Paris and clarified in a number of speeches by senior spokesmen through the late 1980s, was much less radical.[84] Through 1989 President Gorbachev and Soviet Foreign Minister Eduard Shevardnadze appeared to view both alliance structures, and specifically the US military presence in Europe, as stabilizing.[85]

Western observers are often sceptical of the view that Soviet leaders welcome a US presence on the continent, but it has in fact been a recurrent theme in Soviet policy since the late 1950s. When President Gorbachev first raised the notion of a common European home, many Western commentators assumed that it was part of a campaign to push the USA out of Europe and guarantee Soviet domination of the continent.[86] This has been a familiar refrain in the West since the end of World War II. While it is true that Soviet Foreign Minister Andrei Gromyko often voiced anti-US sentiments, the record shows that since the early 1950s Soviet proposals for alternative security systems for Europe almost always included the USA as a treaty partner on the same basis as other members of NATO and the WTO. In the mid-1950s, when the USSR first broached the idea of an all-European Security conference, the goal was to prevent the accession of the FRG to NATO by proposing an all-European security system that would obviate the need for opposing military blocs. In the first of these proposals, presented on 10 February 1954, only European states were envisaged as parties to a general treaty on collective security in Europe, with China and the USA invited as observers by virtue of their permanent membership of the UN Security Council.[87] The following year, however, two draft versions of the treaty, submitted on 20 July and 28 October, 1955, included the USA as a party on the same basis as European states.[88] This was repeated in a note from the Soviet Deputy Foreign Minister Vladimir Kuznetsov to US Ambassador Thompson in Moscow on 15 July 1958.[89]

After the fall of General Secretary Nikita Khrushchev in 1964, Poland took the initiative in reviving the proposal for a European Security Conference, specifically 'a conference of all European states with the participation, of course, of the both the USA and the USSR'.[90] The Political Consultative Commission of the WTO endorsed the Polish initiative the following month, as did General Secretary Brezhnev on a visit to Warsaw in April 1965 and again in an address to the 23rd Congress of the Communist Party of the Soviet Union in March 1966. At the same Congress, Foreign Minister Gromyko blamed the US military presence in Europe for the failure to reduce East–West tension on the continent, and called for 'the settlement of European security problems by the Europeans themselves'.[91] However, in July 1966, when the WTO issued its seven-point Bucharest declaration, the proposal for a European security conference included 'North Atlantic Treaty members and Neutrals', thereby including the USA and Canada. This did not, however, prevent Western publications from referring to 'the pointed exclusion of the United States' from the Bucharest invitation.[92]

While the USSR might not wish an enhanced US role in Europe, Soviet actions suggest, especially after the events in Czechoslovakia in 1967–68, that policy makers in Moscow believe that the US military presence has a stabilising effect, not least in balancing the growing power of the FRG and providing the rational for maintaining Soviet troops in Eastern Europe. Thus, not withstanding Western reports to the contrary, Soviet and WTO invitations to pan-European conferences on European security during the 1950s and 1960s consistently included the USA and Canada. Further evidence of Soviet interest in maintaining a US presence in Europe came in early 1971, when General Secretary Brezhnev chose to respond positively to NATO's invitation to negotiate mutual and balanced forced reductions, thereby tipping the balance of votes against one of Senator Mike Mansfield's amendments calling for unilateral US withdrawals.[93]

Since the early 1970s the USSR has increasingly accepted the US commitment to NATO, and specifically the US military presence in Europe as a norm of international relations. Former US Senator Eugene McCarthy, testifying before the Senate Foreign Relations Committee in September 1974, even suggested that the USSR should be asked to pay half the cost of keeping US troops in Europe.[94] Recognition of the right of USA and Canada to be in Europe is best exemplified in the inclusion of both states as participants in the 35-state CSCE.

In late 1989 and early 1990, with the unification of Germany on the political agenda and Hungary and Czechoslovakia calling for complete Soviet troop withdrawals, the cohesion of the WTO was under severe pressure.[95] Soviet policy makers were thus less preoccupied with formulating a post-bloc pan-European structure and more with how to stabilize the two alliances in order to survive the immediate future. Speaking in London in late November 1989, Sergei Karaganov, Deputy Director of the Institute of Europe in Moscow, emphasized the importance of both NATO and the WTO as the bases on which to build a new security structure, and judged that the European countries had only two to three years in which to formulate a new security concept before events would be out of control.[96]

At the Malta summit meeting in early December 1989 between presidents Bush and Gorbachev, both leaders agreed that NATO and the WTO were needed as stabilizers while Europe was in flux, albeit less as militarily competitive alliances than as politically co-operative organizations. President Gorbachev apparently showed as much interest as President Bush in keeping a US military presence in Europe.[97] This view was reinforced later in the month when Foreign Minister Shevardnadze visited both NATO headquarters and the European Parliament.[98] As President Gorbachev had done in Strasburg the previous July, Shevardnadze acknowledged Western security concerns, specifically NATO's reliance on a US nuclear guarantee. Shevardnadze said that while the Soviet goal was to denuclearize Europe, it would be possible to negotiate nuclear disarmament gradually, first assessing the requirements of minimum deterrence, only later moving to complete disarmament.[99] The Foreign Minister also laid down seven questions the USSR would like to have addressed before the unification of Germany could proceed. These clearly envisaged Soviet and US troops remaining on German territory. In early March, before the elections in the GDR, Soviet Communist Party Central Committee members Valentin Falin and Alexandr Yakovlev endorsed a suggestion from the Social Democratic Party in the GDR that the all four of the World War II allies should all be able to deploy 30 000 troops each in a neutral united Germany.[100] After the Christian Democrats won the election, however, it was clear that the USSR had little leverage over the unification process.

If President Gorbachev appeared to view both alliance structures as stabilizing through 1989, his position changed in early 1990 when it was clear in a number of East–West forums that the WTO was rapidly disintegrating.[101] At the CSCE seminar on military doctrine in Vienna in January–February, and at the 'open skies' meeting of the 23 CFE foreign ministers in Ottawa later in February, the non-Soviet WTO countries acted independently of Moscow. In Ottawa, for example, the foreign ministers of Poland, Hungary and Czechoslovakia agreed with NATO ministers that a united Germany should be anchored in NATO.[102] The Soviet position, expressed on a number of occasions by President Gorbachev, Foreign Secretary Shevardnadze and senior military officers, was that a united Germany must be neutral or non-aligned.[103] The new East German Prime Minister, Lothar de Maizière, and Defence Minister, Rainer Eppelmann, took a middle position, namely that the GDR should remain loyal to the WTO in the short term, but that a united Germany could belong to NATO if NATO renounced its flexible response and nuclear first-use doctrine, and if all nuclear weapons were removed from German soil.[104]

Civilian analysts and officials in the Foreign Ministry in Moscow acknowledged, parallel to the 'two plus four' discussions in early 1990, that it was as much in Soviet as in Western security interest that a united Germany be anchored in NATO rather than become a separate power in central Europe.[105] However, the Soviet leadership found it difficult to back down from its oft-stated position that a united Germany in NATO would destabilize the balance of power in Europe. The first signs of a softer position came in April, when Shevardnadze suggested that a united Germany could for a transition period, pending the establishment of a new pan-European security system, be a member of both two alliances.[106]

US perspectives on Europe: a 'New Atlanticism'?

For US policy makers trying to define a new role in post cold war Europe several new features were apparent in the distribution of power by early 1990. The WTO had ceased to be a cohesive military alliance and an attack on Western Europe was virtually unthinkable. The USSR nevertheless remained a formidable military power, which must either be absorbed in a pan-European structure or balanced by a Western alliance of some sort. A united Germany would soon achieve political and economic dominance in the region and would need to be anchored in both West–West and pan-European institutions. NATO would likely survive as an important anchor for Germany in the short term, but the CSCE would become more important over the long haul, primarily as a way to reassure the USSR and meet the security interests of the former WTO powers. A continuing role for the USA in Europe seemed desirable, but there was no guarantee that Congress would be willing to support this, absent a manifest Soviet military threat.

The two primary candidates to serve as building blocks for a new pan-European system appeared to be NATO and the CSCE. Reporting to NATO heads of state and the Brussels press corps after the Malta summit meeting in December 1989, President Bush and Secretary of State Baker spoke of a 'New Europe and a New Atlanticism'. They both praised the CSCE as a possible building block for a pan-European security system, but the emphasis was on the triumph of Western liberalism, on new missions for NATO and on the need to institutionalize links between the USA and the EC. The proposals included three new NATO agencies: a multilateral verification agency to oversee compliance with a CFE agreement, a centre for the resolution of regional conflicts and a centre to co-ordinate aid and expertise to the new democracies in Eastern Europe.[107]

By early May, US policy on Europe seemed to shift in favour of institutionalizing the CSCE.[108] The new-found US enthusiasm for the CSCE process was as welcome in Western Europe as it was in Moscow, and especially in the FRG, where Foreign Minister Genscher had long espoused building upon the CSCE structure.

In Washington, Senator Nunn endorsed a parallel-path approach that covered an optimistic and a pessimistic future for Soviet *perestroika*.[109] The pessimistic path assumed that reforms would not take hold, that conservatives would regain power in Moscow and that the West would continue to face a serious adversary. For this contingency the policy prescription for the USA was obviously to maintain its leadership of NATO and to maintain a military presence in Europe, albeit at lower levels. The optimistic path assumed that reforms would succeed in the USSR, and that the 15 republics would form a more democratic union and become worthy partners in a pan-European security structure. The policy prescription for this contingency was a serious investment in joint ventures with the Soviet republics and the emerging democracies in Eastern Europe, and rapid institutionalization of the CSCE. For the short term Senator Nunn urged that both paths be pursued simultaneously. As NATO's Supreme Allied Commander, US General John Galvin, noted in late April, the creation of a common security system in Europe does not eliminate the necessity for the NATO alliance.[110] The need to invest in the CSCE process is especially important to head off potentially virulent expressions of nationalism in the weak and impoverished East European states. The CSCE could also serve to anchor a united Germany in a security structure that is reassuring in both halves of Europe.

European attitudes to the US role in Europe

European attitudes towards a US military presence depend on the extent to which US forces are deemed necessary to balance Soviet and German power. As noted above, the 1990s began with a perception that the WTO had virtually collapsed as a military coalition and that the risk of a WTO attack westwards was minimal.[111] Nevertheless, even after a CFE agreement has been concluded and the scheduled unilateral WTO cuts have been accomplished, the USSR will retain powerful nuclear and conventional military forces, remaining a military superpower that many believe will require balancing by US forces stationed in Europe. Furthermore, even if the Soviet threat fades altogether, the prospect of German unification convinced many policy makers in both halves of Europe that a US presence would still be necessary in the 1990s. In March 1990 for example, in an address to the Assembly of the WEU, the Polish Foreign Minister, Krzysztof Skubiszewski, said that the unification of Germany made the stabilizing role of the USA in Europe more important than ever.[112]

Europeans differ about the best way to anchor a united Germany. France tends to accord a higher priority to the European Community than to NATO, but President Mitterrand was energetic in early 1990 in forging links with President Bush, and clearly believes a US presence continues to be important.[113]

The new security arrangements in Europe will likely be a mix of three institutions: a modified NATO, an institutionalized CSCE and a wider and deeper EC. Americans naturally look to those institutions in which they play a role as the best bases for a new European security system. In addition to NATO and the CSCE, these include the 23-state CFE forum, the Economic Commission for Europe (ECE), the Organization for Economic Co-operation and Development (OECD) and the new European Bank for Reconstruction and Development (EBRD).

Most Europeans want to retain a US presence but they also look to institutions to which the USA does not belong, notably the EC and the Council of Europe, as building-blocks for a new pan-European system. Of these the EC is deemed the best foundation for developing a stable set of pan-European economic relations. EC members vary, however, as to what additional tasks the EC might perform, and when. All agree that the EC could perform a democratizing function for the newly emerging pluralist states in Central Europe, much as the EC has already done for Spain and Portugal.

One school of thought, represented by Jaques Delors, President of the European Commission, favours deepening political and economic integration among current EC members before admitting any new members or associates to the EC club. Former President of France Giscard d'Estaing outlined the 'deeper EC' position in an address to the Royal Institute of International Affairs in London in July 1989. Giscard described Europe as five concentric circles. In the centre he saw the 12 West European states in the process of creating a political and economic union; second came the group of the European Free Trade Association (EFTA) states; third the 6 non-Soviet WTO countries; fourth the isolated counties Yugoslavia, Albania, Malta; and fifth the European part of the USSR. Giscard dismissed as unfeasible President Gorbachev's concept of a common European home. Giscard's first priority was consolidation of a union of West European sovereign states that can deal on equal terms with other centres of world power—the USA, the USSR, China and Japan. Closer ties between the EC and the countries of Eastern Europe were a lower order priority.[114]

By contrast, the 'wider EC' school, represented by Prime Minister Thatcher, favours slower integration and earlier widening of the EC to embrace the East European states struggling to develop market economies.[115] The 'deepeners' tend to want the EC to develop competence in defence so that the EC becomes the European pillar of NATO; the 'wideners' want to retain as much national sovereignty in defence matters as possible, not least to make the EC more appealing to North and East Europeans of a neutralist bent.

To the extent that West Europeans interpreted the 'New Atlanticism' pronounced by the Bush Administration in December 1989 to mean that the USA saw the EC as the European pillar of NATO, it was welcomed by the 'deepeners'.[116] However, France was less enthusiastic about new institutional ties between the EC and the USA and about plans to politicize, and demilitarize, NATO.[117] At the NATO foreign ministers meeting in December 1989, French Foreign Minister Roland Dumas suggested that NATO, still dominated by the USA, was not the appropriate forum through which Western Europe should co-ordinate relations with Eastern Europe.[118] In January 1989 President Mitterrand tried to bridge the gap between 'deepeners' and 'wideners' with his proposal for a pan-European confederation. EC Commission President Delors, however, suggested that the EC would first have to to achieve political union.[119]

New enthusiasm for the CSCE

Predictably then, there was a warmer and wider welcome in Europe for the new US enthusiasm for the CSCE process that emerged in the spring of 1990; especially from those who remembered how lukewarm previous administrations had been about CSCE in the 1970s and 1980s.[120] In Bonn, Foreign Minister Genscher had long espoused building upon the CSCE structure. In London, Prime Minister Thatcher's enthusiasm for CSCE was relatively recent, but at the annual Anglo-German Konigswinter Conference in late March 1990 she offered seven proposals for building on the CSCE. These included provisions to ensure free elections, the rule of law and human rights in all CSCE countries; extension of political consultation among CSCE ministers; a conciliation role for the CSCE; provisions to ensure the right of private property; provisions that reaffirmed European CSCE frontiers of all the CSCE states; and provisions for future arms control negotiations after CFE.[121] These proposals were similar to recommendations offered by WTO foreign ministers in Prague earlier in March.[122]

VI. Summary

While this study does not anticipate a total withdrawal of US forces and facilities from Western Europe, it identifies a number of reasons why the US commitment to Europe is being reassessed on both sides of the Atlantic.

In Washington, there have been recurrent questions about the size of the US military presence in Europe ever since President Truman committed four extra divisions to the continent in 1950. Force levels have fluctuated, increasing in response to early crises in Berlin and decreasing when the USA was militarily engaged in other trouble spots. In general, however, the US presence has been remarkably stable and has withstood recurrent congressional attempts to 'bring the boys home' and let the allies take over their own defence. As NATO enters the 1990s, however, the rationale for maintaining several hundred thousand troops and hundreds of military facilities in Europe is being seriously questioned.

In part this is because the EC countries are moving steadily towards economic and political union with the expectation that greater integration must eventually mean integration of defence policies. On the other hand the prospect of German unification tempers European desire for complete autonomy from the USA, as US forces on the continent are seen not only as a necessary counterweight to Soviet power but also as a hedge against German domination.

The main reason why the US Congress may in the end be unwilling to fund a continued US military presence in Europe, however, is the radical change in Western perceptions of a military threat from the WTO. In the USSR, President Gorbachev has reduced the offensive capability of the military and has embarked on a conciliatory foreign policy designed to create optimum conditions for trade with and technology transfer from the West. His renunciation of the Brezhnev Doctrine revealed the illegitimacy of communist party rule in the region, and all six non-Soviet WTO countries are now looking westward for political and economic assistance as they move towards multiparty democracies and market economies. The USSR still retains a powerful military machine, but the WTO as a cohesive military alliance seems dead, and a Soviet advance into Western Europe is now almost as unthinkable as a war between France and the FRG.

Events in the USSR and Eastern Europe have thus undermined the rationale not only of the WTO but also of NATO. Yet, with Europe still in flux, the alliance structures are also recognized as stabilizing forces. As the cold war draws to a close the general mood on the continent is optimistic, but great uncertainties remain, especially about the cohesion of the USSR, the ability of the new governments in Eastern Europe to cope with their enormous economic and political problems, and how to design a new architecture for Europe that can absorb a prosperous and powerful unitary German state.

In both halves of Europe and on both sides of the Atlantic there are strong incentives to build a new security system that will be more co-operative and less adversarial than the post-war two-bloc system of NATO and the WTO. The scaffolding for such a Europe seem to be a stronger European Community and a rejuvenated NATO to anchor a united Germany in the short term while work proceeds to institutionalize the CSCE. The USA obviously will play a role in this new Europe. It is hoped that this volume will contribute to the debate about what that role should be.

Notes and references

[1] In response to such pressures President George Bush has stated: 'We will maintain forces in Europe—ground, sea and air, conventional and nuclear—for as long as they are needed and wanted, as I have pledged. Our forces in Europe are not tied exclusively to the size of the Soviet presence in Eastern Europe, but to the overall Alliance response to the needs of security', *National Security Strategy Report* (US Embassy: Stockholm, 20 Mar. 1990), p. 13.

[2] Kennan, G., 'Considerations affecting the conclusion of a North Atlantic security pact', *Foreign Relations of the United States* (FRUS) (US Government Printing Office: Washington, DC, 1974), vol. 3 (1948), pp. 284–85.

[3] Ireland, T. P., *Creating the Entangling Alliance* (Aldwych Press: London, 1981); McGeehan, R., *The German Rearmament Question: American Diplomacy and European Defense Diplomacy After World War II* (University of Illinois Press: Urbana, Ill., 1972).

[4] US military personnel levels in Europe and the North Atlantic 1950–66 are taken from *US Forces in Europe*, Hearings before the US Senate Foreign Relations Committee, 25-27 July 1973 (US Government Printing Office: Washington, DC, 1973), pp. 150–97. Force levels in 1972–86 are taken from *Selected Manpower Statistics for Fical years 1973–1986* (US Department of Defense, Washington, DC, 1974) See also Baldauf, J, *The American Critique of the US Troop Presence in Europe* (Stiftung Wissenschaft und Politik: Ebenhausen, Dec. 1987).

[5] For accounts of burden-sharing debates within the European alliances between the world wars, see Posen, B. *The Sources of Miltary Doctrine: France, Britain and Germany Between the World Wars* (Cornell University Press: Ithaca, N.Y., 1984; Christensen, T. and Snyder, J., 'Chain gangs and passed bucks', *International Organization*, vol. 44, no. 2 (spring 1990).

[6] For a NATO view, see NATO's Defence Planning Committee, *Enhancing Alliance Collective Security: Shared Roles, Risks and Responsibilities in the Alliance* (NATO: Brussels, Dec. 1988). See especially paragraphs 103–38, which rate the 15 members of the DPC according to their contributions to common NATO projects. France is not a member of the DPC.

[7] Morgan, R., *The United States and West Germany 1945–1973: A Study in Alliance Politics* (Oxford University Press: Oxford, 1974), pp. 82–84.

[8] Mendershausen, H., 'Troop stationing in Germany', Rand report no. RM 5881 (Rand Corp.: Santa Monica, Calif.), pp. 73–76.

[9] Morgan (note 7), p. 126; Treverton, G., *The Dollar Drain and American Forces in Germany* (University of Ohio Press: Athens, Ohio, 1978).

[10] Morgan (note 7), pp. 201–3.

[11] *The Security of the Federal Republic of Germany*, White Paper 1983 (FRG Ministry of Defence: Bonn, 1983) p. 26.

[12] President Johnson's hard bargaining on offset payments is thought by some analysts to have contributed to the political downfall of Chancellor Erhard in 1966. See, for example, Morgan (note 7), p. 148.

[13] Treverton (note 9).

[14] *Mutual and Balanced Forces, Declaration adopted by Foreign Ministers and Representatives of Countries Participating in the NATO Defence Programme* (15 June 1968), *NATO Final Communiqués: Text of Final Communiqués 1949–74* (NATO Information Service: Brussels, undated), pp. 209–10.

[15] Sagan, S., 'Nuclear alerts and crisis management', *International Security*, vol. 9, no. 4 (spring 1985), pp 122–28.

[16] *Congressional Record*, 24 Jan. 1970, p. S-493.

[17] Yost, D. S., 'Western Europe and the U.S. Strategic Defense Initiative', *Journal of International Affairs*, vol. 41, no. 2 (summer 1988), 269–323.
[18] Kissinger, H. A., 'A plan to reshape NATO', *Time*, 5 April 1984.
[19] Kissinger, H. A., 'Alliance cure: redeployment', *Washington Post*, 13 May 1986.
[20] Brzezinski, Z., 'A partial US pullout could benefit Europe', *International Herald Tribune*, 10 June 1987; see also by the same author, *Gameplan: How to Conduct the US–Soviet Contest*, (The Atlantic Monthly Press: Boston, 1986).
[21] Krauss, M., *How NATO Weakens the West* (Simon & Shuster: New York, 1986).
[22] US Deparment of Defense, *Report of the Secretary of Defense Caspar W. Weinberger to the Congress for FY 1988* (p. 258) and *Report of the Secretary of Defense Frank Carlucci to the Congress for FY 1989* (p. 74), annual reports (US Government Printing Office: Washington, DC, 1987 and 1989).
[23] The Stevens amendment became Public Law (P.L.) 97-377.
[24] Williams, P., *US troops in Europe*, Chatham House Paper 25 (Royal Institute for Foreign Affairs: London, 1984), p. 29; see also by the same author, 'The United States commitment to Western Europe', paper presented at the annual International Studies Association—British International Studies Association Conference, London, April 1989.
[25] The European Troop Strength ceiling (ETS) is laid down in P.L.98-525. P.L.99-145 authorizes the ETS to be exceeded by up to 0.5% in certain contingencies. See US Department of Defense, *Manpower Requirements Report for FY 1988* (DOD: Washington, DC, 1987), pp. ix–26.
[26] US General Accounting Office, *Military Presence: US Personnel in NATO Europe* (GAO: Washington, DC, 6 Oct. 1989), p. 2.
[27] US House of representatives, *Report of the Defense Burdensharing Panel of the Committee on Armed Services*, 100th Congress, 2nd Session (US Government Printing Office: Washington, DC, Aug. 1988).
[28] Barber, L., 'US lists 86 bases for closure to cut defence costs', *Finanacial Times*, 30 Dec. 1988.
[29] Fitchett, J., 'Experts say closures of bases will not harm US efficiency', *International Herlad Tribune*, 30 Jan. 1990; Mills, M., 'Cheney's plan for shutdowns a new salvo in long fight', *Congressional Quarterly*, 3 Feb. 1990, pp. 340–42.
[30] Riddell, P., 'Democrat leaders urge deeper cuts in forces', *Financial Times*, 5 Feb. 1990; Mills (note 29).
[31] Anderson, H., Barry, J., Warner, M. and Coleman, F., 'Trimming the troops', *Newsweek*, 12 Feb. 1990, pp. 16–18.
[32] Tyler, P. E., 'US Army outlines plans to cut quarter of troops by 1997', *Financial Times*, 16 Apr. 1990; Reuters, 'Cheney says Soviet crisis could delay military cuts', *International Herald Tribune*, 26 Apr. 1990; Tran, M., 'US ready to cut troops strengths', *The Guardian*, 16 Apr. 1990.
[33] Gordon, M. R., 'Nunn sets the terms for military debate', *International Herald Tribune*, 21–22 Apr. 1990.
[34] On the dispute between the USA and the FRG over the WINTEX exercises in 1987 and 1989, see: Pond, E., 'War games brought NATO rift into the open', *Boston Globe*, 4 June 1989; Fouquet, D., 'Exercise raises NATO and US nuclear debate', *Jane's NATO Report*, vol. 40, no. 28, 4 Apr. 1989, p. 6; Catterall, T., 'War game idiocy fires Kohl resolve', *The Observer*, 30 Apr. 1989.
[35] Kelleher, C. M., *Germany and the Politics of Nuclear Weapons* (Columbia University Press: New York, 1975); Kelleher, C. M., SIPRI, 'The debate over the modernization of NATO's short-range nuclear missiles', *SIPRI Yearbook 1990: World Armaments and Disarmament* (Oxford University Press: Oxford, 1990), pp. 603–22; Schwarz, D. N., *NATO's Nuclear Dilemmas* (Brookings Institution: Washington, DC, 1983).
[36] Sharp, J. M. O., 'After Reykjavik: arms control and the allies', *International Affairs*, vol. 63, no. 2 (spring 1987), pp. 239–57.
[37] Housego, D., 'Soviet arms proposal splits French cabinet', *Financial Times*, 6 Mar. 1987; Amalric, J. 'L'Affaire des euromissiles divise la majorité', *Le Monde*, 6 Mar. 1987;

Marshall, D. B., 'France and the INF negotiations: an American Munich', *Strategic Review*, vol. 15, no. 3 (summer 1987), pp. 20–30.

[38] For example, at the annual Wehrkunde meeting in Munich, US Secretary of Defence Frank Carlucci stated that European reluctance to deploy a nuclear follow-on to the Lance missile would erode the US security commitment to West European security, and that if NATO scrapped its tactical nuclear weapons, the USA would withdraw its forces from Europe. See Markham, J., *New York Times*, 10 Feb. 1988; Lunn, S., 'Current SNF structure and future options', ed. O. Bosch, *Short-range Nuclear Forces: Modernization and Arms Control* (Council for Arms Control: London, 1989), pp. 1–12.

[39] *The Independent*, 26 May 1989.

[40] *Frankfurter Rundschau*, 27 Apr. 1989.

[41] NATO's *Comprehensive Document on Arms Control* (NATO: Brussels, 1989).

[42] White, D., 'NATO searches for a new short-range strategy', *Financial Times*, 4 May 1990; AP, 'Kremlin welcomes US arms proposals', *International Herald Tribune*, 5–6 May 1990; 'Bush says united Germany should belong to NATO' (Bush Oklahoma State University Commencement Address), official transcript, US Information Service, US Embassy, London, 8 May 1990. See also British–American Security Information Council, *NATO Nuclear Planning After the Cold War* (BASIC: London, 1990).

[43] Millward, D., 'Military links with France to be stepped up', *Daily Telegraph*, 5 May 1990; Nundy, J., 'UK and France to co-ordinate security', *The Independent*, 5 May 1990.

[44] Marsh, D., 'Air missile plans expected to provoke NATO dispute', *Financial Times*, 4 May 1990.

[45] Sloan, S. R., *NATO's Future: Toward a New Transatlantic Bargain* (National Defense University Press: Washington, DC, 1985), p. 53.

[46] Ireland (note 3).

[47] Diebold Jr., W., *The Schuman Plan: A Study in Economic Cooperation 1950–1959* (Praeger: New York, 1959).

[48] *Final Act of the Nine Power Conference held in London, September 28–October 3, 1954* (Her Majesty's Stationery Office: London, Oct. 1954), Cmnd 9289, pp. 17–18; cited in Fursdon, E., *The European Defence Community: A History* (Macmillan: London, 1980), p. 321.

[49] *Paris Protocols Amending the Brussels Treaty and Establishing the Western European Union, October 23, 1954*, in *Documents on Germany 1944–1961*, Committee on Foreign Relations, US Senate (US Government Printing Office: Washington, DC, Dec. 1961), pp. 155–75.

[50] See Bertram, C., 'European security and the German problem', *International Security*, vol. 4, no. 3 (winter 1979–80), pp. 105–16; Joffee, J., 'Europe's American pacifier', *Foreign Policy*, no. 54 (spring 1984), pp. 64–82. For a typical British view of the German problem, see Tugendhat, C., *Making Sense of Europe* (Pelican Books: Harmondsworth, 1987), pp. 222–23.

[51] Named for Belgian Foreign Minister Léo Tindemans.

[52] Cited in Kaiser, K. and Roper, J. (eds), *British–German Defence Co-operation: Partners Within the Alliance* (Jane's: London, 1988), p. 162.

[53] Gambles, I., *Prospects for West European Security Cooperation*, Adelphi Paper no. 244 (International Institute for Strategic Studies: London, autumn 1989); Sloan (note 45); Roper, J., 'European defense cooperation', eds C. M. Kelleher and G. A. Mattox, *Evolving European Defense Policies* (Lexington Books: Mass., 1987).

[54] Minc, A., *La Grande Illusion* (Bernard Graset: Paris, 1989).

[55] Grosser, A., 'German question, French anxiety', *New York Times*, 26 Dec. 1089.

[56] Schmidt, H. and Giscard d'Estaing, V., 'Ties that bind the new Europe', *The Times*, 14 Feb. 1990.

[57] Marsh, D., 'Franco-German declaration on European political and monetary union: Bonn initiative behind ambitious target', *Financial Times*, 20 Apr. 1990.

[58] Palmer, J., 'Ministers scupper a federal Europe', *Guardian Weekly*, 27 May 1990.

[59] *New York Times*, 10 Sep. 1950, cited by Sloan (note 45)

[60] *Congressional Record*, 92nd Congress, 1st session, 18 May 1971, pp. S-7215–16.

[61] Enthoven, A. C. and Smith, K. W., *How Much is Enough? Shaping the Defense Program 1961–1969* (Harper Colophon: New York, 1971).

[62] The Eurogroup now has 12 member countries: Belgium, Denmark, the FRG, Greece, Italy, Luxembourg, the Netherlands, Norway, Portugal, Spain, Turkey and the UK.

[63] Cleveland, H., *The Transatlantic Bargain* (Harper & Row: New York, 1970), p. 128.

[64] See the testimony of Kenneth Rush, Deputy Secretary of State, to the US Senate Foreign Relations Committee in *US Forces in Europe* (note 4), p. 46.

[65] Sandler, T. and Forbes, J., 'Burden sharing, strategy and the design of NATO', *Economic Enquiry*, vol. 18 (July 1980). See also chapters 8 and 13 in this volume.

[66] Stürmer, M., 'Is NATO still in Europe's interest?', ed. S. R. Sloan, *NATO in the 1990s* (Pergamonn–Brasseys: London, 1989), p. 109.

[67] Voigt, K., Rapporteur, Defence and Security, North Atlantic Assembly Report on Alliance Security, Brussels, Oct. 1989.

[68] Taylor, T, 'Alternative structures for European defence co-operation', eds Kaiser and Roper (note 52), pp. 170–83; Sloan (note 45).

[69] Taylor (note 68), pp. 173–75.

[70] *Platform on European Security Interests: The Hague, 27 October 1987*, in Western European Union, *The Reactivation of WEU: Statements and Communiques 1984–1987* (WEU: London, 1988), pp. 37–45.

[71] For a comprehensive review of developments in the WEU during the late 1980s, see Clarke, M., 'Evaluating the New Western European Union: the implications for Spain and Portugal', ed. K. Maxwell, *Democracy and Foreign Policy in Portugal* (Duke University Press: Durham, N.C., 1990); van Eekelen, W., 'Security: for a strong West European pillar', *International Herald Tribune*, 8 Mar. 1990.

[72] Mauthner, R., 'European pillar for NATO meeting today', *Financial Times*, 23 Apr. 1990.

[73] 'Ashdown proposes European military', *The Independent*, 10 Mar. 1990.

[74] Gow, D., 'The man who would be Chancellor', *The Independent*, 10 Mar. 1990.

[75] Becher, K. and Kolbloom, I., 'A job for the French in East Germany', *International Herald Tribune*, 18 Apr. 1990.

[76] Evangelista, M. A., 'Stalin's postwar army reappraised', and Lebow, R. N., 'The Soviet offensive in Europe: the Schlieffen Plan revisited', eds S. M. Lynn-Jones, S. E. Miller and S. Van Evera, *Soviet Military Policy* (MIT Press: Cambridge, Mass, 1989).

[77] 'Statement of the Warsaw Pact Defence Ministers Committee "on the Correlation of Warsaw Pact and North Atlantic Alliance Force Strengths and Armaments in Europe and Adjacent Waters"', in 'Warsaw Pact releases figures on force strenghs', *Pravda*; 30 Jan. 1989, p. 5, in *Foreign Broadcast and Information Service–Soviet Union (FBIS-SOV)*, FBIS-SOV-89018, 30 Jan. 1989, pp. 1–8; Mishin, Y., 'Warsaw Pact forces, budget reduction figures', *Argumenty i Facty*, no. 6 (11–17 Feb. 1989), p. 8, in *FBIS-SOV-034*, 22 Feb. 1989, p. 3; Gorbachev speech, FBIS-SOV-89-103S, 31 May 1989, pp. 47–62; Kornilov, Y., 'Facts Behind the military budget figures', *Soviet Weekly*, 17 June 1989.

[78] Speech by President Gorbachev to the Council of Europe in Strasburg, 6 July 1989. See also Tass, 'Statement on 1968 invasion of Czechoslovakia', 4 Dec. 1989, in BBC *Summary of World Broadcasts–Eastern Europe (BBC-SWE-EE)*. On renunciation of the Brezhnev doctrine, see Tass, 'Communiqué of Warsaw Treaty Political Consultative Committee', 8 July 1989, in *BBC-SWB-EE*, 10 July 1989.

[79] Speech by President Gorbachev at the UN General Assembly, 7 Dec. 1988, *Soviet Diplomacy Today* (Soviet Ministry of Foreign Affairs), 1989, pp. 40–47.

[80] For details of WTO unilateral cuts in 1989, see Sharp, J. M. O., 'Conventional arms control in Europe', SIPRI, *SIPRI Yearbook 1990: World Armaments and Disarmament* (Oxford University Press: Oxford, 1990), pp. 459–74.

[81] Reuters, 'Soviet Army said to suspend pullout from East Germany', *New York Times*, 17 May 1990.

[82] Fisher, M., 'At German talks, Soviet face saving is critical', *International Herlald Tribune*, 5–6 May 1990.

[83] 'Germany sets the sylabus for the new maths of Europe', *The Economist*, 5 May 1990. See also the nine-point plan on Germany that President Bush proposed to President Gorbachev at the Washington summit meeting in June 1990. Hoffman, D. and Oberdorfer, D., 'US offer Soviets a nine-point plan on Germany', *International Herald Tribune*, 4 June 1990.

[84] Leonid Brezhnev first used the phrase Common European Home in Bonn in 1981. Malcolm, N., *Soviet Policy Perspectives on Western Europe*, Chatham House Paper (Royal Institute for Foreign Affairs/Routledge: London, 1989). Mikhail Gorbachev first used it in a speech in Prague in April 1987. *Pravda*, 11 Apr. 1987; see also Gorbachev, M., *Perestroika* (Collins: London, 1987), pp. 194–95.

[85] See, for example, the Joint Soviet–FRG Declaration (Gemeinsame Erklärung von Bundeskanzler Kohl und Generalsekretär der KPdSU Gorbatschow) of 13 June 1989, in *Bulletin der Bundesregierung*, no. 61 (15 june 1989), p. 542. For the text in English, see Tass, 'Gorbachev, Kohl issue joint statement 13 June', *Pravda*, 14 June 1989, pp. 1–2, in *FBIS-SOV-89-113*, 14 June 1989, pp. 16–18.

[86] Lellouche, P., 'Architect Gorbachev has designs on Europe', *International Herald Tribune*, 12 Feb. 1988.

[87] The Soviet proposal for a 'General European Treaty on Collective Security in Europe' is reprinted in *Documents on Germany* (note 49), pp. 152–54.

[88] *Documents on Germany* (note 49), pp. 181, 195 and 202.

[89] *Documents on Germany* (note 49), p. 327.

[90] 'Address by the Polish Foreign Minister Rapacki to the UN General Assembly 14 December 1964', US Arms Control and Disarmament Agency, *Documents on Disarmament 1964* (US Government Printing Office: Washington, DC, 1965), p. 527.

[91] *Pravda*, 3 Apr. 1966.

[92] See, for example, the *Military Balance 1966–1967* (International Institute for Strategic Studies: London, 1966), p. 2.

[93] Address by General Secretary Leonid Brezhnev at Tbilisi, 14 May 1971, in *Documents on Disarmament 1971* (US Government Printing Office: Washington, DC, 1972), p. 93.

[94] *Détente: Hearings on Relations With Communist Countries*, Hearings before the Committee on Foreign Relations, US Senate, 93rd Congress, 2nd Session (US Government Printing Office: Washington, DC, 1974), p. 144.

[95] Karpov, V. and Oszi, I., 'Warsaw Pact: to be or not to be?', *Komsomolskaya Pravda*, 20 Dec. 1989, *FBIS-SOV-89-245*, 22 Dec. 1989; Trainor, B. E., 'Turmoil's offspring: Warsaw Pact seen losing its purpose', *International Herald Tribune*, 21 Dec. 1989; Freedman, L., 'How can the Pact survive?', *The Independent*, 15 Dec. 1989.

[96] Pick, H., 'Remodel or fragment, Soviet analyst warns Europe: USSR no longer world superpower', *The Guardian*, 22 Nov. 1989.

[97] 'Disarming', *The Economist*, 9 Dec. 1989, pp. 53–54.

[98] 'NATO Diary', *NATO's Sixteen Nations*, Dec. 1989–Jan. 1990, p. 52; 'Eduard Shevardnadze's Address to the Political Commission of the European Parliament', *Pravda*, 20 Dec. 1989, *FBIS-SOV-89-243*, pp. 28–29.

[99] 'Eduard Shevardnadze's Address to the Political Commission of the European Parliament' (note 98); see also Gorbachev's speech to the Council of Europe (note 79) and Shenfield, S., 'Minimum nuclear deterrence: the debate among Soviet civilian analysts', Brown University Center for Foreign Policy Development, Providence, R. I., 1989.

[100] 'Germany outlines 4-power proposal', *International Herald Tribune*, 2 Mar. 1990.

[101] Hoagland, J., 'As the Germans look to NATO, Soviets urge end to Europe's alliances', *International Herald Tribune*, 30 Apr. 1990.

[102] Almquist, P., 'The Vienna military doctrine seminar: Flexible Response vs. Defensive Sufficiency', *Arms Control Today*, vol. 20, no. 3 (Apr. 1990), pp. 21–25. See also Apple, R. W., 'Gorbachev bares a strategic weakness', *International Herald Tribune*, 16 Feb. 1990.

[103] Gorbachev interview in *Pravda*, 20 Feb. 1990, in *FBIS-SOV-90-039*, 27 Feb. 1990, pp 6–7; Shevardnadze speech at the 'open skies' meeting of the 23 CFE foreign ministers in Ottawa, 12 Feb. 1990, reprinted in *Vestnik*, Apr. 1990, pp. 60–61; Shevardnadze, E., ['Germany and the law'], *Neue Berliner Illustrierte*, reprinted in *Vestnik*, April 1990, pp. 56–57; Akhromeyev interview, 'Military aspects of German unification cited', *Bratislava Pravda*, 30 Mar. 1990, *FBIS-SOV-90-064*, 3 Apr.1990, pp. 1.

[104] Tomforde, A., 'Two Germanys get down to nitty gritty of unity', *The Guardian*, 19 Apr. 1990; Remnick, D., 'East Germany, Moscow, at odds on NATO',*Guardian Weekly*, 6 May 1990.

[105] Danilov, D., 'The search for the ideal compromise', *Moscow News*, no. 15 (15 April 1990). See also Schmidt-Hauer, C., 'Poker zum Beginn', *Die Zeit*, 4 May 1990, citing Soviet Foreign Ministry Chief of Planning, Sergei Tarassenko, to the effect that if a united Germany were eventually anchored in a pan-European system, it would not be destabilizing to have a united Germany in NATO for a transition period.

[106] Steele, J. and Gow, D., 'Soviet plan for Germany in both pacts', *The Guardian*, 12 Apr. 1990. See also inteviews with Shevardnadze in *NATO's Sixteen Nations*, May 1990.

[107] *A New Europe, A New Atlanticism: Architecture for a New Era*, address by Secretary of State Baker to the Berlin Press Club, Berlin, 12 Dec. 1989, *Current Policy* (US State Dept.), no. 1233.

[108] Hoffman, D., 'Bush urges broadened political mission for NATO', *International Herald Tribune*, 5–6 May 1990; 'The common European interest: America and the new politics among nations' (Baker address to the National Committee on American Foreign Policy), offical trancript, US Dept. of State, Washington, 14 May 1990; 'Remarks by the President in University of South Carolina Commencement Address', official transcript, White House, Office of the Press Secretary, Washington, 12 May 1990.

[109] Sloan, S. R., Congressional Research Service, 'The United States and a new Europe: strategy for the future', CRS Report no. 90-245 RCO (US Library of Congress: Washington, DC, 14 May 1990).

[110] 'Das Bündnis in Wandel: Ein Gespräch mit NATO Oberbefehlshaber John Galvin', *Frankfurter Rundschau*, 28 Apr. 1990, p. 6.

[111] At a lecture given at the Swedish Institute of International Affairs in Stockholm on 4 May 1990, NATO SACEUR General Galvin referred to 'the former Warsaw Pact'.

[112] A transcript of the speech was supplied by the Polish Embassy, Stockholm. Skubiszewski's views were echoed by Christoph Bertram in *Die Zeit* the following day; Bertram, C., ['America is determined to prevent Europe from reverting to nation-state roles of old'], *Die Zeit*, 23 Mar., 1990.

[113] Dawkins, W. and Barber, L., 'European security tops Mitterrand, Bush agenda', *Financial Times*, 10 Apr. 1990.

[114] Giscard d'Estaing, V., 'The two Europes: East and West', *International Affairs*, vol. 65, no. 4 (autumn 1989), pp. 653–58.

[115] Rogaly, J., 'Steadying hands on wobbly Europe', *Financial Times*, 26 Jan. 1990.

[116] See, for example, Eberle, J., Kaiser, K and Moisi, D., 'After Baker's speech, the ball is in Europe's court', *International Herald Tribune*, 22 Dec. 1990.

[117] Lemaitre, P., 'Time for EC to grasp leadership of Europe', *The Guardian*, 14 Jan. 1990.

[118] Dumas is cited in *Defense News*, 18 Dec. 1989.

[119] Dawkins, W., 'Mitterrand call for European confederation', *Financial Times*, 2 Jan. 1990. See also Dumas, R., 'Together a greater Europe', *International Herald Tribune*, 15 Mar. 1990.

[120] On US scepticism about CSCE in the Nixon Administration, see Maresca, J. J., 'Helsinki accord 1975', eds A. George, P. J. Farley and A. Dallin, *US–Soviet Security Co-operation* (Oxford University Press: Oxford, 1988), pp. 106–22.

[121] Foreign and Commonwealth Office Arms Control and Disarmament Research Unit, 'Prime Minister's Speech to the Anglo-German Konigswinter Conference, Cambridge, Thursday 29 march 1990', *Quarterly Review*, 17 Apr. 1990, pp. 25–30.

[122] Speech by Foreign Minister Jiri Dienstbier of Czechoslovakia at the meeting of the WTO in Prague, 17 Mar. 1990; English translation from the Czechoslovak Embassy in Stockholm, 3 Apr. 1990.

2. Summary and conclusions

Jane M. O. Sharp
SIPRI

I. Economic issues

A primary purpose of this study was to assemble the most reliable data publicly available on the economic costs and benefits (to all the NATO countries) of US forces and facilities in Europe. This has not been an easy task. Despite Western criticism of the Warsaw Treaty Organisation (WTO) states for lack of openness with military budget data, this project demonstrates that it is also difficult to obtain reliable data about NATO expenditures. Some NATO countries are more forthcoming than others. The USA is more open than Britain, for example, but also data that used to be publicly available in Washington are now classified, not only for the usual security reasons (i.e., to keep information from potential adversaries) but also to defuse domestic opposition to high overseas defense costs. In addition, different NATO allies in Europe do not wish to disclose the bilateral arrangements they have made with the USA about offset payments.

Views from the United States

Alice Maroni (chapter 3) outlines these analytical constraints in her overview of the costs and benefits to the USA of basing forces and facilities in Europe.

The US Department of Defense (DOD) puts the cost of the US commitment to Europe at 60 per cent of the US defense budget, on the order of $180–200 billion in 1987. As Maroni explains, however, this is too all-encompassing a figure, covering all costs associated with Europe, including equipment and training for forces based in the continental USA and earmarked for reinforcement of Europe. Maroni estimates that the direct and indirect budgetary costs of forces and facilities actually in Europe are closer to $50 billion ($47 billion in FY 1987). This includes costs for all forward-deployed general purpose forces and support elements forward deployed in Europe, and corresponds to approximately 15 per cent of the total DOD budget. Of these the direct costs constitute approximately $15 billion. The incremental costs of supporting these forces in Europe, as opposed to supporting them in the continental USA, is $2 billion.

Maroni estimates that withdrawal option A, in which returning US troops are demobilized, could save $11 billion per year, although such savings could be offset by additional expenses incurred in restructuring forces; naval forces might have to be improved, for example. Under option B, in which forces withdrawn would be relocated in the USA, these savings would be more than offset by the considerable incremental costs of building new facilities and providing extra prepositioned equipment in Europe.

Maroni assumes that withdrawal under option A, the 'gone for good' option, would seriously undermine the USA commitment to a 'two-way street' in transatlantic defence procurement, since the economic arguments in its favour have never been compelling. Under option B, the incentives for defence industrial co-operation would remain high, but this in turn will depend on the degree of protectionism that enters into West European transactions after a single EC market is established in 1992.[1]

Joseph Higdon and Paul Friedrich (chapter 4) suggest that the impact of 1992 is likely to be a more significant determinant of transatlantic investment than either of the hypothetical withdrawal options presented in this study. Higdon and Friedrich, both investment analysts, see the fundamental transatlantic relationship as primarily an economic one that is not greatly dependent on the US military presence in Europe. They argue that three factors determine US interest in business growth in Europe: the substantial commitment of US companies already in Europe, the increasingly constructive climate between Western Europe and its Eastern neighbours, and the scramble to obtain a sure foothold in Europe before the single EC market is established in 1992. In their view policy makers in the USA will have to adjust to Europe not needing it militarily, as the US business community has already done.

West European perspectives

David Greenwood (chapter 5) agrees with Maroni's basic findings that the cost generally attributable to forces actually deployed in Europe is approximately 16 per cent of the DOD budget, or approximately $50 billion in 1987–88. While this may seem like a small sum in the context of the total DOD budget, Greenwood believes that it would be a daunting sum to West European governments trying to compensate for a US withdrawal.[2]

Greenwood believes the purely economic impact of a US withdrawal would mean that European NATO would suffer a loss of income of up to $20 billion and job losses of approximately 225 000. Spread among the 14 nations this would pass virtually unnoticed, a faltering in the growth rate or a blip in the unemployment curve. Certainly withdrawal would pose no insuperable problems of macro-economic adjustment. The problem is rather that the impact would not be evenly spread, with the result that micro-economic adjustment in specific areas could present serious difficulties.

For local government officials the principal preoccupation would be the loss of jobs and income from the US bases; for NATO defence ministries it would rather be the costs of replacing the lost military capability. Greenwood sees the possibility of compensation by national forces as unfeasible, but is more optimistic that a US withdrawal could stimulate a new transatlantic bargain and encourage West European co-operation. This judgement is based first on the new political situation created by President Gorbachev and the expectation that force levels for both NATO and the WTO will decrease substantially as a result of both unilateral WTO cuts and the Conventional Armed Forces in Europe (CFE) Negotiation. In addition, Greenwood asserts that there is ample scope in Western Europe for more rational application of military capability and industrial capacity.

The Federal Republic of Germany hosts by far the largest US presence in NATO, with some 209 000 Army, 43 000 Air Force (USAF), 400 Navy and 100 Marine Corps personnel, plus 28 000 dependents. This large contingent represents a heavy burden for the federal budget since under the terms of the basing agreements, US troops along with other forces stationed in the FRG may use all necessary immovables free of charge. But if the federal costs are rather high, the local economic benefits are often considerable. This is especially so for the six small rural communities that host USAF squadrons. USAF bases provide a relatively high percentage of the jobs available for the local population and spending by USAF personnel and their dependants tends to boost the local economies. Since 1985, however, the fall in the value of the dollar has curbed spending considerably by all US personnel and dependants in Europe.

US Army personnel seem to be younger, less well educated, and less well off than their USAF counterparts. They therefore pump less money into local economies and tend to be more disruptive. Local populations find the costs of hosting US Army bases more obvious than the benefits. Despite periodic opposition to low flying exercises, they find the USAF easier to absorb.

Hartmut Bebemeyer and Christian Thimann (chapter 6) assess the economic impact of option B as significantly worse for the FRG than option A, since all the benefits of the US presence would vanish but the costs of keeping the bases running to receive returning troops in a crisis and for regular training purposes would soar. This assessment does not, however, factor in the possible differential costs of replacing US troops withdrawn with extra West German forces.

Belgium, Luxembourg and the Netherlands emerged from the World War II convinced of the need to align themselves with the major Western powers. All were enthusiastic founding members of the 1948 Brussels Treaty and of NATO and supported the effort to establish a European Defence Community in the mid-1950s. Although their military contributions are modest, the Benelux countries form the administrative heartland of NATO, and Belgium and the Netherlands provided three of the first five Secretary Generals of the alliance.

Sami Faltas (chapter 7) notes that during the Viet Nam War in the late 1960s and early 1970s and during the Euromissile crisis of the early 1980s there were signs of anti-Americanism in the low countries, but never any serious questioning of membership in NATO, nor any pressure to send the US troops back home. On the contrary public opinion polls indicate overwhelming approval of the US presence as tangible evidence of the US security guarantee. By NATO standards the US military presence in the Benelux is relatively small, with only 3418 military personnel in Belgium and 3130 in the Netherlands.

SUMMARY AND CONCLUSIONS 37

Withdrawal of these forces and their dependents would therefore have a minimal economic impact. Faltas sees the extra defence effort that might be required of the Benelux countries to make up for the loss of US forces not only from the low countries but also from Western Europe as a whole, in particular from the FRG, as a more serious costs.

In assessing the costs and benefits of the US presence in Britain, Keith Hartley and Nicholas Hooper (chapter 8) suggest that any adverse effects of a US withdrawal, for example on US investment, are likely to be temporary and short term. Given the historical links between the USA and the UK, they suggest that in the long run a stable UK political environment offering profitable opportunities will continue to attract foreign investment. After all, British firms continue to invest abroad in countries from which they have withdrawn military bases.

US forces and facilities based in Britain are air and naval assets rather than ground forces. As such they are obviously there not primarily for the defence of British territory but for the alliance as a whole. In terms of the economic contribution to Britain, US forces added some £800 million in foreign exchange earnings to the UK balance of payments account for 1987–88 and provided some 30 000 jobs both directly and indirectly. Against these benefits were the costs associated with host nation support for the US forces in the form of free or subsidized support, such as for land, housing, schooling and recreation.

In assessing the likely impact on the British economy of a US withdrawal, Hartley and Hooper assume that even if Britain could increase its defence budget, it could not afford and would not choose to replace or reproduce all the US forces being withdrawn. Compensation for combat air assets lost, for example, would obviously have to be done on a NATO-wide basis. Assuming that UK defence budgets would remain essentially unchanged, Hartley and Hooper assume that Britain would be more likely than under current circumstances to align itself with a collective European defence effort on the basis of specialisation by comparative advantage. Under withdrawal option B, in which US forces are committed to return to Europe in a crisis, they assume that extra funds would have to be spent on the care and maintenance of bases to receive returning forces; perhaps at the expense of investing in common European programmes. Finally, assuming a cut in British defence spending to a percentage of the GNP more in tune with the continental allies, the UK would have to look more seriously at the world market to purchase low cost equipment, with possible adverse effects on British defence industry.

Costs and Benefits in the North

Unlike the central region, NATO's Northern Flank does not host many US military personnel but provides bases for important command, control, communication and intelligence (C^3I) facilities. Simon Duke (chapter 9) notes that the economic impact of the US peacetime presence is minimal in the case of Denmark and Norway, but substantial in the case of Greenland and Iceland.

US forces at the Keflavík and Helguvík bases in Iceland play an important strategic anti-submarine warfare role in the north Atlantic. They also substitute for national defence forces. Despite some local opposition, the economic benefits of the US presence in Iceland are considerable, especially in terms of maintaining and running an international airport.

Norway is one of the NATO countries for which it is difficult to obtain data on US facilities since many of them are sensitive intelligence-gathering facilities. While Norway (like Denmark) does not permit the stationing of foreign military forces in peacetime, three important sets of agreements between Norway and the USA established facilities to be used by US forces in wartime. These include the Collocated Operating Base (COB) agreements, the first of which was signed in 1952 and updated in 1974, and the Invictus Arrangements of 1971 and 1980. Both Invictus and COB provide for wartime use by US aircraft of Norwegian airfields. The third agreement provides for the storage of equipment in Norway for a US Marine Corps brigade.

Mediterranean perspectives

Bilateral basing arrangements between the USA in Spain and Portugal have been politically sensitive because the arrangements were originally negotiated with the dictatorial Salazar and Franco regimes. Not surprisingly, the centre–left governments that later came to power in both countries sought to distance themselves from their predecessors and to seek increased independence from the USA. In the event, however, both Spain and Portugal negotiated new base agreements with Washington in the late 1980s, as both countries needed the economic and military aid that came with the basing agreements. This is not to say the agreements were perfectly satisfactory. Both Portugal and Spain complained about decreased levels of US aid and both remain nervous that the US Congress might step in and dilute the aid commitments in the latest agreements. In particular, with the democratization of the non-Soviet WTO countries in 1989–90, Spain expressed concern that investment in Eastern Europe might pre-empt investment in Spain, while Portugal expressed similar concerns about economic aid being siphoned off to Eastern Europe.

As a new member of the EEC and the Western European Union (WEU) as well as NATO, Spain appeared to enter the 1990s with a serious interest in moving beyond bilateral defence co-operation with the USA to broader co-operation with the rest of Western Europe.

Spain and Portugal joined the WEU at the same time but with somewhat different goals in mind. Spain's attitude appeared close to that of France in regarding the WEU as a possible successor to NATO, while Portugal leaned towards the British model which values WEU only insofar as it does nothing to prejudice the transatlantic relationship in NATO.[3]

José Molero (chapter 10) emphasizes that although the terms of trade have improved for Spain with each new bilateral agreement, the relationship remains asymmetric, with the USA still getting the better deal. Spaniards complain that Spain has received rather low-grade US equipment, for instance. On the other hand, the current arrangements obviously represent a net benefit for the Spanish industrial base, and withdrawal would constitute a serious loss that Spain would find difficult to compensate without assistance from the other Western allies. The likely trade-off would be a reconsideration of the current restrictions on integrating Spanish forces with those on the Central Front.

Portugal is one of the founding members of NATO and, as Jane Sharp (chapter 11) notes, is of interest to the USA primarily for the Lajes base complex in the Azores. These facilities serve several important functions, communications, re-fuelling air traffic *en route* to the Mediterranean and points east, and surveillance of Soviet submarine traffic in the Atlantic. Portugal also hosts a major NATO naval command near Lisbon.

Conventional wisdom places most of the smaller allies in the category of 'free riders' but Greece and Turkey both rate highly as burden sharers in NATO's Defence Planning Committee report published in December 1988. Greek and Turkish defence inputs are among the highest in the alliance. Greece devotes the highest percentage of GNP and the highest percentage of manpower to defence, has the longest period of conscription and makes by far the greatest contribution to the alliance pool of available merchant shipping. Turkey fields the largest army in NATO, and increased defence spending during the 1980s by well over the three-per cent NATO target. Neither country, however, has a strong industrial base and both remain dependent on external assistance to maintain modern equipment.

Athanassios Platias (chapter 12) draws a fine line between the costs and benefits for Greece of membership in NATO and of the US presence on its territory. Greece faces a unique problem in NATO in that it feels more threatened by an ally, Turkey, than by its 'adversaries' in the WTO. Greece enjoys far better relations with Bulgaria than it does with Turkey, for example. During 1988 the Greek and Turkish prime ministers began a series of meetings designed to improve relations, but they did not get much beyond defining the four key problems that divide them.[4] Platias reflects the resentment widely felt in Greece that the USA not only favours Turkey with a more generous aid package, but also takes the Turkish side in territorial disputes in the region. The picture is mixed of course. Greeks condemn the USA for malign neglect over the Turkish invasion of Cyprus in 1974, but admit that US pressure was instrumental in defusing the crisis in the Aegean in March 1988. As Platias suggests, one reason that Greece will never leave NATO, for all its ambivalence about the value of alliance membership, is fear that the USA would then give Turkey even more support and thus further undermine the Greek position in the region.

Through the 1980s Greece complained both about the paucity and the nature of US military aid. Much as Molero argues for Spain, Platias claims that such aid programmes make Greece too dependent on US equipment at the expense of building a national defence industrial base, lead to purchases of weapons unsuitable for Greek security needs, and allow the USA too much influence on Greek military procurement policy.

If Greece regards NATO and the US presence as a mixed blessing at best, the Government nevertheless takes a damage-limiting approach to NATO, recognizing that neither neutralism nor removal of all the US bases seems feasible in the short term. Platias estimates the annual economic benefit of US bases to Greece on the order of $105 million, with the annual costs approximately $45 million, giving a net annual benefit of $60 million. Platias sees the impact of option A (the gone for good option) as having a minimal effect economically. Option B, in which the USA makes a commitment to return in a crisis would be more expensive if Greece had the responsibility to maintain the bases in good order. Platias believes that a US withdrawal could produce major benefits if it stimulated better co-operation with other West European allies. To this end, regardless of any withdrawal prospects, in 1987 Greece expressed interest in joining the newly expanded WEU. There seemed little chance of acceptance by the other WEU members in the late 1980s, not least because the Papandreou government was perceived as too anti-American in European Political Co-operation discussions and as a difficult negotiating partner in other aspects of European Community bargaining.[5] It is an open question whether the WEU members would be more welcoming in the event of a US withdrawal, but the incentives for Greece to join would undoubtedly increase.

Turkey is regarded as among the most loyal and robust of the allies and does not produce the kind of headaches (on nuclear policy for example) that Greece causes in Brussels. Saadet Deger (chapter 13) notes that Turkey is essentially a Third World economy, with the lowest per capita income of any NATO country, making its contribution to the alliance all the more remarkable. Deger documents the economic relationship between the USA and Turkey and reflects similar dissatisfaction with the terms of trade that are expressed in the chapters on Spain, Portugal and Greece. US 'aid' programmes tend to dump inferior, outdated military equipment, make Turkey too dependent on the USA as supplier at the expense of domestic production and, especially serious in the case of Turkey, create crippling debt repayment problems. Through the 1980s 'aid' received from the USA did not even cover the outflow back to the USA in debt repayment. If the economic relationship is a net benefit to the USA and a net cost to Turkey, as Deger argues, why does Turkey continue to uphold it?

The answer of course is that Turkey believes that the security benefits justify the economic costs. Despite relatively cordial economic and trade relations with the USSR, Turkey will always be wary of the superpower security threat on its borders and is likely to see NATO membership, and especially the bilateral relationship with the USA, as a vital insurance policy providing deterrence in peacetime and military protection should war occur. At the same time Turkey feels crippled by its dependency on the USA and is struggling to develop an indigenous defence industry in order to bring its armed forces up to date. This need is likely to become even more pressing if NATO as a whole relies less on nuclear deterrence, and if Soviet nationality problems continue to erupt uncomfortably close to the Turkish border.

Turkey needs trade rather than aid in order to develop its economy in general and its defence industrial base in particular. It needs to find a more equitable relationship with Washington as well as better defence and industrial co-operation with the other NATO countries. As member of neither the EC nor the WEU, Turkey feels at a double disadvantage *vis-à-vis* its European allies. Through the 1980s neither organization was unwilling to endorse Turkish membership, although the FRG and the Netherlands provided some help bilaterally.

On the impact of the two withdrawal options, Deger notes that except for the US Tactical Fighter Wing based at Incirlik, the US bases in Turkey are not primarily to assist in the defence of Turkish territory but are geared to peacetime listening and surveillance of Soviet military activity. Deger suggests that if the USA no longer deemed the bases worthwhile, Turkey would probably shut them down. Under withdrawal option B, if Turkey took on the responsibility to maintain the bases it would cost approximately one per cent of the total military expenditure of the country; a minor additional burden. Under the 'gone for good' option, Deger judges that the direct economic benefits lost would easily be compensated for by the recovery of land and buildings for other uses.

Nevertheless, even if the USA chose to withdraw its forces from Europe, it would be unlikely to voluntarily relinquish its valuable intelligence gathering facilities in Turkey, especially if new arms control agreements were to be concluded with the USSR, leading to increased verification requirements.

Italy rates well below average in its defence inputs to the alliance but as Jennifer Sims (chapter 14) argues, it can hardly be described as a NATO 'free rider', since it hosts a substantial US presence. This includes two army bases, three naval bases and five air bases, as well as a number of C^3I facilities.[6] Italy also hosts several important NATO commands. Moreover, Italy is one of the few countries that increased its contribution to NATO through the 1980s.

The economic (as distinct from the military) benefits to Italy of the US presence is hard to gauge as information proved more difficult to obtain than in most NATO countries. Sims suggests that the income from rents, base employment for Italian nationals, construction contracts and support services would be lost in the event of an option A withdrawal. Additional costs of a US withdrawal from Italy would include the commitment by the Italian Government to guarantee employment for its nationals that had taken jobs on US bases. This represents some 3.2 per cent of the Italian defence budget in 1988.[7] While this may seem a small increment, as in the case of the UK and the FRG the impact would be disproportionately high in some regions; notably in the south where the local economies are least able to cope. Sims documents the likely impact of the closure of the USAF base at Comiso. Unless a US withdrawal were accompanied by a reduction in the Soviet presence in the Mediterranean, the impact on Italian national security would also be felt most keenly in the south.

Italians are frequently frustrated by the preoccupations of their alliance partners with the Central Front, and complain that Italy, together with the other Mediterranean powers, is by-passed when commands are assigned.[8] Italian leaders also complain about being left out of important four-power alliance decisions, notably the Guadeloupe meeting at which President Jimmy Carter, Prime Minister James Callaghan, Chancellor Helmut Schmidt and President Giscard d'Estaing approved the double-track Intermediate-range Nuclear Forces (INF) decision in January 1979, and the more recent 'two plus four' talks on German unification. The INF decision was especially galling to Italy since it had always willingly shared the risks of basing US nuclear weapons on its territory, in contrast to the ambivalence of other allies. This was as true when it came to accepting Jupiter missiles in the late 1950s as it was accepting ground-launched cruise missiles in the late 1970s. Italy was also willing to accept the USAF 401st Tactical Fighter Wing (including 72 nuclear-capable F-16s with crews and support personnel) that Spain expelled in 1988, although work was interrupted at the designated base at Crotone (Calabria) immediately after the May 1989 NATO summit meeting, at which US President George Bush agreed to include combat aircraft in the CFE Negotiation.[9]

The perceived economic and military costs of a US withdrawal easily outweigh the benefits for Italy, but of the two withdrawal scenarios, option A would be the easier one to plan for. Under option B, Italy would have little control over the bases to which US troops would return, yet to the extent Italy had to contribute to the running costs of US bases, it would have less to invest in national force improvements. Sims echoes the authors of the other Southern Flank country studies when she observes that the price of alliance membership for Italy has been the debilitating dependence on the priorities of the USA at the expense of Italian national security interests and capabilities. Withdrawal could prove stimulating to Italy's defence industrial base if it meant that national industries, in co-operation with other West European companies, could develop equipment more appropriate to their needs instead of purchasing overly sophisticated US systems. An option A withdrawal could stimulate more Italian interest in co-operative ventures with Spain and France (perhaps via the WEU) in pursuing their common interests in defending the Southern Flank.

II. Assessing the military mission gaps

In assessing the impact of the two withdrawal options on the Nordic countries, Steven Miller (chapter 15) notes the stark contrast between US involvement in NATO's Northern Flank, including its Baltic approaches, and the Central Front. To a large extent the north is already free of US military forces in peacetime, since for the USA this represents a maritime rather than a land theatre. Nevertheless, NATO plans for the defence of the north depend in large part on US

SUMMARY AND CONCLUSIONS   43

ground, sea, and air reinforcements in a crisis. In this sense, as noted earlier, the region could serve as a model for an option B withdrawal for the rest of Europe.

The crucial task for NATO is to prevent Soviet control of north Norway. Miller notes that while NATO policy calls for the forward defence of the FRG at the WTO border, Norway has to be willing to trade space for time—for allied increments in military capability—in the event of a Soviet attack on Finnmark. These include mobilized Norwegian Home Guard and conscript forces from the south as well as external reinforcements from the allies. The latter include British, Dutch, Canadian and West German forces as well as a US Marine Expeditionary Brigade (MEB) which stores prepositioned stocks in the area.

Of far more importance is the incremental airpower that the USA would provide to the Northern Flank in wartime. Miller documents the NATO–Soviet air balance in the region and the plans for incremental allied air power in a crisis. This includes 15 squadrons (300 aircraft) of which 11 squadrons are US. In addition, some US airpower based on aircraft carriers might be available.

Miller assumes that the US Navy would likely operate in northern waters under either withdrawal option and thus be a problem for Soviet planners contemplating an attack on the north. On the other hand, if withdrawal meant that collaboration between the US and its European allies were abandoned, the effectiveness of the US fleet would certainly be reduced. If the US Atlantic Fleet were not available to assist NATO, the European allies obviously could not afford to replace that capability in its entirety. However, Miller argues that such a replacement would be unnecessary, because in the 1980s the US Maritime Strategy assigned the Atlantic Fleet a set of demanding and offensive missions that the Europeans would be unlikely, and unwise, to duplicate.[10]

With respect to the impact of a US withdrawal on defence of the Baltic approaches it is again the loss of incremental US air power in a crisis that would be most keenly felt, since defending the Baltic approaches on sea and on land is the responsibility of Denmark and the FRG.

In sum, for the northern region as a whole, option B represents little change from the status quo, while the most serious impact of option A would be the loss of incremental US airpower that might be brought to bear in a crisis.

Unlike the Northern and Southern flanks, the US contribution to NATO's ground forces in the central region is considered crucial by military planners. In assessing the impact of withdrawal in this region, Hilmar Linnenkamp (chapter 16) finds it impossible to ignore the political element of the US presence in Europe, since the fact that any conflict would thereby immediately engage the awesome power of the USA is the primary message these troops are supposed to convey to the USSR. Replacing the same number of troops would be feasible, if difficult, but any other multinational combination of forces could hardly convey the same political message.

In assessing the military gaps that the two withdrawal options would cause in the central region, Linnenkamp employs a game theory matrix that considers also two different kinds of WTO attack: a deliberate surprise assault from a standing start without reinforcement, and a well prepared attack across a broad front employing all WTO divisions. To these four hypothetical situations Linnenkamp addresses three questions: what is the general political message that a US withdrawal would send to the USSR, what would be the overall strategic message in peacetime and crisis, and how would the USSR perceive the military risks of an attack?

On this last question, Linnenkamp comes to the somewhat paradoxical conclusion that a WTO attack against the 'gone for good' withdrawal option would require well prepared and more sustainable potential, whereas an attack under the circumstance of withdrawal option B would require pre-emption by a smaller force of combat-ready divisions already in place in Eastern Europe.

Linnenkamp suggests that the USA should aim to leave a residual force of one brigade-size unit (normally 3000 men) for each of its three Army corps, to signal to the USSR that the USA intends to stick to its commitment to defend NATO Europe. These combat forces would require a similar number of support troops to receive troop reinforcement from the continental USA in a crisis. This suggests a total residual Army strength of some 20 000 men; considerably fewer than the 75 000 to 100 000 men proposed by US Senator Sam Nunn in April 1990. The discrepancy is reduced if the Nunn proposal is assumed to include several thousand airmen. Specific numbers aside, however, the Nunn and Linnenkamp proposals are consistent with earlier suggestions that US divisions in Europe should be hollowed out, rather than disbanded, in such a way that they could be easily reconstituted in a crisis.

Whereas NATO planners see the northern and central regions as more or less unified theatres of operation to be defended with integrated alliance forces, they view the southern region as five geographically separated theatres to be defended by national forces alone or with some kind of bilateral assistance from the USA.[11] Of the five theatres four are land (northern Italy, Greece/Greek Thrace, Western Turkey/Turkish Thrace and eastern Turkey. The fifth is the entire maritime sweep of the Mediterranean, Adriatic, Aegean and Black seas.

George Thibault (chapter 17) sees two strategic 'glues' binding the southern region together. One is the potential threat to NATO's free passage in the Mediterranean, the other is the US strategic capability. A major problem in assessing and coping with the gaps that a US withdrawal would cause in the southern region is that the US presence in essence consists of a series of bilateral alliances between each local NATO ally and the USA, and it is hard to see one of the other NATO junior partners assuming a similar protector role.

Thibault notes that any NATO partner or group of partners could replace the forces and infrastructure currently provided by the USA, given the same commitment of funds and effort. What would be more difficult to replace would be the cross-regional leadership and the advanced military technology that the USA provides. This contrasts somewhat with the analyses of Molero, Platias and Deger who suggest that one of the benefits of withdrawal could be that

SUMMARY AND CONCLUSIONS 45

greater autonomy might provide the necessary incentive for defence industrial co-operation among the European NATO partners.

Assessments of the NATO–WTO balance in Europe, and of NATO burden-sharing, rarely take account of the US contribution to NATO's command, control, communication and intelligence capabilities. For this study, however, John Pike and Paul Stares (chapter 18) identify the US C^3I systems that contribute to the defence of NATO, evaluate the consequences of the withdrawal of these systems and explore how Western Europe might cope thereafter.

Pike and Stares find that whereas the European allies provide adequate command, control and communication capabilities, they are dependent on US facilities for most of their intelligence on WTO capabilities. US satellite systems provide unparalleled coverage of Eastern Europe and the USSR, territory that would otherwise be largely hidden from European surveillance. Pictures taken by French SPOT satellites, for example, do not yet provide the high resolution imagery available from US systems.

Under both withdrawal options, the most serious C^3I mission gap would be the loss of intelligence capabilities. This would be most severe under option A, if it is assumed that this option precludes the use of any US C^3I facilities in Europe and the withholding from European allies of satellite-derived imagery and signals intelligence. Under option B, Pike and Stares assume that after the withdrawal of a US peacetime presence, European personnel could operate some of the fixed C^3I facilities that the US would leave in Europe. In this case the intelligence gap would be easier to fill, although some US equipment might be considered too sensitive for operation by non-US personnel.

By contrast NATO appears to enjoy a redundancy of communication, command and control assets such that the loss of US facilities would in this regard not be as damaging to NATO security. The European response to the two withdrawal options would vary, but Pike and Stares suggest that the allies would either have to redress their current intelligence-gathering capabilities or adopt new strategy and tactics less reliant on sophisticated C^3I.

The study ends with an examination of the problems NATO would encounter as it tried to compensate for the two withdrawal options. Robert Harkavy (chapter 20) examines how the loss of European bases would affect the USA, and Martin Farndale (chapter 19) offers suggestions as to how Western Europe could cope without the US presence.

Harkavy focuses on how the loss of European bases would affect US strategic nuclear capability and the ability to project conventional power to intervene in contingencies outside the NATO area. Since the 1950s the USA has relied on European bases for all three legs of its strategic triad. After implementation of the Treaty between the USA and the USSR on the elimination of their intermediate-range and shorter-range missiles (the INF Treaty) there will no longer be any land-based US strategic systems in Europe.

To the extent that loss of shore-based facilities would curb the operation of the 6th Fleet in the Mediterranean, a US withdrawal could also mean the loss of some 200 of the carrier-based aircraft that can deliver nuclear weapons to Soviet targets. As Harkavy note, however, many war scenarios envisage the 6th Fleet withdrawing from the Mediterranean at the outset of any conflict in any event. Given the redundancy in US strategic capability, all three legs of the strategic triad could easily survive the loss of European bases. All the ICBMs, and most of the strategic bombers are based in the USA, and the new fleet of Trident submarines will be less dependent than the Poseidon fleet was on Holy Loch and Rota. A more serious loss to the US would be European facilities that support anti-submarine warfare P-3 aircraft.

The USA would also lose tactical nuclear capability as short-range land-based nuclear forces would be withdrawn under both options, but many of these systems are anyway slated for unilateral withdrawal as part of the nuclear modernization plans.[12] Separate negotiations on short-range nuclear forces may drastically reduce these during the 1990s with or without a US withdrawal.

Harkavy believes that the most serious result of a US withdrawal would not be in the nuclear field but rather the opportunities lost to project conventional power into the Middle East and North Africa. Harkavy worries that this at worst might encourage unfriendly nations to attack US allies in the region, or encourage the allies themselves, especially Israel, to adopt more aggressive military postures. It should be remembered, however, that during the late 1980s several prominent US spokesmen proposed withdrawal from Europe precisely to have US troops more easily available for Third World contingencies.[13]

Harkavy suggests that a withdrawal from Europe could also seriously degrade the United States' arms control monitoring capability. Depending on the political climate between East and West this could discourage the US from engaging in arms control negotiations, if the Senate deemed a treaty unverifiable without listening posts in Europe.

As a former Commander of NATO's Northern Army Group, who believes that the INF Treaty seriously undermined NATO capabilities, Martin Farndale finds it difficult to contemplate removing US forces altogether. Nevertheless, despite serious misgivings, Farndale offers several suggestions on how NATO's military planners might restructure forces and command posts to make up for the loss of US forces, if this were decreed by their political masters.

In general he believes that an option B withdrawal would be more difficult for NATO to cope with than option A. This is because West Europeans would always worry that option B was about to become option A, and this would introduce too much uncertainty into NATO planning. At least with option A NATO planners would know where they stood and could plan accordingly. This is in line with Linnenkamp's view that WTO planners would need more substantial forces to contemplate military action against Western Europe after an option A withdrawal than after option B.

For Farndale perhaps the most serious consequence of a US withdrawal would be the need for Western Europe to rely even more than now on nuclear weapons. West European economies would be hard pressed to allocate suffi-

SUMMARY AND CONCLUSIONS   47

cient resources to make up for lost US airpower, and more nuclear missiles off shore would probably be assigned to targets currently covered by the USAF. Since under both withdrawal assumptions there would be no US nuclear weapons in Europe, NATO's nuclear deterrent role would devolve to Britain and France. In Farndale's view this would require increased nuclear assets for the British and French navies, as well as air-launched stand-off cruise missiles and a new nuclear warhead for the Multiple Rocket Launcher System (MRLS).

Obviously NATO planners would also have to think about new doctrine, strategy and operational tactics, but Farndale is sceptical about getting any additional capability from technical fixes and concepts like 'non-offensive defence'. He does, however, believe that the prospect of a US withdrawal could have a salutary effect on moving Western Europe towards a more integrated European defence. He sees a European parliament with a co-ordinated foreign policy as the necessary pre-requisite for a truly integrated system, but in the meantime believes it could be useful to create a European Army composed of independent national contingents, in order to advance conceptual thinking about the problems of a more integrated force.

In the current NATO structure, the US dominates the command structure, but both withdrawal options assume that all NATO commands would be led by Europeans. Under option A, Farndale sees no difficulty in allocating the different commands on the principle that as far as possible nations command in those areas where they have most national concern. He sees three possible options to select SACEUR: give the post to one of the four major allies (UK, France, FRG and Italy); rotate the post among these four; or select the best man for the job, as is currently done for the Chairman of the Military Committee. Deputy SACEUR should obviously come from a different country than SACEUR. For option B, allocation of command to Europeans would present real problems since returning US forces might be reluctant to serve under a European commander. Farndale argues, however, that it would be impossible to change high level command structures once any fighting had started. Serving under European command would be the price the US would have to pay for an option B withdrawal.

Henry Kissinger and other prominent US spokesmen have floated the idea of a European SACEUR on a number of occasions. When asked specifically about a French SACEUR in April 1990, President Bush said he saw no reason to change the present arrangement in NATO.[14] However, with German unification on the agenda, President Bush and Secretary of State James Baker were clearly exploring, with President Mitterrand and other West European leaders, arrangements that would integrate France more closely into the alliance. In Berlin in December 1989 Secretary Baker also proposed new institutional arrangements between the USA and the EC.[15] In France, however, these initiatives were seen primarily as a means to retain US supremacy in the Western alliance. For his part, President Mitterrand appears to have a short-term goal of closer West European integration on defence matters, and for the longer term (as announced in a New Year message in January 1990) a pan-European confederation.[16]

Hubert Vedrine, an adviser to President Mitterrand, emphasized that a stronger European axis on defence need not diminish the US role in NATO. Nevertheless, Vedrine, Foreign Minister Roland Dumas, Mitterrand and others in France tended to emphasize the EC as a more appropriate model for a new European security regime than forums like NATO and the Conference on Security and Co-operation in Europe (CSCE) that include the USA and Canada.[17]

III. Conclusion

While the authors of this study disagree on many details about the costs and benefits of the US presence in Europe and the likely impact of withdrawal, all agree that the benefits of the US presence outweigh the costs, and the costs of withdrawal outweigh the benefits. This finding was hardly surprising, given the broad level of satisfaction in the West with the way that NATO (with a high level of US forces in Europe) has performed the twin tasks of containing the USSR and absorbing the FRG.

With the prospect of German unification, foreign ministers in the non-Soviet WTO states also expressed the view that NATO would play an even more important role in anchoring a unitary German state. These views were expressed at the 'open skies' CFE foreign minister's meeting in Ottawa in February 1990 and especially by the Polish Foreign Minister, Krzysztof Skubiszewski, in a speech to the WEU Assembly on 22 March 1990. Skubiszewski emphasized the US military presence in Europe as a stabilizing factor and noted that 'we all want to move away from Yalta and are afraid to come back to Sarajevo'.[18]

More surprising, however, was the almost unanimous agreement that of the two withdrawal options posited here, West Europeans would find option A ('gone for good') easier to cope with than option B (skeletal peacetime US presence but a commitment to return in a crisis). As Martin Farndale noted, the uncertainties associated with option B and the constant anxiety that it was in reality option A would make contingency planning a nightmare for West Europeans. This judgement is supported by the finding that option B would be more expensive for the USA than option A, and hence the President would always be under pressure from the Congress to convert option B to option A.

Although maintaining the status quo might be the preferred option of most, there was general recognition that the US presence in Europe may not be sustainable for much longer if the perception of a Soviet military threat continues to fade and Congress continues to cut US defence spending. This suggests that it would be prudent for planners on both sides of the Atlantic to prepare for option A, to think through whether NATO without a US presence in Europe could continue to perform the twofold task that NATO with a US presence has performed for the past 40 years, or whether some alternative arrangement would better serve European security interests.

In order to continue to deter Soviet aggression would Western Europe have to rely even more than before on nuclear deterrence, as Martin Farndale suggests? If this were the case, on whose nuclear weapons would they rely? This study shows a steady decrease in the credibility of the US guarantee to initiate the use of nuclear weapons on behalf of Europe, as well as increasing reluctance on behalf of the FRG to share the risks associated with nuclear basing. An important question for further study is whether an explicit first-use nuclear guarantee is necessary for Western Europe to retain confidence in the US commitment to NATO. Alternately, could the West European nations begin to rely on existential deterrence, that is, on the premiss that as long as the two superpowers have intercontinental strike forces, and as long as both wish to retain their spheres of influence in Europe, deterrence would hold?

Another issue, only lightly touched upon in this study, is what role British and French nuclear weapons would play if the USA withdrew all its forces and nuclear weapons from Europe. Must they play a role at all, or could Western Europe feel safe with a force posture based on conventional deterrence alone?

To what extent could European NATO provide for its own conventional defence? This study explodes the myth that the US bears a disproportionate share of the NATO burden, especially in terms of the contribution of conventional forces. Despite continued complaints about 'free riders', European NATO countries provide 90 per cent of the manpower, 85 per cent of the tanks, 95 per cent of the artillery and 85 per cent of the combat airpower in the Atlantic-to-the-Urals area covered by the CFE Negotiation. Thus, a CFE agreement based on ceilings 10–15 per cent below 1989 NATO levels could in theory at least be as dramatic in its effect on NATO's military capability as a US withdrawal. Western Europe would then face a new set of problems. Instead of worrying about how to compensate for the withdrawal of US forces, NATO would be working out how to absorb excess capacity in its defence industries. Would this require more or less co-operation among the European allies, and would it generate more or less incentive to export to the Third World?[19]

This study shows that it would be difficult for West Europeans to compensate for a US withdrawal by purely national efforts. The four wealthier states (France, the FRG, Italy and the UK) would have to provide the lion's share of the funding and would need to co-ordinate their efforts in Central Europe far more effectively than hitherto. Even more important would be the need to co-ordinate plans to compensate for the loss of the US forces and facilities in the Mediterranean countries. The newly revived Western European Union could provide the vehicle for such co-operation but only if the current membership would be willing to admit Greece and Turkey as full members.

However, while the Soviet threat is fading, new manifestations of the German question is moving into centre stage. Other West Europeans, especially in London and Paris, grew increasingly nervous through the late 1980s at the assertiveness of the FRG in its pursuit of improved relations with the East. Perhaps the biggest question that emerges from this study is whether Britain and France will temper their dependence on Atlantic links to join the FRG in forging stronger links with the East, or whether they will continue their traditional role of pulling the FRG back into the Western fold? How this trilateral relationship develops will determine whether some form of Western military alliance can survive without the stabilizing effect of the USA.

Notes and references

[1] On the single market, see Cecchini, P, *The European Challenge: 1992—The Benefits of a Single Market* (Wildwood House: London, 1988).

[2] At the April 1988 'Europe After an American Withdrawal' SIPRI workshop John Roper suggested that an extra $50 billion could be raised if each European ally allocated another 1.5% of its GNP to defence.

[3] Cutileiro, J., 'A Portuguese view: keep the Atlantic Alliance', *International Herald Tribune*, 14 Apr. 1989.

[4] The four issues are Cyprus, the continental shelf, the use of airspace and the military status of the two Greek islands Lemnos and Samothrace.

[5] Clarke, M., 'Evaluating the new Western European Union: the implications for Spain and Portugal', ed K. Maxwell, *Defence and Foreign Policy in Portugal* (Duke University Press: Durham, N. C., 1990).

[6] In January 1990 US Defense Secretary Richard Cheney announced that the USAF base at Comiso is scheduled for closure.

[7] SIPRI, *SIPRI Yearbook 1989: World Armaments and Disarmament* (Oxford University Press: Oxford, 1989), p. 183.

[8] Italian Defence Minister Valerio Zanone complained that no officer from Italy, Turkey, Greece or Spain has been tapped for the senior military post since its creation in the 1960s. 'Italian Minister sees slight by NATO', *International Herald Tribune*, 14 Feb. 1989.

[9] Smith, R. J. and Wilson, G. C., 'Bush proposal puts Italy base in doubt', *International Herald Tribune*, 2 June 1989.

[10] In his first defence budget request to the Congress, US Defense Secretary Richard Cheney reduced the number of US carrier battle groups from 15 to 14. See 'Cheney is doing fine', *International Herald Tribune*, 28 Apr. 1989. In April 1990 Senator Sam Nunn suggested that 10–12 carrier battle groups would be adequate. See Gordon, M. E., 'Nunn sets the terms for the military debate', *International Herald Tribune*, 21–22 Apr. 1990.

[11] Spain has already committed a brigade to serve in Italy.

[12] Barber, L., 'Bush postpones Lance successor', *Financial Times*, 20 Apr. 1990; Gordon, M. R., 'New US plan would cancel German missile deployment', *International Herald Tribune*, 20 Apr. 1990.

[13] Brzezinski, Z., testimony before the House Foreign Relations Committee January 13, 1987; Admiral William Crowe testimony to Senate Foreign Relations Committee, as reported in Fourquet, D., 'Crowe on NATO forces', *Jane's NATO Report*, 25 Apr. 1989, p. 5; Turner, S. (Adm.), 'Arm for the real threats', *International Herald Tribune*, 11 Apr. 1989.

[14] Reuters, 'Bush is against proposals for French chief of NATO', *International Herald Tribune*, 18 Apr. 1990.

[15] *A new Europe, A New Atlanticism: Architecture for a New Era*, address by US Secretary of State Baker to the Berlin Press Club, Berlin 12 Dec. 1989, *Current Policy* (US State Dept.), no. 1233.

[16] President Mitterand's message was reprinted as 'Les voeux de M. Mitterrand', *Le Monde*, 2 Jan. 1990, p. 5. See also Trean, C., 'M. Mitterrand souhaite une "confederation" europeenne avec les pay de l'Est', *Le Monde*, 2 Jan. 1990, p. 1.

[17] Dumas, R., 'A greater Europe', *International Herald Tribune*, 15 Mar. 1990; Davidson, I., 'Western alliance begins to march to a different tune', *Financial Times*, 19. Apr. 1990; Desaubliaux, P.-H., 'Chevenement suggere un "mole europeen" de defense', *Le Figaro*, 6. Feb. 1990.

[18] Skubiszewski speech, English text from the Polish embassy in Stockholm, p. 3.

[19] Gummett, P. and Walker, W., 'The industrial and technological consequences of the peace', *RUSI Journal*, spring 1990, pp. 46–52.

Part II
Economic implications of a US military withdrawal from Europe

Part II

3. US perspectives on the economic costs and benefits of a withdrawal of US troops and facilities from Europe

Alice C. Maroni *

Congressional Research Service, Washington, DC

I. Introduction

The US defence policy debate takes place largely in the context of the annual cycle of congressional action on the President's funding request for national defence programmes and activities. In 1987 and 1988 the defence budget debate for fiscal years (FYs) 1988 and 1989 was characterized by a resurgence of congressional interest in the costs and benefits to the United States of maintaining troops and facilities around the world.

During the 100th Congress, the House Committee on Armed Services established a panel devoted to the exploration of military burden-sharing issues, and hearings were held throughout 1988. Frustrated by both the intransigence of the federal budget deficit and the burgeoning trade deficit, some members of Congress, as in years past, entertained the notion of reducing US European-deployed troops.[1]

In recent years, US perspectives on the economic costs and benefits of withdrawing US troops and military facilities from Europe have been influenced in large part by varying views of the federal budget deficit and the trade deficit. The economic implications of withdrawing US troops from Europe, however, are generally poorly understood and, consequently, often misrepresented. There is neither a simple answer to the question of how much the United States spends on its European defence commitment, nor a commonly accepted way of measuring possible savings associated with a withdrawal of troops and alternative defence policies and commitments.

Purpose and scope

This chapter provides, in strictly economic terms, an overview of the costs and benefits to the United States of withdrawing all US troops and military facilities from Europe. It assesses the impact of such a decision on the US defence budget and US–European defence trade relations, in particular, defence industrial co-operative programmes.

---

* The views expressed in this paper are those of the author and do not represent the views of the Congressional Research Service or any other US Government agency.

The decision to focus exclusively on the economic costs and benefits of a US troop withdrawal from Europe is not intended to suggest that economic considerations are pre-eminent or in any way diminish the importance of evaluating both the military and political costs and benefits to the United States of a US European-deployed troop withdrawal. Furthermore, the author takes no position for or against a US European-deployed troop withdrawal. Arguments for and against the troop withdrawal scenarios used here for the purpose of analysis are beyond the scope of this chapter.

Methodological issues

The almost annually recurring congressional discourse on the costs and benefits of maintaining US troops in Europe is accompanied by remarkably little useful, unclassified budgetary data. The US Department of Defense (DOD) has been reluctant to provide unclassified estimates of US European-deployed troop costs since the early 1980s, because of perceived methodological problems associated with allocating the US defence budget to US commitments on a regional basis and concern about misuse of the data.[2] In 1982, the DOD prepared a detailed analysis of the cost of various US European-deployed troop withdrawal scenarios from which some conclusions may be drawn.[3]

This chapter presents both incremental and total costs associated with the deployment of US troops in Europe. First, estimates of the costs and savings associated with the decision to withdraw US Army and Air Force units from Europe are provided for two possible troop withdrawal options. The discussion of one-time costs associated with each of the options indicates the extent to which a variety of unavoidable costs may offset anticipated savings. The figures provided are relatively crude estimates, based on existing, though somewhat dated, analyses and available cost planning factors.

Second, in the absence of a technically precise methodology for estimating the total cost of individual force elements associated with US European-deployed troops, rough estimates are presented of both the direct and indirect costs of US forces in Europe. While less useful analytically, these statistics have proven to be compelling politically.

Finally, the discussion of US–European trade relations is supplemented by statistics on the defence trade balance which were prepared by the DOD.

Defence burden-sharing from a US perspective

Some Americans have long contended that the European allies fail to do their 'fair share' in support of NATO. Successive administrations and congresses have exhorted the Europeans alternatively to spend more for their collective security in the interest of improving NATO's conventional defence capability and to offset the cost of stationing US troops in Europe.[4]

While no single measure adequately reflects the size and nature of the burden of providing for the defence of Western Europe,[5] the fact that the United States spends more on defence as a share of its resources than any of its NATO allies grates on the sensibilities of those Americans who consider that competing demands for resources in Europe are no greater than in the United States and European economies are sufficiently strong to bear a greater financial burden for their national security. This focus on 'fairness' is a persistent theme in the congressional burden-sharing debate.

While US public opinion polls consistently show high levels of support for maintaining US troops in Europe,[6] the burden-sharing issue is politically popular in the United States where many people continue to harbour some traditional isolationist tendencies and to question both the value of foreign deployments and assistance, in general, and the reliability of US allies, in particular. Since 1981, as the budget deficit and trade deficit have grown, interest in burden-sharing has increased, reflecting the far-reaching search, especially by Congress, for solutions to a set of difficult financial problems.

However, proposals to withdraw US troops from Europe in recent years have generally been associated with either a proposed change in US strategic thinking and defence policy or a plan to encourage greater European defence spending. Troop reduction proposals are seldom heralded as grand money savings options. And when they are, they are often offered under mistaken assumptions.

From the US perspective, the economic cost of maintaining forces in Europe has not been a major determinant of US NATO policy, in spite of the rhetoric. Domestic factions that support the idea of US troop withdrawals have heretofore mainly been concerned about the 'fairness' issue or have preferred alternative strategic schools of thought (perhaps simply questioning the 'over-emphasis' of Europe in US strategic thinking). For example, Earl Ravenal, who advocates a military policy of non-intervention, argues that:

[I]f we are to cut defence spending significantly, we must change our national strategies and our foreign policy . . . Instead of deterrence and alliance, we would pursue war-avoidance and self-reliance. Our security would depend more on our abstention from regional conflicts . . . Over time, we would accommodate the dissolution of defensive commitments, including NATO, that obligate us contingently to overseas intervention.[7]

In defence of his 1987 proposal to withdraw 100 000 US troops from Europe, Zbigniew Brzezinski, former National Security Advisor to President Carter, noted:

[T]he paradox we face is that American conventional forces are weakest where the West is most vulnerable, in Southwest Asia, and are strongest where American allies have the greatest capacity for doing more on their own behalf and where the risk of a U.S–Soviet clash is lowest, in Europe. It simply does not make sense to have the single largest overseas deployment of U.S. forces concentrated in Western Europe.[8]

Alternatively, there are those who propose US European-deployed troop reductions in the interest of prompting increased allied defence spending and improving allied conventional defence capability, but it remains uncertain whether those who want to influence allied behaviour would actually allow sizeable troop withdrawals to occur. The 1984 Nunn–Roth Amendment, ultimately rejected by the US Senate (by a vote of 55 to 41) after a contentious debate, is an example of an incentive scheme that linked US European-deployed troop strength directly to increases in European defence spending and improvements in conventional capability.[9] However, Senator Sam Nunn maintained, 'we do *not* want troop cuts, we want more effective conventional defences . . .'[10]

The wave of troop reduction proposals drafted by members of Congress in 1987 and 1988 is attributable largely to the frustration many Americans feel about the growth of both the federal budget and trade deficits, the size (and perceived 'fairness') of the US financial contribution to West European security (and the protection of Persian Gulf oil), and the continued reliance on the USA by both European and Asian allies for their national security. Members of Congress, including, importantly, many who are neither members of the House or Senate Committees on Armed Services nor among those considered to be defence policy experts, have been challenged by their constituents to explain why the US tax-payer assumes a large share of the burden of Europe's defence. The decision to empower a congressional panel on burden-sharing was made, in part, to deflect errant troop reduction proposals and provide a forum for debating the status of US overseas commitments and troop deployment policy.

II. Economic implications of withdrawal of US European-deployed troops

In recent years, alternative views of the economic costs and benefits of withdrawing US troops and military facilities from Europe have been shaped largely by varying perspectives on the financial burden imposed on the United States. A decision to withdraw all US troops and equipment from Europe would have budgetary implications for the US Department of Defense. In addition, such a move would affect US–European trade relations, in particular NATO defence industrial co-operative programmes. These economic issues are reviewed below.

The US defence budget

Defence analysts use two approaches when considering the cost of US forces overseas—incremental and total costs. The incremental cost of maintaining forces in Europe is the difference between the cost of operating and supporting US forces in Europe and the cost of operating and supporting the same forces in the United States. The total cost of US forces in Europe is a much broader concept that includes either the *direct costs* (largely operations and support costs, for example, pay and base operations) or both the *direct and indirect costs* (for example, transportation, training, logistics support, weapon procurement and research costs) associated with deployed forces.

The burden-sharing debate in the United States frequently focuses on the total cost of operating and maintaining US forces in Europe. This total cost figure, however, is often misinterpreted as the amount the United States would save if US European-deployed forces were withdrawn.

The most analytically useful way to measure the economic costs and benefits of a withdrawal of US forces from Europe is in terms of the *incremental* costs associated with the decision; however, US perspectives on the issue of troop withdrawals are not limited to this narrow, if more analytically appropriate, approach. Continued interest in the *total* cost of US European-deployed forces, a compelling statistic in the congressional burden-sharing debate, reflects for many Americans the broad economic implication of US defence policy *vis-à-vis* Europe.[11] Consequently, this section reviews both incremental and total cost issues associated with US European-deployed forces.

Incremental costs/savings

The most informative way of looking at the budgetary implications of withdrawing US troops and equipment from Europe is in terms of the costs or savings associated with such a move. However, a useful cost figure would reflect a wide range of activities, not all of which are performed more cheaply in the United States.

For example, table 3.1 shows the incremental cost of operating US Army and Air Force units in Europe. The total provided—$1963 million in FY 1987 dollars—is a net figure (costs and savings) reflecting the added annual cost of supporting US troops in Europe over and above the cost of supporting them in the United States. The cost associated with rotating people and transporting supplies to Europe over and above the cost of operating Army and Air Force units in the United States is the largest component of the added cost calculation.

The incremental cost reflected in table 3.1, however, does not represent the net amount that would be saved if European-deployed forces were restationed in the United States. Also, naval implications are omitted. Furthermore, the one-time costs associated with the large-scale movement of troops and equipment that would be required to withdraw all US forces from Europe cannot be overlooked. The extent to which these one-time costs offset projected savings depends on subsequent procurement, operations and basing decisions made by the US Government.

Costs/savings of withdrawing US troops

It is not enough to begin with the simple assumption that all US Army and Air Force units and their equipment have been withdrawn from Europe. An analysis

Table 3.1. Incremental cost of operating US Army and Air Force units in Europe, as of 1986[a]

Figures are in US $m., at constant (FY 1987) prices.

Activity/function	Description	Cost[b]
Permanent change of station	Added cost of rotating people to Europe over and above the cost of a similar number of rotations in the continental United States	856
Transportation	Extra cost of transporting supplies to Europe when compared to transport costs within the continental United States	887
Family housing/quarters	Cost of providing shelter in Europe that would not be incurred in the continental United States	436
Dependant education	Cost of operating schools in Europe for DOD dependants	322
Overseas allowances	Outlays for cost-of-living allowances authorized in Europe	37
Medical care	Savings incurred because medical care and facilities in Europe cost less than in the continental United States	−229
Civilian pay	Savings incurred through lower European wages	−346
Total		**1 963**

[a] No adjustment has been made to account for the decline in the value of the dollar since these estimates were prepared in 1986. The costs and savings shown are therefore understated. Costs of similar activities associated with naval forces afloat are excluded as it is difficult to attribute them to specific regions.

[b] A negative cost represents a savings.

Source: Department of Defense, Office of the Assistant Secretary of Defense (Comptroller), testimony in *Defense Appropriations for Fiscal Year 1984*, Hearings before the Committee on Appropriations, US Senate, Senate Hearing 98-293, 98th Congress (US Government Printing Office: Washington, DC, 1983), pp. 72–73.

of the cost to the United States of such a move cannot be made without establishing at the same time whether the United States, for some reason, elects to retain withdrawn forces on active duty. If one assumes that redeployment to Europe may be necessary, the one-time costs of both basing troops and equipment in the United States and prepositioning duplicate sets of equipment in Europe are sufficiently large (even excluding the very substantial cost of added airlift) that it may be more than 20 years before the added cost of withdrawing troops is offset by the recurring annual savings associated with closing military facilities in Europe and withdrawing troops.

Consequently, it is necessary to consider two scenarios: one (option A) in which forces and equipment moved back to the USA are deactivated on the assumption that they will not be needed to respond to threats to US security in Europe or elsewhere, and one (option B) in which US troops withdrawn from

Europe remain part of the active US force structure and have the capability to redeploy to Europe where they would find prepositioned stocks of equipment.

Option A. If the USA were to withdraw forces and equipment from Europe and deactivate them, thereby reducing force structure, there would be relatively few expenses associated with the withdrawal of forces to offset the savings a reduction in forces would generate. The one-time operation and support (O&S) cost of permanently reducing Army and Air Force end-strengths in option A would be sizeable, but would be considerably smaller than the one-time O&S, basing and procurement costs that would be incurred in option B, to keep 310 000 restationed Army and Air Force personnel on active duty in the USA.

The decision to withdraw troops from Europe and reduce active-duty force strength would be expected to produce defence budget savings in the near term depending on the speed with which the deactivation plan was implemented. This economic 'benefit', however, cannot be measured fairly independent of the related political and military costs and benefits of a decision to withdraw US European-deployed troops. The budgetary savings generated by the withdrawal of troops may misstate the 'real cost' of the redeployment decision when political and military factors are taken into consideration, for example, the cost-effectiveness of US European-deployed forces in terms of their deterrence value.

In addition, while no cost figures are available, the calculation would be different if withdrawn troops were shifted to the reserves rather than demobilized altogether. Indications are that it would be considerably more expensive to shift returning forces to the reserves than to demobilize them, and it would probably be somewhat less expensive to shift forces to the reserves than to retain them on active duty.

Option B. If a decision were made to withdraw all US Army and Air Force units from Europe while maintaining their readiness for redeployment to Europe, the DOD would be forced to rebase withdrawn units in the United States and preposition additional divisional equipment in Europe.

The cost of restationing forces and re-establishing headquarters in the United States would include the one-time cost of refurbishing or constructing required base facilities. While existing excess facilities (and temporary structures) may be adequate to accommodate some brigade-size units, new construction would be necessary to provide permanent base facilities for the return of the required units. Excess shipyards, after all, could not be used by combat and support troops returning to the United States.

The decision to preposition divisional equipment in Europe would necessitate the procurement of extra equipment for the training of forces restationed in the United States. Duplicate sets of equipment for Army and Air Force units would be quite costly but indispensable if US-based forces are to retain any flexibility and remain ready for combat. The size of this one-time cost would depend on how much and what kind of equipment were to be bought. Furthermore, the cost of expanded training with European forces in Europe would be incurred.

In addition, investment in strategic mobility forces (airlift and perhaps sealift) for redeployment in the event of hostilities in Europe and the cost of their operation and maintenance further increases the cost of retaining withdrawn forces on active duty. According to former Secretary of Defense Weinberger, 'the airlift capability we would need to get [100 000 troops] back to Europe in 10 days (to keep our NATO commitment) would be $20 billion to $25 billion with prepositioned equipment, and about $100 billion without'.[12]

The one-time costs associated with the decision to withdraw all US Army and Air Force units from Europe in option B would consist of operation and support costs, basing costs and procurement costs and would be quite large compared to those of option A. The required O&S and basing costs would be incurred relatively quickly, if the withdrawal proceeded quickly, while the added cost of procuring greater strategic mobility assets and additional equipment would probably be incurred slowly over time depending on the investment decisions made.

Cost estimate and analysis. Table 3.2 summarizes the budgetary impact of withdrawing US troops and equipment from Europe. The figures are rough estimates, based on existing, unclassified analyses and available cost planning factors. They represent the cost of returning to the United States all US Army and Air Force European-deployed personnel and headquarters.[13] The figures include the cost of returning dependents as well. In the case of option A, the figures include the cost of withdrawing all equipment. In the case of option B, the figure include the procurement cost of extra equipment for force training in the United States. The figures do not include the cost of withdrawing the Marine Corps' prepositioned equipment in Norway or naval infrastructure ashore elsewhere in Europe.

The withdrawal of US troops from Europe, where they have been stationed since 1945 on a peace-time basis, will cost the US tax-payer money unless returning troops are deactivated.[14] The one-time cost of military construction to accommodate returning forces retained on active duty and the procurement of duplicate sets of equipment for training and sufficient strategic mobility to redeploy forces in the event of hostilities in Europe far outweigh the savings that would be generated in the near term by closing US-operated facilities in Europe. However, a decision to deactivate returning troops, about a one-third reduction in active US Army and Air Force tactical forces, might generate as much as $11 billion in recurring annual savings. Notably, a savings of this magnitude would come at the expense of a decline of about 15 per cent in US active-duty military personnel.

Unless a reduction in force structure is contemplated, it seems clear that a decision to withdraw US troops from Europe would have to be made for other than strictly economic reasons. The large one-time costs associated with re-equipping, rebasing and supporting troops restationed in the United States (not

Table 3.2. Estimate of costs/savings of withdrawing all US Army and Air Force units from Europe[a]

Figures are in US $b., at constant (FY 1988) prices.

Option	One-time costs	Recurring annual savings: operation & support[b]
A: Reduce force structure	+ 1.9 – 3.0[c]	– 11.0
B: Retain in active force	+ 31.1 – 47.7[d]	– 1.6

[a] The figures presented could fluctuate up or down by a considerable amount as a result of changes in the value of the dollar. However, even a 15% fluctuation in the costs and savings shown would be small in comparison to the cost of some items for which estimates are unavailable, for example, the added cost of required airlift and sealift.

[b] Operation and support (O&S) costs include operation and maintenance costs, military personnel costs and other direct, indirect or marginal additional costs associated with weapon support or force change, for example, aircraft replenishment spares and training munitions. Figures are expressed as net savings.

[c] As Option A assumes no restationing costs, the range includes one-time O&S costs only for the five-year period FY 1988–92, expressed in constant FY 1988 dollars, with the cost largely incurred in the first year.

[d] The figure includes one-time O&S costs (including transportation), basing costs for the five-year period FY 1988–92, expressed in constant FY 1988 dollars, and military construction and family housing costs for the construction of new facilities for all withdrawn troops. Costs would be lower to the extent that space is available at existing facilities. Figures also include a rough estimate of the procurement cost of extra equipment required for force training in the United States. No estimate is provided for the added cost of required airlift and sealift.

Sources: Author's estimates based on US Air Force Cost and Planning Factors, AFR 173-13, updated 15 May 1987, and the Air Force Cost Oriented Resource Estimating (CORE) model; Lussier, F. and Myers, B., 'Costs of withdrawing Army troops from Europe', Congressional Budget Office Memorandum, 8 May 1987; and 'Cost of U.S. defense commitment to NATO', prepared by the Department of Defense, submitted for the record by Senator Charles Percy, *Congressional Record*, 12 May 1982, p. S4991.

counting the cost of additional airlift and perhaps sealift required for redeployment) are not likely to be readily accepted by the people of the United States. While there is no question that the prepositioning of equipment in Europe and the procurement of duplicate sets of equipment for training in the United States would improve the combat-readiness of US troops redeploying to Europe in time of crisis, the cost would probably preclude the option.

It is important to note that the figures provided do not reflect the costs to the United States of changes in force structure (in particular, with respect to the Navy) and related procurement and force deployment decisions that would be associated with US Government efforts to compensate militarily for the loss of its 'forward defences' under either troop withdrawal option. The withdrawal of US troops from Europe would effectively shift the first line of US defence closer to the United States. A US decision to establish a less forward, 'open Atlantic' defence that, for example, stops short of the Azores could require a substantially larger not smaller fleet.[15] Whether withdrawn forces are deactivated (option A) or restationed in the United States (option B), a bigger navy may be needed to protect the United States from a Soviet naval threat greatly dispersed in open waters, if European efforts to compensate for a US troop withdrawal fail. The cost of expanding the US Navy and altering its composition to meet new requirements would be sizeable.

In addition, other research and procurement decisions precipitated by the change in US defence policy might be costly. The US Government cannot be expected to execute a meaningful Atlantic defence without trying to compensate for lost sound-surveillance, communications and navigation facilities in Europe. Furthermore, the implementation of a revised US defence policy might include other efforts to improve US defences.

Foreign currency fluctuations

The decline of the dollar relative to European currencies increases the cost to the United States of operating and maintaining forces in Europe and may alter somewhat the calculations of the cost of withdrawing US European-deployed forces. The figures provided in table 3.2 could fluctuate up or down.

Overseas housing allowances and the cost-of-living allowances paid by the DOD to servicemen stationed abroad are increased or decreased based on changes in currency rates as monitored by the DOD Per Diem Travel and Allowance Committee. These allowances are adjusted every two weeks, if warranted. The amount paid to US military personnel abroad depends on rank, time in service, size of family and country of deployment.

In recent years, the purchasing power of the dollar has declined markedly since its peak in the autumn of 1985. It dropped in value by about 12.5 per cent relative to the Deutsche mark during 1987. In early December 1987, it was reported that 'the dollar's value against the Japanese yen and West German mark [was] already roughly 19 per cent below what the Pentagon had estimated at the start of fiscal 1988', two months prior.[16]

The DOD has no choice but to pay these mandatory allowances. When Congress fails to appropriate enough in advance to cover the cost, for example, of overseas station allowances and cost-of-living allowances, the DOD is forced to request supplemental funding or to reprogramme funds from other sources to fund the shortfall. The shortfall (the difference between actual spending and appropriated amounts) for FYs 1987 and 1988 is projected by the DOD at more than $600 million annually.

These allowances represent operation and support expenses that would be counted as recurring annual savings, if US troops were withdrawn from Europe. The more expensive it becomes to operate and maintain US forces in Europe compared to the United States as a result of the decline of the dollar relative to European currencies, the more congressional support US troop withdrawal options may garner.[17]

Cost of deployed forces

US perspectives on the economic costs and benefits of withdrawing US Army and Air Force units from Europe, however, are not limited to the more narrow focus on estimates of one-time costs and recurring annual savings. The cost of the long-standing US military commitment to defend Europe, which includes funding for weapons, technology and training not included in the cost estimates provided in the preceding section, more often defines the parameters of the congressional debate on the economic impact of withdrawing US troops.

The DOD has placed the cost of the US commitment to Europe at about 60 per cent of the US defence budget. In dollar terms, the cost would be on the order of $180 billion, reflecting much more than just the cost of US European-deployed troops.[18] While sometimes misunderstood to be the budget savings associated with a troop withdrawal scenario, the figure includes the cost of forces available in the United States for early and later reinforcement of Europe, as well as the cost of US European-deployed forces. For the purposes of this analysis, it is much too all-encompassing a figure.

In order to understand the economic costs and benefits to the United States of maintaining forces in Europe, it is useful first to examine the major elements that comprise the *direct cost* to the United States of European-deployed forces. In the absence of a technically precise methodology for estimating the total cost of individual force elements associated with US troops in Europe, table 3.3 presents estimates of the cost of US European-deployed forces in terms of four functional categories—force operations and support, management headquarters, military construction, and DOD-operated schools for dependents. The figures reflect US spending for both US and foreign goods and services.

While the figures shown in table 3.3 do not reflect the budgetary costs or savings associated with a decision to withdraw US troops from Europe, the figures establish the distribution of funding between various US non-naval force activities. By far and away the largest component (almost 80 per cent) of direct US European-deployed troop costs is force operations and support—about $12 billion—of which 49 per cent is for military personnel and 43 per cent is for operations and maintenance activities. Overall, investment funding—replenishment spares and military construction—represents only 17 per cent of the total direct cost of US forces in Europe.

The US balance of payments is affected by US direct expenditures abroad for *foreign* goods and services, a somewhat smaller figure. '*In the absence of changes in any other accounts* during a particular time period, increases or decreases in direct defence expenditures abroad will result in an equal increase or decrease in imports of goods and services'.[19] Table 3.4 shows the distribution of US direct defence expenditures for foreign goods and services in NATO Europe. Almost 30 per cent of the total consists of personal spending in Europe by US military and civilian personnel employed by the DOD.

The direct cost of US troops deployed in Europe, however, does not represent all costs logically associated with overseas force deployment. An

Table 3.3. Direct cost of forces deployed in Europe, fiscal year 1989[a]
Figures represent total obligational authority; in US $m.

Activity	Procurement	Military construction	Military personnel	O&M[b]	Total
Force operations and support[c]	972	..	6 284	5 399	12 655
Management headquarters[d]	..	..	102	97	199
Military construction and family housing	..	1 537	..	..	1 537
DOD-dependent schools	..	..	..	345	345
Total direct cost	**972**	**1 537**	**6 386**	**5 841**	**14 736**

[a] Original FY 1984 figures recalculated by the author into FY 1989 dollars using DOD February 1988 deflators. However, no adjustment has been made to account for the decline in the value of the dollar since these estimates were prepared in 1983. Consequently, the costs shown may be understated.

[b] Operation and maintenance costs.

[c] The figures reflect the cost of operating European-deployed combat units and supporting units, which includes the cost of replenishment spares, petroleum, oil, and lubricants (POL), supplies, pay and other training and support costs for operating units and bases. Repair and maintenance of real property, maintenance of POMCUS stocks, and first and second destination transportation costs are included as well in base operating costs along with the cost of fuel, supplies and other personnel support costs.

[d] The figures reflect the cost of operating USAFE, USAREUR, and USEUCOM headquarters.

Source: Department of Defense, Office of the Assistant Secretary of Defense (Comptroller), testimony in *Defense Appropriations for Fiscal Year 1984*, Hearings before the Committee on Appropriations, US Senate, Senate Hearing 98-293, 98th Congress (US Government Printing Office: Washington, DC, 1983), pp. 72–73.

alternative view of the cost of US European-deployed forces includes, in addition to the direct cost of deployed forces, an allocated share of the costs of new equipment, a proportionate share of US-based training and logistics support, research, development, test and evaluation (RDT&E) and DOD administration—so-called *indirect costs*. Reported to Congress by the DOD in the 1970s and early 1980s in unclassified form (and subsequently in classified form) as the cost of the US commitment to NATO,[20] this expanded definition of the cost of US European-deployed forces is considered by some to be a more complete reflection of US troop costs (including naval forces).[21]

Based on extrapolations from funding data for prior years, the estimated cost of US European-deployed forces represents approximately 14–16 per cent of the total annual US defence budget or roughly $47 billion in FY 1987. This figure includes all general-purpose force elements and support elements that are forward-deployed in Europe and reflects both the direct and indirect costs associated with these forces.[22] Indirect costs represent the largest share of the total.

Table 3.4. Direct US defence expenditures abroad for foreign goods and services in NATO Europe, 1987

Figures are in US $m., in constant (1987) prices.

Category	Expenditure[a]
Department of Defense expenditures	
Personnel and related expenditures	2 785
Foreign nationals (direct and contract hire)	1 599
Contractual services[b]	3 107
Construction[b]	542
Major equipment[b]	443
Other materials and supplies[b]	471
Petroleum products	314
NATO infrastructure	263
Military assistance program services	22
Subtotal	**9 545**
Coast Guard expenditures	**3**
Total	**9 548**

[a] Items may not add up to total due to rounding.
[b] Includes foreign expenditures in fulfillment of reimbursable contractual activities by the US Department of Defense on behalf of foreign governments and international organizations.

Source: US Department of Commerce, Bureau of Economic Analysis, from information made available by operating agencies; prepared Sep. 1988.

Over the period FYs 1974–81, investment funding made up roughly 34 per cent of total funding for US European-deployed forces (while total DOD investment funding made up about 35 per cent of the total defence budget). With funding for all DOD investment in FY 1988 at about 43 per cent of the total defence budget, it is logical to surmise that investment funding strictly for US forces in Europe as a percentage of total funding for US European-deployed forces has increased as well, perhaps by a commensurate amount.

Table 3.5 provides data on the cost of US forces deployed in Europe. More recent data are not available in unclassified form.

In broad economic terms, the cost of US European-deployed forces, including both direct and indirect costs, does not represent an especially large share of the total US defence budget. If 60 per cent of the US defence budget is spent on the US military commitment to NATO and only 14–16 per cent of the defence budget is spent on US European-deployed forces, it would appear that the decision to base a large number of US troops in Europe is not the primary contributing factor to the overall cost of the US commitment to help defend Europe. However, the decision to base US forces in Europe continues to be identified as a possible source of budget savings in the context of the congressional burden-sharing and deficit-reduction debate.

Table 3.5. Percentage cost of US European-deployed forces, FYs 1974–82[a]
Figures are percentages.

Fiscal year	Investment	Operations & support	Share of total DOD budget
1974	30	70	16
1975	28	72	15
1976	32	68	14
1977	33	67	14
1978	35	65	15
1979	37	63	16
1980	36	64	16
1981[b]	38	62	16
1982[b]	38	62	16
1974–82 average	**34**	**66**	**15**

[a] Calculated using total obligational authority.

[b] The figures are estimates based on the defence budget submitted by the Carter Administration (January 1981). They do not reflect congressional action on the revised defence budget request submitted by the Reagan Administration (March 1981). Consequently, the figures on which the percentages shown are based may be understated.

Source: *Department of Defense Authorization for Appropriations for Fiscal Year 1982*, Hearings on S. 815 before the Committee on Armed Services, US Senate, 97th Congress (US Government Printing Office: Washington, DC, 1981), Part 1, pp. 61–68.

Defence industrial co-operative programmes

As long as the United States and the countries of Western Europe are allies with a commitment to fight side-by-side if provoked, military industrial co-operation between the Western democracies is in the interest of the United States, whether or not US troops are deployed in Europe. However, a decision to withdraw *and* deactivate US European-deployed troops and allow the Europeans to take full responsibility for their defence would likely alter the incentives that drive US defence trade relations with Western Europe. From a US perspective, interest in equipment standardization and inter-operability and the 'two-way street of transatlantic arms trade', and more specifically, defence industrial co-operative programmes, could well give way to the protectionist sentiment that is felt on both sides of the Atlantic.

However, the economic costs and benefits to the United States of a withdrawal of European-deployed troops in terms of defence trade relations with Western Europe are impossible to quantify. (See table 3.6 for a defence trade balance summary and table 3.7 for a summary of the US balance of trade.)

Standardization and cost

Defence industrial co-operative programmes are generally designed to redress the problem of duplicate or redundant weapons research and production, to facilitate equipment standardization and inter-operability, and to promote economies of scale. Such programmes allow participating Governments to

Table 3.6. US–European NATO defence trade balance summary, 1987
Figures are in thousands of current dollars as of 30 September 1986.

Country	Purchases NATO[b]	Purchases DOD[c]	DOD computed ratio[a] FY 1986	DOD computed ratio[a] FY 1983
Belgium	44 488	84 987	0.52 : 1	8.92 : 1
Denmark	27 293	27 797	0.98 : 1	1.10 : 1
France	65 567	256 978	0.26 : 1	6.03 : 1
FRG	574 600	460 104	1.25 : 1	1.72 : 1
Italy	117 524	89 507	1.31 : 1[c]	3.71 : 1
Luxembourg	5 101	3 070	1.66 : 1	5.75 : 1
Netherlands	224 200	74 154	3.02 : 1[c]	11.29 : 1
Norway	206 256	46 319	4.45 : 1	2.88 : 1
Portugal	241 024	34 938	6.90 : 1	3127.74 : 1
Spain	209 641	58 811	3.56 : 1	97.94 : 1
Turkey	480 974	2 110	227.95 : 1	663.56 : 1
United Kingdom	1 050 562	860 600	1.22 : 1[c]	4.91 : 1
Total	**3 247 230**	**1 999 375**	**1.62 : 1**	**8.21 : 1**

[a] Fiscal year end.

[b] US sales to NATO Europe, including foreign military sales (FMS) and commercial exports licensed under the Arms Export Control Act. Totals for commercial exports represent the dollar value of estimated deliveries made against direct foreign government purchases from US manufacturers of munitions-controlled items.

[c] Figures include prime contract and subcontract awards. Data do not include subsistence, petroleum, construction and support service contracts.

[c] The negotiated ratio is for Italy 2.82 : 1, for the Netherlands 2.9 : 1, and for the United Kingdom 1.50 : 1. These ratios are the result of comparing DOD data with data generated by allies and represent the ratios agreed upon to reflect defence trade.

Source: US Department of Defense, unpublished tables, 10 June 1987.

make the most of limited defence resources and capitalize on available technology whether the programmes are introduced, for example, under the auspices of the NATO Cooperative Research and Development Program in the United States (spawned by the 1985 Nunn and Quayle Amendments) or the NATO Conventional Defense Improvement Program.[23]

However, without a common defence commitment and objectives, the US Government might be less motivated by the need for NATO equipment standardization and inter-operability and less concerned about NATO programme duplication. In addition, the US military might become interested in an altogether different force structure, with many fewer ground forces and much less heavy equipment. US interest in allied armaments co-operation among those in government and industry would likely persist to the extent that (*a*) there is a perceived advantage technologically, (*b*) US defence investment funding is constrained, (*c*) potential markets exist and (*d*) sizeable budget savings seem possible; however, the military and political incentives that truly drive current co-operative arms initiatives would be less compelling, if US and

Table 3.7. US–European NATO balance of trade summary, 1984–86
Figures are in $US m., at current prices.

Country	Deficit (+) or surplus (−)		
	1984	1985	1986
Belgium/Luxembourg	+ 2 014	+ 1 351	+ 1 208
Denmark	− 913	− 1 090	− 1 111
France	− 2 479	− 3 864	− 3 370
FRG	− 8 726	− 12 183	− 15 567
Italy	− 4 129	− 5 756	− 6 474
Netherlands	+ 3 225	+ 2 901	+ 3 485
Norway	− 1 145	− 583	− 233
Portugal	+ 442	+ 97	+ 38
Spain	− 67	− 249	− 341
Turkey	+ 785	+ 650	+ 470
United Kingdom	− 2 834	− 4 300	− 4 615

Source: IMF, *Direction of Trade Statistics Yearbook, 1987*, prepared by the General Economy Division of the Bureau of Statistics (International Monetary Fund: Washington, DC, 1987), p. 404.

allied forces were not expected to have the capability to fight together against a common enemy.

Co-operative weapons programmes have been accorded a relatively high priority in the US defence acquisition community. Since 1985, the DOD has heartily endorsed a variety of legislative proposals facilitating such programmes and has moved quickly to institutionalize its approach to co-operative arms initiatives, including the placement of specialists in US embassies in Western Europe who provide support for arms co-operation projects.[24] If the US Government withdrew its troops from Europe *and* deactivated them, one would also expect it to withdraw its active support for armaments co-operation. Defence industrial co-operative programmes would likely suffer.

To begin with, the bureaucratic infrastructure would likely atrophy. Second, in the absence of a military commitment to Europe and the military and political incentives that gave rise to existing co-operative weapon programmes in the first place, the commitment of federal funds for military research and development (R&D) would be expected to decline, leaving European partners in joint weapon development projects to look elsewhere for resources. While it is evident that a good deal of the funding for co-operative R&D goes to US rather than to European firms, the withdrawal of US research funds from Europe would represent some relatively small added cost to European governments interested in filling the gaps left by the United States. Furthermore, well-established co-operative programmes, such as the NATO Infrastructure Fund, would suffer from the loss of the US financial contribution, currently 27 per cent of the annual total.[25]

In the absence of a military commitment to Europe, it remains uncertain whether economic incentives will ever be sufficient (that is, whether US budgetary savings and income generated by co-operative R&D programmes will ever be large enough) for successive administrations and congresses to continue to encourage and fund defence industrial co-operative programmes at current relatively high levels. As a percentage of total US defence funding for research and development, defence industrial co-operative programmes currently represent a relatively small investment, only 3 per cent. In 1988 the DOD projected that the US funding for co-operative R&D programmes would rise to 25 per cent of total R&D funding by the year 2000.[26]

Implications for the 'two-way street'

From a US perspective, transatlantic arms trade or the 'two-way street' is not an end in itself. It is a reflection of allied co-operation and, in terms of equipment standardization and inter-operability, a means to an end. Absolute equity in the direction of arms traffic has never been a US goal.[27]

The economic arguments in favour of the 'two-way street' from a US perspective have never been especially compelling. While US procurement of European military equipment may entail some budgetary savings,[28] reported savings have never appeared to be especially large from the perspective of the US Government. Furthermore, such a transaction helps neither the US trade balance nor the defence industrial base. A decision to withdraw and deactivate US European-deployed troops, together with the associated erosion of political and military incentives that drive current co-operative arms initiatives and interest generally in transatlantic trade, is likely to undermine the US Government's commitment to the promotion and improvement of the 'two-way street'. In recent years, the defence trade balance, while still favouring the United States, has improved dramatically from 8.2:1 in 1983 to 1.6:1 in 1986 largely as a result of both a conscious effort on the part of the US Government and European efforts to supply its own forces (see table 3.6).

In the absence of active Government support, US industry might still be prompted to pursue co-operative arms programmes (for example, joint production by partners) in lieu of losing marketing opportunities and sales altogether. It is more than likely that US industry would continue to try to sell to the European military, and the 'two-way street' in defence trade might grow more lop-sided. Of course, US weapons would presumably become less desirable to European governments as European self-reliance grows and the Europeans develop a resilient, indigenous defence production base.

However, a worsening US trade balance may lead to tighter restrictions on imports, further limitations on the transfer of technology and other 'red tape' discouraging US industry from competing internationally. Such protectionism would be likely to increase the cost of doing business. Furthermore, the decline of the dollar would consume what little savings some off-shore purchases offer to the DOD. General, non-defence, transatlantic trade might be affected as well, depending on the nature and extent of the trade limitations adopted.

III. Concluding observations

There is neither a simple answer to the question of how much the United States spends on its European defence commitment nor a commonly accepted way of measuring possible budgetary savings associated with a total withdrawal of US troops and alternative defence policies. In addition, there is no way to estimate the economic cost and benefit in terms of trade relations of a dramatic change in US international defence commitments such as the total withdrawal of US European-deployed troops.

European governments interested in making up for the gaps left by the withdrawal of US forces cannot hope simply to spend what the United States has spent and make up, in terms of deterrence and military strength, for the loss of the US contribution to Europe's defence. The economic cost of maintaining US forces in Europe represents only a portion of the total cost of the US military commitment to Europe and does not begin to reflect the value of the US military contribution in terms of the quality of the weapons and ongoing research. Furthermore, the one-time cost of replacing the US contribution will require, for European governments as it does for the US, recurring real annual increases in defence spending to sustain and modernize the established force structure.

The withdrawal of US troops from Europe, where they have been stationed on a peacetime basis since 1945, will cost the US tax-payer money unless returning troops are deactivated. The one-time cost of military construction to accommodate returning forces retained on active duty and the procurement of additional equipment for training and sufficient strategic mobility to redeploy forces in the event of hostilities in Europe far outweigh the savings that would be generated in the near term by closing US-operated facilities in Europe. In addition, the critical economic variable that is not reflected in this analysis is the very substantial cost to the United States of changes in force structure (in particular, with respect to the Navy) and related procurement and deployment decisions that would be associated with efforts of the US Government to compensate militarily for the loss of its 'forward defences' following the withdrawal of US troops from Europe.

From the US perspective, the economic costs and benefits of maintaining forces in Europe have not been major determinants of the pattern of US troop deployment. While a decision to withdraw troops *and* reduce active duty force strength might generate as much as $11 billion in recurring annual savings, this economic 'benefit' cannot be assessed fairly without also considering the related political and military costs and benefits of a decision to withdraw and deactivate US European-deployed Army and Air Force units, raising the question of the relative importance to the United States of economic considerations associated with such a move. The 'real cost' of a decision to withdraw US troops from Europe may be much more a function of political and military considerations than economic factors.

In addition, a decision to withdraw and deactivate US European-deployed troops would likely reduce interest in equipment standardization and interoperability and the 'two-way street', and more specifically, defence industrial co-operative programmes, giving way to more protectionism. Without compelling economic incentives to promote co-operative research, such a dramatic change in the US defence commitment to Europe would probably cause the US Government to withdraw its funding support for weapon research and development in Europe and lend support to European efforts to develop a stronger independent defence production base. A probable effect over time of a strong co-operative European defence effort would be to reduce US arms sales to Europe and generate greater competition between US and European weapon producers for alternative markets in other countries.

Notes and references

[1] In 1987, however, Congress adopted a non-binding expression of the 'sense of the Congress' in favour of maintaining US military personnel in Europe in support of NATO at FY 1987 levels during FY 1988. See Section 1002 of Public Law 100-180, in National Defense Authorization Act for Fiscal Years 1988 and 1989, with an explanation contained in House Report 100-446, 17 Nov. 1987, p. 671. See also *Report of the Defense Burdensharing Panel of the Committee on Armed Services*, US House of Representatives, 100th Congress (US Government Printing Office: Washington, DC, Aug. 1988); Thomas, R. W., 'Alliance burdensharing: a review of data', Staff Working Paper, Congressional Budget Office, June 1987.

[2] Maroni, A. C., *The U.S. Commitment to Europe's Defense: A Review of Cost Issues and Estimates*, Congressional Research Service Report no. 85-211 F (US Library of Congress: Washington, DC, 7 Nov. 1985).

[3] The unclassified version was included in: US Department of Defense, *Report of the Secretary of Defense Caspar Weinberger to the Congress for FY 1984* (annual report) (US Government Printing Office: Washington, DC, 31 Jan. 1983), pp. 186–90. The classified version, submitted to the Senate Appropriations Committee on 24 Sep. 1982, in the form of a letter to Senator Ted Stevens, was entitled 'Cost of Withdrawing U.S. Forces from Europe'.

[4] Eurogroup and NATO officials traditionally respond to critics by representing the European contribution in terms of available forces rather than in dollars. According to their calculations, in the event of war, the majority of the manpower and equipment engaged in conflict on the NATO side would be European—some 90% of NATO's manpower, 95% of its divisions, 85% of its tanks, 95% of its artillery and 80% of its combat aircraft. Van Eekelen, W. F., 'Transatlantic understanding: Eurogroup seminar looks at some of the problems', *NATO Review*, vol. 35, no. 5 (Oct. 1987), p. 19. These figures are somewhat more dramatic than those presented by US writers interested in making the same point. For example, see Galvin, J. R., 'NATO after zero INF', *Armed Forces Journal International*, Mar. 1988, p. 58.

[5] The annual DOD publication *Report on Allied Contributions to the Common Defense* is intended to deal with this problem. In the absence of a 'single, universally accepted formula for calculating each country's "fair share" of the collective defense burden', DOD has developed a way of measuring national performance in terms of a ratio, known as the 'prosperity index', that compares a country's contribution with its ability to contribute. However, the Apr. 1987 and Apr. 1988 editions of the annual 'burden-sharing report' did not include such a measure.

[6] Reilly, J. E. (ed.), *American Public Opinion and U.S. Foreign Policy 1987* (The Chicago Council on Foreign Relations: Chicago, Ill., 1987), especially p. 21.

[7] Ravenal, E. C., *Defining Defense: The 1985 Military Budget* (CATO Institute: Washington, DC, 1984), pp. 36–37.

[8] Brzezinski, Z., 'Choosing where to put our forces', *Washington Times*, 6 Feb. 1987, p. 1-E.

⁹ The Nunn–Roth Amendment proposed a permanent ceiling of 326 414 on US European-deployed troops and called for an annual 30 000 reduction in the ceiling, if the allies failed to meet a set of criteria designed to improve NATO's conventional defence capability. The proposal was offered in the US Senate in 1984 as an amendment to the FY 1985 Defense Authorization Act. Upon tabling the amendment, the Senate went on the approve the substitute offered by Senator William Cohen establishing a ceiling on US troops in Europe and providing funding for side-by-side testing of European weapon systems. For further discussion, see Sloan, S. R., *Defense Burdensharing: U.S. Relations with NATO Allies and Japan*, Congressional Research Service Report no. 88-449 F (US Library of Congress: Washington, DC, 24 June 1988).

¹⁰ Nunn, S., 'Improving NATO's conventional defenses', *USA Today*, May 1985, p. 21; emphasis in original.

¹¹ Depending on which military costs and forces are counted, the annual cost figures used to reflect the overall cost of the US commitment to help defend Europe will vary. For a look at the range of figures used in the debate, see Maroni (note 2).

¹² Weinberger, C., 'We need those troops in Europe', *Washington Post*, 23 June 1987, p. A-19. The DOD argues that the cost of procuring a sufficient amount of strategic mobility to accommodate a complete withdrawal of US European-deployed Army and Air Force units is prohibitive. In its 1982 estimates of restationing-costs associated with illustrative US troop withdrawals from Europe, the DOD estimated the annual cost of additional mobility assets for one Army Corps and five Tactical Fighter Wings rebased in the United States from Europe at $2.2 billion in FY 1983, with a five-year cost (FY 1983–87) of $11.2 billion (current dollars). See 'Cost of U.S. defense commitment to NATO', prepared by the DOD and submitted for the record by Senator Charles Percy, *Congressional Record*, 12 May 1982, p. S4991.

¹³ In particular, the management headquarters that would be returned to the United States include US Army Europe (USAREUR), US Air Forces Europe (USAFE) and US European Command Headquarters (USEUCOM).

¹⁴ According to the DOD: 'If a sizeable portion of our force were withdrawn from Europe and retained on active duty in CONUS [continental United States], we would reduce operating costs but incur additional costs associated with refurbishing or constructing new facilities [in CONUS]. And if we intend to maintain the capability to defend against aggression by the Warsaw Pact, additional mobility forces would have to be purchased to transport these troops back to Europe in an emergency'. Response to a question by Senator Mark Andrews prepared by the DOD for the record, *Supplemental Request for Department of Defense*, Hearing before the Defense Subcommittee on Appropriations, Fiscal Year 1983, Senate Hearing No. 98-293 (US Government Printing Office: Washington, DC, 1983), p. 74. The DOD also submitted to the Committee estimates of the total cost differential for selected units in a classified DOD study entitled 'Cost of withdrawing U.S. forces from Europe', 24 Sep. 1982.

¹⁵ O'Rourke, R., 'The Maritime Strategy and the next decade', US Naval Institute *Proceedings*, Apr. 1988, p. 34.

¹⁶ Dorsey, J., 'Money crunch sends Pentagon's overseas costs soaring', *Norfolk Virginian-Pilot*, 2 Dec. 1987, p. D-4.

¹⁷ On 16 Dec. 1987 Representative Jack Davis introduced a bill (H.R. 3771) requiring the Secretary of Defense to reduce US troops in countries that do not reimburse the USA for certain costs associated with negative currency fluctuations. It was jointly referred to the House Committees on Foreign Affairs and Armed Services. See also Treverton, G., *The Dollar Drain and American Forces in Germany* (Ohio University Press: Athens, Ohio, 1978).

¹⁸ Unclassified introduction, *United States Expenditures in Support of NATO* (annual update pursuant to P.L. 98-525, Sec. 1002, Apr. 1987) (US Department of Defense: Washington, DC, Apr. 1987), p. 11. No other details regarding this figure are available on an unclassified basis.

¹⁹ Department of Defense report to Congress pursuant to Section 8042 of the FY 1988 Department of Defense Appropriation Bill, transmitted in the form of a letter to Senator John C. Stennis from Deputy Secretary of Defense William Howard Taft, IV, dated 25 Apr. 1988. See especially enclosure 2; emphasis added.

²⁰ Prior to 1983, cost estimates of the US NATO commitment were prepared on an unclassified basis by the DOD and appeared regularly in congressional testimony. In 1983 and 1984, Congress required that the DOD report on the costs the United States bears in support of

its commitment to NATO's defence (P.L. 98-94 and P.L. 98-525). The DOD report *United States Expenditures in Support of NATO* (note 18) is classified, along with the cost estimates it contains. For a discussion of the two methods used by the Department of Defense to estimate the cost of the US commitment to NATO, see Maroni (note 2).

[21] One drawback associated with apportioning a *pro rata* share of the investment costs in this manner, however, is that funding for investment may 'rise and fall for reasons that have nothing to do with the specific units committed to NATO . . . In a period of disinvestment in [US] military forces' this calculation may improperly reflect a decline in the cost of US European-deployed forces. US Department of Defense, unclassified introduction, *United States Expenditures in Support of NATO* (note 18), p. 9.

[22] No further details are available on an unclassified basis.

[23] 'The NATO Cooperative Research and Development Program was established in 1986 to support joint weapons and material development work on an equitable cost-sharing basis by the US and one or more NATO Allies . . . To date, $445 million has been appropriated for this program'. Department of Defense, *Support of NATO Strategy in the 1990s*, report submitted to Congress pursuant to P.L. 99-180, Sec. 1001, 25 Jan. 1988, reprinted in the *Congressional Record*, 27 Jan. 1988, p. S128. The NATO Conventional Defence Improvement (CDI) initiative, endorsed by NATO Defence Ministers in 1985, is designed to eliminate critical deficiencies in NATO's conventional force posture.

[24] Dennis Kloske, Special Advisor for NATO Armaments, serves as the DOD's trade and defence co-operation advocate and chairs the Defense Cooperation Working Group, the executive arm of the DOD's Steering Group on NATO Armaments Cooperation which sets overall policy. 46 new staff positions have been established in 15 US embassies abroad—40 of them in Europe—to support all aspects of defence co-operation. Schemmer, B. F., 'Army, USAF, step up side-by-side test; 15 cooperative arms programs under way', *Armed Forces Journal International*, Oct. 1987, pp. 32–33. See also Odorizzi, C. D. and Schemmer, B. F., 'Taft: "side-by side testing a key solution" to two-way street issues', *Armed Forces Journal International*, Dec. 1986, p. 24; Ganley, M. and Stephan, H. J., 'Taft says NATO cooperative programs have made substantial progress', *Armed Forces Journal International*, July 1987, p. 18.

[25] The NATO Infrastructure Fund, established in the early 1950s, represents a formal cost-sharing programme to provide for the financing of facilities, services and programmes to be of common benefit to alliance members.

[26] US Department of Defense, *Report of the Secretary of Defense Frank C. Carlucci to the Congress on the Amended FY 1988/FY 1989 Biennial Budget* (US Government Printing Office: Washington, DC, 18 Feb. 1988), p. 99.

[27] The current two-way street balance is 1.6 : 1 in favour of the United States. Of the countries listed in table 3.6, only three countries currently enjoy a defence balance of trade in their favour: Belgium, Denmark and France. See also Schemmer, B. F., 'Breaking loose from NATO's hang-up over numbers', *Armed Forces Journal International*, Dec. 1985, p. 71.

[28] For example, instead of placing an order with General Dynamics, Northrop Corporation or Grumman, the Navy purchased off-the-shelf jet trainers from the McDonnell Douglas–British Aerospace partnership for $4.5 billion and claimed a savings of about $600 million. Lachica, E., 'Pentagon turns more to European arms: cost factor changes U.S. defense procurement', *Wall Street Journal*, 10 Feb. 1987, p. 42.

4. The impact of a withdrawal on US investments in Europe

Joseph R. Higdon
Capital Strategy Research, Washington, DC

Paul J. Friedrich
Bonn, Federal Republic of Germany

I. Introduction

As we approach a study with such an overwhelming—and not necessarily realistic—premise as a total pull-out of US forces from Western Europe, we would like to introduce a perspective that, at the time of writing (autumn 1988), in the context of this discussion is perhaps unique. While the premise suggests that the United States' primary relationship with Western Europe is military and political, it is our conviction that the fundamental relationship between the United States and Europe is economic, and that it is not one dependent upon the military presence of US forces. The US business community views Europe as an economically safe and promising place to be and will continue to operate there long after US forces have either been reduced or withdrawn. We think the cases of Portugal in the early 1970s and France in the early 1980s, where parties of the left came to power across the board, have very interestingly revealed that the major concern of the United States in Europe was economic. The main issue for the United States in these two instances had to do with what companies would be nationalized and what trade union demands on US businesses would be created. Throughout the debate surrounding these events, there was little convincing discussion of issues concerning military security. It is necessary to point out in this regard that from the perspective of the United States, the military presence in Europe is not an ideological issue. For decades, actors and observers across the political spectrum have viewed a certain reduction of US forces in Europe in a favourable light. Aside from those US companies that are in Europe primarily to serve US military installations, we think it could be fairly argued that the sensitive political nature of the presence of US forces in Europe could serve more as a public relations hindrance to US business operations there than as an asset.

We are reluctant to generalize on the opinion of the US business community, given the extraordinary range of the opinions represented by such a diverse group, but the arguments we put forth are those that have been presented to us most frequently by large US-based multinational companies with long experi-

Table 4.1. Majority-owned European affiliates with greater than 50 per cent ownership by a US parent,[a] 1985–86
Figures are in US$ m.

	1985	1986	Change (%)
Total assets	295 407	349 337	18.3
Total liabilities	189 309	223 039	17.8
Owners' equity	106 099	126 298	19.0
Sales	358 716	396 550	10.5
Net income	16 697	21 661	29.7
Employee compensation	41 364	52 000	25.7
Number of employees (thou.)	2 142	2 084	−2.7

[a] A US parent is a US person that owns or controls, directly or indirectly, 10% or more of the voting securities of an incorporated foreign business enterprise or that owns or controls an equivalent interest in an unincorporated foreign business enterprise.

Source: US Department of Commerce, *Survey of Current Business* (US Government Printing Office: Washington, DC, June 1988).

ence of doing business in Europe and that are included in the *Fortune 100* list of US companies.

There are three main reasons compelling US businesses to continue to view Europe as an area for business growth. One, the considerable commitment of US companies already in place is substantial. Two, the geopolitical climate between Western Europe and its eastern neighbours has rarely been more constructive or promising. Three, and perhaps most compelling, Western Europe's march towards a more unified business entity symbolized in the concept of 1992 has US business aggressively scrambling to secure a more permanent and stable place in Europe.

II. US investment in Europe

Between 1977 and 1986 investment in non-bank foreign affiliates by US multinational companies more than doubled. In 1986 total world-wide assets of these US companies increased 10 per cent to $4746 billion; the assets of their foreign affiliates increased 12 per cent to $932 billion. The largest increase was in Europe. The total assets of majority-owned European affiliates (greater than 50 per cent ownership by a US parent) increased 18 per cent to $350 billion (table 4.1). With US multinational company investments in Europe accounting for 37 per cent of total foreign assets, and with investment continuing, the probability of a business pull-out seems extremely unlikely.

Although these numbers no doubt are crude, their sheer magnitude is significant. In all certainty, sales are now well above $400 billion per year, and owner's equity and net income should also be up proportionally. If the asset value of European affiliates with greater than 10 per cent ownership by US

Table 4.2. European affiliates with greater than 10 per cent ownership by a US parent, 1985–86

Figures are in US$ m.

	1985	1986	Change (%)
Total assets	355 582	426 504	*19.9*
Sales	439 209	486 204	*10.7*
Net income	19 484	25 644	*31.6*
Employee compensation	51 677	64 646	*25.1*
Number of employees (thou.)	2 760	2 662	*– 3.5*

Source: US Department of Commerce, *Survey of Current Business* (US Government Printing Office: Washington, DC, June 1988).

parents (table 4.2) are included, along with US investors' ownership of European debt instruments, the magnitude of US investment in the European Community (EC) becomes very large. The resulting number does not even reflect the market value of those investments.

Of the 1986 majority-owned European affiliate sales, only $8.6 billion worth of goods and services were imported from Europe to the United States. This leaves roughly US $388 billion in sales that were generated in predominantly European markets. Since half of those sales were from the manufacturing sector (table 4.3), there is a clear commitment as well as an incentive to remain in these markets.

The lack of US military presence in Europe will have little impact on US investment practices. The combination of US multinational company participation in the EC market after 1992, and potential sales to new markets in Eastern Europe, means that the United States will continue to buy into Europe.

III. Gorbachev's challenge

Western Europe's most pressing need is not to meet a military threat from the East, but to face the political challenge of the USSR under President Gorbachev. This is a fundamental change in the international situation. Will it last?

How long is the security of Europe to depend on the United States to the extent it has depended on it in the past? It seems to be a fair assumption that the political and security relations in the Atlantic Alliance are likely to change. The structure of Europe as a whole will be a major issue in the 1990s.

Looking at the country in the heart of Europe, with the largest contingent of US troops on its soil, it is clear that the public awareness in the Federal Republic of Germany of a threat from the Soviet Union has drastically changed. A majority of the West German population no longer perceives such a threat, although a majority still believes that the Bundeswehr and membership in NATO are necessary. A poll taken in October 1988 (by the Friedrich-Ebert-Stiftung) showed that only 11 per cent see the Soviet Union as a threat to world peace, as compared to 70 per cent in 1970.

Table 4.3. Majority-owned European affiliates with greater than 50 per cent ownership by a US parent, 1986
Figures are in US $ m.

	Total assets	Sales	Net income	Employees (thou.)
Manufacturing	146 380	191 380	10 724	1 443.7
Finance (except banks), insurance, real estate	77 509	9 762	3 463	42.4
Petroleum	57 796	74 992	3 119	70.1
Wholesale trade	46 437	95 519	3 702	254.4
Services	14 986	16 025	560	137.7
Other industries	6 229	8 873	94	135.8
Total	**349 377**	**396 550**	**21 661**	**2 084.1**

Source: US Department of Commerce, *Survey of Current Business* (US Government Printing Office, Washington, DC, June 1988).

This change of opinion is due to many factors, among them the *rapprochement* of Western Europe and the Soviet Union under President Gorbachev in recent years, the fact that US foreign policy (as it has been formulated with regard to Viet Nam in the early 1970s, Israel or Latin American under President Reagan) has not always been appreciated in Europe, as well as domestic politics such as the re-orientation of socialist and social-democratic parties, which have given up their anti-Sovietism and taken position against nuclear weapons (and thus in one sense also against the United States).

IV. 'Europe 1992' demands US business presence

There are several reasons compelling US business interests to be more involved in Europe by the time of the EC market consolidation in 1992. One is that Europe will have a bigger market after the consolidation. Two, it will be easier to do business in a less barrier-prone market-place. Three, there is a risk that exclusionary barriers will be raised against companies that are not involved in the EC by that time. Four, the EC will probably require direct investments before allowing US businesses open access to the market-place. Five, the general interdependence of the world in financial transactions fosters a greater demand on the part of corporations to be 'on the ground' rather than operating on an absentee basis. What follows is a brief discussion of these points.

As European governments press forward towards their goal of market unification by 1992, US businesses cannot afford to sit on the sideline and wait for the outcome before deciding about their level of involvement. Clearly there are compelling reasons why they will have to be involved early to take maximum advantage of the opportunities that 1992 will present to corporations that are fully operating in Europe at that time. Most US businesses in the past operating in Europe have related to Europe on a country-by-country basis. They have picked and chosen between markets depending on ease of language, working conditions and market size. A unified Europe should present corporations operating in Europe a much bigger market-place in which to sell their goods. 1992 provides, if things go as planned, unified standards of doing business throughout the continent. Those unified standards will make business infinitely easier to transact than at present. Europe has in the past used standards as a means of protecting markets (as in the case of the colour television market). In this new environment, companies that are on the ground and in place will run virtually no risk of exclusion on the basis of standards. There is some indication that Europeans are going to require direct investment before permitting people to do business in this new unified market-place. If this is true, then it behoves US corporations to be fully involved now with direct investment plant facilities in order to take advantage of the 1992 environment.

As part of the 1992 environment, US corporations anticipate that there will be local requirements for description of contents for most manufactured goods throughout Europe. This again will be another subtle means of excluding those who are not fully participating members of the unified market-place.

V. Why US troops in Europe?

A US troop withdrawal from Europe would reduce Washington's influence generally in Europe and would, if the troops were to be kept activated at home, hardly bring any savings. To abandon a role of the United States in Europe would amount to a self-inflicted wound. Keeping troops in place, on the other hand, will provide pragmatic political advantages, support industrial co-operation with the European allies and serve as a bargaining chip for the United States in maintaining access to the European 'market 1992'.

Improving Soviet–European Community relations obviously pleases West European firms, but prompts security worries in some parts of the US Administration. Moscow's continuing charm offensive towards the West increases pressure to relax COCOM restrictions. A reduced US commitment to NATO would even further diminish Washington's leverage. The primary motivation for Europe 1992 is economic as well. To fare best, the USA should keep intact its political weight.

In the final analysis, however, we are all looking at the development of a more global political, social, economic and even cultural environment. As this process continues, the tensions associated with military security will, of necessity, play a less and less determinate role in the affairs of states, which have become highly developed economic and social entities. The US political leadership will have to adjust, as the US business community has done, to a Europe that no longer needs it militarily.

5. The economics of US bases and facilities in Western Europe

David Greenwood

University of Aberdeen, Aberdeen, Scotland

I. Introduction

In a farewell address delivered on the occasion of his retirement in 1796, George Washington invited his fellow Americans to reflect on the potential benefits of their country's 'detached and distant situation' and the potential costs of policies requiring US troops to 'stand upon foreign ground'. He also made a specific reference to Europe. 'Why', he asked, 'by interweaving our destiny with that of any part of Europe, entangle our peace and prosperity in the toils of European Ambition, Rivalship, Interest, Humour and Caprice?'[1]

For the best part of two centuries US politicians—and the public-at-large across the United States—have intermittently pondered Washington's question. At times the sentiment it expresses has been decisively rejected, as it was with the signing of the North Atlantic Treaty in 1949 and the subsequent commitment of US forces to Western Europe's defence under the aegis of the North Atlantic Treaty Organization. But the issue will not go away. In the past few years it has been addressed in several quarters: by senators and editors, by national security advisers whose time has gone and by ambitious representatives who think their time is coming.

Many participants in this most recent debate have concluded that while it would be inappropriate to abandon the Atlantic Alliance altogether, a thorough re-examination of the scope and scale of the US military contribution to NATO is now in order, if not overdue. Exasperated by what they regard as capricious European behaviour, out of sympathy with the mood (or 'humour') of certain European parties and pressure groups, suspicious of Europeans' attention to Europe's own interests—including old rivalries and new ambitions—some have even urged the complete withdrawal of US forces from Europe. This course finds support not only among 'new isolationists' (people who argue that the United States has no need of alliances, and can certainly do without allies who will not pay their way) but also among new internationalists (people who believe that the United States does need strong allies, but contend that nowadays NATO is a source not of strength but of weakness).[2]

To be sure, the option of complete withdrawal has not found favour with recent administrations. In fact, the importance of a continuing presence in Europe of both US nuclear weapons and US conventional forces has been reasserted regularly.³ In Congress and in the country, however, the mood is different. The dominant perception is that the United States has for too long borne a disproportionate share of the burdens (and risks) of NATO defence; and the accompanying prescription is that, since allies are evidently reluctant to do more, then Uncle Sam should do less.

The formula is obviously appealing, especially at a time when the Pentagon's budget is being cut to help eliminate the federal deficit, and when there are pressures to trim overseas spending in particular to help alleviate the nation's trade and payments difficulties. Indeed, some perceive a direct link between the burden-sharing disparity and economic distress. According to one influential congresswoman: 'Americans . . . believe we spend a much greater portion of our wealth on the common defense, and that our allies use the money they save on defense to subsidize their trade, creating [sic] our enormous trade deficit'.⁴

None of this has yet crystallized into an insistent or irresistible congressional and popular demand to 'bring all the boys home'. However, such a demand appears to be taking shape in the crucible of US politics, and could well materialize in the not-too-distant future.

Given this possibility, it makes sense to consider the implications for Western security, and specifically for Western Europe's security. Leaving aside speculation about the precise circumstances surrounding a withdrawal decision (and ignoring, for the time being, the effects of a simultaneous repatriation of some US and Soviet contingents as part of a conventional arms deal), what would a total withdrawal of US forces in Europe and European waters *mean*: psychologically, politically, strategically, militarily? Would it precipitate demoralization and defeatism in Western Europe, a headlong rush to find political accommodation with the Soviet Union, and the concession to Moscow of effective hegemony from the Urals to the Atlantic based on uncontested military superiority? Or would it relieve Western Europe of an enervating dependence syndrome, accelerate Western Europe's political integration and, perhaps, prompt the establishment of what de Gaulle would have called 'a defence for Europe which is a European defence'?

To some extent the answers to these questions depend on the answer to another one: What would withdrawal entail *in economic terms*? To be more precise: If, taking the extreme case, the United States vacated all its bases and facilities in Europe, what sort of economic impact would this have, and what extra resources would the West European members of NATO have to find to make good resultant gaps in the alliance's capabilities (either by replacing like with like or by making different, but essentially equivalent, provision)? To be more positive: Is it economically realistic to contemplate some 'new transatlantic bargain' based on exclusively European provision for first-line deterrence and defence in Europe, under European command, with the United States cast as a 'detached and distant' partner under the North Atlantic Treaty, but for all that still an ally?⁵

In this chapter, these questions are taken as the basis for a tentative exploration of the economics of 'Europe After an American Withdrawal'. The word 'exploration' is used advisedly, because the economic dimensions of the current US presence in Western Europe are not very well documented (and not very well understood), while any attempt to gauge the resource requirements for 'exclusively European provision of first-line deterrence and defence' must of necessity be speculative. The word 'tentative' has to be added too, because much of the analysis rests on fragments of data (or no hard data at all).

II. The US presence in Europe

Scope

The obvious starting-point for this analysis is a general appreciation of the scope and scale of the current US presence in Western Europe and European waters.

In functional terms the scope of the US presence is remarkably comprehensive, encompassing strategic and tactical forces, contingents of all three (or four) services, nuclear and conventional capabilities, front-line units and support elements, plus important command, control, communications and intelligence (C^3I) facilities. This is apparent from the official Department of Defense (DOD) listing of front-line units in table 5.1.

Having said that, the 'strategic forces' component amounts to just one submarine depot ship which may or may not feature in Washington's longer-term plans, while such ground-launched cruise missiles (GLCMs) as became operational in the mid-1980s are being decommissioned under the terms of the 1987 Treaty between the United States of America and the Union of Soviet Socialist Republics on the elimination of their intermediate-range and shorter-range missiles (the INF Treaty). Thus the principal components of 'the presence' are either (*a*) elements of the US contribution to deterrence and defence on NATO's Central Front, or (*b*) forces for operations on NATO's Southern Flank.

In the first category are the ground forces which form the two US corps in the Central Army Group (CENTAG) area of the Allied Forces Central Europe (AFCENT) command plus the skeletal III Corps in the Northern Army Group (NORTHAG) area; the tactical air power provided for operations on the Central Front, comprising a dozen squadrons located in the CENTAG area—forming part of the 4th Allied Tactical Air Force (4ATAF)—together with others located in the UK and the Netherlands; and the logistics and other support units for these 'teeth' formations. In the second category are the 6th Fleet, with its Marines, plus the ground and tactical air forces stationed along NATO's 'southern tier' from Portugal to Turkey, most of which are counted in the order of battle of the Allied Forces Southern Europe (AFSOUTH) command.

Table 5.1. US front-line units in Europe, 1987–88

Unit	Location
US strategic forces	
Navy	
1 Submarine tender (with 10 SSBNs)[a]	Holy Loch, Scotland
US tactical/mobility forces	
Army divisions	
1st Armored Division	FRG[b]
3d Armored Division	(various locations)
3d Infantry Division	
8th Infantry Division	
Bde, 1st Infantry Division	
Bde, 2d Armored Division	
Special mission brigades	
Berlin Brigade	Berlin (West)
Armored cavalry regiments	
2d Armored Cavalry Regiment	FRG[b]
11th Armored Cavalry Regiment	(various locations)
Navy ships/aircraft	
6th Fleet	Mediterranean[c]
(2 carriers, 18 surface combatants/attack submarines, 6 auxiliaries, 1 amphibious ready group, 2 ASW patrol squadrons)	
Air Force tactical forces	
15 squadrons	United Kingdom
12 squadrons	FRG
1 squadron	Netherlands
3 squadrons	Spain
1 squadron	Iceland
27 Flights GLCMs[a]	United Kingdom (10), Italy (7), FRG (6), Belgium (3), Netherlands (1)
Air Force mobility forces	
2 squadrons	FRG

[a] Decommissioned under the INF Treaty.
[b] The US force structure in the FRG incorporates artillery formations, surface-to-surface missile battalions (Pershing, Lance), and surface-to-air missile battalions (Patriot).
[c] Units of the 6th Fleet may be 'forward assigned' to the Indian Ocean.

Sources: Manpower Requirements Report FY 1988 (US Department of Defense: Washington, DC, Feb. 1987), chapter 9; supplemented by information from IISS, *The Military Balance 1987–88* (International Institute for Strategic Studies: London, autumn 1987), pp. 24–26.

The corollary is worth noting. There are no US combat formations 'in place' within the Allied Forces Northern Europe (AFNORTH) area. Because of the 'no bases' policy to which Denmark and Norway subscribe, NATO's concept of operations here postulates defence by indigenous troops (standing forces and mobilized reserve units) augmented by allied reinforcements. The United States has national commitments to contribute to such reinforcement—to deploy a Marine Expeditionary Group plus air power in the event of a threat to northern Norway, for example—and also subscribes to the Allied Command Europe (ACE) Mobile Forces which have deployment options throughout the theatre. Moreover, to facilitate fulfilment of these obligations some equipment has been pre-positioned in the north, and collocated operating base (COB) agreements have been concluded covering the use of airfields. But there is *no* permanent US presence on this flank (unless one judges that regular naval deployments to far northern waters amount to that).

Scale

To the military professional (or competent defence analyst)—someone who knows the meaning of '10 SSBNs', the make-up of an armoured division and a mechanized infantry brigade and an armoured cavalry regiment, the structure of an 'Amphibious Ready Group' and the size of a typical tactical fighter squadron—the listing of front-line units in table 5.1 conveys a picture of not only the scope but also the scale of the US military effort in Europe and European waters. Even the lay observer can appreciate that the 6th Fleet is in effect a small Navy and, if told (for example) that the US Army in Europe has an inventory of around 5000 main battle tanks and that the US Air Force in Europe has a roster of over 700 combat aircraft, that what the United States fields and flies constitutes a considerable military capability. (By way of comparison the United Kingdom has approximately 1200 main battle tanks and some 600 combat aircraft all told.)

The other measures of 'scale' available are, of course, manpower and money. According to DOD statistics, over 320 000 US active-duty military personnel currently 'stand upon foreign ground' in the territory of the European members of NATO. Personnel in non-NATO countries plus those afloat in warships deployed to European waters bring the total located in a more broadly defined Europe to nearly 340 000. In addition more than 40 000 US-based civilian employees of the DOD and defence-related agencies work at European bases and facilities. (How many defence contractors' staff are similarly employed—on a permanent or semi-permanent basis—is an unknown quantity, but the figures must run into the thousands.)

The 'European NATO' military strength represents almost 70 per cent of all US uniformed personnel stationed abroad or afloat (i.e., outside the continental USA, Alaska, Hawaii and other US territories), and nearly 16 per cent of all US forces world-wide. The country-by-country figures in table 5.2 show that these servicemen and -women are overwhelmingly concentrated in the Federal Republic of Germany and West Berlin (almost 250 000 active-duty military, more than four-fifths of them Army personnel), the next three countries in the host nation ranking being the United Kingdom (with just under 30 000 personnel, mainly Air Force), Italy (nearly 15 000) and Spain (around 9000).

Table 5.2. US active duty military personnel in Europe, 31 December 1987

Country/Region	Army	Navy[a]	Air Force	Total
Belgium	1 434	165	1 949	3 548
Denmark	18	33	25	76
France	18	47	17	82
FRG[b]	206 773	424	41 249	248 446
Greece	481	532	2 343	3 356
Greenland	–	–	215	215
Iceland	3	785	1 366	3 154
Italy	4 177	4 843	5 747	14 767
Luxembourg	5	6	–	11
Netherlands	825	27	2 691	3 543
Norway	32	63	123	218
Portugal	61	383	1 219	1 663
Spain	22	3 916	5 151	9 089
Turkey	1 192	134	3 558	4 884
United Kingdom	250	2 715	26 432	29 397
Subtotal	**215 291**	**15 073**	**92 085**	**322 449**
Other Europe/afloat	28	16.571	13	16 612
Total	**215 319**	**31 644**	**92 098**	**339 061**
Total foreign countries	**258 177**	**113 493**	**134 314**	**505 984**
Total worldwide	**777 680**	**783 602**	**605 329**	**2 166 611**

[a] Includes Marine Corps.
[b] Includes Berlin Brigade.

Source: US Department of Defense military manpower statistics, 31 Dec. 1987.

The data in table 5.2 record the number of *uniformed* US personnel in 'European NATO' (which for this purpose, it should be noted, includes Greenland, Iceland and the Portuguese Azores). At most bases and facilities these personnel—and the 40 000 or so US-based civilian employees—have their families with them, an estimated total of well over 300 000 dependants in the countries listed. At virtually every installation there are also locally engaged civilian workers, probably between 75 000 and 80 000 in 'European NATO' as a whole. In the more broadly defined Europe 'the presence' is thus represented by a 'US population' of almost 700 000 and by a 'working population' exceeding 450 000.

In the absence of authoritative and consistent data on the *non-uniformed* people (civilian employees and dependants), producing a precise country-by-country breakdown of these aggregates is impossible. However, by assembling fragments of information from various sources (including other chapters of this book) and making outright guesses where information is lacking altogether, the summary tabulation of population statistics in table 5.3 has been prepared. Because the numbers in columns B, C and E are estimates—and, in some instances, very rough-and-ready estimates—they are given in round figures. So

Table 5.3. US bases and facilities in Europe: population statistics 1987–88 (estimates)

Country/Region	A Active-duty US military personnel	B US civilian personnel[a]	C US dependants[a]	D Total US population (A + B + C)	E Locally recruited civilian personnel[a]	F Total personnel[b] (A + B + E)
Belgium	3 548	1 500	5 000	10 000	750[c]	5 750
Denmark	76	–	50	150	–	75
France	82	–	50	150	–	75
FRG[d]	248 446	34 000	228 500	511 000	60 000[e]	342 500
Greece	3 356	150	3 500	7 000	2 000	5 500
Greenland	215[f]	100	50	350	800	1 100
Iceland	3 154	50	2 500	5 800	250	3 550
Italy	14 767	2 000	16 250	33 000	2 000	18 750
Luxembourg	11	–	10	20	–	–
Netherlands	3 543	200	3 500	7 250	750[c]	4 500
Norway	218	–	100	300	50	250
Portugal	1 663	100	2 000	3 750	250	2 000
Spain	9 089	1 200	7 500	17 700	2 000	12 200
Turkey	4 884	400	3 750	9 000	4 000	9 250
UK	29 397	2 500	38 000	70 000	5 000	37 000
Subtotal	**322 449**	**42 000**	**311 000**	**675 500**	**77 800**	**442 000**
Other/afloat	16 612	–	–	16 500	–	–
Total	**339 061**	**42 000**	**311 000**	**692 000**	**77 800**	**459 000**

[a] Figures are estimates.
[b] Figures are rounded.
[c] Sami Faltas (chapter 7) estimates that there are approximately 1750 locally recruited civilians in the Benelux countries all in all.
[d] Includes Berlin.
[e] If employees of institutions indirectly linked to the US troops are included this figure rises to over 70 000 according to Bebermeyer and Thimann (chapter 6).
[f] This is the official DOD figure. The number should be over 500 according to Simon Duke (chapter 9), who also has a higher figure for US-based civilian employees than is shown here.

Source: Table 5.1 and author's estimates.

too, therefore, are all the totals in the tabulation (except that at the foot of the first column).

Despite its limitations, table 5.3 is instructive. It underscores the point that there is more to the US presence in Western Europe than the service units and the armed forces (tables 5.1 and 5.2). It also brings out the true scale of the presence in the FRG where, if these estimates are right, there is a 'US population' of well over half a million and a 'working population' of over 340 000 sustained by DOD's outlays.

Turning to these outlays, the question arises: What does all this cost? What is the annual bill to the US tax-payer for their country's European deployed forces and their support?

Here too, unfortunately, analysis is hampered by lack of authoritative information, because not all the relevant official data are in the public domain. The obvious source—the DOD's annual Report to the Congress entitled *United States Expenditures in Support of NATO*—is a classified document. This is not so much for reasons of national security, although these doubtless enter the reckoning, but because of the problems of definition and attribution which arise in costing the US commitment to NATO and the consequent danger that whatever figures were published might confuse matters rather than clarify them.[6]

What is known, however, is that in the early 1980s (and, indeed, for a decade before that) the costs attributable to forces actually deployed in Europe—the value which is of interest at this juncture—accounted for around 16 per cent of the DOD's total budget. Assuming that the comparatively long-standing stability of this proportion has been maintained (deliberately or otherwise), a plausible estimate for 'the bill' in 1987–88 is, therefore, $50 billion or thereabouts.

It is important to clarify what 'costs attributable' means. It embraces *(a)* the pay-roll costs of European-deployed military and all civilian employees (see table 5.3 above), *(b)* operation and maintenance outlays (O&M for short, or 'other running costs' in everyday language), *(c)* capital or investment costs (new equipment, replenishment spares and military construction), and *(d)* an overhead element (i.e., a share of the expense of the Pentagon's research and development (R&D), logistics, administrative and training base).

A breakdown of the $50 billion estimate among these cost categories, and among the individual European host nations, would be illuminating. With the use of some heroic—perhaps even irresponsible—assumptions, one can produce such analyses.

Drawing reasonable inferences from overall DOD cost data, the ratios among personnel costs, O&M outlays and capital plus overhead elements are estimated to be (very approximately) 30:30:40. In other words, of the $50 billion, personnel costs probably amount to $15 billion all told, with 'other running costs' coming to about the same, while the remaining $20 billion represents equipment and construction plus overhead expenditure.

The country-by-country cost distribution is harder to assess. However, piecing together fragments of data—much as was done to compile table 5.3—yields the breakdown shown in table 5.4, a summary tabulation of financial statistics which complements the earlier 'population' table. The highlights here are the estimate that the budgetary expense of the labour-intensive US presence in the FRG is probably around $35 billion, or 70 per cent of the European NATO total, while that for forces in the other important host nations—the United Kingdom, Italy and Spain—may well exceed $9 billion, a further 18 per cent.

The obvious comment prompted by table 5.4 is that in Washington's fiscal arithmetic the $50 billion for the annual upkeep of European-*deployed* forces is a comparatively small sum, just one-sixth of the Pentagon's budget. Note,

Table 5.4. US bases and facilities in Europe: financial statistics, 1987-88 (estimates)
Figures are in US$ m.

Country/Region	A Personnel costs	B Operations & maintenance costs	C Equipment & construction costs	D Total budgetary cost (A + B + C)
Belgium	225	175	350	750
Denmark	–	–	–	–
France	–	–	–	–
FRG	10 750	10 750	13 500	35 000
Greece	150	150	300	600
Greenland	50	50	100	200
Iceland	150	150	300	600
Italy	850	950	1 000	2 800
Luxembourg	–	–	–	–
Netherlands	150	150	200	500
Norway	–	–	–	–
Portugal	75	75	100	250
Spain	400	350	500	1 250
Turkey	200	200	400	800
UK	1 500	1 500	2 500	5 500
Subtotal	14 500	14 500	19 250	48 250
Other/afloat	500	500	750	1 750
Total	15 000	15 000	20 000	50 000

Source: Author's estimates.

however, that the United States has directly declared to NATO, for early and later reinforcement of Europe, forces which are costed at around three times this figure. Thus the share of all DOD expenditure ascribed to European-*committed* forces—and perceived as the expense of the NATO commitment in the US popular and political debate—is more like three-fifths or two-thirds.

Note, too, by way of preparation for later discussion of the challenge that 'compensating' for US withdrawal would pose, that while $50 billion may be a small sum in the US budgetary context it represents a considerably greater allocation of resources than any one of Western Europe's 'big spenders'— France, the FRG and the United Kingdom—is in the habit of making; and it exceeds the *combined* defence spending of all the other European members of the alliance. (Put another way, the sum which the US authorities claim as the annual cost of all their forces directly committed to NATO —approaching $200 million in current dollars—exceeds the grand total for defence spending by the whole of European NATO.)

Although the $50 billion attributable cost of European-deployed forces is less than the full cost to the DOD's budget of the European commitment it is also greater than the cost to the US balance of payments of the presence in Europe. That is because the sum includes considerable expenditure—in each of

Table 5.5. DOD direct defence expenditures abroad for goods and services, 1983–87
Figures are in US$ m.

Country/Region	1983	1984	1985	1986	1987
Belgium/Luxembourg	165	121	121	117	184
Denmark	55	57	64	89	100
France	64	96	59	67	78
FRG	4 376	4 587	5 377	6 165	6 384
Greece	537	250	184	204	221
Iceland	85	48	53	60	97
Italy	568	440	516	484	585
Netherlands	83	137	86	99	116
Spain	166	191	168	197	254
Turkey	44	48	75	51	68
United Kingdom	905	925	898	874	1 105
Other NATO	293	227	172	233	357
European NATO	**7 342**	**7 128**	**7 772**	**8 630**	**9 548**

Source: *Survey of Current Business*, vol. 68, no. 6 (June 1988).

the cost categories distinguished earlier—which 'remains' in the United States, that is, does not entail foreign currency outlays. This is true of the whole of the overhead element, most capital spending (the obvious exception being military construction using local labour), and some O&M expenditure. Moreover, part of the pay of military personnel and US-based civilians is saved or remitted home and part is spent on the US goods sold at on-base facilities: not all, therefore, is disbursed as 'local currency expenditure' to the benefit of the economy of the host nation (or locality).

In short, the *foreign exchange cost* of the US presence in Europe is only a fraction of its *budgetary cost*; and, by the same token, so are the related dollar earnings of host economies. The exact fraction is obviously of considerable interest and importance for assessing the economic impact of 'the presence'; and, happily, this is a matter on which there are official data available. The Bureau of Economic Analysis (BEA) of the US Department of Commerce periodically publishes figures on 'direct defence expenditures abroad for goods and services', giving details of outgoings across the exchanges in respect of *(a)* personal spending by US military and civilian personnel employed by the DOD, *(b)* payments to locally engaged civilians, *(c)* outlays by the DOD for foreign goods (including those purchased abroad for resale in post exchanges (PXs) and commissaries) and *(d)* overseas outlays by US and foreign (typically local) contractors engaged for construction and O&M work at installations abroad plus payments to the commonly funded NATO infrastructure programme. The BEA's figures for such expenditures over the five-year period 1983–87 are presented in table 5.5, with country-by-country detail for Western Europe.

According to these numbers, for European NATO as a whole 'across the exchanges' expenditure in 1987 was just over $9.5 billion, or less than one-fifth of the (estimated) budgetary cost of US forces in Europe. Most of this money was spent in the FRG, the $6 billion worth of direct expenditures there including $1.3 billion for locally engaged employees, $2.1 billion for contractual services provided by West German firms (mainly for the operation and maintenance of property), and over $2.0 billion in respect of 'local currency expenditure' by US-based civilians and their dependants.[7]

III. The economic impact of European-deployed forces

The summary data in tables 5.3 and 5.5 provide a point of departure for broadbrush assessment of the impact of the US European-deployed forces on host nation economies.

Military installations create jobs and generate income. They create jobs *directly*, by employing local workers to fulfil a variety of technical and administrative support functions; and *indirectly*, in the local or regional economy, through the extra-mural economic activity that is prompted and sustained by their presence. They generate a flow of income *directly*, through official local purchase of supplies and services plus the off-base private spending of their personnel; and *indirectly*, because such expenditures contribute to further income creation in the local or regional economy (by way of a so-called 'multiplier' process).

For US bases in Europe the direct element in both employment and income generation can be observed from the tabulations. The jobs directly created at US facilities are the 'civilian employees (locally engaged)' of column 5 in table 5.3, an estimated 75 000–80 000 in Europe as a whole. The directly created income is the across-the-exchanges expenditure recorded in table 5.5, amounting to almost $10 billion in 1987.

To estimate the indirect elements it is necessary to know the values of the relevant 'multipliers'—that is, the local employment multiplier and the local income multiplier. What these might be is a matter for speculation (or, better, for large-scale empirical research, going beyond the scope of the present exercise). However, it seems reasonable to assume that they probably lie within the range 1.8–2.0—meaning that every five direct jobs (or dollars of direct spending) generate between four and five indirect jobs (or dollars of indirect local income). Obviously, the exact figure will vary from location to location, depending on the precise nature of the direct employment and income created and on the particular make-up of the local economy.[8] In addition there is local employment creation associated with the local purchases by US personnel to be taken into account, perhaps amounting to one indirect job for every five active duty military plus US-based civilians.

If such reasoning is correct the indirect employment and income associated with the US European-deployed forces come to at least 150 000 jobs and an annual income flow of around $9 billion.

Taken together, the direct and indirect elements yields an overall assessment: one way and another the US presence is estimated as providing employment for around 225 000 and as boosting incomes in Western Europe by almost $20 billion annually.

Income and employment benefits to the host nations in European NATO are the most readily quantifiable (at least in theory) and also the most obvious aspect of the economic impact of the US presence. There may be other—unquantifiable, even intangible—benefits. For instance, is the US willingness to purchase military equipment from Europe, to participate in transatlantic co-development and co-production ventures, and to give aid to several countries, rooted in the formal affiliation under the North Atlantic Treaty (and hence something which troop withdrawal might not affect); or is it attributable, in part anyway, to the practical, concrete ties associated with the physical presence of nearly 700 000 Americans on the continent (and therefore something which might wane if that presence were to end)? More generally, does all-round commercial and financial advantage accrue to West Europeans because the United States is, in an important sense, not just *with* but actually *in* Europe?

This is not the place to pursue such matters. Suffice it to say that they ought to have some place in any comprehensive calculus on 'Europe After an American Withdrawal' (and are, indeed, addressed in other chapters).

There is another side to this coin, however: the difficult-to-quantify, unquantifiable and intangible economic costs that Western Europe bears on account of the US presence. Even the DOD's own annual Report to the Congress on *Allied Contributions to the Common Defense* notes the 'loss of rents and tax revenues caused by the unusually large amount of real estate dedicated to defense purposes'.[9] This cost falls most heavily on the FRG, of course, as host nation to three-quarters of the US troops in Europe—not to mention the around 150 000 troops of other allies—to whom installations, training areas and thousands of housing units are made available without charge. According to the FRG Government, 'the value of the no-cost provision of real estate for stationed forces amounts to more than 40 billion marks (18 billion dollars)'.[10] As the largest foreign contingent in the Federal Republic of Germany, the US forces there probably account for over 60 per cent of such provision, representing a real 'cost' to Bonn which—expressed at 1987/88 values—is likely to exceed $12 billion. (The quoted figures relate to 1985/86.) Resource use such as this must also enter any comprehensive reckoning of the economic impact of European-deployed forces.

Such costs as these fall under the general rubric of *peacetime* host nation support for US forces; and each host country in Europe incurs some expense of this nature, the exact form varying from place to place. Later chapters in this book illustrate the diversity and the often considerable costs involved.

Throughout Western Europe, defence ministries—and civil departments too, in some cases—also have to make budgetary appropriations to cover *wartime* host nation support of deployed forces and their reinforcements, none more than the FRG Ministry of Defence. Under the Wartime Host Nation Support Agreement of 1982 between the Governments of the Federal Republic of

THE ECONOMICS OF US BASES AND FACILITIES IN EUROPE   91

Germany and the United States—which covers the provision of both personnel and facilities to assist US forces—the host nation has undertaken, among other things, to field and equip a 90 000-strong 'supplementary force' for this purpose, which is equivalent to an additional corps.[11]

Clearly this commitment would not exist if there were neither forces in place nor expected reinforcements. Whether it represents an avoidable expense—and hence a true cost of the US presence—is less apparent. Troop withdrawal might or might not be accompanied by abrogation of the reinforcement obligation. If it were not, the FRG would presumably retain its 'supplementary force' in some shape or form. If it were so accompanied, that is to say if the United States 'abandoned' Western Europe totally, there would be much agonizing about the consequences but emphatically not a decision to axe an entire corps from the remaining order of battle.

What all this amounts to is self-evident. Gauging the overall economic impact of the US European-deployed forces is anything but straightforward, especially if the aim is to expose and elucidate all the economic benefits and costs of their presence. More important even than data limitations are the technical and contextual assumptions that must surround any assessment. What is known for certain is that just under $10 billion flow directly across the exchanges to Western Europe's advantage each year (table 5.5); and there is likely to be secondary income generation of, say, $8–10 billion as a result. As for employment, US bases and facilities provide work for between 75 000 and 80 000 Europeans directly (table 5.3); and it is estimated that, across European NATO as a whole, these installations probably sustain indirectly a further 150 000 or so local jobs. Beyond this, however, there are innumerable 'grey areas'—and not a few 'black holes'—making an aggregated cost–benefit calculus impractical. For some countries, however, a national evaluation is worth attempting; and in several of the chapters which follow authors have done just that.

IV. The economics of withdrawal

It is time now to transpose the argument. Having explored the economics of the current US *presence* in Europe, it is necessary to ask what *withdrawal* would entail, in economic terms.

Recalling a formulation offered earlier in this chapter, 'if, taking the extreme case, the United States did vacate all its bases and facilities in Europe, what sort of economic impact would this have?'

The question can be dealt with fairly briefly. Leaving aside the practical problems that would be involved in rapidly repatriating all the front-line units listed in table 5.1 (with their support) and summarily abandoning the various C^3I facilities which the United States has in Western Europe, and provided always that the rough-and-ready analysis of the preceding section is fundamentally sound, the host nations of European NATO would suffer an immediate loss of income (or demand) of up to $20 billion and job losses of 225 000 or thereabouts.

Spread evenly among NATO's 14 European nations, such losses would pass virtually unnoticed, like a faltering in the growth rate or a blip on an unemployment curve. They would certainly pose no insuperable problems of macro-economic adjustment. In fact, the scale of change might well lie within the normal margin of error in governmental demand and employment forecasting.

The US *installations* are not evenly spread, however: there is the national concentration already noted (in the FRG and, less conspicuously, in the UK, Italy and Spain). More important, there is a regional or local concentration (in the southern FRG and eastern UK and in quite specific localities within all the economies of European NATO). Thus there would be problems of micro-economic adjustment at the regional and local level, perhaps quite acute problems in particular instances, such as the closure of a major air base in a relatively remote and/or sparsely populated rural area. Not that these difficulties would necessarily endure—how to deal with military base closures so as to avoid lasting economic distress is a comparatively well-researched and well-understood subject.[12] In any case, much would depend on what the erstwhile host nation might opt to do in order to compensate—to the extent possible—for the US exodus, a theme taken up below.

It is not so easy to predict the consequences of withdrawal for Western Europe's defence *industries*. One can imagine the hypothetical 'retreat to Fortress America' occurring in an atmosphere of all-round disenchantment with the Allies (perhaps with a burden-sharing pretext), accompanied by strident protectionism (born of obstinately intractable 'dual deficit' problems). In such circumstances it is hard to envisage continued US enthusiasm for transatlantic arms deals, especially co-development and co-production ventures. Certainly no one would expect the continuation of recent efforts to promote such activity, by fencing off funds for joint endeavours and encouraging the side-by-side testing of European and US products in open competition for the Pentagon's dollar. The effective barring of the US market to European defence products might even be on the cards. On the other hand, it is equally possible to imagine withdrawal taking place as part of a search for some 'new transatlantic bargain' based on notions of a rational division of labour between Europe and America in provision for Western security, ideas which have received fresh attention lately in Alliance deliberations on novel approaches to the sharing of roles and responsibilities.[13] In these circumstances, a decision to redeploy forces from Europe to the continental United States—perhaps even to disband some or all of them (for budgetary relief)—would not necessarily foreshadow the collapse of NATO, termination of transatlantic trade and other apocalyptic consequences. In fact, Washington might well be at pains to prevent this.

As to whether all the explicit and hidden costs of wartime host nation support would disappear with the departing troops, that too would depend on the precise circumstances of the hypothetical pull-out. If the United States were acting under the terms of—or were simply to prompt—a 'new transatlantic bargain' and, in that spirit, retaining the willingness and ability to reinforce, they would not disappear. They might well rise, depending on the form and scale of the new reinforcement plan.

V. The economics of compensation

The possibility of extra costs on the host nation support account would, however, be among the least of the West European nations' worries, if total withdrawal occurred. Their principal preoccupation would be how to make good the loss of considerable military capability. In that connection the question is: What extra resources would they have to find to fill the gaps?

The annual expense of the current European-deployed forces of the United States has been put, in US budgetary terms, at around $50 billion at today's values. Once an initial investment had been made in replacement infrastructure and hardware (radars, signal processing, data processing equipment, ships, tanks and other fighting vehicles, artillery pieces and missiles, trucks and transporters, fixed-wing aircraft and helicopters, plus munitions for all of these—a daunting list) the resource cost of operation, upkeep and modernization, year in year out, would be $50 billion, *if* Western Europe sought to put in place an identical force structure and identical force levels, and *if* European costs were the same as US costs (item-by-item or on the average).

Unlike businesses, navies, armies and air forces do not keep balance sheets. Therefore one would have to guess at the expense of the initial investment. More practically, one might make the entirely reasonable assumption that the cost would be beyond the reach of NATO's European members, at least within any acceptable time-scale, as would the $50 billion recurring annual outlays for 'operation, upkeep and modernisation' for that matter. (Nor would any likely difference between US and European costs invalidate that judgement, even allowing for the big difference between the personnel bill for an all-volunteer force compared with that for part-conscript forces.)

In other words, replicating present US provision, replacing like with like, is not an option for Western Europe. Nor is the option—superficially appealing in some cases—of what might be called 'national in-fill', that is, each country 'replacing' at its own expense the US forces it had previously hosted. For one thing this would saddle the FRG with *(a)* the financially impossible burden of equipping and maintaining over two corps including nuclear, dual-purpose and conventional missile and artillery batteries, together with a tactical air force of over 300 combat planes (half the size of the present Luftwaffe); and *(b)* the demographically impossible task of finding 250 000 extra personnel. No less significant, the implicit rationale of such an option is that the existing pattern of US deployments somehow corresponds to the particular security needs of the host country in which each element of the force structure happens to reside.

This is manifestly absurd. The US corps and tactical air assets in the FRG are there not to defend the host country as such but because that is where NATO's front-line is located. Similarly, the fighter wings based in the United Kingdom are there not to protect the United Kingdom itself but because it would be militarily unwise to have all of NATO's tactical air power based on the continent.

Put simply, replication and 'national in-fill' are not practical propositions, which raises the question: Is 'compensation' in any form actually pie in the sky, and in cloud-cuckoo-land at that?

The short answer is: perhaps not. It may be economically (and politically) feasible to contemplate a 'new transatlantic bargain' founded on what was defined earlier in this piece as 'exclusively European provision for first-line deterrence and defence in Europe, under European command'.

The basis for this judgement is twofold. In the first place, there is clearly scope for revision of the pattern of NATO's military dispositions, even within the framework of a strategy founded on forward defence and flexible response; and the opportunity for undertaking such revision, moving NATO in the direction of a lower-cost posture, has been provided by President Gorbachev's recent actions. In the second place, there is abundant scope for better use of the resources allotted to military purposes by the European allies, through exploitation of more efficient approaches to fielding forces and acquiring arms.

As regards the first of these themes: slow though NATO bodies may be to recognize it, amendment of concepts of operations plus adjustment of force structures and force levels are inevitable in the 1990s as the requirements of first-line deterrence and defence are reassessed, to take account of *(a)* the unilateral reductions in Soviet and other Warsaw Pact forces announced over the turn of the year 1988/89 (which the East has actually begun to make), *(b)* the reorganization of the remaining Warsaw Pact forces that has been promised, involving their assumption of an explicitly defensive orientation (which has also begun) and *(c)* force reductions resulting from the Conventional Armed Forces in Europe (CFE) Negotiation. Obviously it is too early to state categorically what precise 'amendment plus adjustment' will find favour on the NATO side. But one can confidently predict, among other things, greater interest in the 'defensive defence' notions which have already received considerable analytical attention and a greater preparedness to contemplate substitution of mobilized reserves for ready forces in NATO's order of battle. In sum, making 'compensation' for a withdrawal of US troops in Europe ought to become progressively more manageable as time goes by.

Turning to the second theme, the circumstances have never been more propitious for new approaches to 'fielding forces and acquiring arms'. So far as weapon acquisition is concerned, progress towards improved co-ordination of procurement among European NATO states has gathered considerable momentum already. The key developments here have been the revitalization of the Independent European Programme Group (IEPG) and the pursuit of numerous bilateral initiatives (of which the Franco-British and the Franco-German are perhaps the most noteworthy). Prospects for greater collaboration in upstream R&D and for more competition in arms development and production during the 1990s are extraordinarily good. This is partly because the IEPG has prepared, and begun to implement, an innovative blueprint for collaborative R&D and for a more open European market generally. Also relevant is the fact that European industrialists are increasingly choosing to work in harness for their own hard-headed commercial reasons. A truly pan-European approach to weapon acquisition is in the making, and will undoubtedly help drive costs down.[14]

However, it is in 'fielding forces' that NATO's European members have most to gain. 'Exclusively European provision for first-line deterrence and defence' might be affordable if nations were prepared to practise some degree of role specialization, each country contributing to the collective defence in accordance with its comparative advantage in different forms of military provision. Practical studies have identified a host of attractive options here: concentration by Western Europe's smaller navies on the shallow- and sheltered-water tasks to which they are best suited; generation of extra ground formations by improved reserve mobilization practices among continental conscript armies; and rationalization of provision for tactical air power and for air defence (especially missile air defence in the FRG, which is a mission tailor-made for locally recruited regular personnel augmented by specialized reservists). The possibilities, if not endless, are certainly numerous.[15]

Detailed examination of the potential of judicious role specialization, and of how it might bring 'compensation' for Europe's loss of US forces within reach, lies beyond the scope of this chapter. Two final thoughts are in order, however. First, the thrust of this concluding argument is that the necessary precondition for a viable 'new transatlantic bargain' is an imaginative 'new intra-European bargain'. Second, if that could be struck—on the twin pillars of role specialization and armaments co-operation—it would finesse the burden-sharing argument and pressure for massive US troop withdrawals might recede. In short, find the solution and there might not be a problem in the first place.

Notes and references

[1] Gilbert, F., *To the Farewell Address* (Princeton University Press: Princeton, N. J., 1961), p. 145.

[2] There are lots of 'new isolationists'. The leading spokesmen for this school are Irving Kristol and Melvyn Krauss (author of *How NATO Weakens the West*). See Krauss's contribution to Weinrod, B. (ed.), *The US and NATO: Should the Troops Stay?* The Heritage Lectures, no. 118 (The Heritage Foundation: Washington, DC, 8 May 1987) for a concise statement of his position.

[3] See for example *National Security Strategy of the United States* (The White House: Office of the Press Secretary, Jan. 1988), p. 18.

[4] Representative Patricia Schroeder (Democrat–Colo.), quoted in the *Chicago Tribune*, 24 Mar. 1988.

[5] The expression 'new transatlantic bargain' is taken from Sloan, S., *NATO's Future: Towards a New Transatlantic Bargain* (National Defense University Press: Washington, DC,

1985), following Cleveland, H., *NATO: The Transatlantic Bargain* (Harper & Row: New York, 1970).

[6] For useful clarification see Maroni, A. C. and Ulrich, J. J., *The US Commitment to Europe's Defense: A Review of Cost Issues and Estimates*, Congressional Research Service Report no. 85-211F (US Library of Congress: Washington, DC, 1985); Williams, P., *US Troops in Europe*, Chatham House Papers no. 25 (Routledge/Royal Institute for International Affairs: London, 1984), pp. 25–27.

[7] Details from *Survey of Current Business*, vol. 68, no. 6 (June 1988). For a detailed breakdown of the $9548 million direct expenditures in 1987, see table 3.4 in Alice Maroni's chapter in this volume.

[8] The calculation of local income and employment multipliers is discussed in Greenwood, D. and Short, J., 'Military installations and local economies—a case study: the Moray air stations', *Aberdeen Studies in Defence Economics (ASIDES)*, no. 4 (Dec. 1973); and in Short, J., Stone, T. and Greenwood, D., 'Military installations and local economies—a case study: the Clyde submarine base', *ASIDES*, no. 5 (Aug. 1974). See also Terner, I. D., *The Economic Impact of a Military Installation on the Surrounding Area: A Case Study of Fort Devens and Ayer, Massachusetts*, Research report no. 30 (Federal Reserve Bank of Boston: Boston, Mass., 1965).

[9] Cited in *The German Contribution to the Common Defense* (Press and Information Office of the Government of the Federal Republic of Germany: Bonn, 1986), p. 7.

[10] *The German Contribution to the Common Defense* (note 9), p. 21

[11] *The German contribution to the Common Defense* (note 9), p. 14.

[12] See, for example, Lynch, J. E., *Local Economic Development After Military Base Closures* (Praeger: New York, 1970); *Economic Impact of Military Base Closings* (US Arms Control and Disarmament Agency: Washington, DC, Apr. 1970), 2 volumes.

[13] The key document here is the NATO Defence Planning Committee Report *Enhancing Alliance Collective Security* (NATO Defence Planning Committee: Brussels, Dec. 1988).

[14] On the IEPG's recent activities see the Report of the Independent Study Team entitled *Towards a Stronger Europe* (Independent European Programme Group: Brussels, 1986) and the Communiqué issued at the end of the IEPG Ministerial meeting in Luxembourg on 7 Nov. 1988, with the appended Action Plan. For a summary of the 'innovative blueprint' which this Action Plan represents, see Greenwood, D., 'Collaboration and competition in West European arms procurement', *International Defense Review*, Apr. 1989, p. 512.

[15] For more on this see Greenwood, D., 'Reshaping NATO's defences', *Defence Minister and Chief of Staff*, no. 5 (1984), pp. 9–19; 'Towards role specialization in NATO', *NATO's Sixteen Nations*, July 1986, pp. 44–49 (both summaries of work done with Steven Canby) and 'Making better use of resources in NATO', *Defence Minister and Chief of Staff*, no. 2 (1988), pp. 5–8; and Volten, P. M. E., *Voor hetzelfde geld meer defensie* (Clingendael: The Hague, 1987).

6. The economic impact of the stationing of US forces in the Federal Republic of Germany

Hartmut Bebermeyer and Christian Thimann

European Study Group on Alternative Security Policy, Bonn, Federal Republic of Germany

I. Introduction

In the West German discussion of the US troop presence in the Federal Republic of Germany, the focus has mainly been on economic considerations, not the least as these apply to the issue of a partial or complete withdrawal.

As the economic impact of the US force presence can be expressed both in terms of costs and benefits, a cost–benefit analysis seems to be the most appropriate way to approach the subject.

The economic ties between the US forces and the Federal Republic of Germany are extensive, and include orders placed by US installations with West German firms, wages and salaries paid to West German civilian workers and money spent by US soldiers, civilian employees and family members on West German products and services. The economic impact of the US troop presence also has an indirect effect on defence expenditures, social welfare, public transport and environmental protection.

To date, there has been no comprehensive study of these economic aspects by US or FRG government agencies or research institutes, nor has any complete set of statistics been compiled on which to base such a study. The only available data are scattered and fragmentary. Forecasts on this basis can therefore only be approximations.

The legal basis for the US presence

The legal basis for the stationing of the US forces in the Federal Republic of Germany is set down in the Agreement between the Parties to the North Atlantic Treaty Regarding the Status of their Forces of 19 June 1956, and the Supplementary Agreement of 3 August 1959 as amended on 21 October 1971.[1]

Economically of greatest importance is Article 63 of the Supplementary Agreement, which states that the US troops—along with other NATO forces stationed in the FRG—may use all necessary immovables free of charge. They are only required to assume the costs for maintenance and repair. This provision puts a heavy burden on the Federal budget of the FRG.

Table 6.1. US active forces in the Federal Republic of Germany, 1986–87
Figures are in thousands.

Year	Army	Air Force	Navy	Marine Corps	Total
1986	208.9	40.3	0.3	0.1	249.6
1987	208.6	42.4	0.4	0.1	251.4
1988	209.0	43.0	0.4	0.1	252.5

Source: *Manpower Requirement Report, Fiscal Year 1988* (US Department of Defense: Washington, DC, Feb. 1987).

The number of US forces in the FRG

In 1988 there were a total of 252 000 US servicemen stationed in the FRG. Of these, 209 000 are Army and 43 000 Air Force personnel. The numbers of Navy and Marine Corps servicemen (400 and 100 respectively) were so small as to be negligible. The total includes US servicemen stationed in Berlin (mainly 4300 Army personnel). Despite the ongoing debate in the United States about a reduction of US forces in the FRG, the number of US Army personnel in that country has essentially remained unchanged in the past three years, while the number of US Air Force servicemen increased over the same period by more than 1000 per year (table 6.1).

As far as the evaluation of the economic impacts is concerned, therefore, the number of US forces in the FRG can on the whole be regarded as constant. Given the lack of significant numbers of Navy and Marine Corps personnel, only Army and Air Force data need be considered.

The US forces relevant to this investigation—those units of the US Army Europe (USAREUR) and the US Air Force Europe (USAFE) stationed in the FRG—are under the command of the US European Command (USEUCOM) in Stuttgart, central headquarters for all US forces in Europe. The main units of USAREUR are the V Corps in Frankfurt/Main and the VII Corps in Stuttgart. Both have two divisions, one brigade, one regiment and numerous supplementary battalions. The units of the V Corps are deployed in the densely populated metropolitan Rhine-Main area (Frankfurt, Mainz, Wiesbaden, Offenbach and Darmstadt), in the surrounding Main-Kinzig area and in Rhine-Hesse. The units of the VII Corps are mainly located in the cities of Franconia and the Upper Palatinate, namely Würzburg, Ansbach, Augsburg and Nuremberg.

There are three air forces under the command of USAFE, one each in the United Kingdom, Spain and the Federal Republic of Germany. The latter, the 17th Air Force with headquarters in Sembach, has six squadrons stationed in Bitburg, Hahn, Spangdahlem, Ramstein, Zweibrücken and Sembach itself. In contrast to the US Army garrisons, all of these bases are located in rural, structurally weak regions of the Eifel and the Palatinate.

The United States has 308 military installations in the entire FRG: 170 barracks, housing complexes, headquarters and schools, as well as 138 depots,

Table 6.2. US facilities in the Federal Republic of Germany, 1974

Type of facility	Number	Type of facility	Number
Barracks	170	Swimming pools	15
Schools	130	Gymnasiums	152
Shops/stores	340[a]	Bowling halls	114
Theatres/cinemas	306	Golf courses	21
Clubs	95	Churches	147
Libraries	103		

[a] 78 food stores; 262 goods and service outlets.

Source: *Facilities Maintained by U.S. Department of Defense in the Federal Republic of Germany* (Friends Committee on National Legislation: Washington, DC, 1974).

radio stations and training camps. These installations are distributed geographically as follows: 81 in Bavaria, 75 in Hesse, 74 in Baden-Württemberg, 61 in Rhineland-Palatinate, 14 in West Berlin and 3 in other states.[2]

As these distribution figures show, the US forces represent a significant economic factor for the middle and southern parts of the FRG only. They do not contribute to reducing the economic disparity between the north and the more prosperous south.

Costs and benefits

Costs for the FRG accruing from the stationing of US forces appear in many parts of the federal budget, to a lesser extent also in state and municipal budgets. Aside from direct costs covered by the federal defence budget, costs associated with the US force presence include expenditures for social welfare, public transport and environmental protection. On the benefit side for the FRG economy there are the payments made by US authorities, units and personnel, including dependants, for West German goods and services.

These economic aspects are described in detail and—as far as possible—quantified in the following sections. On the basis of these data, the last section concludes with a cost–benefit analysis from which two important conclusions may be drawn:

1. The US force presence is a significant factor in the FRG economy.
2. While most costs are paid out of public funds, drawn either from federal, state or municipal budgets, the benefits mainly go to FRG private firms and civil employees.

II. The US force presence as an economic factor

As for the economic effects of the stationing of US troops in the Federal Republic of Germany, there are several reasons why a clear distinction should be made between US Army and US Air Force units. First, in terms of personnel, there are great differences in their respective number, income, age, years of service, education, training and ratio of officers to non-commissioned officers (NCOs). Second, in terms of location, the two services predominate in regions that differ in terms of of economic infrastructure and development. Third, the procurement and investment programmes of the two services are organized differently. Finally, the two services are not affected in the same way by the debate about a withdrawal of US forces. The presence of the US Air Force is much less disputed than that of the US Army. In the West German debate, only a reduction of Army personnel is discussed; the US Air Force is often not even mentioned.

Army garrison/Air Force base characteristics

The US Army garrisons in the FRG are with few exceptions located in towns. At the end of World War II the USA took over the barracks of the wartime German *Wehrmacht*, which were often located in town centres. As most US soldiers and civil employees, along with family members, live in or near the garrisons, so-called US 'ghettos' have developed in these centres. An average US Army garrison in the FRG quarters between 5000 and 10 000 soldiers, the same number of family members and about 2000 civilian employees, of which 25–50 per cent come from the USA. The presence of one or several garrisons in a given area will therefore have a marked impact on local demography. To take one example, four cities in the Frankfurt area—Hanau, Wiesbaden, Darmstadt and Gießen—with a total population of less than 600 000, host more than 77 000 US soldiers, civil employees and family members. Statistically, one out of every eight persons residing in this area is a US citizen.[3]

Such concentrations of US soldiers and dependants are often associated with high social costs, although this is something that can be quantified only to a very limited extent. Garrison towns commonly suffer from higher levels of criminality, juvenile delinquency and prostitution than surrounding areas, with implications, not the least, for the local investment climate. Furthermore, as the average income of the USAREUR personnel is comparatively low and only few industries benefit from the small share of private consumption that in this way flows into the West German economy, the social and economic integration of the garrison troops with the host community is often at a disadvantage. This has been the case in particular in the cities of the Rhine-Main area, where local opposition has often been voiced against the large US troop presence.

The situation of the US Air Force is markedly different. All six bases of the USAFE in the FRG are located in small towns in rural areas, where they provide vital impetus to the local economies. A good example is the Eifel town of Bitburg, with 12 000 West German inhabitants.[4] The town is also the home of 11 000 US citizens, whose private expenditures in 1986 amounted to an estimated $14.3 million, or DM 33.2 million, assuming a dollar/D-mark exchange-rate of 2.32 (the annual average for 1986). Of the local West German work-force, one out of seven persons works on the base. In the town itself, that is, outside the base area, US citizens own 2700 apartments and houses, to which

380 newly built homes were added in 1986. For these and other construction programmes the base invests about $150 million annually. On the average, roughly 80 West German firms are working on the base at any given time. Three-quarters of these firms come from Bitburg itself, the rest almost exclusively from the surrounding area. In 1986, the base paid $6 million for electricity and $1.4 million for sewage to the municipality. The benefits to the local economy were almost entirely made in the services sector, however. Virtually all equipment on base, military as well as civilian, is purchased from the United States.

The principle of self-sufficiency

The US soldier [in the FRG] lives in an entirely American world, in which . . . he gets paid in dollars and buys US products in US stores . . . his bank account is with an American bank, the post office is American and it charges a domestic rate for mail delivery to the USA. Officers bring their own cars from America, and those who cannot afford one drive free of charge in US military buses from one US installation to another. Insurance policies are American, and in case of illness one sees an American physician in a US hospital. The children go to US kindergartens and schools, . . . in their spare time the US soldiers watch American movies, eat their hamburgers and steaks in their own snack bars and clubs, work out in their own gymnasiums, . . . heat with imported US coal and make telephone calls with their own communication facilities. In brief, the Americans in Germany live in America. The few things they require from us are electricity, gas and water; they use our roads and garbage dumps . . . The life of our American allies is organized in such a way that they live completely autonomously, independent from us and meet all of their needs themselves.[5]

This quotation, taken from a 1985 investigation of the social situation of the US soldiers in the FRG, depicts a typical characteristic of all US bases in the country: the exceptionally high degree of self-sufficiency. With free access to all these facilities, there is almost no need for the US serviceman, civil employee or family member to leave the base in order to satisfy the requirements of daily life. Nearly every garrison or base provides a complete range of facilities covering everything from basic needs to entertainment.

In addition, a low dollar/D-mark exchange-rate has resulted in a rapid increase in sales at all US military commercial installations. For example, from December 1986 to December 1987, when the dollar declined from DM 2.00 to DM 1.63, sales at all US commercial installations amounted to more than $1.2 billion, 20 per cent more than in the year before. Sales at all PX stores increased over the same period by 12 per cent to $673 million (1987).[6] These figures can be broken down with the help of data provided by the Army and Air Force Exchange Service (AAFES), the supplier for all US military commercial installations such as PX stores, clubs, restaurants, gas stations, and so on (table 6.3).

Out of these $1.34 billion only $353 million, or about one-quarter, flow into the FRG economy (table 6.4).

Table 6.3. Army and Air Force Exchange Service (AAFES) direct sales in the FRG, FY 1987

Sales are in US $m., at constant (1987) prices.

	Sales	Percentage share
Retail	1 111	*83*
Food	146	*11*
Services	41	*3*
Vending	42	*3*
Total	**1 340**	*100*

Source: Department of the Army and the Air Force, Headquarters Army and Air Force Exchange Service, Europe (AAFES-EUR-PA) (written reply to letter, 8 Apr. 1988).

Assuming a low dollar/D-mark exchange-rate (less than DM2/$1), AAFES expects a strong increase in direct sales and a slight increase in dollar payments to West German suppliers. At the same time, however, the low exchange-rate means a decrease in revenues for the suppliers. Furthermore, the weak dollar will result in a reduced share in the AAFES sales of West German goods (DM 706 million in 1987, or 26 per cent), simply because West German products become increasingly expensive relative to imports from the United States.

The economic situation of US servicemen in the FRG

The force presence allows US defence dollars to enter the West German economy in three ways: through on-base employment of West German citizens, through orders placed with West German firms for goods and services, and through the private consumption of US personnel. As regards the latter, a distinction must again be made between Army and Air Force personnel.

For the members of the US Air Force, USAFE in Ramstein estimates that those who live on-base (a share of 30 per cent) spend about 30 per cent of their income in D-mark, that is, in the West German economy, whereas those who live off-base (a share of 70 per cent) spend about half of their income in D-mark.[7] For the latter, there often are no nearby US shops. In addition, Air Force personnel living in West German houses must pay their rents in West German currency. Averaging these figures, the estimated share of D-mark spendings of all USAFE members is 44 per cent, with rent, tourism, restaurants and West German 'quality products' (cars, household machines, souvenirs and luxury goods) as the main items of expenditure.

US Army soldiers and US civil employees spend a much smaller share of their income in D-mark, largely because of their lower average incomes. But their consumer behaviour is also to some extent determined by their average age and level of education, which for the Army is lower than that of that of Air Force personnel. Among Army personnel the demand for cultural events is negligible, the interest in tourism is smaller and the demand for goods is restricted to daily necessities. In the rather blunt formulation of a local West

Table 6.4. Army and Air Force Exchange Service (AAFES) payments to West German suppliers, 1988
Payments are in thousand US $, at constant (1987) prices.

	Payments	Percentage share
Retail merchandise	146 000	*41*
Concesssionaires	77 458	*22*
Plant contractors	18 454	*5*
Support services	31 660	*9*
Gasoline	79 290	*23*
Total	352 862	*100*

Source: Department of the Army and the Air Force, Headquarters Army and Air Force Exchange Service, Europe (AAFES-EUR-PA) (written reply to letter, 8 Apr. 1988).

German official, in times of a weak dollar, only three trades are supported: restaurants, taxis and prostitution.

Tables 6.5 and 6.6 give a breakdown in rank and income for USAREUR and USAFE personnel in the FRG. Table 6.5 provides an average income for US Army personnel in 1987 of $17 734 per year or $1478 per month. By contrast, table 6.6 gives an average income for the USAFE in 1987 of $32 080 per year or $2673 per month. There are two main reasons for the considerable difference in income between the two services: (*a*) the substantially higher proportion of officers and higher-ranking soldiers within USAFE, and (*b*) the higher average income of each USAFE rank as a result of the greater length of service and the many supplementary payments, that is, for hazardous work or weekend and night shifts.

The 33 000 US civilian employees receive an average annual income of $20 000 (1987); here differences between Army and Air Force personnel as well as between posting regions may be disregarded.[8]

With the help of the US Department of Defense (DOD) deflators, which are lower for civil personnel than for military personnel, the average per capita income can be aggregated for previous years and forecasted for 1988, as shown in table 6.7. Personnel strengths are here regarded as constant.

The annual gross income of all US personnel amounted to $5.46 billion in 1986. In a larger investigation, USAREUR has calculated that out of this amount, $1.536 billion (or 28 per cent) is spent in the West German economy on rent and individual consumption.[9] However, from 1986 to 1988 the decline of the dollar has caused a loss of purchasing power in the FRG of 25 per cent. To compensate this loss, US personnel have continually cut their expenses in D-mark, shifting their purchases to the US commercial facilities, as reflected in the AAFES sales figures presented above. There are no official figures yet, however, that show how much FRG sales have decreased in 1987 and 1988 from the 28 per cent share in 1986. According to press reports, FRG sales figured at around 10–12 per cent in the autumn and winter 1987/88.[10] The yearly average for 1987 will therefore have been around 18 per cent. This corresponds to a sum of $1.09 billion, or DM 1.95 billion, for 1987. In view of

Table 6.5. Average 1987 income of USAREUR military personnel, by rank

Rank	Average annual income ($)	Percentage of total personnel
Generals	75 132	0.02
Staff officers	52 884	1.73
Officers	30 600	5.63
Warrant officers	33 708	1.86
Master Sergeants	31 248	7.24
Sergeants	20 604	28.83
Privates	11 424	54.68

Source: Department of the Army, Headquarters V Corps, Frankfurt am Main (written reply to letter, Mar. 1988).

the dollar's development in 1988, falling to an annual average of DM 1.74, US personnel expenditures that flow into the FRG economy will have amounted to approximately $0.84 billion or DM 1.47 billion in that year. Compared to the situation in 1986, in other words, the economic benefit to the FRG of the private consumption of the US personnel in 1988 would have been reduced by 60 per cent, or almost two-thirds. For 1989, however, a further reduction does not appear to be plausible as the dollar has significantly recovered and—in the first four months—stabilized at a level clearly beyond DM 1.85.

The US forces as employer

The employment of US civilians

According to official US sources, there were 33 000 US citizens employed by the US forces in the FRG in 1987.[11] This number will have grown further. US civilians are replacing West German civilian employees in increasing numbers, mainly because the latter have become considerably more expensive (see below), but there are other reasons as well: with living costs rising for US service families as a result of the falling dollar, a growing number of dependants are seeking employment. For others, work serves as an escape from isolation and boredom. As very few US dependants speak sufficient German to enter local labour markets, on-base employment remains the only option for most. In addition to these categories, a growing share of US soldiers chose to stay in the FRG after retirement, putting additional pressure on the on-base labour market.

The living style and consumer behaviour of these US civil employees are not very different from that of US military personnel with comparable incomes, with the provision that a higher proportion of the civilian personnel are married. Furthermore, as civilian personnel incomes do not differ greatly from one service or region to another, data provided by the US Air Force[12] can be aggregated to provide an estimate of the total outlay for wages for US civilian personnel in the FRG. In 1987, an average annual income of $20 000, or DM 46 400, meant a total outlay of $660 million, or DM 1.53 billion.[13] Assuming

Table 6.6. Average 1987 income of USAFE military personnel, by rank

Rank	Average annual income ($)	Percentage of total personnel
Generals	77 590	1
Staff officers	53 601	19
Officers	37 126	32
Master Sergeants	29 646	5
Sergeants	21 330	16
Privates	15 461	27

Source: US Department of the Air Force, Headquarters United States Air Force, Europe (Office of the Commander-in-Chief), Ramstein (written reply to letter, 5 Jan. 1988).

that D-mark expenditures remain at 28 per cent, the US civilian employees spent $185 million, or DM 428 million, for private consumption in the West German economy. However, also here the decay of the US currency has caused a sharp reduction in expenditure, to probably 18 per cent in 1987 and to about 11 per cent in 1988

The employment of FRG citizens

In 1986 the US forces employed 71 000 FRG civilians, according to official sources in Bonn.[14] US official sources mention 59 445 West German employees (on 31 March 1987),[15] but this figure only includes on-base personnel and not West German civilian employees of AAFES, American Express Bank and other institutions indirectly linked to the US troop presence. However, these jobs are also part of the economic impact—in this case benefit—of the US force presence in the FRG. It would therefore seem more appropriate to use the West German figure in this context.

The professional spectrum of the West German employees includes guards, drivers, crafts- and tradesmen, engineers and physicians. Their annual salary, which flows entirely into the West German economy (taxes are only paid in the FRG), adds up to DM 3 billion.[16] This corresponds to an average gross income of DM 42 250, including incidental wage costs and fringe benefits. These sums are paid by the United States according to West German legislation.[17]

For a variety of reasons, the number of West German employees is currently being reduced considerably (by 4000–5000 by the end of 1988).[18] Local employees are comparatively expensive to the US employer, as no part of their salaries flows into the US economy, not even through tax payments. In addition, the decay of the US currency means rising costs for wages paid in D-marks. The constant unemployment rate in the United States also affects US employment policies for positions abroad, with pressure mounting to keep them available for US citizens.

As a consequence of the heavy concentration of US military installations to certain regions and communities, a cut-back of locally recruited personnel would create large social problems and costs in specific host areas. This applies

Table 6.7. Per capita incomes and D-mark expenditures as shares of total income of US military and civil personnel in the FRG, 1986–88

	1986	1987	1988
Average per capita annual income			
Army	16 780	17 750	18 730
Air Force	30 250	32 080	33 760
US civil employees	19 730	20 000	20 650
Annual average DM/$ exchange-rate	2.32	1.79	1.76
Total income (in billions)			
US $	5.46	5.73	6.05
DM	12.66	10.26	10.53
Expenditures of US military and civil personnel in the West German economy (in billions)			
US $	1.54	1.09	0.84
DM	3.56	1.95	1.47
Percentage of total income	*28*	*18*	*14*

Sources: Calculations by authors; deflators from *National Defense Budget Estimates for FY 1986* (US Department of Defense: Washington, DC, Mar. 1985), p. 50.

in particular to the USAFE bases in Rhineland-Palatinate, with 2300 and 800 West German civilian employees in the small towns of Ramstein and Bitburg, respectively. Also in the Rhine-Main area as well as in Baden-Württemberg (Stuttgart) would a dismissal of West German employees lead to a sharp rise in local unemployment. Increased unemployment benefit and social welfare expenditures would be inevitable for the municipalities.

US orders of West German goods and services

'The US Forces are a big economic factor in Germany, indeed', USAREUR concludes in a major 1986 study.[19] American Express Bank, which handles practically all currency transactions for the US forces in the FRG, that year handled transactions worth $6.113 billion billion, or DM 14.182 billion. This figure is corroborated by a report of the West German *Bundesbank*, which lists a total of DM 15.4 billion under 'revenues for goods and services provided to US military installations'.[20] The difference is mainly due the fact that the *Bundesbank* report includes figures for the US installations in West Berlin in the total figure. The half-official periodical *Wehrtechnik* gives a comparable sum of DM 14.6 billion for the expenditures of the US forces in the FRG in 1984.[21]

The USAREUR figures for 1986 of about DM 14.2 billion can be broken down as follows: DM 2.73 billion for salaries to West German blue and white collar personnel; DM 3.56 billion for private consumption of US soldiers, civilians and family dependants; DM 1.64 billion for purchases made by US contracting agencies; DM 3.06 billion for goods and for services purchased by

US installations; DM 1 billion for construction work; and DM 2.2 billion for general orders (e.g., to *Bundespost*, *Bundesbahn* and for private consumption and rents). Calculating from the official West German figures for the West German civil employees (71 000 instead of approximately 60 000), which give a total expenditure for salaries of DM 3 billion, the expenses of the US forces in the West German economy rise to DM 14.46 billion. This is the figure included in the cost–benefit analysis.

III. Costs to the Federal Republic of Germany

Real estate and infrastructure

The real estate used by the US troops free of charge (immovables, houses, training camps, depots) is valued at DM 28 billion.[22] Most of it (DM 20 billion) is owned by the Federal Republic, the rest belonging to various states and municipalities. Assuming an annual user value of 5 per cent, the present arrangement means an annual loss of income for the FRG of at least DM 1.4 billion. US estimates are even higher, with a calculated income loss of DM 1.86 billion.[23] At a rough estimate, DM 1.5 billion would seem a reasonable average.

FRG costs also include expenditures for the upkeep of infrastructure used by the US forces. Only a small share of these costs is reimbursed (e.g., for the use of the *Bundesbahn*). In general, the FRG bears the costs for all infrastructure programmes (such as road construction), even when the US forces are the prime or even exclusive user. These payments are made either directly from the federal budget or over state and municipal budgets, with the latter receiving reimbursement from the Federal Government.[24] There are no official West German estimates for these expenditures. USAREUR assumes $58 million, or DM 134 million, for infrastructure costs per annum.[25]

The NATO Infrastructure Programme and Wartime Host Nation Support Agreement

FRG payments to the NATO Infrastructure Programme (NIP) amount to DM 2.62 billion for the period 1986–89, with an annual average of DM 654 million.[26] There are no figures indicating what share of this sum is allocated to US troops, but 50 per cent seems to be a moderate estimate. This would mean somewhere around DM 327 million per year. Most of this US-tagged infrastructure funding actually returns to the FRG through contracts with FRG firms, mainly in the construction sector. In 1986, orders worth an estimated DM 200 million were placed in the West German economy. It would be misleading, therefore, to see these NIP expenditures on the cost side of the balance only.

The agreement between the United States and the FRG on Wartime Host Nation Support (WHNS) from 1982 provides West German logistics support for the US troops coming to the FRG in times of crisis or war. For the period 1986–89, the FRG will have spent about DM 153 million annually to meet

WHNS commitments.[27] These payments refer directly or indirectly to the US troops; principal for those orders is the West German *Bundesamt für Wehrtechnik und Beschaffung* (BWB). Consignees are West German firms, paid by the Federal Government (out of departemental budget—or 'section'— 14 of the federal budget) via the BWB to provide facilities (mainly road, depot and telecommunications construction) for use by the US troops under WHNS. Consequently, these DM 153 million also involve benefits as well as costs to the FRG.

Maintenance and manœuvre damage costs

Section 35 of the federal budget (Ministry of Finance) covers costs pertaining to the stationing of foreign forces in the FRG and for which funding is not specifically provided elsewhere in the federal budget. This includes costs for communications, road construction and maintenance, supply and disposal facilities and damages. Under this title, expenditures of DM 143 million were associated with the US forces in 1986, West Berlin excluded.[28]

As far as damages caused by US troop manœuvres in the FRG are concerned, only 75 per cent of costs are covered by the United States; the rest is paid for by the FRG.[29] For this purpose a sum of DM 19.5 million was drawn out of the federal budget in 1986.[30] Section 35 does not cover costs associated with West German real estate and infrastructure used by US forces (discussed above).

Exceptional costs

In general, on-base construction programmes are paid for by the US forces themselves. In 1986 there were three exceptions, however: on two occasions the FRG paid DM 300 million to modernize US barracks, and on one occasion DM 171.2 million for the construction of a US installation in Garlstedt.[31] However, as such payments are not made on a regular basis, they are not included in the cost–benefit balance presented below in section V.

West German expenditures in the United States

Between 1982 and 1986 the FRG spent on the average DM 1.68 per year on *Bundeswehr* procurement in the United States. This represents 48.7 per cent of West German military procurement in foreign countries and 9.3 per cent of the entire *Bundeswehr* procurement budget. However, this immense sum includes the purchase of the Patriot System in 1985 (price: DM 4.78 billion).[32] The total *Bundeswehr* procurement in the United States is considerable. Contrary to common belief, however, there is no legal requirement for the FRG to purchase US arms in order to offset the costs to the United States of stationing troops in the FRG. Therefore, there is no formal link (e.g., in form of a hidden foreign exchange balance) between arms purchases and the US presence in the FRG. Nevertheless, the high level of procurements attests to a 'good-will attitude' on the part of the Federal Republic.

The costs for the FRG accruing from West German training camps in the USA in 1986–89 have been estimated at DM 746 million, giving in an annual average of DM 186 million.[33] However, since there is no direct correlation between these costs and the presence of US troops in the FRG, these expenditures are taken up in the final balance of this analysis.

Social welfare and other social costs

Considerable social service costs are associated with the US troop presence in the FRG. Since 1949, 200 000 marriages have been contracted between US military personnel and West German citizens. It is not uncommon for these marriages to end in divorce or separation, with wives or husbands (along with any children) either left in the FRG after the husbands have been transferred to postings in the United States or elsewhere, or returning from the United States after a divorce. In either case, many of these women come to rely for their support on the West German social services. Also the illegitimate children of US servicemen—as many as 30 000 since 1946—are often raised on West German social security payments alone.[34] Experts in the municipalities and cities concerned estimate that a 5 per cent share of the local social budget is spent on problems associated with the US force presence.[35] According to computer forecasts, these expenditures amount to DM 60 million per year.

There are numerous social costs linked with the US forces' presence in the FRG which cannot be quantified, even though they represent an economic loss. There is no doubt that problems such as criminality, prostitution and drug addiction are enhanced by the US presence (particularly in larger cities where a high number of US soldiers live), often with costly consequences (e.g., AIDS). For all its economic implications, however, this complex problem is not expressible in figures and remains merely a qualitative element in the equation.

IV. Benefits to the West German economy

The US forces benefit the Federal Republic of Germany in three ways (figures refer to 1986): (*a*) as purchasers of West German goods and services (DM 7.9 billion); (*b*) as employers of West German civil personnel (DM 3 billion; DM 2.73 billion according to US estimates); and (*c*) as private consumers on West German markets (DM 3.56 billion).

This amounts to a total of DM 14.46 billion spent by the United States in the West German economy in 1986.

The 50 per cent decrease in the value of the US currency relative to the D-mark from 3.30 in March 1985 to 1.65 in January 1988 has meant a corresponding cut in the purchasing power of US servicemen stationed in the FRG.

They have compensated this loss by continuously reducing their expenses in D-mark and by shifting to dollar purchases in US shops. In the course of 1988, however, this input into the West German economy decreased by about DM 2 billion. In 1988 the decay of the US currency apparently reached a turning-point, with the dollar value levelling out an at annual average of DM 1.74, and showing an upward tendency at the beginning of 1989. For lack of available data, it is not possible to give a more detailed forecast.

US orders placed with West German firms

Of US defence dollars spent in the West German production sector, the construction branch is the largest single beneficiary, with orders worth DM 1 billion placed in 1986.[36] In accordance with US legislation, bids must be invited for all construction orders exceeding DM 48 000.[37] The biggest items here are the housing programmes and the construction of new bases. Because of the decay of the dollar, the US orderers have begun importing prefabricated houses and construction parts from the United States. The US Air Force even flies such parts to Europe in order to keep the dollar/D-mark exchange burden at a low level, a clear illustration of the the extent to which US purchasing power in the FRG has dropped. Current construction programmes include the building of new facilities in Wildflecken, as well as in Ansbach and Würzburg, as part of a 'Master Restationing Plan' in which three brigades of the 8th US Infantry Division (17 000 soldiers) are to be moved from the urban Rhine-Main area to the area around Fulda (Hohenfels, Wildflecken and Grafenwöhr), closer to the FRG–GDR border. The West German construction industry will benefit, locally as well as regionally, from orders placed through 1995.

For years the West German automobile industry was a favoured supplier to the US forces. In 1984 alone, US military units purchased cars at a value of $87 million,[38] or DM 247 million, given the 1984 exchange-rate of DM 2.84 to the dollar. In terms of private consumption, US servicemen also gave preference to West German automobiles, in particular to Porsche, Mercedes Benz, BMW and Audi. Since the decline of the dollar in 1985, however, US purchases have declined drastically. A poll of dealers of West German automobiles in the stationing areas has proven that purchases by US soldiers have decreased to zero. As a direct consequence of the falling dollar, therefore, the West German automobile industry has lost an important group of customers.

Retailers in the stationing areas have also benefited from the private expenditures of US servicemen. Furniture, electrical appliances, HiFi-equipment, antiques and clocks ('Schwarzwald-Uhren', or cuckoo clocks) have been in great demand. This sector was also heavily stuck by the falling dollar. All retail federations consulted have confirmed that revenues accrued from sales to US servicemen have fallen drastically. Although no detailed figures are available, it is evident that US personnel and dependants in the FRG have since 1987 to an increasing extent made their daily purchases in US shops. The rapid increase in the sales of the PX shops—12–13 per cent in 1987—is a clear indication of this trend.[39]

The demand for services

Service contracts account for a considerable share of the D-mark expenditures of the US forces in the Federal Republic. Reliable data are only available, however, for contracts involving the port facilities in Bremerhaven and the housing of US personnel in West German-owned houses.

Since 1945 the port facilities in Bremerhaven have served one of the most important supply centres for US forces in Europe. In 1987 the turnover for US troops (in- and outgoing) reached a total of 450 000 tons.[40] About 60 per cent of these shipments were of household goods of US servicemen stationed in the FRG and of general, non-military supply goods (the latter another indicator of the high measure of self-sufficiency of the US forces). The economic return to the FRG is an estimated DM 69 million in that year.

The West German housing sector has also benefited considerably from the US presence. More than 30 per cent of all US servicemen live in rented West German-owned houses and apartments. This estimate is based on figures collated from numerous sources, there being no comprehensive set of figures to draw from.[41] The share is higher for Air Force personnel than for Army personnel. According to official West German sources, 80 per cent of the rent is reimbursed by the employer (DOD); the remaining 20 per cent as well as additional costs for electricity, water, gas, heating and telephones is borne by the tenant. These costs have also risen sharply as a result of the falling dollar, making living in West German houses too expensive for many. The 30 per cent average is diminishing as a consequence.

Calculating from 1986 figures, a total of 70 000 US servicemen and civilian employees living in West German houses and paying an average annual rent of DM 8500[42] provides the West German lessors with an income of roughly DM 595 million per year.

Prior to 1985 the US force presence provided a steady source of income for the West German tourist industry, not the least for local entrepreneurs in the resort areas of Upper Bavaria and the Black Forest region. Inevitably, this relationship has also suffered drastic change as a result of the currency fluctuations. Hard figures are not available, however, that could illustrate the importance of this group for the tourism industry or give a measure of the change that has taken place.

The same is true for the entertainment industry and the prostitution trade, where it is also only known that the times of 'quick returns' are past.

It has been estimated that until 1985 almost 50 per cent of the revenues of the taxi trade in stationing areas derived directly from the US troop presence. The demand for taxis is very elastic; the trade reacts quickly to changes in the dollar/D-mark exchange-rate. The benefit of the US force presence for this business sector has clearly diminished in the past three years.

By contrast, US expenditures to skilled craftsmen, particularly for repairs and maintenance of barracks and housing, have not been reduced. The US Army Contracting Agency Europe (USACAE) estimates for 1984 list expenditures of $360.6 million for such services.[43] Given an annual average exchange-rate of DM 2.84 for 1984, this corresponds to DM 1.02 billion. Since then, D-mark expenditures in this sector have remained at roughly the same level. According to information provided by the trade federations concerned, however, and in keeping with the general trend, private expenditures by US personnel have diminished.[44]

Consequences for the job market

The West German job market has benefited greatly from the US force presence. It provides not only direct employment for 71 000 West German civilians, but also indirect employment through expenditures in the West German economy. Through orders placed by US installations in the West German production and service sectors and through the private consumption of servicemen, civilian employees—US as well as West German—and dependants, DM 14.46 billion flowed into the West German economy in 1986. According to USAREUR estimates, these expenditures correspond to the indirect employment of 100 000 persons.[45] This is comparable to the personnel figures of large firms in the FRG with sales volumes on a par with the US expenditures, which range between 70 000 and 120 000 employees.

Of a total work force of 28 million persons in the Federal Republic (1986), 0.62 per cent are directly or indirectly employed through the US forces. In other words, a total withdrawal of the US forces would—in theory—leave 174 000 West Germans out of work. In 1986, 2.23 million persons were unemployed. Without the US force presence, their number would be significantly (7.8 per cent) higher.

V. Cost–benefit analysis

The balance

A comparison of costs (DM 2.337 billion) and benefits (DM 14.813 billion) shows a positive balance of DM 12.476 billion for the FRG in 1986 (table 6.8). Three important points must be kept in mind, however:

1. The net benefit of DM 12.48 billion does not represent profits or net income, only turnover or the proceeds of sales.
2. The balance is an expression of monetary costs and benefits only. Such non-quantifiable costs to society as are incurred through crime, drugs, social conflicts, etc., are not included. Also not reflected in the balance is the cost

Table 6.8. Cost–benefit balance of the US force presence in the Federal Republic of Germany, 1986

Figures are in million DM, at constant (1986) prices.

Costs		Benefits		
Real estate (5% user value)	1 500	*US purchases of goods and services*		
NATO Infrastructure Federal budget	327	Construction	1 000	7 900
		Rental costs	600	
WHNS Federal budget	153	Repairs	1 020	
		Bundesbahn/-post	1 020	
Maintenance Federal budget	143	AAFES	353	
Infrastructure Federal budget	134	*NATO Infrastructure Fund* Orders placed with FRG firms		200
Barracks, housing Exceptional payments	(771)	*WHNS* Orders placed with FRG firms		153
Manœuvre damages Federal budget	20	*Wages to FRG citizens* 71 000 employees (DM 42 250 each)		3 000
Military procurement in the USA Federal budget	(1 680)	*Private consumption* Share of 28% from total income (DM 12 670)		3 560
FRG training camps in the USA Federal budget	(186)			
Social welfare Municipal budgets	60			
Social costs (crime, prostitution, drug abuse, etc.) Municipal budgets	..			
Total	2 337			14 813
Balance				12 476

Source: Authors' calculations.

share for staff and material expenses accrued by federal, regional and local public authorities affected by the US presence (the Ministry of Defence, municipal authorities, the police, etc.).

3. The balance does not reflect the economic and social implications of the regional distribution of costs and benefits. In social welfare terms, US expenditures are more valuable to the national economy in rural regions with high levels of unemployment than in economically more prosperous areas. In this study, however, they were rated according to the value of the gross income throughout. Similarly, in the context of a social welfare-oriented economic policy, it makes a great difference where US orders are placed—whether in the armament industry or in the construction or agricultural sector, for instance.

Such distinctions have not been made here, however, as this falls outside the realm of the strict economic analysis.

Appraisal

The net benefit of the US force presence corresponded in 1986 to a share of 0.62 per cent of the West German gross national product (GNP). This positive balance were significantly reduced in 1987 and 1988, as a consequence of the fall of the dollar. Data for these years are not yet available but estimates point to a decrease of almost DM 1.5–2 billion. Given an average growth of the GNP of about 2 per cent per year, this would mean a reduction to about 0.52 per cent.

At first sight, the benefit the FRG derives from the US force presence—hardly more than half of 1 per cent—seems almost negligible. Measuring only against the GNP is misleading, however, for two reasons:

1. As the net benefit represents a return on payments in the defence sector, a more suitable reference than the GNP or the gross domestic product (GDP) might be the defence budget, which in 1986 amounted to DM 50.3 billion (section 14 of the federal budget). Here, the share would be 24 per cent.

2. While the economic impact of the US force presence may be small in terms of the national economy, as implied by its share of the GNP, this does not hold true for the regional economies involved. In urban areas where US garrisons are concentrated, such as in the Rhein-Main and Main-Kinzig areas and in the cities of Bavaria and Baden-Württemberg, these installations play an important economic role. In the rural areas of the Eifel and the Palatinate, the trade and employment opportunities offered by the Army garrisons and Air Force bases in the region are even vital to the local economies.

VI. Economic implications of a withdrawal

In the light of the decisive political changes in Eastern Europe since autumn 1989, representatives of the NATO and WTO countries agreed upon an upper limit for US and Soviet troops in Central Europe of 195 000 each. This result of the Ottawa 'open skies' CFE foreign minister's meeting in early 1990 could mean a decrease in the number of US military servicemen in the FRG of up to 80 000. This figure could rise if the Conventional Armed Forces in Europe (CFE) Negotiation in Vienna results in further troop reductions.

Independent of these developments, the DOD decided in early 1990 to close down Zweibrücken AB, one of the six USAFE bases in the FRG. By the end of 1993 the base, with 3500 US servicemen, a corresponding number of dependants, 360 US and 340 West German civil employees, will be abandoned. The total benefit of this base to the FRG economy has been estimated by USAFE at DM 140 million per annum.[46] The Government of Rhineland-Palatinate set up in early 1990 a special commission to investigate ways of offsetting the local and regional economic impact of this and possible future closures.

There are numerous scenarios for a withdrawal of US forces from the FRG, and for each scenario there are manifold reactions and possibilities of compensation (regarding the number of Army/Air Force troops to be withdrawn, whether they are to be replaced by other NATO/FRG troops or whether there is to be no compensation, etc.). Hence, economic forecasts must contain a high share of uncertainty and come close to mere speculation. Consequently, only some facts shall be provided:

1. A critical factor is the dollar/D-mark exchange-rate. The impact of the falling dollar on US expenditures in the FRG is illustrated by the concomitant decrease in the private consumption of US personnel in D-mark by more than DM 2 billion.

2. Growing political and economic pressure in the USA is likely to lead to reduced D-mark expenditures on the part of the US installations, as well as to cuts in the number of West German civilians in favour of US personnel.

3. While the loss of revenue associated with a withdrawal of the US forces would have negligible effect on the West German national economy, the host regions would suffer great economic damage. In particular rural communities with strong economic ties to nearby USAFE bases would suffer severe losses, with unemployment figures rising by 10 per cent or more.

4. A withdrawal would in the first hand lead to significant reductions in public expenditure. In the federal budget alone, outlays of DM 2–3 billion per annum, corresponding to roughly 5 per cent of the defence budget, would disappear. On the other hand, enormous subsidies, which could well amount to several billion D-mark, would have to be paid out of the federal budget to offset the economic consequences to host regions. In other words, a withdrawal would entail a huge re-distribution of federal resources. Additionally, the Government would have to pursue extensive and costly job creation programmes in the affected regions. Measures such as these are generally not very efficient.

5. The economic consequences for the military, in terms of filling gaps created by a US withdrawal, are not predictable in detail. It is clear, however, that the resources required in terms of personnel and funds would not be available to fully replace the US forces.

Option A: the US forces do not return in times of crisis

In the event of a total withdrawal of US forces from the Federal Republic of Germany, with no provisions made for their return in times of crisis, practically all costs and benefits for the West German economy vanish completely. The real estate used by the US forces (currently valued at DM 20 billion for federal lands and property and DM 8 billion for lands and property owned by individual states and municipalities) may either be sold or rented out for approximately DM 1.5 billion per year. All other costs for the federal budget accruing from the US presence, including infrastructure, maintenance, NATO Infrastructure Fund and WHNS costs as well as other exceptional costs, would disappear. Social welfare costs would at first remain constant and then decrease with a certain time lag, and no further social welfare costs would arise. Social costs (criminality, prostitution, etc.) would also diminish.

Furthermore, if the US forces are 'gone for good', West German military procurement in the United States would most probably decrease as well, as there would no longer be any need to demonstrate a 'good-will attitude' nor any formal or informal obligation to offset US expenditures in the FRG. In addition, a withdrawal might lead to reduced co-operation between the US and European armament industries, simply because the need for standardization would be reduced if the idea of Europe as a common defence area were abandoned.

West German expenditures for military training in the United States would most probably not be affected by a withdrawal, however, as it includes elements (such as the Hawk, Nike and Patriot missile training programmes) that cannot easily be transferred to the Federal Republic, even if more facilities were available there.

The FRG would be deprived of virtually all direct economic benefits of the US force presence in the event of a total withdrawal. These include US orders placed with West German firms, employment opportunities at the US installations, private consumption by personnel on US pay-rolls, as well as a share of such federal expenditures as are linked to the US presence through the WHNS and NIP. The cost–benefit balance would therefore point to a slight deficit, with (residual) social costs of DM 60 million (showing a downward trend) and zero benefits.

Option B: the commitment to return in times of crisis remains

If the US forces withdrew completely but remained committed to return in times of crisis or war, the economic consequences for the Federal Republic of Germany would be very negative. Practically all benefits would disappear, while costs would remain almost unchanged. Base real estate would not be available for sale or lease, and it might even have to be maintained by the FRG, adding costs of several billion D-mark to the federal budget. The costs for the WHNS and NIP programmes would probably increase. Only running costs for maintenance, infrastructure (section 35 of the federal budget) and for manœuvre damages would vanish or at least be significantly reduced. West German expenditures for military procurement and training in the United States would most probably show the same tendency as discussed under option A.

West German WHNS and NIP expenditures linked to the US commitment to return would remain. These involve costs as well as benefits for the FRG, however, adding to both sides of the balance.

It is clear that if the United States remains committed to return to Europe in times of crisis or war, it is because it is in its interest to do so. In contrast to the situation in option A, it is therefore reasonable to suppose that the USA would share the costs accrued in keeping the return option open.

To simplify, the cost–benefit balance for the return option would include costs for real estate, NIP, WHNS as well as residual social welfare costs (total: DM 2.04 million). These are to be contrasted with the benefits derived from federal WHNS and NIP orders placed with West German firms (total: DM 353 million). This balance would be significantly altered—depending on the contribution—if the US commitment to return included a financial commitment towards the maintenance of facilities and infrastructure in the FRG.

Notes and references

[1] *NATO-Truppenstatut und Zusatzvereinbarungen* (Beck: Munich, 1987); see also preface to section 35 in *Bundeshaushalt 1987* (Bundesminister der Finanzen: Bonn, 1987), p. 2.

[2] *Facilities Maintained by U.S. Department of Defense in the Federal Republic of Germany* (Friends Committee on National Legislation: Washington, DC, 1974).

[3] See Schmidt-Eenboom, E., *Die militärischen Strukturen im Main-Kinzig-Kreis. Gutachterliche Stellungnahme im Auftrag des Kreistages, 1. Version* (Politisches Informations- und Nachrichtenkontor: Weilheim, Dec. 1987), p. 132–34.

[4] 'Millionen Dollar fließen in die heimische Wirtschaft/US Base Bitburg', *Mitteilungen der IHK Trier*, no. 1/86 (Industrie- und Handelskammer Trier: Trier, 1986).

[5] Seiler, S., *Die GIs, Amerikanische Soldaten in Deutschland* (Rowohlt: Reinbek bei Hamburg, 1985), p. 23–24 (authors' translation).

[6] Kohl, H.-H., 'Es reicht nur für die Ansichtskarte', *Frankfurter Rundschau*, 4 Jan. 1988 (calculation and forecast by authors).

[7] US Department of the Air Force, Headquarters United States Air Force in Europe, Ramstein (reply to letter, 5 Jan. 1988).

[8] *Information, Flugplatz Ramstein* (316th Air Division/Cost Branch, USAFE: Ramstein, 30 Sep. 1988). See also US Department of the Air Force (note 7).

[9] *Financial and Employment Impact of the U.S. Forces on the German Economy in FY 1986*, report no. AEAHN-GR 5-5f (US Department of the Army, Headquarters United States Army, Europe and Seventh Army: Heidelberg, 31 Aug. 1987) (reply to letter, 2 Feb. 1988).

[10] See Kohl (note 6).

[11] *Worldwide Manpower Distribution by Geographical Area. March 31, 1987*, report no. DIOR/M 05-87/02 (US Department of Defense: Washington, DC, 1987), p. 11.

[12] See US Department of the Air Force (note 7).

[13] The exchange-rate in 1987 was DM 1.79 on annual average. See *Monatsbericht der Deutschen Bundesbank*, vol. 40, no. 2 (1988), p. 80.

[14] Bundesminister der Finanzen, Bonn (reply to letter, 4 Nov. 1987).

[15] Department of the Army (note 9).

[16] See Bundesminister der Finanzen, Bonn (note 14).

[17] *NATO-Truppenstatut und Zusatzvereinbarungen* (note 1), articles 9 and 56 respectively.

[18] Gen. Glen Otis, Commander in Chief, US Forces Europe, quoted in Kohl (note 6).

[19] See Department of the Army (note 9).

[20] 'Die Zahlungsbilanz der Bundesrepublik Deutschland', *Monatsberichte der Deutschen Bundesbank*, no. 7, 1987, p. 23 (Reihe 3: Zahlungsbilanzstatistik).

[21] Heckmann, E., 'Die USCAE beschafft für die US-Army in Europa', *Wehrtechnik*, vol. 17, no. 5 (1985), p. 88.

[22] Bundesminister der Verteidigung, Bonn (reply to letter, 23 Dec. 1987).

[23] See Department of the Army (note 9).

[24] Section 35 of the federal budget covers most of these costs, with the exception of road construction costs, which are covered by section 12.

[25] See Department of the Army (note 9).

[26] See Bundesminister der Verteidigung (note 22).
[27] See Bundesminister der Verteidigung (note 22).
[28] See Bundesminister der Finanzen (note 14).
[29] In 1986, a total of DM 202 million were drawn from section 35 of the Federal budget for the regulation of damages.
[30] See Department of the Army (note 9).
[31] See Bundesminister der Verteidigung (note 22).
[32] See Bundesminister der Verteidigung (note 22).
[33] See Bundesminister der Verteidigung (note 22).
[34] A very thorough description and analysis of the social situation of the US soldiers in the Federal Republic can be found in Seiler (note 5).
[35] No official figures are available. The estimates presented here are based on data from telephone surveys of social welfare offices in the Rhine-Main area and in Rhineland-Palatinate.
[36] Estimate from Department of the Army (note 9).
[37] Hessische Friedens- und Konfliktforschung, Frankfurt (reply to letter, 16 Dec. 1987). The main US construction programmes in the Federal Republic in 1988–89 are those in: Ansbach, tagged at DM 25.6 million; Darmstadt, at DM 29.2 million; Frankfurt, at DM 18.7 million; Hanau, at DM 42.8 million; Pirmasens, at DM 14 million; Schweinfurt, at DM 17.8 million; Wildflecken (near Fulda), at DM 20million; Würzburg, at DM 48 million; and Ramstein, at DM 11.5 million. *Department of Defense Budget for Fiscal Year 1988/89, Construction Programs (C-1)* (US Department of Defense: Washington, Jan. 1988.)
[38] See Heckmann (note 21).
[39] See Kohl (note 6).
[40] Bremer Lagerhaus-Gesellschaft (BLG), Bremerhaven (reply to letter, 29 Dec. 1987).
[41] Figures for Wiesbaden, for example, can be found in Schmidt-Eenboom, E., *Wiesbaden: Eine Analyse der militärischen Struktur in der hessischen Landeshauptstadt* (Institut für Friedensforschung: Starnberg, 1987), p. 62.
[42] In Bitburg, where the tenants are all USAFE personnel, the average annual rent amounts to DM 10 800. See 'Millionen Dollar fließen in die heimische Wirtschaft/US Base Bitburg' (note 4). In US Army garrison towns, however, the average rent paid is considerably lower, because of the lower average incomes of Army personnel.
[43] See Heckmann (note 21).
[44] The shoe-maker federation may be cited as an instance: of an overall turnover of DM 1.4 billion in 1986, about DM 13 million were proceeds from contracts with the *Bundeswehr*, police and fire-brigades and with foreign forces. Of these, the US forces account for an estimated share of DM 4–5 million. Bundesinnungsverband des Deutschen Schumacher-Handwerks (reply to letter, 27 Nov. 1987).
[45] See Department of the Army (note 9).
[46] Kirbach, R., 'Wenn die Amis gehen, gehen die Arbeitsplätze mit', *Die Zeit*, 23 Feb. 1990.

7. Economic consequences of a withdrawal of US forces from the Benelux states

Sami Faltas
Eindhoven University, Eindhoven, Netherlands

I. The political setting

In the course of the two World Wars, Belgium, Luxembourg and the Netherlands learned that their traditional policies of neutrality and appeasement had become totally inadequate as a means of maintaining peace and security in their region. Since the end of World War II, there has been a strong consensus in the Benelux countries that the only way to provide security for their territory, and indeed for Western Europe, lies in international arrangements that would minimize the risk of war and provide a guarantee from the major powers for the security of the smaller states of the region.

This is not to say that the governments of these countries opted for a Western military alliance as soon they regained their freedom in 1945. As in many other countries, they initially hoped that the United Nations would achieve universal disarmament and provide a world-wide system for the prevention of war and the defence of all states against armed aggression. The tradition of seeking peace through international law is particularly strong in the Low Countries.

It soon became clear, however, that the United Nations would not be capable of providing adequate protection for Western Europe or any other part of the world. The new organization was severely handicapped by the rivalry between the two new superpowers. Furthermore, the rapid consolidation of Soviet control in Eastern Europe led many West European statesmen to believe that the USSR also posed a direct threat to their own region. They were therefore in a hurry to reorganize their defence, and they could not do it alone.

Under these circumstances the Benelux states soon abandoned their policy of neutrality and formed an alliance with the major Western powers. In fact, their governments played an active role in the conclusion of the Treaty of Brussels and the attempt to establish a European defence organization.

There were also domestic factors which contributed to the formation of a Western alliance. Most governments in Western Europe were in favour of restoring the pre-war political order and pursuing conservative economic policies, but they faced strong domestic opposition from the labour movement and various left-wing parties and movements. West European governments and the United States agreed that one of the reasons why an alliance was desirable was that it would serve as a mainstay against radicalism in Europe.[1] Even today it is probably correct to claim that the presence of US forces in European countries serves to maintain the domestic status quo. In the case of Belgium, Luxembourg and the Netherlands, where such radicalism today plays an insignificant political role, this function is no longer important.

Finally, the decision to enter into a Western alliance was part of a broader drive towards international integration. Statesmen from the Benelux countries pursued this with great enthusiasm, not only because they foresaw that the creation of a European market would benefit their economies, but also because they felt that the unification of Europe would put an end to war between the countries involved and strengthen them collectively against external threats.

The Benelux states are in several respects NATO's heartland. Historically, they helped lay the foundations of the alliance, joining the United Kingdom and France in concluding the Treaty of Brussels in 1948. Geographically, Belgium forms the administrative centre of NATO, as host to the civil headquarters of the organization and the seat of its oldest and most important military command. Politically, Belgium and the Netherlands have in the past been closely associated with the leadership of the alliance, having supplied three of its first five secretaries-general.

Militarily, the Low Countries play only a minor role, but their limited defence resources are almost entirely assigned to NATO. With the exception of the Federal Republic of Germany, no other states have so closely identified their security with NATO's system of collective defence. In fact, the principal objective of the Benelux armed forces is not so much to defend their national territory, but rather to contribute to the collective defence of Western Europe.

For many years the Benelux states enjoyed a reputation of being particularly loyal and compliant allies.[2] The fact that Dutchmen and Belgians have held high posts in the organization more frequently than the size of their countries would warrant suggests that these countries have been seen as a cohesive force in the alliance. This began to change as a result of popular opposition to the Viet Nam War in the 1960s and the massive anti-nuclear-weapon campaigns of the 1970s. The change was more pronounced in the Netherlands than in Belgium, while Luxembourg seems hardly to have been affected.

How did the anti-nuclear-weapon campaigns, which were supported by millions of Dutchmen in the late 1970s and early 1980s, affect public opinion concerning collective NATO defence in general and the US military presence in particular? At the time, many domestic and foreign observers believed that the Netherlands was slowly but surely sliding out of NATO. If it was—which is debatable—it was not doing so consciously, and certainly not intentionally.

The campaigns, commonly grouped under the loose heading of the 'peace movement', were in fact led by a coalition of highly diverse groups which agreed on a small number of main issues. It was precisely because they focused narrowly on these issues that they were able to maintain their unity and mobilize large numbers of people. Their principal objective was to prevent the deployment of new nuclear weapons in Europe. For practical reasons, they focused on the basing of US cruise missiles in the Netherlands. Soviet nuclear armaments received little attention, and even less was said about British and French nuclear weapons.

For practical reasons, again, the campaigns aimed to influence the Netherlands Government and Parliament, rather than NATO or the US Government. The ideological mood of the campaigns was diffusely pacifist. Unlike the anti-Viet Nam War movement, the campaigns against nuclear weapons did not broaden into a general opposition against US foreign policy or the NATO system. Anti-Reagan sentiments found strong expression but there was little anti-Americanism. When asked, if obliged to choose, whether they would describe themselves as pro- or anti-American, two-thirds of the respondents in a 1983 poll labelled themselves as pro-American. Less than 15 per cent opted for 'anti'.[3] At no time did the presence of US military forces in the Netherlands become a campaign issue.

In the end, the peace movements failed to push through a unilateral decision to ban cruise missiles from the Netherlands soil.[4] Had they succeeded, this might have provoked an exasperated US Government to withdraw its forces from the Netherlands. Although unlikely, it was conceivable. What is important to note is that such a withdrawal was not widely discussed, it was not the objective of the anti-nuclear-weapon campaigns, and public opinion would not have welcomed it. It would certainly have dismayed the Netherlands Government and all the major political parties.

For all their enthusiasm for unilateral nuclear disarmament, the Dutch never ceased to support NATO and the alliance with the United States. In 1984, 60 per cent of the respondents in a public opinion poll wanted the Netherlands Government to defer or reject a decision to approve the basing of cruise missiles in the Netherlands, and 89 per cent wanted the Netherlands to remain a member of NATO.[5]

The early 1980s also saw a small blossoming of academic and political debate on alternative concepts for the defence of Europe, such as the formation of a European 'pillar' within NATO, the creation of a Common Security system for the whole of Europe and the introduction of alternative military doctrines, sometimes described under the heading of 'defensive defence.' However, most of these concepts assumed that superpower guarantees, by necessity nuclear, would still be required for the security of Europe. Although such alternative defence concepts often implied the withdrawal of foreign forces from European countries, no one suggested that the US forces should start packing their bags.

In the Benelux states, the presence of US troops in Western and Central Europe is seen as: (*a*) tangible proof of the US commitment to the collective defence of Western Europe, seen as essential to the security of the Benelux states; (*b*) a military asset which would be extremely costly to replace; (*c*) A living reminder of the leading role played by the United States in the liberation of Western Europe from Nazi occupation; (*d*) a symbol of US hegemony which is accepted as a fact of life; and (*e*) a symbol of nuclear armaments and nuclear politics, which are detested by a large part of the population.

In the minds of the Benelux public and policy-makers, the benefits clearly outweigh the drawbacks of the US presence. Under present conditions, there is no significant public support for a withdrawal of US forces.[6]

II. The US military presence

In March 1987 there were 3418 US military personnel in Belgium, 3130 in the Netherlands, and a few hundred in Luxembourg.[7] By NATO standards, this is not a very large US military presence. In 1987, six other European NATO countries, including Iceland, each had more US troops on their territory than either Belgium or the Netherlands.

On a national scale, the approximately 17 000 US military personnel and dependants do not have a significant social impact in the Benelux countries. Their presence is felt on a local level, but not strongly. Reports in local newspapers suggest that relations with the local population are marginal, but that such contacts as do exist are good. The US personnel tend to spend most of their time during their two- to three-year tours of duty within their own community and generally to follow instructions to be discreet and polite in all contacts with the local population.

III. The impact of withdrawal

This study presents two possible options for a withdrawal of US forces. In both cases it was assumed that the posture of the Warsaw Treaty Organization remains as it was in early 1988 and that the North Atlantic Treaty remains intact. During 1989–90, however, WTO forces decreased substantially.[8]

Option A assumes that the US forces have 'gone for good', and that the US Government has not made a specific commitment to return to Europe in a crisis. Under these conditions it is inconceivable that the US forces would continue to play a leading role in NATO's European commands. The Supreme Allied Commander Europe (SACEUR) would no doubt be a European, and fewer US soldiers and civilians would be assigned to NATO's administrative bodies. However, the United States would remain a member of NATO.

Option B is based on the assumption that the United States has promised to return if a crisis arises. This implies that various facilities and types of personnel would remain in place, in order to facilitate an orderly return of forces from the USA. The US role in NATO's administrative machinery would not change much.

This chapter is mainly concerned with the economic implications for the Benelux countries of a withdrawal of US forces from their territory. This is by no means the same issue as the economic implications for the Benelux countries of a withdrawal of US forces from Western Europe as a whole. As is

ECONOMIC CONSEQUENCES FOR THE BENELUX STATES   123

seen, a withdrawal of US forces from the Benelux countries would constitute a small economic loss to the Low Countries. The loss would be greater in option A than in option B; but the difference between the two scenarios is less significant than the issue of whether the US forces are to be replaced, and, if so, in what way and by whom.

Who would leave?

In 1988, there were 7777 US military and other defence personnel in the Benelux region (table 7.1). This is more than in previous years, as it includes a complement involved in the completion of cruise missile sites at Florennes in Belgium and Woensdrecht in the Netherlands. It is assumed here that these bases are fully manned and equipped in the pre-withdrawal situation.[9]

A US withdrawal according to option A would remove over 7600 US servicemen and other defence personnel from the region. A community of about 16 000 people (including dependants) would leave. A total of 31 US defence facilities would be removed, including three Air Force bases, a large transportation terminal, 11 equipment depots of various types, and a wide range of communications facilities. An estimated 1521 Benelux civilians working for US defence facilities in their region would lose their jobs.

NATO personnel deserve separate mention, as there are about 1439 US citizens in the Low Countries working for NATO or as members of the US delegation to NATO. They comprise about one-fifth of all US defence-related personnel in the Benelux region. Our assumption here is that the United States would remain a member of NATO after its withdrawal, but in this scenario US nationals working for military NATO organizations—SHAPE (Supreme Headquarters Allied Powers Europe), the SHAPE Technical Centre and AFCENT (Allied Forces Central Europe)—would leave, as would some of the US personnel working for, or based at, NATO's civil organizations. It is assumed here that the only US defence-related personnel that would remain in the Benelux countries after a option A withdrawal would be 165 civilians at NATO headquarters in Brussels.

Option B is less drastic. Presumably, it would involve the withdrawal of 4558 employees, especially from the three US Air Force bases, while all US personnel working for NATO would remain in place, and there would be an increase of 230 people working at equipment depots, communications facilities and the Military Traffic Management Command, all of which would face a greater workload supporting US forces coming to Europe for exercises. In a crisis, they would have a Herculean task to perform, facilitating the rapid movement of large US forces to the European theatre. All in all, option B, as defined here, would reduce US defence personnel in the Benelux countries from 7777 to 3219 people. An estimated 1094 Benelux nationals working for US defence facilities would lose their jobs.

Table 7.1. US defence personnel in the Benelux countries (incl. NATO staff) before (1988) and after withdrawal

Figures rounded to nearest five are estimates; abbreviations are explained on pages xx–xxiv.

Country[a]	Location	Unit	No. of personnel 1988	Option A	Option B
B	Barronville	USAr munitions depot	25	0	30
B	Ben-Ahin	USAF DCS troposcatter link	25	0	30
NL	Brunssum	USAr POMCUS depot	25	0	30
NL	Brunssum	NAMPA Center	25	0	30
NL	Brunssum	AFCENT HQ	500	0	500
NL	Cannerberg	NATO 2d ATAF Joint Ops Center	25	0	30
NL	Capelle/IJssel	USAr MTMC, TTCE HQ	200	0	300
B	Casteau	SHAPE HQ	500	0	500
B	Le Chenoi	USAF DCS link	25	0	30
B	Chièvres AB	USAr AB, USAr/NATO comm.	200	0	50
NL	Coevorden	USAr POMCUS depot	25	0	30
B	Daumière Caserne	USAr logistics, munitions depot	25	0	30
B	Evere (Brussels)	NATO TARE main operating hub	339	165	339
NL	Eygelshoven	USAr POMCUS depot	25	0	30
B	Flobecq	USAr communications centre	25	0	30
B	Florennes AB	USAF GLCM base	1 564	0	100
B	Grobbendonk	USAr logistics, munitions depot	25	0	30
NL	The Hague	SHAPE technical centre	100	0	100
NL	Het Harde	USAr TNW storage (custodial)	25	0	30
NL	Havelterberg	USAr TNW storage (custodial)	25	0	30
NL	Heerlen	USAr reserve hospital	25	0	30
NL	Hook of Holland	USAr DCS troposcatter link	25	0	30
B	Houtem	USAr DCS troposcatter link	25	0	30
B	Kester	NATO SATCOM, DSCS comm.	25	0	30
B	Kleine-Brogel AB	USAF TNW storage, DCS link	50	0	60
B	Koksijde AB	USAF logistics depot, MAC terminal	100	0	100
L	Luxembourg	USAr depot, anti-crisis centre	100	0	100
NL	Maastricht	USAr AUTODIN, TARE station	25	0	30
NL	Maastricht	USAr III Corps HQ	300	0	50
B	Mons	NATO DCS, TARE hub	25	0	30
NL	Schinnen	USAr logistics centre	100	0	100
NL	Soesterberg AB	USAF 32d TFS	1 607	0	100
NL	Steenwijkerwold	USAr communications facilities	25	0	30
B	Sugny	USAr/FRG munitions depot	25	0	30
NL	Vlagtwedde	USAr POMCUS depot	25	0	30
NL	Volkel AB	USAF comm, TNW storage	25	0	30
B	Westrozebeke	DCS troposcatter link	25	0	30
NL	Woensdrecht AB	USAF GLCM base	1 517	0	100
Total			**7 777**	**165**	**3 219**

[a] B = Belgium; NL = Netherlands; L = Luxembourg.

Sources: Duke, S., SIPRI, *United States Military Forces and Installations in Europe* (Oxford University Press: Oxford, 1989), appendices 2A–2C, pp. 32–36; Benelux press reports; US, Netherlands and Belgian official documents.

Economic consequences

How would a withdrawal of US forces from Benelux territory affect the economies of these countries? The major items to consider are: (*a*) the facilities provided by the Benelux governments to US forces in the Benelux countries and the payments received for such facilities; (*b*) contracts given by the US Department of Defense to Benelux firms for goods and services to be supplied to US forces stationed in the region; (*c*) the incomes of US Government military and civilian personnel in the Benelux countries and their cost to the host country; and (*d*) the cost to the host countries of increasing their defence effort to compensate for a withdrawal of US forces.

An attempt is made here to assess the first three items. The data used are largely derived from figures for the FRG presented in chapter 6 by Hartmut Bebermeyer and Christian Thimann, to make up for the lack of reliable data for the Benelux countries. Item four cannot be properly investigated in this chapter.

Host country facilities

NATO states provide land and other real estate free of rent to each others' forces. For fiscal year 1981 the US Congress prohibited the payment of property taxes by the US Government for overseas military facilities.[10] It is assumed here that none are paid in the Benelux countries, so that in fact all the real estate used by US forces is paid for entirely by the host country.

All other things being equal, this means that the evacuation of US military facilities in the region would benefit the Benelux treasuries. On the basis of data for the FRG, it is possible to give a rough estimate for the annual income currently forgone by the Benelux governments, by providing real estate free of charge to US forces, at $30 million. A US withdrawal as envisaged in option A would eliminate this loss. However, this only applies if the property is then put to profitable use. If the Belgian or Netherlands armed forces use the property vacated by the US forces, or if it is lent free of charge to another ally, the host countries would continue to forego the potential income as before.

Option B could, in theory, reduce the loss of potential income by some $10 million.[11] As the US forces would probably need to keep the real estate available for immediate use in a crisis, however, the property would in fact probably not be vacated, and therefore no income would be generated. In other words, the property income the Benelux governments are likely to gain from a US military withdrawal under the conditions defined by options A and B is next to nothing.

Contracts with local firms

US defence organizations and facilities in Western Europe receive all their military equipment from the USA, but purchase most of the civilian goods and services they need locally. For instance, nearly 80 per cent of the contractors of Woensdrecht Air Force Base in the Netherlands are Dutch firms.[12]

In the FRG, the institutions of the US forces in 1986 spent about $3.4 billion on contracts with West German firms and utilities for a large variety of goods and services. This represented an expenditure of about $13 650 per US defence employee in the country. From 1986 to 1988, the value of the dollar in D-Marks fell so low that the US forces would in 1988 have needed about $4.8 billion to buy the same package of West German goods and services as in 1986. However, US institutions and individuals cut back their purchases in expensive European currencies.[13] Assuming that the US institutions reduced their purchases of DM by 15 per cent from 1986 to 1988, their expenditure would have totalled approximately $4 billion in 1988, or $15 850 per US defence employee in the FRG.

Extrapolating from Bebermeyer and Thimann's figures, this suggests that US defence institutions in the Benelux countries spent about $87 million in 1986 on goods and services provided by local firms and utilities. The estimate for 1988 would be about $101 million.[14] Almost all of this income would be lost if the US forces withdrew according to option A. In option B, the economies of the Benelux countries stand to lose about $73 million in revenues.

Personal incomes

Table 7.2 provides fairly accurate information on the personal incomes of US defence personnel in the Benelux countries. Since allied forces in NATO countries are exempt from most forms of taxation under the 1952 Status of Forces Agreement, including tax on real estate and income, the departure of US soldiers would not mean a loss of tax revenues. Likewise, a US withdrawal would not make much difference in terms of the use of public services provided by the host country. US personnel participate only to a very limited extent in the social life of the host countries, largely relying on facilities and services provided by the US Government. In addition, unlike the US military presence in the FRG, the US defence community in the Benelux countries is small. US servicemen do pay some local duties and levies. Presumably these payments cover the small cost of public services provided to US personnel by the local authorities.

Press reports on the US Air Force base Camp New Amsterdam (CNA) in Soesterberg, the Netherlands, suggest that roughly one-half of the take-home wages of CNA personnel remains in the US economy.[15] This money is spent on food, drink and a wide range of other consumer products and services offered by on-base franchises. A part is paid into savings accounts or sent home to families. The remaining half is spent in the host country's economy, on housing, food, recreation, petrol, motor vehicles and durable consumer goods. All in all, the 7777 US defence employees in the Benelux countries spend about $104 million of their own money in the host country's economy. The figure

Table 7.2. Personal incomes of US defence personnel in the Benelux countries, 1988
Wages are in US$.

Personnel Category	Number	Wages Average[a]	Total (m.)
NATO (NATO HQ, AFCENT, SHAPE)	1 439	33 760[b]	48.6
US Army (military personnel)	1 125	18 730	21.1
US Air Force (military personnel)	4 439	33 760	149.9
Other military personnel	140	30 721[c]	4.3
Civilian personnel	634[d]	20 650	13.1
Subtotal	**7 777**	**30 463**	**236.9**
Benelux personnel	1 521[e]	27 915[f]	42.5
US social security payments to Benelux personnel		11 166	17.0
Subtotal	**1 521**	**39 081**	**59.4**
Total	**9 298**		**296**

[a] Wage figures from Bebermeyer and Thimann (chapter 6).
[b] Assumed to be equal to average USAF wage.
[c] Assumed to be equal to weighted average Army and Air Force wages.
[d] Assumed to be 10% of operational US defence personnel (excluding NATO, AFCENT and SHAPE staff); based on figures for FRG.
[e] Assumed to be 24% of operational US defence personnel (excluding NATO, AFCENT and SHAPE staff); based on figures for FRG.
[f] Based on wages given in DM for FRG personnel in 1986, multiplied by the average growth rate of US defence personnel wages in dollars from 1986 to 1988 and converted to dollars as the average rate for 1988 (DM 1.68). The 1986 rate was DM 2.32.

Sources: Personnel: Duke, S., SIPRI, *United States Military Forces and Installations in Europe* (Oxford University Press: Oxford, 1989), appendices 2A–2C, pp. 32–36; wages: Bebermeyer and Thimann (chapter 6).

may be declining in response to the falling rate of the dollar relative to Dutch guilders and Belgian francs. Additionally, the Benelux nationals working for US defence facilities in their region spend almost all of their income in the local economy. If the premiums paid by the US Government for their social security are included, the income of Benelux employees working for the US armed forces amounts to about $59.4 million.

In option A, some $161 million spent by US and local employees of the US defence community in the Benelux economies would be lost. A withdrawal as in option B would cost the Benelux economies revenues of up to $107 million. However, if the exchange-rate of the dollar remains unfavourable, US defence institutions and personnel will increasingly avoid local businesses, and the local economy will have less to lose from a withdrawal of US forces.

Given the present deployment, the departure of US defence employees would affect local businesses in the towns surrounding the major US bases. Landlords in particular would suffer; however, the effects would not be dramatic.

Table 7.3. Economic costs and benefits to the Benelux countries of a withdrawal of US personnel and facilities (options A and B)

	1988	Option A	Option B
Personnel			
Total US personnel	7 777	165	3 219
(Operational US personnel)	(6 338)	(0)	(1 780)
Benelux personnel	1 521	0	427
Total	9 298	165	3 646
Costs (–) and benefits (+) in US$ m			
Sale/lease of base real estate		+ 30	+ 10
Local firms loss of US defence contracts		– 101	– 73
Loss of private expenditure (US personnel)		– 102	– 64
Loss of employment/income (Benelux personnel)		– 59	– 43
Total		– 233	– 169

Sources: Author's calculations, based on Duke, S., SIPRI, *United States Military Forces and Installations in Europe* (Oxford University Press: Oxford, 1989) and official Netherlands documents.

Summary

The rough calculations presented here suggest that a withdrawal of US forces according to option A would cost the Benelux economy $233 million a year, or some $263 million if real estate vacated by the US forces is not used profitably. A withdrawal as in option B would cost $169–179 million (table 7.3). In other words, the economic consequences of a withdrawal of US forces from the Benelux region to the host countries would be negative, but not painful enough to make a political difference.

However, this is true only if the Netherlands, Belgian and Luxembourg armed forces are not required to take over duties now performed by the US forces in Western Europe. If the Benelux states were to shoulder even a small part of the burden which the United States would lay down in option A, a steep increase in defence expenditure would be unavoidable. First, the three governments would have to replace 1274 US NATO officials now working in the Benelux countries, plus an unknown number of US NATO officials based elsewhere in Europe. This would cost at least $43 million a year. Second, it would probably fall to Belgium and the Netherlands to replace the US units now stationed on their territory. The annual cost would be in the order of $1 billion a year. Third, a quarter of a million US soldiers stationed in the Central Sector and some of the 19 000 US sailors serving in European waters would need to be replaced by Europeans and Canadians. The European countries with major responsibilities in the FRG and in the Atlantic Ocean and Mediterranean Sea areas would be expected to maintain their share in the collective defence of these areas. Belgium, which has a relatively large commitment in the Central Sector, and the Netherlands, with a fairly large commitment both in the Central Sector and the Eastern Atlantic, would be strongly affected. They would be among the countries likely to face the largest relative increase in defence expenditure as a result of a US withdrawal. The political impact would be dramatic. Fourth, if the US forces left Western Europe for good, this would affect international relations in a way which is almost certain to be unfavourable to the internationally oriented economies of Belgium and the Netherlands. For instance, US investment in the region is likely to decline.

Military–industrial relations

In the past 20 years the United States has been the largest exporter of military equipment to the Benelux armed forces. During the 1950s and the early part of the 1960s, the United States provided arms under the Military Assistance Program, but since then Belgium and the Netherlands have been buying US defence equipment on commercial terms.

Like many other industrial states, Belgium and the Netherlands insist on offset contracts for their own industries whenever they import military equipment. The US Government considers these arrangements both inconvenient and inefficient, but it also recognizes that giving the smaller West European countries a stake in manufacturing NATO weapon systems in this way strengthens their commitment to the Atlantic Alliance. It is not certain, however, that offset contracts are beneficial to the countries that ask for them. Almost invariably they raise the price of the imported equipment, and this additional cost sometimes outweighs the benefits obtained by industry in terms of jobs and technological innovation.

Experts agree that a rationalization of arms production and procurement in the NATO region which would eliminate the wasteful duplication of effort and introduce a division of labour reflecting comparative advantage would be much better than the current approach. Smaller NATO states such as Belgium and the Netherlands take a particularly keen interest in armaments co-operation, because it represents their only chance to control the cost of modern defence forces and maintain a military industrial capability in the long term. However, as long as real progress in armaments co-operation remains elusive, individual governments continue to demand a quid pro quo when they buy arms abroad.

A withdrawal of US forces from Western Europe according to option B would not necessarily affect US military exports to Western Europe or military–industrial co-operation within the alliance as a whole. However, if the US forces were to leave for good, as is assumed in option A, then this would be sure to have an impact on military–industrial trade in NATO. The nature of this impact would depend almost entirely on the circumstances under which the US forces withdraw. If it were part of a multilateral disarmament agreement, arms production and arms purchases in Europe would slump and arms imports from the United States would fall dramatically. In the new situation, one might expect most of the defence equipment required to safeguard Western Europe's security, or the common security of Europe as a whole, to be developed and produced in the region, rather than in the USA. A decreased dependence on US arms would probably go hand in hand with a decreased dependence on US forces. The political cohesion of NATO as it stands today would be weakened. These effects would be much stronger if the US forces left without the consent of the West European allies. There would be an intense desire to become self-reliant in defence, hence in armaments, which might lead the West Europeans to bury their differences and pool their military–industrial resources in order to defeat US competitors. This response would further undermine an alliance already reeling under the impact of the unilateral withdrawal of US troops.

IV. Conclusion

The economic effects of a US military withdrawal from the territory of the Benelux countries would be unfavourable, but not grave. There is no reason to expect any dramatic effects as long as the USA clearly and credibly commits itself to returning its forces to Europe in a time of crisis. However, if the USA were to withdraw its forces without promising to send them back if they are needed, the indirect and long-term economic effects could be drastic. Failing an agreement on multilateral disarmament in Europe, the Belgian and Dutch armed forces would be obliged to shoulder part of the defence burden laid down by the US forces, and this would be very costly. Moreover, the Benelux governments would be expected to contribute generously to a large variety of armaments projects intended to make Western Europe as self-reliant in defence as possible.

Most people concerned with defence policy in the Benelux states are at least intuitively aware of these implications, which is why very few of them are in favour of a US military withdrawal under the present conditions.

Notes and References

[1] Kolko, J. and Kolko, G., *The Limits of Power: The World and United States Foreign Policy* (Harper & Row: New York, 1972).

[2] For the Netherlands, see van Staden, A., *Een trouwe bondgenoot: Nederland en het Atlantisch Bondgenootschap, 1960–1971* (Uitgeverij In den Toren: Baarn, 1974).

[3] Vaneker, Ch. J. and Everts, Ph. P., *Buitenlandse Politiek in de Nederlandse Publieke Opinie* (Nederlands Instituut voor Buitenlandse Betrekkingen 'Clingendael': The Hague, 1984), p. 41.

[4] It could be argued that some of the credit for the conclusion of the US–Soviet INF Treaty banning intermediate-range nuclear missiles should go to these activists who strongly opposed the modernization of nuclear forces in Europe. Without the pressure of their campaigns, it is very doubtful whether the US Government would ever have proposed the zero option, which eventually served as a basis for the INF Treaty.

[5] Vaneker and Everts (note 3).

[6] A US military withdrawal could under certain conditions become a popular notion in the Benelux countries. One such instance would be a situation where an East–West agreement on multilateral nuclear and conventional disarmament and alternative security arrangements for the European continent as a whole were believed to be within reach.

[7] *Defense 87 Almanac* (American Forces Information Service: Alexandria, Va., Sep./Oct. 1987).

⁸ For detailed estimates of WTO cuts achieved in 1989 and scheduled for 1990, see Sharp, J. M. O., SIPRI, 'Conventional arms control in Europe', *SIPRI Yearbook 1990: World Armaments and Disarmament* (Oxford University Press: Oxford, 1990), pp. 459–74.

⁹ The cruise missile bases had in fact largely been completed when the INF Treaty banning them was signed in Dec. 1987. The host governments wanted the missile units to be replaced by other US units.

¹⁰ Comptroller General of the United States, *Report to the Chairman of the Subcommittee on Defense*, Senate Committee on Appropriations (US General Accounting Office: Washington, DC, 31 July 1984), p. 10.

¹¹ Calculated from parameters for FRG figures given by Hartmut Bebermeyer and Christian Thimann in chapter 6 of this volume and applied to the number of US defence personnel in the Benelux countries.

¹² Personal communication from peace movement activists in Woensdrecht.

¹³ Bebermeyer and Thimann (chapter 6).

¹⁴ Figures for US defence personnel working for NATO organizations are not included in the basis for these estimates.

¹⁵ This is supported by figures for Bitburg Air Base in the Federal Republic of Germany, where 30% of the personnel live on the base and spend 30% of their income in the host country's economy. The 70% that live outside spend half their income in the German economy. On average, USAF personnel at Bitburg spend 44% of their income in the local economy; Bebermeyer and Thimann (chapter 6).

8. Costs and benefits to the United Kingdom of the US military presence

Keith Hartley and Nicholas Hooper
University of York, York, England

I. Introduction

Closure of US bases in the United Kingdom is an issue which is dominated by myth and emotion, lacking independent analysis, critical evaluation and empirical evidence. Questions arise about the effects of withdrawal on British defence policy and budgets, and on the performance of the economy, particularly in terms of employment, price stability, growth and balance of payments targets. At the outset, the magnitudes involved need to known. Are the numbers sufficiently large to have a substantial impact on the protection of the UK and on its economy (e.g., of a kind similar to the oil price rises of the 1970s or the discovery of North Sea oil); or are the impacts so small and marginal that, apart from a few local effects, they can be ignored?

An analytical framework for assessing these issues is outlined in this chapter. Consideration is given to the benefits to the UK from membership of a military alliance and evidence is presented on the actual contribution of US forces to British protection. The assumptions about US withdrawal are explained and evaluated. Finally, evidence is presented on some of the defence budget and economic implications of withdrawal. A distinction is made between defence implications in terms of the effects on military budgets, equipment procurement and national defence industries, and economic implications particularly in relation to arguments about the burdens of defence spending. Throughout, the aim is to develop a conceptual framework to specify the policy-relevant questions, consider the available evidence and identify the gaps in our knowledge, all of which provide 'inputs' into a broad cost-benefit analysis viewed from a British perspective. Informed choices about US withdrawal require information on the likely gainers and losers.

II. US forces and British defence

The economics of alliances

NATO is an international voluntary club specializing in collective defence (a public good). Nations will remain members, paying the club fee, as long as membership is worthwhile. On this basis, NATO survives by offering more protection and/or lower defence costs than separate national defence arrangements. Economic models of military alliances offer a number of policy-relevant predictions:

1. Larger nations in an alliance, by valuing defence more highly, will usually allocate larger shares of their national income to defence than smaller nations. It is in their national interest to do so.[1]

2. Smaller nations will get a free ride by consuming the collective protection offered by the larger allies. The model does not imply that smaller nations *ought* to be required to bear a larger share of the common burden: their contributions reflect their national interests.[2]

3. In the absence of a political union, NATO defence expenditure based on separate sovereign states will be less than the amount desired by the common interests of the group as a whole.[3]

4. In principle, a military alliance offers further potential benefits in the supply of defence equipment and the provision of armed forces. Here there are opportunities for applying the principles of specialization by comparative advantage and for exploiting economies of scale (by standardizing equipment).

5. Recent developments have modified the predictions about burden-sharing and free-riding. A distinction is made between deterrent and protective forces, with the latter subject to 'thinning' (e.g., as more territory has to be protected). It is then predicted that, regardless of income levels, unequal burdens will be incurred by those allies who either provide deterrent forces (e.g., Britain, the USA), or are major providers of private and/or excludable public outputs offering alliance-wide deterrence. Alliances relying on protective and conventional forces are less likely to be characterized by free-riding and are more likely to share burdens in relation to the benefits received.[4] Empirical work supports the modified model. There was evidence of substantial free-riding and support for the original deterrence model during the 1950s and into the 1960s. However, in the 1970s and 1980s there was a shift in NATO away from a purely deterrent alliance to flexible response and a more conventional and protective alliance.

Estimates of the relationship between US and British defence spending suggest that a $1 rise in US military expenditure results in a 1.5 cents cut in British defence outlays, whereas the corresponding reductions for France and the Federal Republic of Germany are 7.5 and 9.2 cent, respectively. Similarly, the USA will respond to a $1 rise in British defence spending by cutting its military budget by 87 cents.[5]

While NATO is a specialist defence club providing security and protection, it might also be seen as offering wider economic and political benefits. For example, the UK might believe that NATO membership offers indirect spin-offs in the form of US investment and export earnings (e.g. US tourists). A US withdrawal might increase uncertainty, with possible adverse effects on the willingness of US firms and shareholders to invest in the UK and for its citizens

Table 8.1. United Kingdom defence expenditures and manpower, 1987
Expenditure figures are in £ m.

Programme	Expenditure	Manpower (thou.) Services	Civilians
Nuclear strategic forces	882	2.0	4.1
Navy general purpose forces	2 491	41.2	11.5
European theatre ground forces	3 035	101.3	25.5
Other Army combat forces	177	15.2	6.4
Air Force general purpose forces	3 416	61.2	9.2
Reserves	385	3.0	3.0
Research and development	2 337	1.0	23.5
Training	1 277	59.7	15.5
Equipment support	881	8.0	38.5
Stocks	588	..	..
Other support	3 373	35.0	43.1
Total	**18 782**	**327.6**	**180.3**
UK jobs due to MoD procurement (direct and indirect)			360
UK jobs due to export of defence equipment			110

Source: British Ministry of Defence, *Statement on the Defence Estimates* (Her Majesty's Stationery Office: London) for 1987 (Cmd 101) and 1988 (Cmd 344).

to visit the country. However, given the historical links between the UK and the USA, any adverse effects are likely to be temporary and short-term. In the long-run, a stable British political environment offering profitable opportunities will continue to attract foreign investment. British firms continue to invest abroad without requiring overseas military bases. Nevertheless, it is recognized that this is an area where it is extremely difficult to make reliable predictions.

US forces in the UK

Viewed as an input into British defence, the US forces based in Britain make a substantial contribution to British protection and security. In 1986 there were some 32 000 US service personnel and about 38 500 dependants based in the UK.[6] The US forces are equivalent to almost 10 per cent of total British service personnel (tables 8.1 and 8.2). Most of the US contribution is in the form of air forces located at bases in East Anglia, Oxfordshire and Gloucestershire (table 8.2). Indeed, the US Air Force (USAF) in the UK has a combat force equivalent to over 50 per cent of the number of Royal Air Force (RAF) combat aircraft, so making a major contribution to the total air capability based in Britain. In additition, there are some 2500 to 4000 US Navy personnel providing communications and support facilities. The US Army presence is nominal and does not involve any combat troops. However, it also has to be recognized that US forces in the UK are for the protection of Europe and not solely the UK, a view which will affect the British response to US withdrawal.

Table 8.2. Major US military installations in the United Kingdom, 1987–88

Location	Military personnel[a]	Civilian personnel[a]	Type of installation
US Air Force			
Abingdon	52		Storage facility
Alconbury	4 567	1 034	Air base
Bentwaters/Woodbridge	4 500	970	Air base
Chicksands	1 331	476	Communications
Croughton	422	62	Communications
Fairford	1 200	450	Air base
Fylingdales			Communications
Greenham Common[b]	1 520	482[c]	GLCM site
High Wycombe	125	25	Communications
Lakenheath	4 732	2 167	Air base
Martlesham Heath			Communications
Mildenhall	2 850	916	Air base
Molesworth[b]	250+		GLCM site
Oakhanger[d]			Communications
Sculthorpe[e]	33	10	Air facility
St. Mawgan[d]	100		Communications
Upper Heyford	4 937	7 641	Air base
Wethersfield	400+		Air base
US Navy			
Holy Loch	51+	38	SSBN port facility
Brawdy	300		Communications
Edzell	767	170	Communications
London	961	363	Communications
Macrihanish	100		Weapons facility
Thurso	180	59	Communications
Other			
Menwith Hill			NSA Sigint facility
Burtonwood	44		Army depot
Welford	200		Ammunition supply
Total	**29 622**	**14 863**	

[a] Numbers of service personnel and civilians are broad estimates and vary according to deployment; it is not possible to obtain accurate data for each base. In the text, total US service personnel is estimated at almost 32 000 in 1987–88.
[b] To be disbanded in accordance with the terms of the INF Treaty.
[c] UK (MoD) personnel.
[d] Joint USAF/USN facility.
[e] Annexe to Lakenheath.

Sources: Duke, S., *US Defence Bases in the United Kingdom. A matter for Joint Decision?* (Macmillan: London, 1987); Duke, S., SIPRI, *United States Forces and Military Installations in Europe* (Oxford University Press: Oxford, 1989).

Table 8.3. Costs and benefits to the UK of the US force presence, 1987–88
Figures are in US$ m., at constant (1986) prices.

	Costs		Benefits
US forces offered land surplus to British requirements	(Market value unknown)	*Injections into the UK* US payments to central Government	327
US forces offered land for housing and recreation	(Market value unknown)	Other US payments plus private expenditures of US personnel	837
Land transferred to US forces since 1973 Cost Sharing Agreement	71	*Trident* Waiver on US R&D costs	48–60
UK subsidized share (86%) of rates on property	..		
UK manning of Rapier air defence of US bases	290		
Total	**361+**		**1 212–1 224**
		Employment	(thou.)
		Total number of civilian jobs dependent on US forces (direct and indirect)	29.6
		Minus US civilians	2.5
		Total UK civilians	**27.1**

There is a danger of double-counting—e.g., employment is the result of spending by US forces. Some of the figures should be regarded as broad orders of magnitude.

Sources: British Ministry of Defence, *Statement on the Defence Estimates* (Her Majesty's Stationery Office: London, 1987), Cmd 101, vol. 2; National Audit Office, *Ministry of Defence: Costs and Financial Control of British Forces in Germany*, report no. HCP236 (Her Majesty's Stationery Office: London, Jan. 1988), p. 31.

The economic contribution of US forces

US forces add foreign exchange earnings to the United Kingdom's balance of payments account, totalling about £800 million for 1987–88. Their spending benefits local economies, providing some 30 000 jobs both directly and indirectly (including new construction projects).[7] In fact, US service personnel and their associated civilian workers add an extra 60 000 people to the military-industrial complex in the UK: the result is a substantial interest group which will oppose public policies likely to adversely affect their incomes (table 8.3).

Against these benefits, there are costs. The UK provides host nation support for US forces. This involves free or subsidized support in forms such as surplus land for facilities, housing and recreation. In principle, such transfers should be priced at their market value (surplus defence land has alternative uses). Also, as part of its contribution to the Trident research and development (R&D) costs, it has been agreed that British forces will be manning the Rapier air defence of

US air bases in the UK. However, with Trident, there is a benefit to the UK reflected in the favourable terms for its payment towards US R&D costs (table 8.3). Finally, complete withdrawal means that the UK would receive a transfer of assets in the form of the value of the capital stock embodied in the US bases.

III. Assumptions

It is assumed that a US withdrawal takes place with an unchanged threat and with or without a specific commitment to return in a crisis. Following the withdrawal, the UK will respond by either increasing defence expenditure, leaving it unchanged or reducing it, depending on its future security needs. Even these apparently simple assumptions are fraught with difficulties. They conceal a wide variety of subsidiary assumptions and complex relationships involving the defence sector, the adjustment period and the operation of the British economy, all in a world of uncertainty about the future. Consider the following examples which are meant to be illustrative rather than comprehensive.

Assumptions about British defence spending

At one extreme, it might be assumed that the UK will increase its defence spending to replace the protection previously provided by US forces (particularly if there is no commitment to return in a crisis). To compensate exactly for the loss of US forces would require the UK to expand greatly its air force and to do so by occupying abandoned US bases. This would preserve current levels of protection for the UK and maintain current levels of spending in areas where US bases are located. However, it does not follow that the spending patterns of increased British forces in ex-US bases would be identical to those of US personnel; and different consumption patterns have different implications for 'leakages' of spending from base areas to other regions in the economy and overseas. Nor does it follow under this assumption that the UK would use its increased defence spending to adopt the same mix of manpower and equipment as the US forces. The UK will have different views and preferences about the appropriate form of protection, and the relative prices of manpower and equipment are likely to differ between Britain and the USA. If relative input prices differ, then the UK will tend to substitute relatively cheaper for more expensive defence inputs. On this basis, the UK might achieve the same level of protection by substituting, say, nuclear for conventional forces, or land and sea forces for air forces. Increased spending is also likely to mean more orders for British defence contractors, raising questions about the mobility of resources between the civil and military sectors and the implications for the international competitiveness of the British economy.

Another assumption might be that British defence spending remains unchanged at current and planned levels. The UK might respond by re-arranging its existing force mix either by adopting a more independent stance (like

Sweden) or by aligning its forces with a possible European defence force, the form of which would need to be specified. For example, would European nations contribute forces on the basis of specialization by comparative advantage and would there be an associated European common market in defence equipment? Of course, the impact on British security might not be so great if the USA withdrew with a commitment to return in a crisis. In this case, the UK would have to allocate some limited resources to the care and maintenance of US bases.

Finally, it might be assumed that the UK reduces its defence spending below current and planned levels. Questions arise about the extent of the reduction, with European allies often suggested as a bench-mark. For example, in 1986 British defence expenditure was 5.1 per cent of the gross domestic product (GDP), compared with 4.1 per cent in France and 3.1 per cent in the FRG. It does not follow that under this option, protection will be correspondingly reduced since the UK might use a smaller budget more efficiently by purchasing equipment from the lowest-cost suppliers in world markets, with adverse effects on the size of the British defence industrial base.[8]

Assumptions about the transition

If the US withdrawal and the British response are not instantaneous, some assumptions have to be made about both the time required for US troops to depart and the associated adjustment costs. For example, recent British experience with a *planned* 3 per cent NATO commitment and the *unexpected* Falklands conflict, suggests that an increase in defence spending of some £2 billion required about 3 years and a rise of £4 billion required about 6 years (1986–87 prices; see table 8.5 below). The adjustment costs for both the military sector and the economy will depend on the choice of force mix and its training inputs, as well as on assumptions about how well and quickly labour markets operate. For example, if increased British defence spending is used to expand the air force, then the Ministry of Defence (MoD) will have to train extra pilots at a cost of over £1.7 million per combat aircraft pilot.[9]

Assumptions about the economy

US forces in the UK resemble foreign tourists. Their injections of spending power into the British economy add to the levels of aggregate demand in the UK, and contribute foreign exchange to the invisible account of the balance of payments. US troop withdrawals will immediately result in lower levels of spending in the UK and the loss of foreign exchange earnings will initially have an adverse impact on the current account of the balance of payments, reducing the size of any surplus or increasing the size of a deficit. The subsequent effects will depend on whether a market or Keynesian view of the British economy is adopted.

In a completely free market, it is assumed that all prices always adjust to clear *all* markets quickly, if not instantly (i.e., markets for goods, services, labour, land, capital and foreign currency). However, some free market economists accept that actual markets are not characterized by instant adjustment and equilibrium, but instead they show continuous change: entrepreneurs will change their plans in response to price signals but they will always be uncertain as to whether the changes are permanent.[10]

Critics of free markets raise doubts about how well they actually work (market failure), questioning the time required to adjust to changes and their implications for the whole economy. Worries are expressed that a free market policy will lead to substantial and continuing unemployment, with further fears of a continuous downward spiral for the economy. Keynesians regard output, quantities and unemployment rather than prices and wages as the means by which the economy adjusts to change. With this approach, US troop withdrawals (lower injections), everything else remaining unchanged, would have an immediate adverse effect on domestic income, output, employment and on the balance of payments. Lower aggregate demand is also likely to dampen down inflationary pressure and the deterioration in the balance of payments might create downward pressure on the exchange-rate. Over time, much will depend on the adjustment and responsiveness of the British economy to the changed circumstances following US withdrawal and whether there will be compensatory spending policies.

If British defence spending is increased to compensate for the loss of income of US forces, then domestic output and employment might remain unchanged. However, there will still be balance-of-payments costs and exchange-rate implications of withdrawing US forces, as well as a need to re-allocate resources from civil to military uses. Critics of increased defence spending will point to defence R&D 'crowding-out' valuable civil research, with adverse effects on the British economy's long-run competitiveness.[11]

If British defence spending remains unchanged or falls following US withdrawal, then the closure of US bases could have adverse effects on local economies, with multiplier effects leading to direct and indirect job losses. The extent to which the spending of US troops 'leaked' into areas outside the vicinity of their bases will determine the geographical distribution of reduced spending following withdrawal. Whether the final result is increased local unemployment depends on job vacancies in each area, on the mobility of labour and the willingness to accept lower wages.

The detailed consideration of assumptions shows that apparently simple propositions about US troop withdrawals require further assumptions and a complex exercise in modelling the British economy. The counter-factual requires analysts and policy-makers to consider what would happen in the absence of US troops in the UK. Sensible and informed public choices on this issue cannot ignore the need for careful and independent analysis, critical evaluation and appraisal of the available evidence. Questions about the gains and losses of US withdrawal require evidence on issues such as the budget costs of compensatory British defence spending and its likely effects on the British economy. Which regions and which industries will be affected, and what are the likely effects on jobs, technology, growth, inflation and the balance of payments?

IV. Evidence

By how much would the UK have to increase its defence spending?

To answer this question, assume that the US withdraws without a commitment to return in a crisis (option A). Much then depends on whether US forces in the UK are viewed as contributing to the defence of the UK or to the protection of Europe. This issue can be resolved by using a two-stage approach. First, estimates will be made of the costs to the UK of replacing all US forces in Britain. Second, consideration will be given to how this burden might be shared between the UK and other European members of NATO. To estimate the costs to the UK of replacing US forces some simple 'back of the envelope' calculations provide broad approximations of the likely orders of magnitude. Two estimates are used based on numbers and forces:

1. The numbers option: assume that the UK replaces all withdrawn US service personnel. In 1987–88, yearly expenditure exceeded £57 000 per Regular soldier. At these unit costs, the replacement of almost 32 000 US personnel would cost the British defence budget an extra £1.833 billion per annum. The transitional costs and the costs of recruiting additional manpower in the 1990s, which will be an era of demographic change, are not taken into consideration. It also assumes that per capita expenditure on British service personnel will provide a level of protection equivalent to that previously provided by the US forces (e.g., British prices are used and not US prices; US forces might be more capital-intensive, operating more modern equipment; and annual defence budgets do not reflect the value of the capital stock).

2. The forces option: assume that the UK adopts a replacement force structure identical to that of the US forces. On this basis, the UK would need to create an extra 25 squadrons of strike-attack and offensive support aircraft (i.e., in excess of 300 combat aircraft). Allowing for both direct and indirect expenditures (e.g., for headquarters, training, stations, aircraft R&D, other support) gives an estimated annual cost of £115 million per combat aircraft squadron. To replace the USAF in the UK would cost some £2.875 billion per annum. In addition, the UK would need to replace some 4000 US Navy personnel. Assuming some £57 000 per person (see 1 above), gives an additional total cost of almost £230 million per annum. Thus, if the UK adopted an identical force structure, it would probably increase the defence budget by over £3 billion per annum (1987 prices).

These crude estimates suggest that to replace withdrawn US forces, the UK would probably have to raise its annual defence spending by between £1.8 billion and over £3 billion: an increase of 10–17 per cent on the 1987 defence budget (1987 prices). Of course, if the US forces are really contributing to European defence, then the UK might argue that this burden should be shared with other European members of NATO.

Various burden-sharing rules can be formulated equivalent to a poll tax or proportional versus progressive income taxes. For simplicity's sake, assume that the costs of replacing the US forces in the UK are shared equally between Britain, France, Italy and the FRG. On this basis, each nation would contribute between £0.45 billion and almost £0.8 billion per annum. For the UK, this contribution would be equivalent to an extra 2.4–4 per cent of its 1987 defence budget. Such a percentage increase is similar to that which the UK willingly accepted in the form of the NATO 3 per cent growth target between 1979 and 1986. However, this extra burden would only fund the replacement of US forces withdrawn from the UK; further costs would be required to replace US forces withdrawn from elsewhere in NATO.

If US bases are closed, which areas will lose?

Most US service personnel are in bases located in East Anglia and the South of England. These are areas where local unemployment rates are usually considerably lower than the national average (see table 8.4). On this basis, closure of USAF bases is less likely to create serious local unemployment problems. Much, however, will depend on the willingness and ability of labour (e.g., skills) to move to different jobs in neighbouring localities and possibly on their willingness to accept lower wages. Elsewhere, a few US bases accounting for over 1000 service personnel are located in relatively high unemployment areas of Scotland, Wales and Lancashire.

Base closures will also have multiplier effects. For British military installations, estimates suggest local income multipliers of 1.25–1.35 and civilian employment multipliers of 0.7–1.5 (i.e., the total civilian employment both directly and indirectly associated with a base is 0.7–1.52 times the number of service personnel in the local base).[12] As a result, a US withdrawal could mean the immediate loss of between 22 400 and almost 50 000 jobs (compare the earlier estimate of civilian jobs). Here, however, it has to be remembered that faced with adversity, firms and local communities can be remarkably resilient and prophets of doom and gloom have often been wrong. There are examples of local economies which have experienced and survived the closure of a major employer (as in the coal-mining and steel industries). Nonetheless, base closures are likely to create problems for some communities, necessitating adjustment policies to compensate the potential losers from US withdrawal.

Table 8.4. UK local (base area) and regional unemployment rates, December 1987
Figures are in percentages.

US base/ Local town	%	US base/ Local town	%
RAF Alconbury		*RAF Lakenheath*	
Cambridge	3.8	Kings Lynn	9.1
Huntingdon and St. Neots	5.8	Norwich	8.1
Peterborough	9.1	*Menwith Hill*	
RAF Bentwaters		Harrogate	5.7
as Alconbury		*RAF Mildenhall*	
Burtonwood		Bury St. Edmunds	4.3
Warrington	10.3	Norwich	8.1
Wigan-St.Helens	16.2	*RAF Molesworth*	
Liverpool	18.7	Corby	11.7
Edzell		Kettering	5.7
Dundee	14.5	Peterborough	9.1
Fairford		*RAF Upper Heyford*	
Gloucester	7.2	Banbury	7.6
Swindon	7.4	Oxford	4.4
RAF Greenham Common		*RAF Wethersfield*	
Newbury	4.0	Colchester	7.6
Reading	4.4	Ipswich	5.7
Holy Loch		*RAF Woodbridge*	
Glasgow	16.2	Ipswich	5.7
Region	%	Region	%
East Anglia	6.6	North West	12.3
South East	6.6	Scotland	13.1
South West	8.1	Great Britain	9.5
Yorkshire and Humberside	11.2	UK	9.7

Source: Department of Employment *Gazette*, vol. 96, no. 2 (Feb. 1988) (Her Majesty's Stationery Office: London, 1988).

Effects of US withdrawal on British defence industries, regions, the balance of payments and the exchange-rate

Changes in British defence spending will affect the size of the defence industrial base. Questions arise as to which industries and which regions are likely to benefit from increased defence spending and which will lose from any cuts. During the 1980s, the MoD was British industry's largest single customer, spending about 75 per cent of its equipment budget in the UK, a further 15 per cent on collaborative projects and 10 per cent on imports.[13] The major suppliers to the MoD are the aerospace, electronics, ordnance and shipbuilding industries, with the Ministry being a dominant customer for these industries. Britain's leading defence contractors include British Aerospace (including Royal Ordnance), GEC, Rolls Royce and Vickers Shipbuilding.

In the mid-1980s, the British defence industries were a major employer, forming an interest group likely to oppose reductions in military spending. Expenditure by the MoD on equipment accounted for some 360 000 direct and indirect jobs with another 155 000 direct and indirect jobs resulting from the Ministry's non-equipment outlays (e.g., food, fuel, clothing). Exports of defence equipment accounted for a further 110 000 jobs. Limited data are available on the geographical distribution of direct employment due to MoD equipment spending. The South East of England accounts for over 40 per cent of defence jobs associated with equipment contracts, with a further 30 per cent of the employment located in the North, North West and South West. Some towns such as Barrow-in-Furness, Bristol, Brough, Preston and Yeovil are heavily dependent on defence business.[14]

If British military spending were reduced following a US withdrawal, there would be an immediate concern over job losses in the defence industries. Estimates were made of the employment impact of re-allocating a given defence expenditure to other industries, with the re-allocation based on the existing distribution of National Health Service, total government and civil government expenditure. The results, which are tentative, indicate that for every job resulting from military final demand, some 1.5 jobs might be created by an equivalent amount of civil government expenditure. On the other hand, if US withdrawal leads to higher British military spending, including defence R&D, there might be crowding-out effects. Worries have been expressed by the Government that spending on defence R&D may crowd out valuable investment in the civil sector, and so impairing industry's ability to compete in the international market for civil high technology products.[15]

Consideration must also be given to the likely balance of payments effects of US withdrawal and the implications for the exchange-rate. A US withdrawal would mean the loss of foreign exchange earnings (£0.8 billion in 1987–88), with an immediate adverse impact on the current account of the United Kingdom's balance of payments. This change was incorporated into a model for forecasting exchange rates. The resulting estimates showed that the loss of foreign exchange earnings from US withdrawal appeared to have little noticeable effect on the United Kingdom's exchange-rate (e.g., a fall of 0.25 per cent or less). However, free market economists would regard the exchange-rate as an adjustment mechanism and not a policy objective!

There are other possible defence balance of payments effects from US withdrawal. Much depends on the assumptions made about continued trade between the UK and the USA, an area which is fraught with uncertainty. In 1978, the balance of trade in defence equipment was 4:1 in favour of the USA; by the mid-1980s the corresponding ratio was just under 2:1.[16] A US withdrawal without a commitment to return might persuade the UK to be more protectionist in its equipment procurement, particularly if there was a move towards a European defence industry and an associated common market in defence equipment. Of course, if protectionism leads the UK to buy defence equipment from higher cost suppliers, it will sacrifice the gains currently obtained from trade with the USA.

V. The economic impact of British military expenditure

The issues

US withdrawal could lead to higher, lower or unchanged British defence spending. This section evaluates and tests the argument that military expenditure is a burden on the British economy. It considers the relationship between military expenditure and the performance of the economy. Two main questions are examined: First, whether military expenditure has any impact on the major economic variables; and second, if so, whether such impact is unique, that is to say different in nature or size from the impact of other types of expenditure.

The argument that military expenditure is a burden on the economy has led to claims that defence spending adversely affects investment and growth. This hypothesis is tested using econometric techniques. The issue of a unique relationship is assessed by comparing the results for defence spending with similar equations embodying other types of civil expenditure, namely, health, education and private consumption.

The facts

Recent trends in defence spending and in the British economy are shown in table 8.5. In relation to arguments about the burdens of defence spending, the years 1978–85 provide a useful case study. During this period, British defence spending in real terms rose by almost 30 per cent. Within this rising total, there was a clear preference for allocating an increased share to equipment expenditure to the obvious benefit of British defence contractors: typically some 10 per cent of MoD equipment expenditure is spent overseas. While defence spending rose, there was no evidence of a general deterioration in all the economic indicators. Indeed, there were some periods of improvement in the GDP, productivity and the inflation rate, and there was a surplus on the current account of the balance of payments for much of the period. However, the unemployment rate rose substantially, suggesting that the economy was operating with spare capacity and within its production possibility frontier. Such casual empiricism, however, is not conclusive and is no substitute for detailed analysis, empirical testing and critical evaluation.

Previous studies

There is a substantial literature on the economic impact of defence spending. Critics claim that military spending imposes a major burden on the economy. It

Table 8.5. Defence and the British economy

	Defence indicators:			Defence spending		Manpower[c]		Defence balance of payments[d]	
Year	Expenditure[a] (£ b.)	D/Y[b] (%)		Equipment (%)	Personnel (%)	Service (thou.)	Civilians (thou.)	Net visibles (£ m.)	Net invisibles (£ m.)
1978–79	14.6	4.5		40.0	44.2	315	286	288	−751
1979–80	15.3	4.9		39.7	42.6	321	276	390	−879
1980–81	15.7	5.0		43.7	40.7	334	265	776	−902
1981–82	16.1	5.2		44.7	40.1	328	252	705	−1 068
1982–83	17.1	5.3		43.7	37.8	321	243	712	−1 348
1983–84	17.6	5.4		44.8	37.0	326	233	572	−1 106
1984–85	18.8	5.2		45.8	34.9	326	206	567	−1 293

	Economic indicators:						Current account of balance of payments[i]	
Year	GDP index[e]	British unemployment (%)	Inflation index[f]	Productivity index[g]	G/Y[h] (%)		Visible balance (£ b.)	Invisible balance (£ b.)
1979	102.8	4.8	84.8	102.2	43.3		−3.45	2.71
1980	100.0	6.1	100.0	100.0	45.1		1.36	1.74
1981	98.5	9.5	111.9	101.8	46.0		3.36	2.87
1982	100.3	11.0	121.5	105.8	46.4		2.33	1.70
1983	103.3	12.1	127.1	109.8	45.9		−0.84	3.99
1984	106.7	12.6	133.4	111.3	45.5		−4.39	5.27
1985	110.7	13.1	141.5	113.6	44.5		−2.07	5.02

[a] 1986–87 prices.
[b] Share of defence spending in GDP at market prices.
[c] Service personnel are for British Regular forces; civilians are MoD civilians.
[d] Defence balance of payments data in current prices, showing net balances. Visibles are for identifiable equipment only; invisibles do not include payments by US forces in UK to British firms or non-central government agencies, nor expenditures by US personnel.
[e] GDP at constant factor cost (1980 = 100).
[f] Retail price index (1980 = 100).
[g] Productivity based on output per employed person (1980 = 100).
[h] General government expenditure as share of GDP.
[i] In current prices.

Source: Authors' calculations.

is accepted that in an economy with spare capacity, the immediate effect of increased defence spending is likely to be greater output and higher employment. However, in the long run, it is claimed that there are adverse effects on the balance of payments, investment, technical progress and, ultimately, on economic growth.[17] Empirical work has been produced to support these claims. There are, for example, econometric studies showing an inverse relationship between shares of defence in GDP and the share of investment.[18]

The view that defence spending is a burden is not universally accepted. It is often pointed out that military expenditure contributes to jobs, exports, import-saving and technological spin-off benefits. One review of the arguments and the evidence found that for the USA, the economic impact of military spending is only marginally different from that of other forms of Federal spending—hence it concluded that 'the dispute over the economics of military spending is political, not economic'.[19]

A satisfactory evaluation and empirical testing of the opposing views about the burdens or otherwise of defence spending needs a fully and properly specified economic model which identifies the causal factors linking different levels and components of defence spending to the major macro-economic variables in the UK (e.g., growth, inflation, unemployment). All too often correlation is confused with causation, and the underlying economic determinants of investment and growth are conveniently ignored. Not much attention is given to the economic impact of alternative government and private expenditure; nor is allowance made for possible changes in the efficiency of markets and their effect on economic performance. In this context, the views of two authorities are worth considering:

1. An analysis of the theoretical relationship between military spending and growth concluded '. . . on *a priori* considerations one cannot give any answer to the question whether disarmament will have positive or negative consequences on economic growth'.[20]

2. 'If the quest for universal propositions about the defence-growth relationship is to continue, let it be purged of prejudice; in particular please could the indisputable fact that high levels of military spending have been associated with high, medium, low (and no) growth be given the critical scrutiny which it merits?'[21]

The empirical results

Consider the hypothesis that military expenditure diverts resources from investment and adversely affects economic growth. This hypothesis was tested using models in which investment shares and growth rates were explained by defence shares and a set of other economic variables which have been used in previous empirical studies. A negative or inverse relationship was predicted between defence spending on the one hand, and investment and growth on the other. The results should be regarded as illustrative and exploratory, showing some possible directions for further work. As with other estimating equations in this field, they can be criticized for their *ad hoc* nature, multi-collinearity, and their sensitivity to the countries chosen, the time-period studied and the equation specification. Table 8.6 presents a sample of empirical results for the UK from a large number of equations which were estimated.

The equations in table 8.6 explained over 80 per cent of the variations in investment shares. The results provided no support for the belief that defence spending adversely affects investment: the coefficients for the defence variable were not statistically significant. Elsewhere, the unemployment and per capita

Table 8.6. The economic impact of British defence expenditure

Dependent variable	1. I/Y[a]	2. I/Y	3. Growth	4. I/Y	5. I/Y	6. I/Y
Coefficient[b]						
Constant	7.2	7.7	−6.5	5.6	6.9	8.2
D/Y[c]	0.12 (0.29)	0.05 (0.18)	0.51 (0.37)	..	..	..
I/Y[a]	..	..	−0.52 (0.71)	..	..	..
Growth[d]	−0.05 (0.71)	..	..	−0.02 (0.23)	−0.05 (0.91)	−0.05 (0.69)
Prices (inflation)[e]	−0.001 (0.04)	..	−0.32[f] (4.12)	−0.05 (1.18)	−0.04 (1.55)	−0.002 (0.69)
Unemployment[g]	−0.60[f] (7.02)	−0.59[f] (7.89)	−0.80 (1.60)	−0.70[f] (7.65)	−0.54[f] (10.05)	−0.59[f] (8.69)
Y/P[h]	0.003[f] (4.35)	0.003[f] (4.72)	0.006 (1.63)	0.003[f] (4.97)	0.003[f] (5.88)	0.003[f] (4.93)
NHS/Y[i]	..	..	..	1.33 (1.66)	..	..
E/Y[j]	..	..	..	..	0.88[f] (3.64)	..
C/Y[k]	..	..	..	..	..	0.002 (0.02)
$\bar{R}^2$[l]	0.86	0.85	0.53	0.88	0.92	0.86
DW[m]	1.15	1.26	2.01	1.51	1.83	1.17

[a] Percentage share of investment (I) in GDP (Y).
[b] Figures in brackets are t-ratios (a measure of the reliability of the estimated coefficients: significance at the 1% level indicates a high degree of reliability).
[c] Percentage share of defence (D) in GDP (Y).
[d] Annual percentage change of rate.
[e] Annual percentage change.
[f] Significant at 1% level.
[g] National unemployment rate (percentage).
[h] Per capita income.
[i] Percentage share of Government spending on the National Health Service (NHS).
[j] Percentage share of Government spending on education (E).
[k] Percentage share of private consumption (C) in GDP.
[l] Measures how well an equation explains the dependent variable (goodness of fit), with a complete explanation shown by a value of 1.0 (100%); here adjusted for degrees of freedom.
[m] The Durbin-Watson (DW) statistic is a further indicator of the reliability of an estimated equation. It measures autocorrelation, such that if there is no autocorrelation, the DW statistic will be near to 2.0, and the further it is from 2.0 the more likely there is autocorrelation.

Sources: Central Statistical Office, *UK National Income and Expenditure, 1986* (Her Majesty's Stationery Office: London, 1986); authors' calculations.

income variables were statistically significant with plausible signs (equations 1 and 2). Nor was there any evidence of defence spending adversely affecting economic growth. However, the growth equations often showed a relatively large unexplained residual and a surprising lack of statistical significance for the investment variable (equation 3).

The empirical results for investment and growth were generally supported by other equations which were tested for the sub-periods 1965–75 and 1975–85, and for equations which included a time-trend. Interestingly, though, for the period 1975–85, the defence variable had a statistically significant but *positive* effect on investment shares.

Questions also arise as to whether the economic impact of defence spending is unique and different from that of other types of expenditure. This view was tested by incorporating various types of civil expenditure into the investment shares and growth equations. Table 8.6 shows that taking private consumption and Government spending on education and health, only education expenditure had a statistically significant and positive effect on investment shares (equations 4–6). Further tests showed that none of the various types of civil expenditure had any statistically significant effect on economic growth.

In total, the empirical results do not support the view that defence spending adversely affects investment and growth. Indeed, the economic impact of defence spending is similar to that of private consumption and government expenditure on the health service. This conclusion also casts doubts on simple, single causal explanations offered for the United Kingdom's relatively poor economic performance (e.g., it is all due to too much spending on defence). Admittedly, military expenditure uses resources which might be used elsewhere, but 'it is a mistake to consider that the military sector is responsible for such macro-economic developments as upswings in prices and unemployment. In particular, the worsening economic performance in the industrial economies during the past decade cannot properly be attributed to changes in military spending'.[22] US withdrawal, however, is likely to focus British and European attention on the efficiency of their existing defence spending. Indeed, given the United Kingdom's preferences in the late 1980s, it seems unlikely that withdrawal would cause Britain to increase its defence spending substantially.

VI. European co-operation

With equipment becoming increasingly expensive and with downward pressure on defence budgets, European nations cannot avoid questioning the efficiency of their defence industries and of the current arrangements for providing armed forces. US withdrawal from Europe will bring these issues into sharper focus. It might encourage greater European co-operation with opportunities for the creation of a European defence industry, more multilateral arms programmes, a common market in equipment and a European defence force. Significantly, substituting a European political union for the current alliance would mean that the unified system as a whole has an incentive to provide the amount of defence which is in its collective (group) interest, with the constituents of the union being required to contribute the amount required in their common interest.[23]

A European defence industry

References are often made to a European defence industrial base (DIB). This is a vague concept that needs to be defined more carefully and clearly. Questions arise about the most efficient structure for a European defence industry. What are the minimum components of a DIB which Europe regards as essential for military purposes? How much is the minimum likely to cost and is Europe willing to pay the price? Structural features as reflected in the number of firms, their size and entry conditions for various industry sectors will affect the conduct of firms (e.g., price and non-price behaviour) and, ultimately, their performance (e.g., profitability[24]). To share costly R&D and exploit the available economies of scale and learning in industries such as aerospace requires large firms and suggests a monopoly solution which, in a protected market, would lack competition from rival suppliers. However, the monopoly problem can be overcome by ensuring that markets are contestable (through the threat of rivalry from other European defence contractors and from overseas).

Economic principles suggest guide-lines for a European defence industry and market. Relevant principles include specialization by comparative advantage, scale economies, gains from free trade and the contestability of markets.[25] However, these principles ignore the real difficulties involved in moving from the existing situation to an ideal state. Inevitably, proposals to rationalize European defence industries will encounter massive opposition from those firms, workers and regions which are likely to lose from change. There are worries that vote-sensitive governments and budget-conscious bureaucracies will create a European defence industry and market which is a highly protected and inefficient cartel, with the worst features of the Common Agricultural Policy. Were this to happen, the Europeans would fail to exploit the available opportunities for obtaining defence equipment at lower cost. The conflict between economic efficiency and the losers from efficiency improvements emerged in the debates of the late 1980s about a stronger European defence industry.[26]

Towards a stronger Europe

An important contribution to these debates was the so-called Vredeling Report[27], prepared for NATO's Independent European Programme Group (IEPG), which in examining Europe's defence industries made a number of proposals for increasing their competitiveness:

1. A policy of competition exposing Europe's defence industries to normal market forces should be pursued, and obstacles to free trade removed, leading to the creation of a single European arms market. In 1986, the European Community members of NATO spent some 20 billion European Currency Units (ECUs) on defence equipment. The introduction of competition into previously protected monopoly defence industries can lead to cost savings of between 10 per cent and over 30 per cent,[28] and further savings are available from exploiting economies of scale and learning.

2. Governments should not distort the market directly or indirectly and should be more willing to award defence contracts to suppliers in other European nations.

3. There should be greater user of competing consortia which would be subject to fixed price contracts for all stages of the project, with no restrictions on profit levels.

The Vredeling proposals are, however, subject to a number of constraints resulting in a substantial departure from a competitive market solution, with adverse effects on equipment costs. Constraints take the form of restrictions on the range of suppliers invited to tender, the need to maintain and develop the industrial base and national technologies, and the need to support nations with less developed defence industries. Further constraints on competition arise from the desire to share work on the basis of equity rather than efficiency criteria, and the concern with employment and balance of payments objectives.[29]

The situation is a classic economic problem showing that there are no costless solutions. The current organization of European defence industries is characterized by 'unnecessary' duplication of costly R&D and by inefficient production which fails to exploit available scale economies. At the same time, to achieve a competitive European common market in defence equipment will involve losers, and constraints on competition are not costless. Clarification of the issues in this debate requires that the aims of policy be specified. If the aim is to protect Europe through the efficient provision of defence equipment, then a competitive market solution, including a willingness to purchase from outside Europe, is required. In contrast, however, collaboration is often suggested as the means of achieving a European defence industry.

The benefits and costs of European collaboration

Governments are fond of claiming major cost savings and benefits from international collaboration, but they have provided little empirical support for their claims. Indeed, it is likely that some of the alleged benefits might have been exaggerated by groups of politicians, bureaucrats, scientists, military staffs and industrialists with an interest in starting a project.[30] In other words, European arms co-operation might not result in all the benefits suggested by the model of the ideal case.

Collaboration has its disadvantages and costs. Bargaining between partner governments, their bureaucracies and armed forces together with lobbying from

Table 8.7. Immediate effects of a US withdrawal

		Employment in			
Option	UK defence budget	area of US bases	other areas	Balance of payments	Technology
Complete replacement by UK	↑	↔	↔	↓	↑↓ (?)
No replacement by UK	↔	↓	↓	↓	↔

interest groups of scientists, engineers and contractors can lead to inefficiencies. These are claimed to arise from work-sharing, duplication, compromises in operational requirements and substantial administrative costs. Nevertheless, even with inefficiencies collaboration can lead to cost savings for each partner compared with an identical national programme. Two equal partners on a joint aircraft project incurring a collaboration penalty of 30 per cent on R&D and 5 per cent on production will each save some 35 per cent on its R&D costs and 5 per cent on unit production costs. Other estimates have suggested that on a combat aircraft with substantial collaboration inefficiencies (33 per cent on R&D and 5 per cent on production), a doubling of output from 200 to 400 units due to collaboration leads to savings on unit costs of about 20 per cent.[31]

VII. Conclusion

Compared with the adjustments following the end of World War II and the oil price shocks of the 1970s, the economic effects of US withdrawal are likely to be relatively small, with some possible local problems. Furthermore, if the US withdrew completely without a commitment to return in a crisis, then it is possible to envisage a range of European responses and solutions. At one extreme, the European part of NATO could be retained, everything else remaining unchanged. At the other extreme, there could be a move towards some form of unified European defence force, a European defence industrial base and a European common market in defence equipment. Experience with collaborative weapons acquisition suggests that agreement on a unified European defence force and associated industrial arrangements will be difficult to say the least. There are major questions of national sovereignty, burden-sharing and role specialization.

Proposals for the withdrawal of US troops from the UK will also be opposed by those interest groups likely to lose from withdrawal. Critics will point to the implications for British defence policy and budgets, for the survival of NATO and for the United Kingdom's international prestige. US forces might oppose the loss of a foreign posting. Reference will also be made to the loss of jobs following the closure of US bases and to adverse effects on the United Kingdom's balance of payments. This chapter has presented a framework for analysing these issues and some of the evidence needed for assessing the implications of withdrawal. Ideally, informed public choices on the issue require that all the trade-offs be identified and quantified using an information framework of the type shown in table 8.7.

Table 8.7 is illustrative of the type of information which can be presented as a basis for more informed choices. Directions of change are shown for the immediate effects of US withdrawal, all other factors being equal or unchanged Ideally, the response of other European nations needs to be included. Further parameters can be added, such as expected categories of unemployed (identified by wage, skill, sex, age, or full- versus part-time positions) and local unemployment rates. Even if this approach does no more than raise a number of unanswered questions, it identifies what we know, do not know and need to know about the defence, budget and economic effects of a US withdrawal.

Notes and references

[1] Olson, M. and Zeckhauser, R., 'An economic theory of alliances', *Review of Economics and Statistics*, vol. 48, no. 3 (Aug. 1966), p. 268.

[2] Olson and Zeckhauser (note 1), p. 278

[3] Olson and Zeckhauser (note 1), p. 279

[4] Sandler, T. and Forbes, J., 'Burden sharing, strategy and the design of NATO', *Economic Inquiry*, vol. 18, no. 3 (July 1980), pp. 425–44.

[5] 1980 prices. See Sandler, T. and Murdoch, J., 'Defense burdens for northern European allies', ed. D. Denoon, *Constraints on Strategy* (Pergamon/Brasseys: London, 1986), p. 102.

[6] In 1983 there were 27 000 service personnel and 32 500 dependants. The 1986 estimate of dependants is based on the 1983 ratio of dependants to service personnel. See British Ministry of Defence, *Statement on the Defence Estimates* (Her Majesty's Stationery Office: London, 1983), Cmd 8951, vol. 1, p. 19.

[7] Based on the 1983 ratio of jobs to US service personnel.

[8] Hartley, K., Hussain, F. and Smith, R., 'The UK defence industrial base', *The Political Quarterly*, vol. 58, no. 1 (1987), pp. 62–72; Hartley, K. and Hooper, N., 'Defence procurement and the defence industrial base', *Public Money*, vol. 7, no. 2 (Sep. 1987), pp. 21–26.

[9] 1978–79 prices. See *RAF Pilot Training*, First report of the Defence Committee, 1980–81 Session, House of Commons Paper 53-649 (Her Majesty's Stationery Office: London, Feb. 1981), p. xi.

[10] Hayek, F. A., *Knowledge, Evolution and Society* (Adam Smith Institute: London, 1983), p. 22.

[11] British Ministry of Defence, *Statement on the Defence Estimates* (Her Majesty's Stationery Office: London, 1987), Cmd 101, vol. 1, p. 48.

[12] Short, J., Stone, T. and Greenwood, D., 'Military installations and local economies: a case study: the Clyde submarine base', University of Aberdeen Centre for Defence Studies *ASIDES*, no. 5 (Aug. 1974), pp. 47 and 55. The local income multipliers for UK military bases are similar to some local tourism income multipliers, e.g. 1.34; see Archer, B., 'Tourism multipliers: the state of the art', *Bangor Occasional Papers in Economics*, no. 11 (University of Wales: Bangor, 1977). They are all the same high in relation to similar multipliers for the UK economy which have been estimated at 1.1–1.35; see Kennedy, M. C., 'The economy as a whole', ed. M. J. Artis, *The UK Economy*, 11th ed. (Weidenfeld and Nicolson: London, 1986). US base income multipliers may also differ if US service personnel send remittances to the

USA, which will increase the leakages from the UK economy, so reducing the size of the multiplier.

[13] British Ministry of Defence, *Statement on the Defence Estimates* (Her Majesty's Stationery Office: London, 1988), Cmd 344, vol. 1, p. 37.

[14] British Ministry of Defence (note 13), vol. 2, p. 62.

[15] British Ministry of Defence (note 11), vol. 1, p. 48.

[16] British Ministry of Defence (note 11), vol. 1, p. 49.

[17] Chalmers, M., *Paying For Defence* (Pluto Press: London, 1985), ch. 6; Smith, D. and Smith, R., *The Economics of Militarism* (Pluto Press: London, 1983), ch. 4.

[18] Smith and Smith (note 17).

[19] Adams, G. and Gold, D., 'The economics of military spending: is the military dollar really different?', eds C. Schmidt and F. Blackaby, *Peace, Defence and Economic Analysis* (Macmillan: London, 1987), p. 268.

[20] de Haan, H., 'Military expenditures and economic growth: some theoretical remarks', ed. C. Schmidt, *The Economics of Military Expenditure* (Macmillan: London, 1987), pp. 95–96.

[21] Greenwood, D., 'Note on the impact of military expenditure on economic growth and performance', ed. C. Schmidt (note 20), p. 102.

[22] Blackaby, F., Introduction: 'The military sector and the economy', eds N. Ball and M. Leitenburg, *The Structure of the Defence Industry* (Croom Helm: London, 1983), p. 20.

[23] Olson and Zeckhauser (note 1), p. 279.

[24] Hartley, K., 'The European defence market and industry', eds P. Creasey and S. May, *The European Armaments Market and Procurement Co-operation* (Macmillan: London, 1988).

[25] Hartley, K., 'Public procurement and competitiveness: a Community Market for military hardware and technology?', *Journal of Common Market Studies*, vol. 25, no. 3 (Mar. 1987), pp. 237–47.

[26] Vredeling, H., *Towards a Stronger Europe*, Independent European Programme Group (North Atlantic Treaty Organization: Brussels, 1986), vols 1 and 2.

[27] Vredeling (note 26), pp. 3–7.

[28] *The Institutional Consequences of the Costs of Non-Europe*, Committee on Institutional Affairs, report no. PE 118.040 (European Parliament: Brussels, Feb. 1988), p. 12.

[29] Vredeling (note 26), pp. 4–5.

[30] Hartley, K., 'Defence, industry and technology: problems and possibilities for European Co-operation', ed. G. Hall, *European Industrial Policy* (Croom Helm: London, 1986).

[31] Edwards, Sir G., *Partnership in Major Technological Projects,* seventh Maurice Lubbock Memorial Lecture, 14 May 1970 (Oxford University Press: Oxford, 1970), p. 24.

9. The economic impact of a US military withdrawal from NATO's Northern Flank

Simon Duke
Pennsylvania State University, University Park, Pennsylvania

I. Introduction

Denmark, Norway and Iceland play an important role in NATO's security co-operation as hosts to a number of important early-warning, anti-submarine warfare (ASW) and other facilities and units guarding the alliance's Northern Flank. Within the context of this study they constitute a special case because of the various bi- and multilateral arrangements permitting the extensive reinforcement of the Northern Flank in times of crisis or emergency. The general theme of this book is to examine the effects of a withdrawal of all US forces from Europe to see how the European NATO countries might cope. Of the two withdrawal scenarios described in the introductory chapter, the option B, allowing for the pre-positioning of supplies and the return of forces from the United States to Europe in time of tension, is of particular interest when considering the case of the Northern Flank countries, as such arrangements already form the basis for the current NATO roles of Norway and Denmark. Close scrutiny of these countries is warranted as they provide models for possible future arrangements in other NATO countries on, for instance, the Southern Flank, where US basing is a highly contentious issue and where pre-positioning of supplies coupled with a diminished peacetime presence may well be less politically divisive.

In Norway's case, there are some particular problems associated with distinguishing between economic and political or technological factors. Norway is for political reasons very sensitive towards issues of US financing of defence installations, and this is reflected in the paucity of sources. Another general problem is contained in the very concept of 'withdrawal' as laid out in this book. What does it include? Again, to use Norway as an example, the US Department of Defense (DOD) spends a substantial portion of its overseas research budget in Norwegian universities and advanced research institutes. Should it be assumed that, as part of this withdrawal, part or all of this funding would cease? If so, the economic impact could be appreciable. The cost or benefit of basing is often assessed not only in economic but also in socio-economic and political terms. Thus, it could be that a given base or group of bases may offer benefits to all parties concerned, while the actual cost of basing may be too high for one nation, in particular for a small would-be host. This chapter deals with some specific instances—Greenland offering the prime example—in which a foreign military presence has a large, and largely positive, economic impact, while having little or nothing to do with the defence of the host country itself.

Cost-sharing within NATO of various projects after a US withdrawal is a further problem. The construction of hardened air shelters (HAS) in Denmark as part of the US Air Force Europe (USAFE) programme to upgrade survivability may serve as an example. During time of NATO alert, USAFE would use five air bases in central Denmark prepositioned with supplies and facilities for logistics support. Growing concern for the potential vulnerability of aircraft that would be based at these and similar bases prompted the initiation of the HAS programme. The programme for the construction of 88 HAS was paid for by the Danish Government, which in turn recouped that money from the NATO Infrastructure Fund.

Another major problem in the Northern Flank concerns the US military presence on Greenland. The US facilities in Greenland are not for the defence of Europe but rather for the defence of the North American continent. These facilities could therefore be seen as falling outside the scope of this study. However, since the issue at hand is the hypothetical withdrawal of all US forces from Europe, regardless of whose interests they serve, and since Greenland–US relations are maintained through Copenhagen, Greenland is included in this chapter. It is difficult to define adequately the economic costs or benefits accruing from the US presence in Greenland because of the problems of separating the economic issues from political, technical or even sociological ones. An under-emphasized aspect in this regard is that the military partnership with the USA in Greenland has effectively served Denmark's economic interests. Many of the functions carried out by US personnel at Greenland's second main base at Søndre Strømfjord, such as air traffic control, airport operations and helicopter ambulance and search and rescue services, have direct civil applications that are essential to Denmark's efforts to develop the island. These functions, while primarily developed to serve military operations, are crucial to the all-important air traffic between Denmark and Greenland and between the isolated Greenlandic communities. The cost of assuming this role would be considerable and would strain Denmark's already weak economy.

Another factor that merits attention is the geographic location of the Northern Flank countries. None of the countries have a sufficiently developed economic infrastructure to support large-scale basing operations on their own, in part because of their small populations, but also because of the high cost of constructing facilities in relatively remote locations. The adverse climate has also tended to discourage large-scale military land operations in the northern part of the region, and more emphasis has therefore been put on the support and early-warning roles for other parts of NATO and of the continental United States. Two hypotheses arise out of these observations: (a) because of the small populations of Iceland and Greenland, and because a relatively high degree of dependence on services provided by the US forces has been built up over the years (USAF responsibility for air traffic control in Greenland serves as an example), the economic impact of basing in this region is bound to be higher than in other locations (although early US non-fraternization policies and political opposition to US bases have diminished the potential impact); and (b) the running costs of basing in this region are relatively high, as the local economies can provide little in terms of services and infrastructure, and as most supplies have to be flown in.

As regards the economic impact of the US military presence, the Nordic NATO countries can be separated into two distinct groups. The first, in which the impact is low, includes Denmark and Norway. The prime consideration here is that there are no substantial forces deployed to these countries in peace-time, although there has been considerable expenditure on wartime deployment bases for air and ground forces. These are one-time costs, however, mainly involving the construction of support facilities that require very little in the way of maintenance and manning. The second group includes Greenland, Iceland and the Faeroe Islands. All three are characterized by small populations and the absence of indigenous armed forces. The deployment of US forces to Greenland and Iceland in particular has had a significant impact on the local economies, although much of this impact has been associated with major capital projects such as the construction of a new military airport and a new civilian air terminal at Keflavík. This impact will become more apparent as automation continues and systems such as the Distant Early Warning line of air defence radars (the DEW line) in Greenland are phased out. Moreover, many of the functions carried out by US forces in Iceland and Greenland are highly technical and require highly trained personnel that, security considerations aside, cannot be recruited locally.

II. The base agreements

A helpful starting point for an evaluation of the economic costs entailed by the US military presence are the base agreements, which give an idea of the financial burdens encountered by the United States and the respective host nations. The various agreements signed by the United States and the various Nordic countries do not specify fixed amounts and, unlike arrangements made in the Southern Flank, there is no explicit trade-off linking base access to economic assistance. Beyond the letter of the agreements there are *ad hoc* arrangements that apply to specific projects. For instance, the agreement covering the NATO infrastructure funding of construction costs for the HAS at various Danish airfields was made more palatable to Denmark, economically as well as politically, by linking it to the sale of a certain amount of Danish bacon to the USA.

The base agreements, as they apply to the economic impact of the US presence, are discussed below on a country-by-country basis. Greenland and Denmark are discussed together, as all base agreements concerning Greenland were signed with Denmark, which also retains control over all matters pertaining to defence and foreign affairs after the establishment of Greenlandic Home Rule in 1979.

Denmark and Greenland

The basic documents covering Danish–US defence co-operation are the North Atlantic Treaty and a bilateral treaty on the defence of Greenland signed in 1951. Under the 1951 agreement the two countries agreed to 'establish and operate jointly such defense areas' as may be deemed necessary to the defence of 'Greenland and the North Atlantic Treaty area'. The treaty was in fact drafted a full 10 years after the US military arrived on the island, and it replaced the wartime agreement which had governed the presence until then. The 1951 agreement also meant that the NATO Status of Forces Agreement could be applied to US servicemen stationed in Greenland in the same way that it applied to US servicemen in other NATO countries.[1]

The agreement between Denmark and the United States on the use of facilities in Greenland was signed in Copenhagen on 27 April 1951 and will remain in force for the duration of the North Atlantic Treaty. Under the agreement the United States and Denmark pledge to 'take such measures as are necessary or appropriate to carry out expeditiously their respective and joint responsibilities in Greenland, in accordance with NATO plans'.[2] The United States is permitted, as a party to the North Atlantic Treaty, 'to assist the Government of the Kingdom of Denmark by establishing and/or operating' 'defense areas' that are necessary for the 'development of the defense of Greenland and the rest of the North Atlantic Treaty area, and which the Government of the Kingdom of Denmark is unable to establish and operate single-handed'.[3]

The actual arrangements for the defence of Greenland are vague. Article II specifies, for instance, that the 'responsibility for the operation and maintenance of the defense areas shall be determined from time to time by agreement between the two Governments'.[4] Where the operation and defence of a 'defense area' is the responsibility of the US Government, the Danish Commander-in-Chief of Greenland has the right to attach Danish military personnel to the staff of the US commanding officer. The former shall be consulted 'on all important local matters affecting Danish interests'.[5] The US Government is further empowered, within the confines of the 'defense area' to 'improve' and 'fit the area for military use', as well as to 'construct, install, maintain and operate facilities and equipment, including meteorological and communications facilities and equipment, and to store supplies'. It may also 'station and house personnel' and provide the requisite facilities to cater for them as well as to provide for the protection and internal security of that area. The US forces may 'improve and deepen harbours, channels, entrances and anchorages'.[6]

Similar regulations apply to those defence areas under Danish control. The US Government may attach US military personnel to the staff of the Danish

Table 9.1. US and Danish areas of financial responsibility for bilateral and joint NATO defence arrangements on Danish territory

Acronyms are explained on pages xx–xxiv.

US areas of responsibility	Danish areas of responsibility
Partial payment of COBs (construction and wartime running costs) through NATO Infrastructure Fund	Partial payment of COBs (construction and peacetime running costs) through NATO Infrastructure Fund
Wages to Danish and Greenlandic personnel working at Thule and Søndre Strømfjord and of US military personnel at various locations	O&M of UKADGE early-warning radar (DEW line) at Thorshavn (Faeroes)
O&M[a] of civil air traffic control at Søndre Strømfjord	Inspection/maintenance of War Reserve Material
Construction/modifications at Thule and Søndre Strømfjord	O&M of NARS facilities in Greenland and the Faeroes
Proportion of cost of Allied exercises	
O&M for Greenland DEW Line segment	
Transport to/from Greenland and the USA	
O&M of Angissoq (Greenland) and Ejde (Faeroes) Loran-C facilities	
Wages to Danish personnel at Duedodde operating Sigint facility (Bornholm)	
Construction costs for Sigint installations	

[a] Operation and maintenance.

commanding officer. The United States may also use Danish 'defense areas' in co-operation with the Danish Government.[7] A reciprocal right exists for the Danish use of US 'defense areas'.[8] The US forces enjoy rights of 'free access to and movement between' the defence areas throughout Greenland, including the territorial waters of Greenland.[9] The United States also agrees to respect all laws, customs and regulations of the local population and pledges that 'every effort will be made to avoid any contact between US personnel and the local population which the Danish authorities do not consider desirable for the conduct of operations under this agreement'.[10]

In addition to the agreements between the Danish and US governments regarding bases on Greenland, there are a number of agreements between the respective governments on the use of Danish facilities in Denmark in time of NATO alert. Denmark has a policy of not allowing the peacetime stationing of foreign military forces on Danish soil except in Greenland[11] and a parallel policy of allowing no nuclear weapons on Danish territory.[12] Following the crash near Thule in 1968 of a B-52 bomber carrying four nuclear weapons, all of which were destroyed in the ensuing fire, it was explicitly stated that the latter policy also applies to Greenland.

Successive Danish governments have recognized that outside reinforcement forms a significant part of Danish security policy. An agreement was signed in

1976 whereby the Danish Government agreed to make available a selected number of air bases as part of the Collocated Operating Base (COB) programme. In the same year an Ordinance Governing the Admission of Foreign Warships and Military Aircraft to Danish Territory in time of peace was signed at Amalienborg Palace on 27 February. The latter agreement contains no significant economic terms since it is mainly an agreement on access to and passage through Danish territorial waters and airspace. This agreement was superseded by a resolution, passed on 14 April 1988 in the Danish Parliament, which required the Government of Denmark to inform visiting naval vessels of Denmark's policy not to accept nuclear weapons on Danish territory. The resolution prompted a crisis within NATO and a souring of Danish–US relations as it posed a direct challenge to the long-standing US policy of neither confirming nor denying the presence of nuclear weapons aboard its ships.

In 1982 the Danish Government accepted SACEUR's Rapid Reinforcement Plan, which includes US air reinforcement as well as British air and ground reinforcement. To implement the plan a number of bilateral agreements were reached with the allied countries participating in the reinforcement of the Nordic region, namely the United States, Britain, the Netherlands and Canada. These agreements cover the prepositioning of logistics support material (fuel, munitions and other support equipment) but no heavy material. The future of the reinforcement plan is currently in doubt since the British announced in 1987 that they would reduce or abolish their reinforcement brigade or possibly redirect it to the Central Front.

Since the main costs incurred by basing in Denmark are those of air travel and associated facilities, a specific agreement was negotiated between the US and Danish governments on the use of aeronautical facilities and services. This agreement, signed at Copenhagen on 7 July 1960,[13] recognized that Denmark could not single-handedly, 'within the resources available to it', afford to implement the International Civil Aviation Organization (ICAO) North Atlantic Regional Plan. The USA agreed, 'subject to the continued availability of funds and authority, to provide certain air navigation, communication, and related services for the benefit of international civil aviation in connection with its operation of its facilities and services' pursuant to the 27 April 1951 agreement on the operation of 'defense areas' in Greenland.[14] In Article I of the agreement Denmark delegates to the US Government, again subject to the availability of funds, 'the responsibility for the establishment, maintenance, and operation of aeronautical facilities and services in Greenland which are required for the air traffic services and the protection of air movements over Greenland necessary to carry out programs of the ICAO'. The US Government is not obliged, however, to provide facilities and services beyond those which it is in a position to furnish as a part of its obligations under the 1951 agreement. The Government of Denmark reserves the right to relieve at any time, in whole or part, the US Government from the responsibility delegated in the 1960 agreement under Article I.[15] In practice there is little reason for the Danish Government to do so as the savings afforded by this 'delegation' are considerable for the country. Thus, Søndre Strømfjord, although an important Strategic Air Command (SAC) base and centre for logistic support for the DEW line and Thule, is also designated as an international aerodrome.[16]

Iceland

An exchange of notes, comprising the agreement of 6 December 1956, forms the central document regulating the US use of bases in Iceland. The 1956 agreement[17] confirms an earlier Defense Agreement of 5 May 1951,[18] confirming that both sides had held discussions concerning the revision of the earlier agreement and the withdrawal of forces, and stating that world circumstances and 'the continuing threat to the security of Iceland and the North Atlantic community call for the presence of defense forces in Iceland . . .'.[19] An Iceland Defence Group consisting of not more than three senior representative of each government was established to review the defence needs of Iceland and the North Atlantic area.

The economic aspects of basing are first mentioned in the 1951 agreement which, in the Preamble, states:

Having regard to the fact that the people of Iceland cannot themselves adequately secure their own defenses, and whereas experience has shown that a country's lack of defenses greatly endangers its security and that of its peaceful neighbours, the North Atlantic Treaty Organization has requested . . . that the United States and Iceland . . . make arrangements for the use of facilities in Iceland in defense of Iceland and thus also the North Atlantic Treaty area.[20]

Under this agreement Iceland is bound to 'make all acquisitions of land and other arrangements required to permit entry upon and use of facilities in accordance with this Agreement, and the United States shall not be obliged to compensate Iceland or any national of Iceland or other person for such entry or use'.[21] The agreement also seems to infer that Iceland also pays for civil aviation operations at Keflavík Airport which, like Greenland's Søndre Strømfjord, is a joint civilian and military aerodrome.[22] The absence of any specific mention of US obligations implies that the United States is financially responsible for all other aspects of basing, such as construction (much of which was already done by British and later US forces during World War II).

The economic aspects of the US military presence were given further attention in 1974 in a Memorandum of Understanding (MOU) and an Exchanges of Notes,[23] which served as a continuation of the 1951 Agreement. The MOU and Exchange of Notes, signed in Reykjavík, entered into force on 22 October 1974. The Exchange of Notes echoes the 1956 exchange in stating that both governments agree that 'the present situation in world affairs as well as the security of Iceland and of the North Atlantic Community, call for the continuation of the facilities and their utilization by the Iceland Defense Force under the Agreement on mutually acceptable terms'.[24] The MOU states that the United

Table 9.2. US and Icelandic areas of finacial responsibility for defence arrangements in Iceland

US areas of responsibility	Icelandic areas of responsibility
Construction/modifications of military facilities at Keflavík	Acquisition of all land for US facilities
Wages to US and Icelandic personnel at Keflavík and other facilities	Share of costs of new civilian terminal for Keflavík
Equipment to upgrade civilian air facilities at Keflavík	Civilian aviation at Keflavík
Construction and supply of Helguvík and Stakksfjördhur fuel depots	
Transport to/from Iceland and the USA	

States 'will seek to reduce its military personnel in the Iceland Defense Force as agreed upon between the two Governments' (see below).[25] It was also agreed that 'within a reasonable time Icelandic personnel with suitable qualifications and training will replace US personnel in certain positions with the Iceland Defense Force'. At the same time, the MOU also states that 'the United States does not seek to employ or retain more United States or Icelandic personnel than circumstances warrant'.[26] The United States pledged to 'seek funds, where available, to construct family housing units on the agreed areas sufficient to accommodate eligible United States military personnel stationed in Iceland' and that pending completion of the housing, the United States should act as 'agent for personnel residing off the agreed areas'. More importantly, following constant Icelandic protest against the base as well as congressional concern about security at Keflavík, where approximately 20 civilian flights per day operate on the same parking apron as US military aircraft,[27] the United States agreed to take 'such mutually agreed measures as will effectively separate the civilian air terminal from base facilities in the Keflavík agreed area'.[28] The United States also agreed to 'seek to provide certain equipment to upgrade the Keflavík Airfield over a ten-year period to meet ICAO standards'.[29]

Added to the Exchange of Notes and the MOU was an agreed minute in which the United States pledges to reduce its military personnel strength by 420, 'to be replaced by qualified Icelandic personnel as such personnel became available. The Icelandic Defense Force will train Icelanders for these positions as required'. Keflavík was again mentioned, this time in connection with the proposed separation of military and civilian facilities. The United States agreed to 'cooperate in the construction of a new terminal complex subject to availability of funds and military requirements'. Discussion ensued about the financing of the taxiways and ramp, roads, including an access road, and the 'rehabilitation of the refuelling system'.[30] The United States also agreed to 'study the feasibility' of purchasing geothermal heat for use by the Icelandic Defense Force in the Reykjavík area.

Norway

Although Norwegians are very sensitive about the basing issue, and in spite of a 1949 ban on the *permanent* peacetime deployment of foreign combat forces on Norwegian soil (thus not excluding temporary deployment, training or infrastructure facilities), Norway would play an important role in the defence of NATO's Northern Flank in the event of a NATO alert, receiving sizeable numbers of reinforcement troops. It is therefore quite consistent with official Norwegian policy on basing that a secret agreement was concluded between the SAC and the Norwegian Government for the wartime use of two airfields at Sola and Gardermoen, where fuel and spare parts are to be stored in peacetime. A more recent agreement has been reached to store equipment for the US Marine Amphibious Brigade (MAB) in peacetime. As long as Norwegian officials can claim that there are no foreign bases on their soil and that all bases are under Norwegian control and manned by Norwegian personnel—even if the equipment they tend is not for their own use in wartime—this is seen as consistent with Norway's stance on foreign basing. The logic of the argument suggests that it is quite acceptable for foreign agencies to fund Norwegian military installations—as the US Central Intelligence Agency (CIA) and National Security Agency (NSA) do—as long as the installations are actually manned by Norwegians. Key facilities such as the VHF-SHF-UHF receivers at Viksjøfjellet and Vardø, which intercept the telemetry of Soviet missile launches from the Barents Sea, White Sea and Plesetsk, are manned by Norwegian technicians on behalf of the NSA.

The economic consequences of a withdrawal, given the nature of the deployments (or non-deployments in peacetime), would be slight compared to the consequences in other NATO countries where appreciable numbers of US forces are present in peacetime. Nevertheless, a considerable amount has been invested by the United States and NATO into facilities in Norway, although the precise amounts are often obscured by the sensitive nature of their functions.

The three agreements which regulate military co-operation between Norway and the United States give only a hint at the economic ties. The secret agreement referred to above, signed in October 1952, was entirely US-financed. This was in 1974 superseded by a COB agreement. The discussion below focuses on the latter and on the Invictus Arrangement[31] of 1980.

The COB agreement with Norway was secretly concluded two years before the agreement with Denmark. There are eight known COBs in Norway (out of around 70 reserved for USAFE use in Europe) which together could host up to 200 aircraft in time of emergency.[32] In a USAFE document that appeared at the same time as the US–Norwegian COB agreement[33] the financial obligations were laid out as follows (the operational terms are discussed in chapter 15):

1. USAFE sponsor units will plan for the operational, logistics and administrative support necessary for the augmentation squadrons to accomplish wartime flying operations from the COBs.

2. Peacetime functions of the sponsor unit include development of required joint national plans for the reception and support of the augmentation forces as well as inspection and maintenance of prepositioned war reserve material (WRM).[34]

3. Wartime functions of the sponsor unit include providing a reception team to assist the host base and the deploying unit's Initial Support Element in receiving and bedding down the US unit at the COB.

The Norwegian sponsor units at the various COB locations continue to provide or arrange logistics and administrative support for US forces at the COB sites even after the deployed units are operationally ready. COBs are specifically intended to augment aircraft on allied airfields for NATO wartime functions. Where possible, US aircraft would share existing allied air bases. The costs may be incidental if the facilities already exist. If construction is required, the costs would be covered by the NATO Infrastructure Fund.

The other instance of defence co-operation of relevance here is the Invictus Arrangement of 27 February 1980,[35] replacing an earlier Invictus Arrangement of 1971, which in turn updated the original Arrangement of 1960. The 1980 Arrangement is designed to provide pre-positioning of supplies at one or more airfields for aircraft from US carriers. In the event of an aircraft carrier being damaged or crippled, these bases could act as alternative repair/deployment platforms. The Invictus sites are still officially classified, but it is common knowledge in Norway that they are at Ørland and Værnes. As a result of Norway's basing policy, the cost of maintaining these sites falls on Norway. The relevant provisions of the Invictus Arrangement are contained in articles III and VI. Article III states: 'Norway will be responsible for physical security of the facilities, custody, and maintenance of prepositioned equipment, including the provision and supervision of guard and maintenance personnel. Norway will also ensure that qualified personnel will be available to operate and maintain the infrastructure facilities and prepositioned equipment both in peacetime and wartime'.[36]

Article VI, perhaps as a wary concession to the sensitivity of basing and nuclear issues in Norway, contains an unusual property clause that would apply in the event of the arrangement being nullified: 'Movable equipment and supplies belonging to the United States will *remain the property* of the United States, and for the duration of the agreement and for a reasonable time thereafter, *may be freely exported from Norway*, or *disposed of* in Norway under terms which may be in effect or agreed upon between the two governments.'[37]

Under the Invictus Arrangement, Norway is the host nation and the Chief of Defence, Norway (CHOD Norway) is responsible for the logistics assistance, operation and maintenance as well as the control and security of facilities.

One other agreement deserves brief mention. On 13 January 1981 the Norwegian Parliament consented to the signing of a framework agreement between Norway and the United States on the storage in central Norway of material for the MAB.[38] The idea behind the MAB site was a Norwegian Government

Table 9.3. US and Norwegian areas of finacial responsibility for bilateral and joint defence arrangements in Norway

Abbreviations are explained on pages xx–xxiv.

US areas of responsibility	Norwegian areas of responsibility
Proportion of construction costs for COBs and MAB, Invictus and NADGE sites	Proportion of construction costs for COBs and MAB, Invictus and NADGE sites
Common funding of ACE HIGH troposcatter communication facilities	Common funding of ACE HIGH troposcatter communication facilities
MAB administration and O&M costs	Training of personnel to run/maintain infrastructure facilities and pre-positioned equipment in peacetime and wartime
Cost of construction and occasional manning of certain C^3I sites	
O&M costs of all Sigint stations	Maintenance/inspection of prepositioned War Reserve Material at COBs
Construction and O&M costs of Loran-C and Omega stations	Peacetime security of infrastructure facilities and pre-positioned equipment
Construction and O&M costs of SOSUS chain	
Part-time operation of P-3 ASW aircraft	
Construction and O&M costs of seismic and navigational stations	
Transport to/from Norway and the USA	

proposal to Parliament. The proposal recognized that 'owing to limited resources, Norway is dependent on Allied support in the event of an emergency' and that 'others cannot be expected to come to Norway's assistance unless Norway herself makes arrangements for receiving reinforcements'.[39] Steven Miller discusses the details of this arrangement in chapter 15. Suffice to note here that the arrangement provides that the control and maintenance of equipment shall be the responsibility of Norway. Norway is also obliged to 'supply sufficient means of transportation for the loading and transport of the MAB's personnel and equipment if they are to be transported to other locations in Norway'. Host nation support for the MAB includes:

– approximately 150 tracked vehicles;
– two transport companies (90 trucks each);
– one ambulance company (35 ambulances);
– one fuel supply section (6 tank trucks);
– engineering and airfield equipment;
– ammunition, provisions and fuel from Norwegian stocks until US supply support is established, in case the MAB were to be moved to other threatened areas in Norway.[40]

The MAB is largely dependent on Norwegian support to function. The extensive construction of warehouses and other storage facilities for equipment, as

ECONOMIC IMPACT ON NATO'S NORTHERN FLANK   165

well as modifications to Ørland and Værnes airfield, is financed through the NATO Infrastructure Fund.

III. The economic impact of a US withdrawal

Section II contains a summary of the financial obligations as laid out in the various treaties between the United States and the host nations. Section III examines the economic impact of a hypothetical withdrawal of US forces from Europe, concentrating on the data available (which do not necessarily correspond to the summary tables in section II). The scarcity of accurate data is a difficulty, in particular in the cases of Denmark and Greenland, and it would seem that data have not been compiled in any systematic manner. While economic factors are undoubtedly significant, they are not necessarily paramount given the small scale of the economic exchange between the host countries and the US forces.

The assumption that all personnel, equipment and facilities have been withdrawn from Europe—including all command, control, communications and intelligence (C^3I) and other shore-based facilities—but that the North Atlantic Treaty remains intact creates additional problems. Since the assumption is that the withdrawal is not one made in anger, the question of what actually prompts it arises. Furthermore, given that it is an amicable split, how does it affect European threat perceptions? In addition to the withdrawal of US forces currently in Europe, are all US-financed facilities also shut down? Assuming that the US military forces in Europe in a sense protect US investments in Europe by creating a secure atmosphere for commerce, would North American business be less inclined to invest in European projects after a US forces withdrawal? The economic ramifications of a withdrawal, perhaps also coupled to a measure of disinvestment, are difficult to grasp and no prediction could be made as to the possible consequences. Similarly a complete withdrawal would have far-reaching consequences for the Nordic countries. The dependency of the Norwegian fishing fleet operating in the Norwegian Sea on the US Loran-C navigation system[41] is a case in point. If the relatively minor (in terms of size and personnel) Loran-C facilities in the Nordic area were shut down, what would be the economic impact on Norway's economy?

The following data are based on the most current figures available at the time of writing—which in most cases means 1986 figures—and for reasons of comparison the figures will be given in US dollars wherever possible. As in section I, the data are presented on a country-by-country basis.

Denmark and Greenland

Military installations in Denmark and Greenland can be divided into two categories: those that are wholly US-financed and those that are paid for through the NATO Infrastructure Fund. The first category includes Thule,

Søndre Strømfjord and a number of communications facilities. The loss of these facilities would mean the loss of approximately 1000 jobs for Denmark which, bearing in mind the small population and small budget, could have serious repercussions. The second category represents a potentially more serious impact for the Danish economy (and those of other NATO countries as well). If the United States were to withdraw the prepositioned stocks from the COBs, the NATO countries would be faced with the choice of either paying for the US share of the Infrastructure Fund (signifying an increase of 32 per cent for each country, assuming the cost is shared equally) or else closing a selected number of facilities. If it were to choose the former, Denmark would be faced with a cost increase of $16 million.[42] For a small country with a debt of approximately $37 billion in 1985, this would be a difficult burden to bear. The alternate course of making adjustments within other spheres of government expenditure would inevitably meet with strong political opposition.

Notwithstanding these difficulties, Denmark is capable of assuming the relative increase in its NATO Infrastructure Fund contributions. There would be few other costs to consider within Denmark, since the facilities used by the US forces are wartime deployment bases. The heavy expenditures incurred by peacetime basing are thus not present. The economic impact of a US withdrawal would not be as traumatic as in Iceland or Greece, since the Danish injunction against the stationing of foreign troops means that there would be no troop withdrawal that could have an effect on local economies.

Greenland poses different problems, to the United States as well as to Denmark. For the United States, a withdrawal from Greenland would constitute a much greater loss than a withdrawal from Denmark proper, both in economic and in military terms. The United States has spent significant amounts upgrading military facilities in Greenland in the mid-1980s (over $18 million in 1986 alone).[43] In particular, it has embarked upon a major investment programme to modernize (or, as the USSR argues, replace) the 20-year old Ballistic Missile Early Warning System (BMEWS) conventional radars at Thule and Fylingdales (UK) with solid-state phased-array radars. Modifications at the Thule site were completed by the end of fiscal year (FY) 1986. Since the effective coverage of all approaches to the United States from the USSR depends on the joint operation of all three radars in the BMEWS chain (the third is at Clear, Alaska), a US withdrawal from Greenland would mean an investment loss to the system as a whole, as well as a significant reduction in US ballistic missile early-warning preparedness. Since BMEWS is directly associated with the defence of North America, and not of Europe, it is difficult to see why European NATO countries would wish to shoulder the costs associated with the continued operation of the chain.

As mentioned above, the US military facilities in Greenland are financed entirely by the United States and are not eligible for NATO funding. There are therefore no reliable figures that indicate how much the European NATO countries would have to pay to maintain these facilities should they decide to do so following a US withdrawal. As a general point there would be considerable short-term costs if the United States' NATO allies did consider gradually taking over the functions of the US personnel, many of whom are highly skilled. To train non-US NATO personnel to carry out many of the C^{3}I activities would be extremely expensive and, once such skills are acquired, there is no guarantee that the private sector might not prove more attractive. Since the study posits a US *military* withdrawal, a case could be made for arguing that US civilians in highly technical jobs could remain, at least for the transitional period encompassing the training of non-US personnel to assume these functions. Many of the existing US personnel are highly skilled specialists, often with postgraduate or advanced qualifications. It could also be questioned whether the United States would want its NATO allies to assume some of these highly sensitive tasks, which properly fall under the aegis of the North American Aerospace Defense Command (NORAD) rather than of NATO.

A 1987 decision by the United States and Canada to modernize the Alaska–Canada segment of the DEW line by constructing a new North Warning System (NWS)[44] is of interest as it noticeably excludes Greenland. Faurby and Petersen, two noted Nordic security experts, have commented that the DEW line is 'obsolescent and costly to operate, while any replacement would obviously involve heavy investments', but they caution that 'the air-breathing threat against which the line was originally built seems to be reappearing in the 1980s with the forthcoming introduction of a new low-flying Soviet bomber (the Blackjack) and especially of air and submarine-launched cruise missiles which could significantly threaten sea lines of communication in the western part of the Atlantic'.[45]

A US withdrawal from Greenland would have serious economic effects on Greenland and Denmark in spheres other than the purely military. There would be an immediate loss of income tax revenue to the Greenlandic Government of around 10 per cent. More serious, the existing air transport system in Greenland would collapse, as a complete US withdrawal would mean a removal of practically the entire infrastructure. There would be immediate repercussions for Denmark, as the Danish Government would have to assume costly security functions as well as air traffic control and other civil aviation costs.

Since the 1950s there has been a reduction in the use of US personnel at the bases and an increase in the employment of Danes, who provide a significant contribution to the Greenlandic economy through the taxes they pay to the Greenland authorities. The revenue accruing from base areas was around 56 million Danish crowns in 1983 and has from 1980 represented between 8 and 10 per cent of the income tax revenue in Greenland.[46] Table 9.4 shows employment figures at the base areas for 1986.

The employment issue has another dimension as well. The Greenlandic trade union, SIK, wants more local workers employed at the US bases, although this would require a reinterpretation of the clause in the 1951 agreement requiring that the US bases have as little impact as possible upon traditional Greenlandic

Table 9.4. Personnel at US military installations in Greenland, 1986

	US	Danish	Total
Military			
US Air Force	572	..	572
US Coast Guard	19	..	19
Danish	..	3	3
Civilian	377	794	1 171
Total	**968**	**797**	**1 765**

Source: the SIPRI data base.

life. There are hardly any Greenlanders employed at Thule AB but an increasing number have successfully found work at Søndre Strømfjord and the associated civilian terminal. In a small-scale and highly overstretched economy such as Greenland's, the bases' potential as places of local employment is a factor that must rank very high in any consideration of the economic consequences of a US withdrawal.

Iceland

The US military presence in Iceland—through the Iceland Defense Force (IDF)—is well documented, particularly since the creation of the Defense Affairs Office (DAO) in April 1985. The DAO's main function is to ensure the smooth operation of the defence agreements between Iceland and the United States. Iceland is not a contributor to the NATO Infrastructure Fund[47] and the apportioning of costs for all projects is done on a bilateral basis. Most programmes currently under way are financed by the US or the NATO Infrastructure Programme. Although Iceland is not part of this programme, construction or modifications at NATO bases are eligible for Infrastructure funding. For example, the 13 HAS being constructed at Keflavík, according to Icelandic estimates at a cost of $2–2.5 million each (making a total of $26–32.5 million), are financed through the Infrastructure Programme.[48]

The most important projects currently under way are the construction of aprons and taxiways for the new civilian terminal at Keflavík, a new road to the terminal, the Ytri-Njardvík fuel dump and work on the new harbour at Helguvík. Preliminary work on radar stations, air shelters and other smaller projects is also being carried out.[49] In addition to current projects, additional IDF projects have been approved (mostly paid for by the United States) which amount to approximately $59 million, the final price depending on contract negotiations between the Navy and Icelandic Prime Contractors (IPC). The projects were scheduled to take place over the next two or three years but it is unclear how much, if any, has actually been paid for. IPC has also been awarded a large maintenance contract for roads, vehicle parkways, runways, waterways and some masts worth an estimated $11 million.[50] The total value of

Table 9.5. IDF personnel and dependants, 1984–87
Figures are as of 31 December of respective year.

	1984	1985	1986	1987
IDF employees	3 104	3 057	3 104	3 171
IDF dependants	2 144	2 045	2 054	2 030
thereof children	1 284	1 230	1 257	1 113

Source: Foreign Affairs Report 1988, A Report to the Althing by Steingrímur Hermansson, Minister of Foreign Affairs (Icelandic Ministry of Foreign Affairs: Reykjavík, 1988), p. 52.

IPC contracts was worth an estimated $25 million in 1985, $54 million in 1986, $56 million in 1987 and $70 million in 1988. The gradual increase in expenditure by United States and NATO countries is mainly due to the decision to separate the civil and military air facilities at Keflavík.[51] Expressed as a percentage of Iceland's GDP, the IPC contracts represent a small but significant portion (1.39 per cent in 1986[52]).

The operating expenses associated with the IDF are substantial. IDF payments to Icelandic parties in 1987 amounted to $153.3 million, of which wages paid to Icelandic IDF employees accounted for $38.8 million. The remaining $114 million were payments to Icelandic firms for wages, contracts and purchases of goods and services.[53] During the same year the IDF paid approximately $101 million to Icelandic firms and individuals for wages, contracts and the purchase of goods and services. The private expenditure of IDF personnel and their dependents on Icelandic agricultural produce was estimated to be $1.1 million, which represented an increase of roughly 15 per cent from 1985. The overall cost of structures currently in use by the IDF has been estimated by the Icelandic Foreign Ministry at $309 million. The estimated rebuilding cost of these structures would be approximately $1384 million at 1987 exchange rates.

At the the end of 1987 all but 63 of the IDF personnel lived on base accommodation. The IDF are all military except for 49 civilian employees, of whom 17 are teachers at base grade schools. The remainder are engaged in technical work, either as employees of DOD or as representatives of US companies doing contract work on various US installations. IDF personnel figures for 1984–87 are given in table 9.5

The figures suggest that there would be considerable long-term savings for the United States in the event of a complete withdrawal. The problem would then be if the remaining NATO allies would be willing to shoulder the US portion of various construction costs as well as man the installations. There would also be a considerable loss of investment. Table 9.6. gives an idea of the portion of costs borne by the United States and NATO respectively for construction at the Keflavík base.[54]

The withdrawal of IDF personnel would certainly find favour in Congress. Funding for projects in Iceland has long been unpopular and budget requests have consistently been cut, often quite drastically. For example, a budget request for a total of $21 780 million was submitted for FY 1986. The House

Table 9.6. NATO and US cost shares for constructions at Keflavik AB, 1984
Figures are in US$ m.

	NATO	USA	Total
Fuel storage	61.0	61.0	122
Fuel pier	25.0	25.0	50
Aircraft shelters[a]	30.0	–	30
Control centre			
Building	12.3	4.78	17.08
Equipment	–	39.8	39.8
Various construction[b]	–	32.5	32.5
Host nation support	–	2.77	2.77
Total[c]	127.3	165.3	293

[a] Total cost for construction at 13 sites.
[b] Excluding construction costs for civil air terminal and personnel housing
[c] Items may not add up to totals due to rounding.

Sources: *Military Construction Appropriations for 1985*, Hearings before a Subcommittee of the Committee on Appropriation, US House of Representatives, 98th Congress, 2nd Session (US Government Printing Office: Washington, DC, 1984), Part 2, pp. 578–99; *Hearings on H.R. 1816*, Committee on Armed Services, US House of Representatives, 98th Congress, 1st Session (US Government Printing Office: Washington, DC, 1984), pp. 292–98, 305; *Department of Defense Budget for Fiscal Year 1985*, Construction Programs (C-1) (US Department of Defense: Washington, DC, 1984).

Armed Services Committee recommended a total of $1 270 million, in effect cutting the request by $20 510 million.[55] Within Congress the prevailing feeling has been that NATO should pay for more of the construction programmes in Iceland, in particular as regards the modifications to the airfield at Keflavík.

Despite what is said above, the main impact for Iceland of a US withdrawal will, to an even greater degree than for other Nordic countries, be strategic rather than economic. Since Iceland has no indigenous defence forces, the country will be faced with the immediate problem of how to defend itself against potential aggressors, while the remaining NATO allies will have to deal with the problem of how to fund, in whole or in part, both such a defence force and the completion of the existing construction programmes. Iceland's role in the overall defence of the Northern Flank is vital, and recent developments in Soviet fighter and cruise missile technology have emphasized the significance of the IDF. Without the IDF the Greenland–Iceland–United Kingdom (GIUK) Gap would be wide open, a situation which would cost NATO a large, but intangible, amount to rectify.

Norway

Norway, like Denmark, has an injunction against the permanent basing of foreign troops on its soil in peacetime. Again, this means that there will be no discernible impact on the local economies as a result of a troop withdrawal. The main economic burden that would fall on Norway, and on its NATO allies, would be that of assuming a higher portion of the NATO Infrastructure Fund. Many of the initial construction costs for facilities in Norway were paid for totally by the United States, but a few more recent projects are paid for out of the Infrastructure Program. In the case of the COB programme, Norway has undertaken to assume a portion of the cost of building HAS and this cost is reimbursed to Norway from the Fund.

Norway's current contribution to the Infrastructure Program amounts to 3.14 per cent. Without US participation that figure would rise to 4.33 per cent, assuming the same relative division of costs among the other NATO countries. Norway currently devotes approximately 2 per cent of its defence budget to the Infrastructure Program; without the United States this would rise to over 2.5 per cent or, expressed in monetary terms, a rise in the defence budget of approximately $13 million. An unspecified amount of money could be raised to meet this increase in the short-term by the sale of facilities left behind by the US military withdrawal. Long-term problems would be serious if there is continued interest in keeping the facilities open. In common with other Nordic countries that host US facilities, Norway is a net receiver of NATO Infrastructure aid, paying only 3.14 per cent of expenses in 1988 but receiving over 8 per cent of the programme's resources. The initiation of construction of COBs and MAB facilities has meant a significant increase in Infrastructure funding and presumably an enhanced profit for Norwegian contractors. It should be noted, however, that while the United States does fund over 27 per cent of the programme, it has enjoyed the exclusive use of approximately 46 per cent of the facilities.[56]

Many of the facilities in Norway that have more sensitive missions do not fall under the Infrastructure programme and are wholly US financed. This applies specifically to the signals intelligence (Sigint) stations at Jessheim, Vetan and Vadsø and other facilities with specialized functions at Karasjok and Kongsberg. Since the tasks carried out at these intelligence installations are technically demanding, a small number of highly specialized personnel are required. In spite of official statements that all personnel at these formally Norwegian locations are Norwegian—more a reflection of political sensitivities than anything else—an undisclosed number of US 'civilian' specialists and advisers are employed at many of these installations.[57] There would be an economic impact if the 'invisible' US specialists left, since the cost of training highly skilled personnel could initially be high. Circumstances that point to the presence in Norway of an undisclosed number of US personnel has been presented in parliamentary debates. The RB-47 incident serves as an interesting case study. On 1 July 1960 a US reconnaissance flight originating from Brize Norton AB (UK) was shot down by Soviet fighter aircraft on the grounds that it was engaged in espionage activities and was violating Soviet airspace. After the RB-47 aircraft was shot down over the Barents Sea, the Norwegian Ministry of

Defence required that all US technicians at Bø be out of Norway within four days so that the Minister could, in the event of a question in Parliament, deny US–Norwegian collusion and state that there are no US personnel connected with any military facility in Norway.[58] In some cases US civilians openly work in Norway on projects that are not wholly military, but nevertheless have strong military connections. One example of this was the 1971 decision of the Norwegian Government to allow the United States to install and operate the Norwegian Seismic Array (NORSAR), data from which are transmitted to the Seismic Data Analysis Centre in Alexandria, Virginia.[59] Norway is also involved in the US Navy Sound Surveillance (SOSUS) programme, operating one of 23 SOSUS sites. According to the DOD, in 1988 there were 206 active duty military personnel in Norway (presumably mainly engaged in official guard duties and advisory roles).[60]

To put figures to the above is difficult, in part because of the paucity of accurate data, but also because of the secret nature of some of the US-financed facilities in Norway. In addition, the routing of funds from the United States is not always obvious since many other agencies, such as the NSA, are involved as well. Some preliminary estimates can be made on the basis of data from the last four to five years, which, while not offering a complete picture, do perhaps provide a feel for the amounts involved.

The construction of facilities for the MAB were paid for by the United States. This also includes payments for in all about 25 Norwegians engaged in the maintenance of equipment and the associated administration of the pre-positioned stock arrangements. The expenses for this small team are estimated to be about $6.2 million for 1988.[61] For the United States this amount would be saved in the event of a withdrawal. The other major project involving Norwegian personnel was the building of COB facilities at eight Norwegian locations. In FY 1988 the expenses for Norwegian maintenance crews, involving a total of approximately 30 men, was estimated to be approximately $8.4 million. The additional expenses for depot maintenance (snow clearance, heating, etc.), again borne by the USA, were approximately $15.6 million.

The Invictus Agreement has also involved extensive construction work at Værnes and has so far cost approximately $125 million. The opportunity cost to the United States of this arrangement is, it has been argued, the addition of several carriers to the Norwegian Sea fleet.[62] Even if Norway is only able to support aircraft from one aircraft carrier, this would represent a saving of roughly $2 billion.[63]

Two additional points need to be discussed at this stage concerning facilities that are common to many NATO countries. The first is the NATO Air Defence Ground Environment (NADGE) network of some 85 radar facilities linked with the ACE HIGH troposcatter communications network. NADGE is funded under the Infrastructure Program and although it is not new it is constantly being upgraded and modified. The withdrawal of US personnel would seriously damage the system, while assuming the costs for operation and maintenance would entail an economic burden for all other NATO countries—particularly if the initial training costs are taken into account. The total manning of NADGE in Europe is roughly 6500 operators and 2300 programmers and an unknown number of maintenance personnel. The number of personnel operating the 13 Norwegian NADGE facilities is estimated to be 1300.[64] The operating costs for Norway of NADGE have amounted to approximately $9.4 million during the period 1984–87. Since there will still be a need for such a system after a hypothetical US withdrawal, the only serious option would be for the European NATO allies to assume the United States' portion of the development and maintenance costs. The other option, that of developing a comparable European warning system, is for economic and technological reasons hardly feasible. Similarly, the Omega submarine navigation system facility at Bratland is totally US-financed, except for the normal provision of ground, water and other services, and it is not clear what the economic or possible security implication might be if Norway took over the tasks associated with this facility. Like the Loran-C system (of which Norway hosts two transmitters), Omega arguably offers some civilian spin-offs, most notably as accurate navigation aids for shipping. Potential costs or benefits are impossible to estimate, however, as the annual US contributions to the running costs of the stations in Norway appear not under the budget of the Ministry of Defence, but rather under that of the Ministry of Communications, usually listed as 'other expenses'. The expenses incurred by Norway in operating the Omega station are refunded in full by the United States and appear under the heading 'other income'.[65]

A second consideration that may have serious economic implications for the European NATO allies involves US funding for research and development (R&D). The assumption would be that the United States would be less inclined to invest in European R&D in the event of a withdrawal. Information exchange between Europe and the USA on technological matters (small as it is) could conceivably also dry up as a result of a withdrawal. The loss of US military R&D contracts would heavily constrain future defence research projects undertaken at European universities and other research establishments. Again, there are no reliable figures to indicate what the economic impact might be, but it is estimated that the majority of US military R&D expenditure abroad goes to Europe.[66] This would have a great impact on Norway in particular, since Norway won over half of all R&D contracts awarded by the Pentagon in 1986 to foreign governments, non-profit organizations and universities. With a total of $12.1 million in DOD contracts, Norway obtained 51 per cent of all such DOD research money spent with NATO allies and almost 44 per cent of all such money spent abroad.[67] The United Kingdom followed closely with $9.5 million (or 40 per cent) of the $23.5 million in contracts that went to NATO countries and the FRG third with $1.05 million. Eighty-six per cent of all contracts for 1986 went to NATO countries. In Norway's case most of the money (93 per cent) went to Norway's Royal Norwegian Naval Material Establishment. The Pentagon's total R&D budget for 1986 stood at $2.7 billion and was split among 318 non-profit organizations, government agencies and universities. Although only one per cent of this was invested abroad, among 47 institutions, these small investments represent an important source of research funds for many institutions struggling with limited budgets.

In the event of a US withdrawal it is unlikely that Norway would save much or any money because of the ambitious infrastructure programme that has been embarked upon, which, if continued for the use of NATO forces, would require considerable outlays in the future as well. A withdrawal would pose serious problems for Norway and its European NATO allies. Since the facilities being constructed are for the exclusive use of the United States, they have a choice of either halting construction and leaving the facilities unused *or* assuming the costs of the US share and finding the available manpower, airlift and equipment to replace, for instance, the MAB. It is in the author's view unlikely that enough funds would be available to do this. Norway has not been able to fill the gap left by the Canadian Air/Sea Transportable (CAST) Brigade group and the chances of filling an even larger gap are thus highly unlikely (assuming there is a need to do so). Few savings are foreseeable for Norway since the US or NATO allies finance virtually all bases and major construction programmes in Norway. Even the savings that could be made by the withdrawal of the MAB, to take one example, are few. The main obligation is the provision of ground support equipment, which includes a minimum of 150 over-snow vehicles, two motor transport companies (90 trucks each), one ambulance company (35 ambulances), a refueller section (6 trucks) and sundry engineering and air base support. Most of this equipment has been drawn out of existing stocks and would probably be re-assimilated by the Norwegian Army. Demobilization of these forces would cause insignificant savings.

It should be stated again that for Norway, as for Denmark and Iceland, the economic implications of a US withdrawal are not only difficult to calculate but they are also perhaps the least significant effect of a withdrawal. The security considerations of a withdrawal would far outweigh the economic implications.

IV. Summary

Although Denmark, Iceland, Norway and their territories host a small peacetime US presence, the implementation of several major capital-intensive programmes since the mid-1970s has meant that a significant amount of US defence dollars have been spent in the Nordic region, mainly to support the US Maritime Strategy. These programmes—the construction of COBs, storage locations, the new phased-array radar at Thule and the new terminal at Keflavík—are important (but temporary) injections into the local economies. As such these programmes are having an impact in particular in Iceland and Greenland. Nevertheless, aside from the Keflavík base, which is a significant employer and actor in the Icelandic economy, the facilities in themselves have little economic significance to the national economies. The most common peacetime type of installation in the Nordic region serves a C^3I role, which demands a small number of highly specialized personnel.

Although the actual presence of US forces has a limited economic impact at the moment, it seems clear that the scenario envisaged in the project—a complete US military withdrawal—would create several immediate problems for the Nordic allies. The central issues are whether the remaining NATO countries would continue to share infrastructure costs, and if so, how. If it is assumed that they would and that cost shares remain proportional, the Nordic region would face a tough economic challenge. The increased defence burden could only be shouldered at the cost of cuts in other areas of government expenditure. All of the countries have been net recipients of aid. Put crudely, in purely economic terms they receive more than they give, and much of this can be accounted for by US expenditures on construction. The economic aspects of the question do tend to distort the real issues, however. It is possible to measure the amount of investment that might be lost, the number of jobs lost and so on, but it would be unrealistic to assume that in the event of a withdrawal there would necessarily be an interest in replicating all or any of the withdrawn US facilities. The loss of Thule would be a heavy blow to the USA, but the fact remains that that base provides no defence for Greenland, and since its facilities are concerned entirely with the defence of North America, there would be no immediate Danish interest in assuming any of the functions now associated with it.

From the US perspective there would obviously be a heavy loss due to the formidable amount that has been invested in the Nordic region, both on a bilateral basis and through the NATO Infrastructure Fund. A vast majority of this investment is not recoupable since it has been invested in physically inert structures—buildings, runways and so on. The United States could save in the long-term, depending on what kind of redeployment patterns are envisaged, since the enormous costs associated with flying provisions through Keflavík would no longer be incurred. The types of US deployment evident in these countries, which stresses low combat force readiness and a high C^3I ability, provides a counterpoint to the US pattern of basing on NATO's Central Front, where logically the bulk of the US manpower in Europe is stationed. In spite of the low profile of the US forces in the region, these still generate considerable controversy, and decisions on their future will tend to be influenced more strongly by political and strategic considerations than by economic costs or benefits of the US military presence. In deciding on their future presence, however, the economic aspects should at least be borne in mind.

Notes and references

[1] Duke, S., SIPRI, *United States Military Forces and Installations in Europe* (Oxford University Press: Oxford, 1989), appendix 3C, p. 55.

[2] TIAS 2202, *Agreement between the Government of the United States of America and the Government of the Kingdom of Denmark, pursuant to the North Atlantic Treaty, concerning the defense of Greenland*, signed in Copenhagen 7 Apr. 1951, Article I.

³ TIAS 2202 (note 2), Article II.
⁴ TIAS 2202 (note 2), Article II, Section 2.
⁵ TIAS 2202 (note 2), Article II, Section 3(a).
⁶ TIAS 2202 (note 2), Article II, Section 3(a) i–vii.
⁷ TIAS 2202 (note 2), Article II, Section 4(a) and (b).
⁸ TIAS 2202 (note 2), Article II, Section 3(c).
⁹ TIAS 2202 (note 2), Article V, Section 3.
¹⁰ TIAS 2202 (note 2), Article VI.
¹¹ For a more detailed description of the evolution of Denmark's 'no foreign bases' policy see Petersen, N., 'Denmark and NATO 1949–87', *Forsvarsstudier*, no. 2/87 (Forsvarshistorisk Forskningssenter: Oslo, 1987), pp. 23–28.
¹² It should be noted, however, that the Danish anti-nuclear stance is contentious and that NATO planning assumes that nuclear weapons could be deployed on Danish territory in time of emergency or war. The Danish Social Democratic Party proposes to change this possibility by establishing a nuclear-free zone in the Nordic region. The Social Democrats do not advocate withdrawal from NATO and want to continue the arrangements for Allied forces to reinforce Danish forces in wartime.
¹³ TIAS 4531, *Agreement between the Government of the United States of America and the Government of the Kingdom of Denmark concerning the establishment and operation of certain aeronautical facilities and services in Greenland,* signed in Copenhagen 7 July 1960.
¹⁴ TIAS 4531 (note 13), Preamble.
¹⁵ TIAS 4531 (note 13), Article IV.
¹⁶ TIAS 4531 (note 13), Appendix, Attachment A.1.1.
¹⁷ TIAS 3716, *Exchange of notes between the Icelandic Minister for Foreign Affairs and the American Ambassador,* 6 Dec. 1956.
¹⁸ TIAS 2266, *Defense Agreement between the Government of Iceland and the Government of the United States of America,* signed in Reykjavík 5 May 1951.
¹⁹ TIAS 3716 (note 17).
²⁰ TIAS 2266 (note 18).
²¹ TIAS 2266 (note 18), Article II.
²² TIAS 2266 (note 18), Article VI.
²³ TIAS 7969, *Exchange of Notes between the American Ambassador and the Icelandic Minister for Foreign Affairs,* Memorandum of Understanding and Agreed Minute, signed in Reykjavík 22 Oct. 1974.
²⁴ US forces in Iceland are referred to as Iceland Defense Forces, largely to assuage local political opposition to the US military presence as well as serving as a reminder of the function of the US forces. Similarly, US bases and facilities in Greenland are referred to as 'defense areas'.
²⁵ TIAS 7969 (note 23), Paragraph 1.
²⁶ TIAS 7969 (note 23), Paragraph 2.
²⁷ *Millitary Construction Appropriations for 1989*, Hearings before a Subcommittee of the Committee on Appropriations, US House of Representatives, 100th Congress, 1st Session (US Government Printing Office: Washington, DC, 1989), Part 2, pp. 586–87.
²⁸ TIAS 7969 (note 23), Paragraph 4.
²⁹ TIAS 7969 (note 23), Paragraph 5.
³⁰ TIAS 7969 (note 23), Paragraph 5, Section C.
³¹ Not *Agreement* as it is sometimes mistakenly called. 'Invictus', from the Latin word for invincible, was the code name for the initially classified project and was later applied to the Arrangement.
³² Gleditsch, N. P., 'The strategic significance of the Nordic countries', *PRIO Report 14/86* (International Peace Research Institute: Oslo, 1986), p. 16.
³³ USAFE Form 137, 'The Collocated Operating Base Concept' (OPR, HQ US Air Force Europe: Ramstein AB, FRG, Jan. 1974), p. 1.
³⁴ WRMs are not normally built-up rounds but are munitions of low-risk and high bulk. Security is supplied by host-nation support groups.

³⁵ See Gleditsch, N. P., 'Invictus agreement declassified', *PRIO Inform 14/84* (International Peace Research Institute: Oslo, 1984) and the Norwegian Defence Ministry's information bulletin *Fakta*, no. 1084 (Nov. 1984), for the full texts of the Invictus Arrangement.
³⁶ Invictus Arrangement (note 35), Article III.
³⁷ Invictus Arrangement (note 35), Article VI; emphasis added.
³⁸ TIAS 9966, *Norway: Defense Prestockage and Reinforcement*, Memorandum of Understanding, signed in Washington 16 Jan. 1981. In Article II the composition of a MAB is defined as two Air Defense squadrons, two close support squadrons and appropriate support aircraft and approximately 75 heavy transport and light support helicopters.
³⁹ 'Storage in Norway of equipment for an American brigade', *Fakta*, no. 0586 (Norwegian Ministry of Defense: Oslo, Aug. 1986).
⁴⁰ Invictus Arrangement (note 35), Article VIII.
⁴¹ For an excellent in-depth discussion of the Loran-C system see Wilkes, O. and Gleditsch, N. P. (in collaboration with I. Botnen.), *Loran C and Omega. A Study of the Importance of Radio Navigation Aids* (Norwegian University Press: Oslo, 1987).
⁴² Author's estimate—refers only to the additional NATO Infrastructure expenditure and does not take into account bilateral projects with the USA.
⁴³ *Military Construction Authorization Act, FY 1986*, Committee on Appropriations, US House of Representatives, 99th Congress, 1st Session (US Government Printing Office: Washington, DC, 1986), pp. 195–96.
⁴⁴ Upon completion in 1992, NWS will include at least 52 sites strung along the 70th parallel, each equipped with both microwave and OTH-B (over-the-horizon backscatter) radars.
⁴⁵ Faurby, I. and Petersen, N., *The Far North in Danish Security Policy*, paper presented at the ISA Conference, St Louis, Mo., Apr. 1988 (available from the Department of Political Science, University of Aarhus, Denmark), p. 20.
⁴⁶ Figures from Taxation Board of Greenland and *Grønland Kalaallit Nunaat 1983* (Ministry for Greenland: Copenhagen, 1984), quoted in Archer, C., 'Greenland and the Atlantic Alliance', University of Aberdeen Centre for Defence Studies *Centrepiece* no. 7 (summer 1985), p. 25.
⁴⁷ The DAO has reportedly carried out a detailed study of the possibility of Iceland joining the NATO Infrastructure Fund.
⁴⁸ *Foreign Affairs Report 1987, A Report to the Althing by Geir Hallgrimsson, Minister of Foreign Affairs* (Icelandic Ministry of Foreign Affairs: Reykjavík, 1984), p. 21.
⁴⁹ For full details of all projects and modifications at Keflavík see Gunnarsson, G., *The Keflavik Base: Plans and Projects*, Occasional Paper no. 3 (Icelandic Commission on Security and International Affairs: Reykjavík, 1986). For details on other projects see *Foreign Affairs Report 1987, A Report to the Althing by Matthiàs Å. Mathiesen, Minister of Foreign Affairs* and *Foreign Affairs Report 1988, A Report to the Althing by Steingrímur Hermansson, Minister of Foreign Affairs* (Icelandic Ministry of Foreign Affairs: Reykjavík, 1987 and 1988).
⁵⁰ Figures from *Foreign Policy Report 1988* (note 49), pp. 48–49.
⁵¹ For a detailed description of US activities at the Keflavík base, see Jonsson, A., *Iceland, NATO and the Keflavik Base* (Icelandic Comission on Security and International Affairs: Reykjavík, 1989).
⁵² OECD statistics for February 1988.
⁵³ Figures from *Foreign Policy Report 1988* (note 49), p. 51.
⁵⁴ Gunnarsson (note 49), quoting *Military Construction Appropriations for 1985*, Hearings before a Subcommittee of the Committee on Appropriations, US House of Representatives, 98th Congress (US Government Printing Office: Washington, DC, 1984), Part. 2, pp. 578–99.
⁵⁵ *Military Construction Authorization Act, FY 1986* (note 43), p. 196.
⁵⁶ Duke, S., 'The US military presence in Europe: economic aspects', *RUSI & Brassey's Defence Yearbook 1988* (Brassey's: London, 1988), pp. 81–97.
⁵⁷ This point is is discussed in Duke (note 1).
⁵⁸ See Wilkes and Gleditsch (note 41), p. 287.
⁵⁹ Edge, K., 'US intelligence in Norway', *Counterspy*, vol. 4, no. 1 (1980), pp. 32–42.
⁶⁰ US Department of Defense, *Manpower Statistics for All Authorized Full Time Military Personnel, 30 September 1988*, table P309A (US Government Printing Office: Washington, DC, 1988). Of the 206, 48 were Army, 42 Navy , 22 Marine Corps and 94 Air Force.

⁶¹ Author's estimate based on Nkr figure and converted at a US$ rate of 6.3754 (Dec. 1987). The value fluctuates little and is normally within the range 6.3–6.7. The figures are from the International Monetary Fund's *International Financial Statistics*, vol. 41, no. 2 (Feb. 1988), pp. 380–81.

⁶² Ausland, J., 'Norways's strategic role', *New York Times*, 19 May 1980.

⁶³ The figure was calculated at 1976 prices and appears in Gleditsch (note 35), p. 7.

⁶⁴ Author's estimate.

⁶⁵ Wilkes and Gleditsch (note 41), p. 273.

⁶⁶ Gleditsch, N. P., Hagelin, B. and Kristoffersen, R., *Some Transnational Patterns of Military R&D*, PRIO publication S-6/82 (International Peace Research Institute: Oslo, Feb. 1982).

⁶⁷ *Armed Forces Journal International*, May 1987, p. 24.

10. Economic aspects of the US military presence in Spain

José Molero
University of Madrid (Complutense), Madrid, Spain

I. Introduction[1]

From a methodological point of view, it is very difficult to distinguish the economic issues from other political, strategic and technological aspects of the United States military presence in Spain. While probably true for many other countries as well, it is particularly the case with Spain because of the nature and the special political implications of the first bilateral treaty signed with the United States in 1953. Specifically, while the treaty and subsequent agreements make reference to several direct economic commitments by the United States in payment for the use of the bases, they also have other, indirect implications for the Spanish economy that, while often difficult to quantify, in the long run are of much greater importance.

Complicating the matter further, this chapter was prepared while negotiations were in process between the Spanish and US governments aimed at reducing the US military presence. As a consequence of these deliberations it has recently become much more difficult to gain access to official information, an unfortunate development only partially offset by the abundance of press reports on the event itself. This difficulty in obtaining information compounds the already much criticized aura of secrecy that surrounds the evolution of the military agreements with the United States.[2]

The study is divided into four sections. The first is a historical account with the twofold aim of gaining a dynamic perspective on the economic issues as they have developed from the moment the first bilateral agreement came into force, and of emphasizing the significance of the US–Spanish defence relationship for the host country in political as well as economic terms. Section II is devoted to the treaty of 1982–83, in which a new element is added as a result of Spain's entry into NATO in 1982. Section III deals with what may be called the indirect economic effect, of paramount importance in evaluating the role of the defence relationship with the United States for the Spanish economy. The concluding section considers some of the implications of the new treaty signed in 1988, in which important changes were introduced.

II. Historical evolution of the military agreements and treaties signed by Spain and the United States, 1953–82

Initiation of the relationship and the 1953 agreements

The origins of the US military presence in Spain can be traced back to the late 1940s. At the time, Spain's internal situation was as desperate as its international standing was low—the latter in no small part due to the political links of the Franco regime with the defeated powers of World War II. In terms of US–Spanish relations, however, this was effectively offset by the political climate of the immediate post-war period, which greatly influenced the US attitude towards the Franco regime.[3]

Spain's internal difficulties were not only political in nature; the economic situation was also precarious. The aftermath of the war, the historical stagnation of the Spanish economy, the international economic outlook and the autarchic measures of the Government led to serious difficulties on all fronts, not the least as regards the country's foreign economic relations. In brief terms, Spain's dilemma was 'the inability to export, within the structure and institutional framework of the forties and fifties, and thereby to obtain sufficient foreign currency to meet the import demand originating from the need for the reconstruction and modernization of the Spanish productivity machine'.[4]

In this context, United States aid linked to the military agreements 'would not only mean the recovery of the regime on an international scale, but would also have important repercussions in eliminating certain productivity bottlenecks, thus permitting Spain to sustain and increase import flows'.[5]

In the early 1950s, as far as the US plans were concerned, the geographical situation of Spain began to be considered of vital importance to US strategic planning for Western Europe, the Mediterranean and the North Atlantic.[6]

Spain's anti-communist attitude permitted the United States to overcome the political obstacles of dealing with the dictatorial Franco regime. However, Spain's *rapprochement* with the United States was not without its problems. The governments of France and England were for different reasons unable to envisage an end to the isolation of the Franco regime. At the same time, radical isolationist elements within the regime were against the idea of such a *rapprochement*. Finally, the Spanish armed forces had neither the strength nor the equipment necessary to participate effectively within the military framework implicit in a block policy.

Notwithstanding these difficulties, the so-called 'Madrid Pacts' were signed by the US and Spanish governments on 26 September 1953. Apart from the defence agreement, they included an economic aid agreement as well as several secret adjuncts pertaining to the classification and implementation of the defence accords. From the time of its signing, three issues dominated the political discussion:

ECONOMIC ASPECTS OF THE US MILITARY PRESENCE IN SPAIN

1. The status and terms of the agreement: in spite of Spain's interest in a formal treaty, an executive agreement was signed that did not require formal ratification. The Spanish Government also felt that the agreement was asymmetrical in that it provided that Spain assume certain risks while not providing any guarantees of US assistance in the event of an attack.

2. The US economic compensation for the use of the bases was considered insufficient in light of what the Spanish Government was offering in terms of military assets.

3. Also an important issue was the role the agreements played in bolstering, and giving legitimacy to, the Franco regime and in breaking Spain's political and economic isolation.

The extent to which these issues overlap in the political debate illustrates the degree to which the economic repercussions of the agreements are inseparable from the political and strategic aspects. This interrelationship remains an element also of future renewals of the agreements.[7]

In order to evaluate the economic importance of the compensation received by Spain, the following factors must first be considered.[8]

1. Prior to the signing of the pacts, Spain had already received aid from the United States. In 1951 it received two loans of $30 million, one from the National City Bank and the other from Chase National Bank.

2. Extraordinary funding from the USA was to increase during the 1950s. In 1956, when the economic aid derived from the agreements was being utilized, these funds amounted to almost 40 per cent of income from exports.[9]

3. Spain encountered some difficulties in processing the aid, especially the loans provided by the US Export–Import Bank (Eximbank). In particular, it had difficulty in obtaining rapid shipments of capital goods for industrial projects. It should here be borne in mind that Spain, as part of the agreement, was under obligation to purchase on the US market.

4. The positive result of the negotiation on the bases was seen from the outset as a way of obtaining economic aid. The condition for this was precisely the granting of military facilities to the United States.

5. The pact was signed at a time when external aid from the USA was in decline, making it difficult to obtain greater compensation for the bases.

6. The figures forecasted in the agreement for the base assets were low. Although the amounts the Spanish Government later received were higher, this was not the result of negotiations preceding the agreement.

US economic aid to Spain in the 1950s developed as shown in table 10.1. Two main points can be made to the figures presented here. First, the total amount is a small sum, both for the *carte blanche* that Spain gave the USA with regard to the use of the bases and in light of the needs of the Spanish economy. Second, for the period in question the figures show a clear decline in total aid.

Table 10.1. US economic aid to Spain based on 1953 pacts, FYs 1953–58
Figures are in US $m.

Source	1953	1954	1955	1956	1957	1958	Total
Congress	75	30	50	50	40	50	295
McCarran Amendment	–	55[a]	–	–	–	–	55
Executive	10	–	10	20	15	–	55
Total	**85**	**85**	**60**	**70**	**55**	**50**	**405**

[a] Of this amount, only US $24 m. is considered donation.

Source: Viñas, A. et al., *Política Comercial Exterior de España (1931–1975)* (Banco Exterior de España: Madrid, 1979).

As indicated, 'beside the aid itself, the Hispano-US agreements would encourage other relations with a beneficial effect on the economy, since additional resources were channelled to Spain from different quarters'.[10] These include for the whole decade: Eximbank loans ($71 million), Public Law 480[11] ($392 million) and Charity from the National Catholic Welfare Conference ($120 million).

In evaluating the impact of the aid directly linked to the agreements, the actual orientation of the aid programme should also be taken into consideration. Under the terms of the US Law of Mutual Security, during the first five-year period 60 per cent of the US Defense Support Program 'aid' to Spain was destined for US projects (mainly construction of bases) and 10 per cent for administrative costs incurred by the United States in Spain. Only the remaining 30 per cent was donated to the Spanish Government.[12] After 1958–59, with the bases practically completed, the percentage of the donation rose to 90 per cent, with the remainder intended to cover expenses incurred by the United States.

Thus, aside from the reduced amount of aid awarded, the small proportion of it that initially made its way directly into the economy was another constant issue of concern for the Spanish Administration. Finally, criticism was also levelled against the provision of US Public Law 480 that the resources made available through it were for the most part intended for the purchase of US agricultural surpluses, despite the more urgent import needs of the Spanish economy for such things as raw materials and capital goods.

It was the virtually unanimous opinion of analysts at the time that the aid was merely a short-term remedy—at most capable of resolving some of the more serious problems of the balance of payments—with a very limited impact on the basic problems of the Spanish economy.

The economic implications of the first agreement may be measured in three dimensions: the total aid received up to the first ratification of the convention in 1963, the importance of the aid with regard to Spain's balance of payments and US aid to Spain as a share of the total US international assistance programme.[13]

The total amount of US resources placed at Spain's disposal is shown in table 10.2. Two important factors should be mentioned here. First, the Spanish Government only considered the funds originating from the Defense Support

Table 10.2. US aid to Spain, 1953-63
Figures are in US $m.

Programme	Sub-total	Total
Economic aid (including technical assistance)		504.5
Defense support program		
Public Law 480		506.5
National Catholic Welfare Conference charity		174.0
Other donations		3.8
Development loan fund		17.1
Wheat operations 1953		20.0
Eximbank loans		297.5
Congressional loans 1953–54	62.5	
Cotton operations 1953–54	24.0	
Congressional loans 1954–59	72.0	
Congressional loans 1960–63	139.0	
Total		**1 523.4**

Source: Viñas, A. et al., *Política Comercial Exterior de España (1931–1975)* (Banco Exterior de España: Madrid, 1979).

Program as donations, since the rest represented bank operations. Second, the quality of the military equipment received was poor (in some cases the shipments even consisted of scrapped material), even though it meant an important resource for the poorly equipped Spanish armed forces.

The impact of the aid on Spain's balance of payments is shown in table 10.3. Put simply, it can be said that the total US resources received in 10 years, counting all accounts, was approximately equal to the income of four year's worth of Spanish exports. As a share of the total US international assistance programme Viñas shows that until 1962, US aid to Spain decreased.[14]

The military agreements with the United States also had an indirect economic effect, notably in the areas of US investments, technology import and the Spanish armaments industry.

With regard to US investments, reference should be made to the agreements on economic aid included in the Madrid Pacts. As a prerequisite for the aid to be received, Spain agreed, among other things, to seek to stabilize its currency and its balance of payments, to create and maintain financial stability, to put an end to cartels and monopolies, to encourage competition and, above all, to liberalize its foreign investment legislation.[15]

It must be remembered that the legal conditions for foreign investment in Spain had since 1939 been particularly unfavourable. It is common knowledge that direct investment was in the early 1950s one of the basic strategies for international expansion of the US economy. For this reason, the relationship between US military and economic interests in the deliberations preceding the 1953 agreements should not be overlooked. It is true that it took several years (until 1959) before the promised liberalization of foreign investment legislation became a reality, mainly because of opposition from Spanish private interests.

Table 10.3. Significance of US aid to Spanish foreign trade, 1953–58
Figures are in US $m.

Income	1953	1954	1955	1956	1957	1958
US aid						
Economic aid:						
Goods	–	24.1	82.4	81.9	58.4	73.8
Donations, net income	–	3.0	44.3	24.1	28.6	20.5
PL 480 imports	–	–	16.0	77.8	76.2	92.0
Eximbank loans	48.1	10.9	3.3	9.8	4.3	8.0
PL 480 loans	–	–	–	9.0	–	–
McCarran Amendment loans	–	–	–	20.0	–	–
Total US aid	**48.1**	**38.0**	**146.0**	**222.6**	**167.5**	**194.3**
Income from exports	405.6	398.4	387.6	402.8	416.8	498.0
US aid as percentage of total income from exports	*11.8*	*9.5*	*37.7*	*55.3*	*40.2*	*39.0*

Source: Viñas, A. et al., *Política Comercial Exterior de España (1931–1975)* (Banco Exterior de España: Madrid, 1979).

As proof of the importance the USA gave to this issue, however, official US representatives have since 1953 insisted on the need for Spain to comply with this part of the agreements.

Although there was some growth during the 1950s, legal restrictions still kept down the volume of US investments in Spain. Indicative of the growing interest in the Spanish economy among US investors, however, between 1956 and 1960 the number of multinational subsidiaries and associated companies rose from 21 to 92, as many large US firms opened offices in or sold equipment to Spain, carefully testing the opportunities offered by the Spanish economic and legal structure.[16]

After liberalization was completed in 1959, US investments showed considerable growth, with the United States in the end ranking first among countries investing in Spain on a permanent basis. Incentives to US investors were a rapidly expanding market, a favourable environment for investment, relatively cheap labour costs, negligible tax pressure and political stability.[17]

With respect to technology import, two phenomena are of particular interest: the acquisition of productive technology by Spanish industrial firms and the introduction of scientific labour methods in Spain. Of these two, the first received little attention among scholars prior to the 1960s. The following conclusions are drawn from a 1979 study.[18]

In the industrialization plans of the 1940s it was envisaged that Germany would be the main source of new technology. These plans were essentially taken over by planners of the 1950s, with the important difference that the Unites States now occupied first place as the source of necessary technology.

This is corroborated by a comparison of technology-transfer agreements of a large number of industrial firms during the period 1939–63. Of a total of 227 agreements, most of them (164) were signed between 1950 and 1963. The USA played an important part in the supply of this technology, with 21 agreements (12.8 per cent) during this period. US dominance is particularly evident in sectors such as agricultural machinery, fuel, chemicals and pharmaceuticals.[19]

The growing preference of Spanish industry for US technology was enhanced by the incorporation of US organization methods. A US Technical Aid Program was set up as part of the 1953 agreements, in which three mechanisms for the introduction in Spain of US organization methods were developed: technical exchange visits, visits of foreign—mainly US—experts and the supply of technical services.

As several previous studies have indicated[20] these mechanisms played an important role in introducing more productive organization methods in Spanish industry. These were seen as an essential supplement to the import of machinery and production technology and the Government gave its firm support also to this method of technological modernization.[21]

The third area referred to above, the impact of the Madrid Pacts on the Spanish armaments industry, is much more difficult to examine, as practically no data are available on the subject. However, there is sufficient evidence to suggest that although the arrival of military material from the USA improved the level of equipment of the Spanish armed forces, it had a negative effect on the development of the national defence industry, as the number of orders to domestic arms producers fell sharply once the US armaments began to arrive.[22]

The agreement renewals of 1963, 1970 and 1976

The first agreement was renewed in 1963. No important modifications were made at this time. The second renewal marked, however, for the first time a growing difference of opinion between the Spanish and US governments. From a Spanish point of view, these differences can be summed up as follows:

1. Spain was becoming increasingly conscious of the asymmetry of the original agreement, in which Spain was given practically no guarantee of any kind of US defence commitment. This was accentuated when the USA refused to support Spain in the latter's colonial conflict with Morocco, at a time when Spain relied heavily on US military equipment. As a results of these difficulties, Spain demanded that the existing agreement be exchanged for a mutual defence treaty.

2. There was a growing awareness in Spain of the need for the USA to declare its position with regard to Spain's dispute with the UK over Gibraltar.

3. The secret clauses of the original agreement (and their implication for Spanish sovereignty) became a political issue when in 1966 a US B-52 bomber crashed and accidentally dropped four atomic bombs over Spanish territory.[23] This event gave rise to strong criticism of the terms of the agreement.

4. The economic compensation was considered insufficient, and the amount and quality of the military equipment received left much to be desired.

As regards the economic aid, in the period 1962–68 this amounted to $59.8 million in donations and $420 million in loans (mostly from Eximbank). In the words of one analyst: 'The US military presence in Spain from 1963 to 1970 was not compensated by any economic benefits of significance to Spain's development. Sooner or later the loans received have to be repaid'.[24]

The military aid was reduced considerably compared to the previous period. From 1962 to 1968 this aid amounted to a total of $163.4 million in direct transfers and $2.3 million for the purchase of equipment in the United States, a reduction of 52 per cent compared to the previous period.

Against this background and in the context of the completely new situation of the Spanish economy after a decade of economic growth, Spanish Minister of Foreign Affairs Fernando María Castiella summed up Spain's demands in the following manner: (*a*) increase in military aid, to compensate for the reductions of the previous years (a request was made for $1 billion over a period of five years); (*b*) transformation of the agreements into a real defence treaty; (*c*) US diplomatic support in the dispute with the UK over Gibraltar;[25] and (*d* liberalization of US restrictions regulating how the compensation is used.

The negotiations were extremely complicated, involving a mixture of economic as well as non-economic issues. Among the many topics discussed were the need for improving Spain's investment climate for US investors, US doubts about the signing of the preferential trade agreement between Spain and the EEC and Spain's entrance into certain technological areas such as nuclear energy research and space exploration.[26]

The negotiations ended abruptly in 1969 when Foreign Minister Castiella was replaced by Gregorio López Bravo. The shift indicated a desire to tone down the original demands, a development linked to the increasingly difficult internal situation. Although the final agreement was far from what Spain originally had envisaged, some improvements, such as the establishment of Spanish sovereignty over the bases and the transfer to Spanish control of the Rota–Zaragoza oil pipeline, were achieved. As compensation for continued US access, Spain received $120 million in Eximbank loans for the purchase of second-hand military equipment, additional loans (up to $60 million, which 'would perhaps be obtained from Congress'[27]) for updating facilities and the purchase of additional military equipment, the loan of various ships and the supply of machinery and tools for the production of ammunition.

Of great importance for the amounts actually received was that this and subsequent agreements contained a provision whereby there was a commitment for the first year only, while the amount for the subsequent periods were subject to approval by Congress. This introduced an element of insecurity in Spanish defence planning. The evolution of aid actually received during the following years can be seen in table 10.4.[28]

When the time again came to renew the agreement, the main points on the agenda (defence treaty level, guarantee of US commitments, greater economic

Table 10.4. US military and economic aid to Spain, 1970–74
Figures are in US $m.

Year	Military aid	Economic aid	Total
1970	131.2	61.2	192.4
1971	149.6	13.4	163.0
1972	103.5	452.4[a]	555.9
1973	119.9	173.0[a]	292.9
1974	187.0	161.7	348.7
Total	**691.2**	**861.7**	**1 552.9**

[a] Mainly Eximbank loans.

Source: Chamorro, E. and Fontes, J., *Las Bases Norteamericanas en España* (Euros: Barcelona, 1976), p. 197.

compensation) remained unresolved. Following the death of Franco, however, expectations rose that the United States might take a new stand on these issues.

Despite the changed circumstances, US resistance to Spain's demands again led to difficult negotiations. Finally in 1976, a treaty of friendship and co-operation was signed which went half-way to meeting Spain's wish for a defence treaty. Spain also succeeded in including a security clause linking the US–Spanish bilateral relationship with NATO as well as a provision restricting (in theory at least) US nuclear activities in and over Spanish territory.

As for compensation, Spain obtained a total sum of $1.2 billion, which breaks down as follows: civil economic aid ($35 million), Eximbank loans ($450 million), military donations ($75 million), donations for an alert and control system ($50 million) and a loan for the purchase of military equipment from the United States ($600 million).

III. The 1982 Agreement

General conditions for a new agreement

It is important to stress the general dissatisfaction in Spain with the terms of the existing agreements. To this must be added two additional factors influencing Spain's position in the new rounds of negotiations: first, the excessive dependence of the Spanish armed forces on US equipment (which, as has been noted, was not always in perfect condition), and second, Spain's growing dependence on the loans (as the percentage share of the donations had fallen), which meant that compensation for the use of the bases steadily decreased in real terms over the years.

Furthermore, Spain had undergone several important changes, political as well as economic, since the 1976 renewal. This was also of significance in setting the context for the new agreement. The consolidation of a democratic regime, following a peaceful transition period, gave an impetus, moral as well as political, to the Spanish position that it hitherto had lacked. At the same time, the newly respectable left was very outspoken in its criticism of the agreements, in part for reasons already mentioned. Among other things it pointed to the danger of the Torrejón base, located only a few kilometres from Madrid.

In 1982 Spain had become a member of NATO, and negotiations for Spain's entry into the EEC were well under way, both developments strengthening Spain's international position.

The country's socio-economic structures had become considerably more modern. In the process, earlier concerns about Spain's balance of payments vanished, as did the need for aid from abroad.

Since the second half of the 1960s important changes had been made in the plans for equipping the Spanish armed forces. One new development was a trend towards diversifying the origin of military equipment, reducing Spain's dependence on the USA. Another was a new interest in modernizing the Spanish defence industry through licensed and joint production ventures. The latter gave an incentive to claims for greater US compensation to the industry.[29]

At the time of the negotiations, the Spanish position, although strengthened by these developments, was weakened by a series of internal circumstances. It become clear that the centre-right Centre Democratic Union (UCD) Government taking part in the talks was about to fall, with real prospects of a Socialist Party triumph in the near future. Beyond its criticism of existing agreements, the latter voiced its reservations regarding the probable content of a new one. At the same time, relations between the military and the civil sectors hit a new low following the attempted *coup d'état* on 23 February 1982.

The contents of the 1982 Agreement: a general description

From the Spanish standpoint, the new agreement had to serve the aim of improving the international standing of Spain and of the new democratic regime. At the same time, it was necessary to eliminate some of the more unfavourable aspects of the former agreements.

The parties faced several difficult unresolved issues: the extent and nature of the US defence commitment, control over the use of the bases, compensation, loan terms, technology transfer and joint production of high-technology military equipment, Spain's difficulty in meeting its balance of payments, the status of the US forces, the nuclear presence and the status of the new agreement.

After many meetings an Agreement of Friendship, Defense and Cooperation was signed on 2 July 1982, entering into force on 24 May 1983.[30] This Agreement was given a new political context, with a prologue affirming the common ideals of freedom and democracy and Spain's full territorial integrity, as well as recognizing the country's incorporation into the Atlantic Alliance framework. The Agreement also called for the establishment of a Hispano-American Council, which was to regulate, through a number of subordinate bodies, US–Spanish co-operation in defence, economic as well as cultural matters.

ECONOMIC ASPECTS OF THE US MILITARY PRESENCE IN SPAIN

It is important to note that the Agreement called for the elimination of the US nuclear presence in Spain, as well as prohibiting overflights of aircraft carrying nuclear arms or equipment. Nuclear-propelled vessels were hereafter required to seek official authorization to dock in Spanish ports.

One of the most controversial issues, and one in which Spain encountered fierce opposition from the United States, related to the control over movements of US forces in Spanish territory, including airspace and waters. A compromise was reached differentiating between bilateral, NATO and other US use of Spanish territory. The concept of threat or attack by third parties was clarified, with the understanding that it only applied to infringements of the territories of the two signatories. Provisions were included by which the Hispano-American Council could be overridden for reasons of legitimate defence or in other national emergency situations. US operational and support installations were described and listed in an Annexe. Spanish sovereignty over the bases and supporting installations was affirmed.

With regard to procurement, it was stated that the price would be calculated under favourable terms for Spain and that top priority would be given to delivery. Loans for the first year would amount to $400 million and donations for military instruction would amount to $3 million.

With respect to defence industrial co-operation, obstacles to purchasing items produced in either country were removed, enabling Spain to make offers on equal terms with other NATO member countries.

In terms of the status of the US forces, the most important change was the introduction of a NATO standard Status of Forces agreement.

With regard to economic co-operation, reference was only made to such general objectives as increased communication and collaboration in fields of common interest.

Scientific and technological co-operation was defined in terms of objectives, as was cultural co-operation. For the first year the United States made donation of $12 million towards the development of these spheres of co-operation.

With regard to the economic compensation, the $415 million already listed for the first year was established not in the wording of the agreement but in an exchange of official notes.[31] A comment on the amount and composition of this sum is warranted.

1. Donations make up a small share, both in terms of the total sum and in terms of what has been received under previous agreements. Until 1982 US donations totalled $22 million for each fiscal year. In fiscal year 1983—the first in which the terms of the new agreement were applied—the donations dropped to $12 million.

2. The loans for purchasing equipment were increased considerably compared to the previous period (see table 10.5). This coincided with Spanish plans for modernizing its armed forces. Most of these $400 million were to be used to pay for new F-18 aircraft purchased from the United States.

Table 10.5. US loans and grants obligations and loan authorizations to Spain, FY 1946–85

Figures are in US$ m.

Programme	Marshall Plan Period 1949–52	Mutual Security Act period 1953–61	Foreign Assistance Act period 1962–81	1982	1983	1984	1985	1962–85	Total loans and grants 1946–85	Repayments and interest 1946–85[a]	Total less repayments and interest 1946–85[a]
Economic assistance											
Loans	52.7	295.2	0.3	–	–	–	–	0.3	336.8	366.6	–29.8
Grants	0.1	618.9	113.2	22.0	12.0	12.0	12.0	171.2	783.2	–	783.2
Total economic assistance	52.8	914.1	113.5	22.0	12.0	12.0	12.0	171.5	1 120.0	366.0	753.4
Military assistance											
Loans	–	–	602.3	125.0	400.0	400.0	400.0	1 927.3	1 927.3	675.2	1 252.1
Grants	–	499.5	509.4	2.0	2.5	3.0	2.9	519.8	1 016.9	–	1 016.9
MAP grants	–	427.3	268.3	–	–	–	–	268.3	694.2	–	694.2
Credit financing	–	–	602.3	125.3	400.0	400.0	400.0	1 927.3	1 927.3	675.2	1 252.1
IMET[b]	–	16.4	27.9	2.0	2.5	3.0	2.9	38.3	54.1	–	54.1
Transfers of excess stock	–	14.4	68.7	–	–	–	–	68.7	82.7	–	82.7
Other grants	–	41.4	144.5	–	–	–	–	144.5	185.9	–	185.9
Total military assistance	–	499.5	1 111.7	127.0	402.5	403.0	402.9	2 447.1	2 944.2	675.2	2 269.0
Total loans	52.7	295.2	602.6	125.0	400.0	400.0	400.0	1 927.6	2 264.1	1 041.8	1 222.3
Total grants	0.1	1 118.4	622.6	24.0	14.5	15.0	14.9	691.0	1 800.1	–	1 800.1
Total loans and grants	52.8	1 413.6	1 225.2	149.0	414.5	415.0	414.9	2 618.6	4 064.2	1 041.8	3 022.4
Other US loans											
Eximbank loans	–	114.3	2 113.5	37.8	–	–	–	2 151.2	1 974.2	2 093.4	–119.2
All other	–	–	72.0	1.0	–	–	–	72.9	10.8	13.3	–2.5
Total other loans	–	114.3	2 185.5	38.8	–	–	–	2 224.1	1 985.0	2 106.7	–121.7

[a] Values are net of deobligations. [b] International Military Education and Training Program.

Source: US Agency for International Development, *US Overseas Loans and Grants and Assistance from International Organizations* (US Government Printing Office, Washington, DC, 1987).

3. As indicated above, the amounts were only established for the first year, for which Congress had approved a sum by the time the agreement was signed. In the future, each additional transfer of funds required new approval. Normal US practice was to approve a similar amount in ensuing years. However, when the negotiation of terms for the new treaty was well under way, Congress reduced the level of its aid considerably. The loans for buying arms, as part of the US Foreign Military Sales (FMS) programme, corresponding to more than 90 per cent of the total aid to Spain, were in fact reduced from $400 million to $105 million in 1987.[32] Cuts in aid to other countries in the region were less dramatic. For example, aid to Greece decreased from $500 million to $345 million, and while in the case of Turkey non-military aid fell from $485 million to $178 million, military aid (MAP) to that country increased from $215 million to $437 million in the same period. This made it difficult for the Spanish Ministry of Defence (MoD) to meet payments, especially for the F-18 purchase. In the end it had to rely on additional funding from the Spanish Government.[33]

4. The conditions of the loans were relatively favourable, with a 30-year amortization period and a 5 per cent interest rate for 1987. It has been claimed, however, that even lower interest rates could be obtained on the international money market, since the US Federal Financial Bank has established for these military equipment loans an interest rate of more than 0.125 per cent above the rate it offers the US Government.[34]

IV. Approach to an economic evaluation of the US military bases in Spain

As stated in the beginning of this chapter, an evaluation of the economic impact of the US bases in Spain must take into account both direct and indirect aspects. This section will first deal with aspects relating to the operation of the bases and then analyse Spain's purchase of US military equipment and how this relates to the development of the Spanish armaments industry. Finally, some of the issues relating to the transfer of technology will be considered.

Economic aspects of the bases

After much effort it has been possible to establish a means of estimating the direct economic benefits (other than aid) derived from the bases. Unfortunately, it has not been possible to find data on what the US presence costs the Spanish Government. This means that a precise balance cannot be given; it is nevertheless clear that in terms of implications for the Spanish economy as a whole, the indirect effects of the US presence are more important than the direct costs and benefits.

The first element of the analysis pertains to the number of personnel assigned to the bases. Table 10.6 illustrates how difficult it is to make an estimate even of this. These figures provide the basis for subsequent estimates of expenditures

Table 10.6. Number of personnel at US bases in Spain, 1987

Category	Torrejón	Others	Total
US personnel			
Military: Authorized	4 507	8 038	12 545
Actual (estimated)	4 093	5 640	9 733
Civilian: Authorized	685	984	1 669
Actual (estimated)	882	1 653	2 535
Total : Authorized	5 192	9 022	14 214
Actual (estimated)	4 975	7 293	12 268
Spanish civilian personnel (estimated)	1 097	1 003	2 100
Temporary personnel (max. no.)	420	935	1 335

Sources: 1982 Agreement (TIAS 10589); Arkin, W. M., 'La presencia militar de los Estados Unidos en España', eds M. Aguirre and C. Taibo, *Anuario sobre Armamentismo, 1987/88* (Debate C.I.P.: Madrid, 1988); SIPRI data base.

and employment effect. As a preliminary estimate, the bases provide direct employment to 2100 and indirect employment to 8600–18 400 Spanish civilians (according to the multipliers David Greenwood uses in chapter 5), giving a total employment effect of 10 700– 20 500 jobs.

It is much more difficult to estimate US expenditures, although here some calculations can be made. Working from available figures on the average annual expenditure for the Torrejón base and taking previous calculations into account, a rough estimate can be arrived at for the whole US base structure.

Spanish authorities have stated that the annual US operation and maintenance (O&M) costs for Torrejón are about 7 billion pesetas ($41 million).[35] According to other sources, O&M costs correspond to about 50–55 per cent of total costs to the USA, giving a total for this one base of around $78 million (see table 10.7). Wages to Spanish civilians amounts to about $13 million.[36]; the rest is attributable to expenditures in Spain on US personnel and other minor costs. For the other bases, the wages of Spanish personnel total an estimated $11.88 million and the expenditures of US personnel $30.12 million. With

Table 10.7. US running costs for installations in Spain, 1985–86
Figures are in US $m., at constant (1985) prices.

	Torrejón	Other installations	Total
Operation and maintenance	41	46.42	87.42[a]
Spanish personnel wages	13	11.88	24.88
US personnel and other costs[b]	24	30.12	54.12
Total	**78**	**88.42**	**166.42**

[a] Yearly totals prone to fluctuate.

[b] Includes only share of total wages spent in Spain on permanent and transit US personnel, including rents for off-base housing and other minor outlays.

Source: Author's calculations.

Table 10.8. Committed Spanish Ministry of Defence purchases of US military equipment, 1983-87.
Figures in US $m.

Purchase	Price
72 F-18 aircraft	2 059
12 AV-8B Harrier aircraft	170.6
6 LAMPS helicopters	188.2
6 Chinook helicopters	94.2
Total	**2 512**

Source: Author's data.

O&M costs at an estimated $46.42 million, this gives a total cost for the remaining US installations in Spain of around $88.42 million. Adding up the two parts, this gives in the last column of table 10.7 an amount of $166.42 million.

Although this is a very simple method, it gives a rough idea of the scale of the economic benefit to Spain of the bases. It must be emphasized that it is in reference to effects for Spain only; the real cost to the United States is higher. The difference is explained by the relatively low level of local US expenditure.

It is common for a large part of the US staff to be housed in residential estates near the bases. Such an area is El Encinar de Los Reyes, an estate 12 kilometres from Madrid housing 886 US families and offering all kinds of services, including a supermarket, a church, a school and even a police station. Apart from the suppliers, which are difficult to quantify, the area provides employment to 110 persons engaged in various maintenance activities. It should not be difficult to convert this type of facility and thus maintain the existing jobs.[37]

The problems related to companies that provide services to the bases, especially all type of construction, are greater. Again, Torrejón offers the best illustration of this. In 1986–87 there were approximately 40 service companies working exclusively for the base, together employing over 2000 Spanish civilians.[38] The situation has been similar for the other important bases, although no specific data are available.

Purchases of US military equipment and the Spanish arms industry

Spain's purchase of US military equipment is without doubt one of the central issues of US–Spanish economic relations. The task here is to see how important the purchase of military equipment from the USA (with or without military aid funding) has been for the development of the Spanish arms industry.

The historical dependence of the Spanish armed forces on US equipment remains. The Ministry of Defence has committed large sums to arms purchases in the United States during the 1983–87 period, as can be seen in table 10.8. To this should be added the purchase, according to media and expert sources, of

Table 10.9. Main Spanish arms procurement programmes involving US compensation, 31 December 1986

Purchase	Supplier
EF-18 fighter/attack aircraft	McDonnell Douglas
AV-8B Harrier vertical take-off aircraft	McDonnell Douglas/British Aerospace
Chinook transport helicopters	Boeing Vertol Company
Harpoon system armaments	McDonnell Douglas
LAMPS MK II anti-submarine system	Sikorsky/IBM/General Electric
C-130 Hercules transport aircraft	Lockheed
Bushmaster 25-mm gun	McDonnell Douglas
M-113 vehicles	FMC
TOW anti-tank missiles	Hughes
Gas turbines/control system for frigates	General Electric
Navy ammunition transport vehicles	BMY (Harsco Corporation)
Super-Puma helicopters	Aerospatiale
Spare parts for CL-215 Canadair aircraft	AIM Enterprises
Roland armaments system	Euromissile
Aspide armaments system	Selenia/Contraves

Source: Rodriguez, A., 'Los programas de compensaciones asociados a las adquisiciones de material de defensa', *Economia industrial*, Jan./Feb. 1987.

spare parts and supporting equipment for the Spanish Air Force and Navy to a value that corresponds to more than one quarter of the overall figure expressed in the table. This means that total payments to the United States from 1983 to 1987 lie between $3.235 billion and $3.529 billion. This figure corresponds to over 60 per cent of the Spanish MoD's total expenditure, national as well as international, for new investments and replacements for that period.

To the above must be added an important qualitative consideration: the items corresponding to spare parts and supporting equipment, on which practically no hard data are available, constitute the basic element of Spain's dependence on US military equipment. This is largely due the fact that the equipment in question cannot be replaced by other suppliers on the international market, either because of its strategic nature, making it subject to (re)export restrictions, or because the technology is simply not available outside the USA. Given the indirect link between US exports and the existing accord, this dependence invariably weakens Spain's position *vis-à-vis* the USA at the negotiation table.

Another element to the procurement side of US–Spanish defence cooperation is the technological and economic compensation to Spanish industry for the purchase of US defence material. These arrangements are the result of direct negotiations between Spanish defence authorities and the US industries concerned. Such arrangements are becoming increasingly common as a means of financing international arms sales. The objectives are usually to improve the balance of payments, gain access to new technology, obtain systems design and development capacity, aid sectors not used in the purchaser's industry and create incentives in certain sectors.[39]

Table 10.10. US compensation to Spanish industry, 31 December 1986
Figures are in US $m.

Category	1986 dollars			1981 dollars	
	Requested	Rejected	Pending	Approved	Accredited
Defence	400.347	162.896	36.407	201.134	181.338
Commercial	322.664	74.051	53.251	195.342	141.084
Total	**723.081**	**236.947**	**89.658**	**396.476**	**322.422**

Figures may not add up due to rounding.

Source: Rodriguez, A., 'Los programas de compensaciones asociados a las adquisiciones de material de defensa', *Economia industrial*, Jan./Feb. 1987.

In the case of Spain, the MoD requires compensation for the purchase of armaments which cannot be supplied by the national industry. To facilitate relations with the United States in this area, a special US compensation directorate (Gerencia de Compensaciones con los Estados Unidos) has been set up within the Spanish Ministry of Defence, with the Ministries of Foreign Affairs, Finance, Industry and Energy represented as well.

Speaking in broader terms—i.e. not strictly limited to defence-related arrangements—compensation usually consist of: (*a*) *Industrial compensation*, direct if the programme enables Spanish industry to manufacture part of the system covered by the main agreement, and indirect if the agreement permits the manufacture of other industrial products in Spain; (*b*) *Technological compensation*, consisting of the transfer of foreign technology (both military and civilian) Spanish companies; and (*c*) *Economic compensation*. Table 10.9 shows the main defence-related compensation programmes existing in Spain.

Because of its volume and complexity, the first of these programmes is the most important. Compensation amounting to $1.543 billion has been obtained for this programme, which corresponds to 100 per cent of the payments for the F-18 aircraft until 1986. The compensation programme requires exporting companies, as the net recipients of these US subsidies, to offer guarantees that: (*a*) the value of the technology purchased from developed countries should be more than 40 per cent; (*b*) the value of the technology transferred should amount to more than 10 per cent of the commitment; and (*c*) compensation for tourism is not to exceed 10 per cent. The Gerencia evaluates and compares incoming requests and expresses the amount as compensation in terms of 1981 dollars. For this purpose, the requests may at all times be approved, rejected or investigated by the Gerencia. Table 10.10 gives the situation as of December 1986. Up to that time the forecasts established had apparently been confirmed.

Table 10.11 lists the programmes existing in December 1986 by sectors. As can be seen, the defence sector tops the list, with 11 participating companies receiving over 50 per cent of the total compensation. In total, the programmes include 160 companies. Of these, 7.5 per cent are public, receiving 55 per cent of the total approved value; less than 70 per cent are multinationally owned,

Table 10.11. Distribution of US compensation to Spanish industry by sector, 31 December 1986
Figures in US $m.

Sector	Approved in 1986 dollars	Per cent of total	Accredited in 1981 dollars	Number of firms
Defence[a]	201.134	50.73	181.338	11
Chemicals	69.655	17.57	44.805	11
Iron and steel	37.231	9.39	28.474	19
Engineering and services (civil)	22.799	5.75	16.874	8
Naval construction	15.673	3.95	11.368	2
Investment and technical transactions	8.818	2.22	7.131	6
Machinery and tools	5.688	1.43	4.161	18
Electronics and information (civil)	5.552	1.40	4.302	5
Transformation metal	5.155	1.30	3.794	28
Food	4.482	1.13	3.614	19
Tourism	3.018	0.76	2.560	9
Other capital goods	1.371	0.35	1.035	16
Construction materials	0.876	0.22	0.652	2
Consumer goods	0.686	0.17	0.508	8
Electrical material	0.349	0.09	0.265	3
Other	13.989	3.53	11.541	6
Total	**396.989**	**100.00**	**322.422**	**160**

[a] Includes aeronautics and electronics industries.

Source: Rodriguez, A., 'Los programas de compensaciones asociados a las adquisiciones de material de defensa', *Economia industrial*, Jan./Feb. 1987.

receiving just over 8 per cent; and more than three-quarters are national companies with less than 500 employees, receiving only 25 per cent.

It should be emphasized that some especially complex projects may involve extra costs for the participating Spanish companies; in such cases the Spanish Government provides special subsidies. As an example, in programmes involving joint production of aeronautical structural sub-units and systems, $117 million (in 1981 dollars) have been set aside as a subsidy to cover any extra costs the Spanish companies might incur.

Technology transfer

In the long term, the technology transfer effect of the compensation programmes is of even greater value to Spain than the immediate economic benefits. Emphasis here is on two areas: the overall importance of US technology to Spanish industry (especially to the defence industry) and the problem of so-called dual-use technologies.

Generally speaking, dependence on US military equipment generates an immediate technological dependence, both as it applies to O&M and to activities related to licensed or joint production.[40]

ECONOMIC ASPECTS OF THE US MILITARY PRESENCE IN SPAIN

Table 10.12. Spanish arms industry technology transfer contracts with foreign firms, 1974–84

Spanish firm	I	II	III	IV	V	VI	VII	VIII: Foreign firms (number/country)
Naval industry								
Bazán	43	10	25	–	4	4	33	13US, 11FRG, 5F, 5UK, 3J, 2I, 2NL, 1CH, 1N
San Carlos	2	–	2	–	–	–	1	1UK, 1FRG
Astilleros Españoles	1	–	1	–	–	–	1	1F
Electronics industry								
Eisa	8	–	8	–	–	–	3	7F, 1S
Eesa	4	2	2	–	–	–	2	3F, 1UK
Cecsa	1	1	–	–	–	–	1	1US
Sener	1	–	–	–	1	–	1	1P
Marconi	1	1	–	–	–	–	1	1FRG
Bressel	1	–	1	–	–	–	–	1US
Aeronautical industry								
Casa	25	3	4	–	5	13	13	10US, 8UK, 4F, 3FRG
Military vehicles								
Enasa	20	2	9	2	2	7	11	5UK, 4F, 4US, 2CH, 2NL, 1FRG, 1I, 1B
Armament industry								
Santa Barbara	3	1	1	–	–	1	–	2F, 1FRG
Llama	1	–	–	–	1	–	1	1UK
Barreiros	3	–	3	2	–	–	1	2US, 1NL
Laguna de Rins	1	–	1	–	–	–	1	1F
Total	**115**	**20**	**57**	**4**	**13**	**25**	**70**	

Legend:
- I: Total Contracts
- II: Final product manufacture
- III: Components manufacture
- IV: Production premises
- V: Use of patent, licence
- VI: Technological support
- VII: Dual-use technology

Country key:
- B: Brazil
- CH: Switzerland
- F: France
- FRG: FR Germany
- I: Italy
- J: Japan
- N: Norway
- NL: Netherlands
- P: Poland
- S: Sweden
- UK: United Kingdom
- US: United States

Source: Ranninger, H., 'La transferencia internacional de tecnología. Teoría y evidencia', unpublished doctoral thesis, Faculty of Economics, University of Madrid (Complutense), 1987.

Table 10.12 offers another way of looking at the importance of US technology for the Spanish defence industry. This table summarizes technology transfer contracts with the Spanish defence industry from 1974 to 1984. As can be seen, the United States is the leading supplier in terms of quantity. Of the 115 contracts included, 30 are US, with emphasis in the aerospace, naval construction and military vehicle industries. If the orders placed under these contracts are considered,[41] the United States also comes out ahead.

The origin of the conflict on dual-use technology is to be sought in US pressures on the Spanish Government to agree to join the Coordinating Committee on Export Controls (COCOM), and thereby participate in the multilateral NATO effort to control the export of strategic materials to the Soviet Union and its allies. As an alternative, the US Government demanded that Spain introduce domestic regulations that would give the Government power to sanction companies re-exporting dual-use technology without permission.

After internal debate, the Spanish Government responded to the US demands by issuing the Ministerial Order of 5 June 1985 declaring its intention to join COCOM. Affiliation was affirmed in a royal decree of 25 March 1988. This legislation applies fully to all methods of technology import, whether through contractual transfer, direct investment or the acquisition of goods incorporating certain technical resources. Its purpose is to set up a control system for commercial operations involving the export of goods incorporating dual-use technologies previously imported through any of these three channels.

Fundamental to this policy is of course the concept of dual-use technology. Paradoxically, this concept is not defined in any of the articles of the Ministerial Order of 1985, nor, for that matter, anywhere else in the Spanish law book. In its implementation, therefore, the Order leaves the task of deciding whether or not any given technology is to be considered dual-use to the country of origin. According to the recent decree, the solution is to keep a list of dual-use technologies, to be made available to all applicants.[42] Through this legislation the Spanish Government offers guarantees to protect the interests of countries granting transfer of technology to Spain in all cases involving dual-use technology. In other words, on this particular issue the function of the Spanish code of law to defend national interests against those of foreign firms and countries involved in technological transfer operations is reversed.

What has been said above of course only refers to what is given in the law texts, as no information has yet been published on how the legislation really works. Whether or not it is followed to the letter will depend on specific government practices and on how well these serve Spanish interests in each case.

V. Conclusion: towards a new phase of US–Spanish military relations

After looking at both the historical background and the present situation, several key elements to understanding the economic implications of the US military presence in Spain emerge. In the first place, the economics involved cannot be grasped without an understanding of the origin and development, in political terms, of the US presence. Second, the Spanish perspective has always been to focus on the scant compensation offered by the USA. The total sum received in direct compensation has been considered insufficient in light of the strategic as well as economic value of the bases and the risks assumed by Spain as host. While the loans have been sizeable and have gradually increased, they cannot really be considered compensation as such, given the heavy repayment burden they entail. A third element is the quality of the military equipment that the USA has provided, especially that received as a transfer or loan under earlier agreements. On occasion great expense has been incurred simply for the maintenance of old and faulty equipment.

Another unfavourable aspect for Spain is the fact that no guarantees have ever been made as to the amount to be received each year from the USA, as decisions are made by Congress on a year-by-year basis. This has become especially evident in recent years, as reductions in aid have directly affected the Spanish Government's ability to meet payments for the F-18 aircraft purchase. As for the costs and benefits associated with the operation of the bases, the impact on the Spanish economy is marginal. In certain sectors, however (in particular for certain special service companies catering to the bases and to surrounding communities), the bases are an important source of direct and indirect income and employment.

The most noteworthy aspects of this military presence have to do with the purchase of armaments and spare parts, and the opportunities for compensation and technology transfer that these purchases have made possible. These opportunities have changed concomitantly with the terms of the renewed agreements, but everything seems to point to the fact that through the mechanisms introduced as part of the compensation arrangements, Spanish industry has benefited, indirectly, from the US military presence. On the other hand, a growing dependence on certain replacement parts and other strategic material has made it increasingly difficult for Spain's armed forces to become less dependent on equipment received from the United States.

This study cannot be concluded without mentioning the main elements of the agreement recently reached between the Spanish and US governments on 28 September 1988. This agreement reflects a new orientation for Spanish defence policy, the basic premises of which were outlined in the Cortes on 23 October 1984. These can be summed up in three main points:

First, Spain will stay in NATO under the terms of the 1986 referendum, which set three basic conditions: Spain is to be nuclear free, Spanish forces are not to be integrated into NATO's joint military structure and the US presence in Spain is to be reduced. In December 1988 NATO's Defence Planning Committee approved the general rules for the Spanish military contribution; the other two conditions were taken into consideration in the new bilateral treaty between Spain and the United States.

Second, Spain has sought to join the Western European Union (WEU). This process culminated on 14 November 1988, when the WEU approved both Spain's and Portugal's applications for membership.

Third, Spain is to seek to negotiate a new agreement with the United States, where consideration is taken both to Spain's new defence policy and the changed socio-political conditions in the country (the strengthening of the democratic system) and to ongoing as well as planned modernizations of the

Spanish armed forces. The latter means a greater Spanish contribution to Western defence, in qualitative and quantitative terms. On 1 December 1988 the new Defense Cooperation Agreement was signed between Spain and the United States. The most important features of this Agreement are:

1. The US military presence will be reduced as a consequence of the withdrawal—within 3 years—of 72 F-16 aircraft from Torrejón AB and of most of the 4000 US personnel servicing the aircraft. In addition, the refuelling aircraft at Zaragoza will be reduced from 15 to 10 units and transferred to Morón AB in Andalusia.

2. From a Spanish point of view the agreement means a number of qualitative improvements. These include the strengthening of the Spanish command's authority over the US bases; improvements in the service contracting system (building, repairs, etc.) and of the contract conditions for local personnel; and the adoption of a new status of forces regime based on the principle of reciprocity for the US forces stationed in Spain.

3. The principle of linking the use of the bases to US aid or other commitments has been abandoned. This has two important consequences. First, no credits are now granted by the USA for the purchase of US military material, which for Spain means that a greater budgetary effort is required in the short term. Second, all other aspects of the bilateral co-operation included in previous agreement are abandoned. A separate agreement on the conditions for cultural, technical and economic co-operation is being negotiated.

4. One of the most contentious issues has been the clause prohibiting the stationing or transit of nuclear weapons, introduced as a result of the 1986 referendum. Technically, under the terms of the new agreement, nuclear-armed vessels of the US 6th Fleet are now prohibited from making use of the Rota naval base, while B-52 bombers with nuclear payloads are no longer allowed to use Spanish airspace. A compromise has been reached, however, establishing that while Spain does not accept nuclear arms on its territory, the Spanish Government will not ask to be informed about what types of weapon are carried aboard visiting or transiting US ships or aircraft.[43]

5. One of the consequences of the new agreement is that Spain assumes a greater economic burden.[44] Specifically, this means assuming the O&M costs for the Torrejón base that previously had been a US responsibility—the base will remain operational in order to receive US reinforcements in a crisis situation—and to pay for Spain's enlarged contribution to NATO.

The general view seems to be that the increased cost burdens are compensated for by the political benefits offered by the new agreement. In particular, the agreement is regarded as improving Spain's international status as well as fostering a more autonomous and more European-oriented Spanish defence policy.

As for the latter, this does not run counter to the shifts in the European security system brought about by the political changes that have taken place in the Warsaw Pact countries in the autumn and winter of 1989–90 The context of the US presence has changed as a result of these events, for Spain as for all the West European host countries, but not necessarily in such a way as to reverse or halt the trend, described in this chapter, away from reliance on the United States and toward greater European co-operation. However, while it is impossible to make a detailed forecast of the implications of these changes to the formulation of Spanish defence policy in the future, two immediate challenges should be mentioned. The first relates to the division that already exists between those allies favouring the maintenance of the current organization of NATO, with its strong emphasis on the trans-Atlantic relationship, and those more inclined to develop an autonomous system in Europe. As NATO re-evaluates its traditional objectives in central and northern Europe, a reorientation of forces and facilities on the Southern Flank is inevitable. As the traditional threat—and defence posture—is dismantled, the US military presence in southern Europe may increasingly be perceived as mainly serving out-of-area operations. The second challenge relates to the provision in the 1988 agreement for the transfer of the F-16 squadron from Torrejón to Italy. Already the subject of strenuous debate over costs, the uncertainty in early 1990 over Europe's future political order and security framework has served to reinforce the view among some in the US and NATO military establishment that the transfer should be delayed.

Notes and references

[1] I am greatly indebted to Dr Herman Ranninger for his aid and comments and to Santiago Lopez for his effective collaboration in the collection of data.

[2] At the time of writing, negotiations are under way on the content of a new treaty, following the Spanish decision to request that the United States withdraw its F-16 squadron from Torrejón. The new talks.have been surrounded by great secrecy. This has been denounced by Viñas (see note 6 below) and others. The situation is given very good treatment in Yuste, J., 'Washington guarda secreto sobre sus pactos con Franco', *El País*, 25 Sep. 1987.

[3] There is abundant literature on this subject. The texts we have used most often are Marquina Barrio, A., *España en la Política de Seguridad Occidental* (Ediciones Ejército: Madrid, 1986); Viñas *et al.*, *Política Comercial Exterior en España (1931–1975)* (Banco Exterior de España: Madrid, 1979); Vasquez Montalban, M., *La Penetración Americana en España* (Edicusa: Madrid, 1974); and Grasa, R., 'España y la política de bloques', *Anuario sobre Armamentismo en España* (Fontanara/Centro de Investigación para la Paz: Barcelona/Madrid, 1986).

[4] Vinas *et al.* (note 3), p. 741.

[5] Vinas *et al.* (note 3), p. 742.

[6] These matters have been analysed in depth in Marquina Barrio (note 3), p. 380, and in Viñas, A., *Los Pactos Secretos de Franco con Estados Unidos* (Grijalbo: Barcelona, 1981). In this context it is worth remembering that the contacts between the US Government and the Franco regime go back to World War II, when the air transport (December 1944) and the Air Transport Command (ATC) (March 1945) agreements were signed.

[7] Marquina Barrio (note 3), Viñas (note 6) and Chamorro, E. and Fontes, J., *Las Bases Norteamericanas en España* (Euros: Barcelona, 1976).

[8] The following considerations are based on the information and analysis contained in Viñas *et al.* (note 3), p. 744.

[9] See Viñas *et al.* (note 3), p. 750.

[10] See Viñas *et al.* (note 3) p. 775. The figures below are taken from the same source.

¹¹ Public Law 480 (of 10 July 1954) provided credits to the Spanish Government for the purchase of US agricultural surplus.

¹² This does not apply to the funds emanating from the McCarran Amendment, of which only 20% was applied to US expenses.

¹³ Viñas et al. (note 3), pp. 798 and following.

¹⁴ Viñas et al. (note 3).

¹⁵ On this subject see Viñas (note 6); Braña, J., Busa, M. and Molero, J., 'El fin de la etapa nacionalista: industrialización y dependencia en España', *Investigaciones Economicas*, no. 9 (May/Aug. 1979).

¹⁶ Dura, J., 'Las inversiones privadas norteamericanas en España', unpublished working paper, University of California, Berkeley (undated mimeo).

¹⁷ See 'Stanford Research Institute', *Las inversiones norteamericanas en España* (US Chamber of Commerce: Madrid 1972, p. 16).

¹⁸ Braña, Buesa and Molero (note 15).

¹⁹ See Braña, Bueso and Molero (note 15).

²⁰ Gil Pelaez, J., 'Los EEUU en el movimiento de la productividad' *Información Comercial Española*, no. 409 (Sep. 1967); Buesa, M. and Molero, J., 'Cambio técnico y procesos de trabajo: una aproximación al papel del Estado en la introducción de los métodos de la organización científica del trabajo en la economía española durante los años cincuenta', *Revista de trabajo*, no. 67–68 (July/Dec. 1982); Herrero, J. L., 'La introducción de la organización científica del trabajo en la España de los años 40 y 50', unpublished doctoral thesis, School of Economics, País Vasco University (Bilbao),1987.

²¹ For this reason the Comision Nacional de Productividad Industrial (the National Commission for Industrial Productivity) was created in 1952. See Buesa and Molero (note 20), p. 253; Herrero (note 20), ch. 9.

²² This is mentioned by Buesa, M. and Braña, J., 'Tecnología y dependencia: el caso de la industria militar', *Informacion Comercial Española*, no. 552 (Aug. 1979). For a more in-depth study of the aeronautical industry, see Buesa, M., 'El Estado en el proceso de industrializacíon: contribución al estudio de la política industrial española en el período 1939–1963' unpublished doctoral thesis, School of Economics, University of Madrid (Complutense), 1982, pp. 173–82.

²³ See SIPRI, *SIPRI Yearbook 1977: World Armaments and Disarmament* (Almqvist & Wiksell: Stockholm, 1977), p. 66.

²⁴ See Vazquez Montalban (note 3), p. 41.

²⁵ The claim in the national press that this demand was made was denied by the Spanish Government.

²⁶ See Marquina Barrio (note 3), pp. 842–43.

²⁷ See Marquina Barrio (note 3).

²⁸ As in all such cases, the political process surrounding the signing opens additional channels for aid not mentioned directly in the terms of the agreement.

²⁹ This point is developed further in Molero J., 'Foreign technology and local innovation: some lessons from Spanish defense industry experience', eds P. Gummet and J. Reppy, *The Relations between Defence and Civil Technologies*, NATO ASI Series (Kluwer Academic Publishers: Dordrecht, 1988).

³⁰ The most important reason for the delay was the electoral triumph of the Socialist Party in Oct. 1982. The new Government added a short protocol as a way of safeguarding (at least in the face of public opinion) its electoral promise to review the agreement. In fact, the protocol added nothing new; see Marquina Barrio (note 3), pp. 930–31 and Grasa (note 3), p. 86.

³¹ See notes from the Spanish Minister of Foreign Affairs to the US Ambassador to Spain, numbers 250/1, 251/1 and 252/2 of 1982, acknowledging the receipt of notes informing of the US Government's funding requests for Spain to Congress for FY 1983; appended to the text of the treaty, *Treaties and Other International Acts, Series* 10589 (US Department of State: Washington, DC, 1982).

³² In FYs 1988 and 1989 the FMS to Spain was eliminated. The total aid in the FY 1989 proposal has been reduced to less than $3 million.

³³ *El País*, 12 Feb. 1988.

³⁴ *El País*, 18 Sep. 1986.

[35] *El País*, 17 Jan. 1988.
[36] This figure is based on a total of 1097 Spanish employees with an average yearly wage of 2 million pesetas ($11 765).
[37] *El País*, 1 Feb. 1988.
[38] *El País*, 17 Jan. 1988.
[39] See Rodriguez, A., 'Los programas de compensaciones asociados a las adquisiciones de material de defensa', *Economia industrial*, Jan./Feb. 1987.
[40] Production under US licence was especially important in the naval sector (frigates and carriers). Also worth mentioning is the use of US engines for aircraft developed by CASA, and the import of electronic subunits and materials by the automation and electronics industries, etc.
[41] See Castells, M. *et al.*, *Nuevas Tecnologías: Economia y Sociedad en España* (Alianza: Madrid, 1986), chapter 4, section 2; Ranninger, H., 'La transferencia internacional de tecnología: teoría y evidencia', unpublished doctoral thesis, School of Economics, University of Madrid (Complutense), 1987; Molero (note 29).
[42] The list of dual-use materials and technologies was published by the Spanish Government in early 1990; see *Boletín Oficial del Estado*, 7–8 Feb. 1990. Government officials have since acknowledged that the system needs to be reorganized to allow for the changing international situation.
[43] This compromise has been criticized by many analysts. See for instance Aguirre, M. J., 'Las funciones de las bases', *Anuario El País* (El País-Aguilar: Madrid, 1989), p. 110.
[44] See Cajal, M., 'Historia de una negociación', *Anuario El País* (note 43), p. 106.

11. Costs and benefits of the US military presence in Portugal

Jane M. O. Sharp
SIPRI

I. Portugal's security needs

Prior to General Antonio de Spinola's *coup* on 25 April 1974, Portuguese defence policy was dominated by the two tasks of preserving the colonial empire and maintaining a separate identity from Spain on the Iberian peninsula. After the uprising the leadership of Portugal was taken over by democratic forces that swiftly put an end to the colonial era. One result of this was that Portuguese defence policy became more focused on Europe. As a member of NATO, prior to 1974 Portugal was primarily interested in exploiting military aid from its allies to support its colonial wars. After 1974 Portugal was anxious to use its alliance membership to demonstrate its Euro-Atlantic credentials, especially to obtain US and West European economic assistance.

Following World War II, Portugal's relations with the Western democracies remained uneasy. The Government of António Salazar refused Marshall Aid when it was first offered in 1947, but was later forced, in view of Portugal's growing national debt in the late 1940s, to apply for a special extension of the plan (subsequently approved in 1949–50). Joining NATO was also controversial in 1949, but geostrategic interests prevailed over ideology. Despite the opposition of many in Portugal to intra-European entanglements, the emergence of the United States as the predominant Atlantic power left Salazar with few options. Ironically, as in the case of US aid to the Franco regime in neighbouring Spain, membership in NATO helped give Salazar a measure of recognition abroad and legitimacy at home. The decision to hold the February 1952 North Atlantic Council meeting in Lisbon proved particularly useful to Salazar in this regard. The export of Portuguese armaments to the newly established Federal Republic of Germany also helped to give new life to Portuguese defence industries in the early 1950s.[1]

II. US interests in Portugal

Although situated on the periphery of Western Europe and making little contribution to the forward defence of NATO territory, Portugal has long been recognized as having great strategic importance to the United States. Portugal's key asset is that it controls the strategic triangle bounded on two sides by the island groups of the Azores and Madeira and on the third by the Portuguese mainland. This area is crossed by vital sea lines of communication between the United States, Europe and North Africa, and includes the waters connecting the northern and the southern Atlantic as well as the approaches to the Mediterranean Sea. Portugal first granted the United States the use of facilities in the Azores in World War I and, in the late 1930s, before the outbreak of World War II, US Admiral Sterling asserted that the Azores were 'an advanced strategic border of the North American continent with a similar position in the north Atlantic to that of the Hawaiian Islands in the Pacific Ocean'.[2]

Portugal has been allied with Britain in some form or other since the 14th century, and joined the European Free Trade Association in 1960 largely because Britain was a member. Despite these traditional ties and the World War I links with the United States, however, Salazar would have preferred to follow a policy of strict neutrality in World War II, since he was ideologically closer to Hitler and Mussolini than to either Stalin or the Western allies. This proved impossible, however. In May 1943, President Franklin Roosevelt and Prime Minister Winston Churchill drew up a contingency plan (Operation Lifebelt) to occupy the Azores if Salazar would not agree to make the islands available to the Western allies. Reluctantly, Salazar acquiesced, although as George Kennan (then Charge d'Affaires in Lisbon) noted in December 1943: 'Salazar fears association with us only slightly less than with the Russians'.[3] US forces arrived in the Azores in early 1944 and have been there ever since.

US facilities on Portuguese territory

During World War II the United States maintained a base on the island of Santa Maria in the Azores. This was returned to Portugal in 1944 in exchange for facilities at Lajes, on the island of Terceira. In January 1951 Portugal signed a Mutual Defence Assistance agreement with the United States, followed in September of the same year by an agreement that formally granted the United States the use of the Lajes air base. This agreement has been reviewed approximately every five years and modified by Technical Agreements in 1957, 1976 and 1984. An Exchange of Notes in December 1983 extended the use of the Azores facilities until 1991. The United States also has the use of facilities at Angra do Heroismo, Pico da Barrosa and Praia da Vitória, all in the Azores, as well as in Lisbon.

The Azores base serves three main purposes for US forces. First, it serves as a stopping-off point and refuelling base for US aircraft crossing the Atlantic to reinforce NATO forces in Western Europe. Lajes has also been used as a staging base for US military activities outside the NATO area, such as the supply of arms to Israel in the 1967 and 1973 wars and the raid against Libya in 1986. Second, the Azores serves as a base for surveillance of Soviet submarine activity in the Atlantic, the western Mediterranean and, in particular, in the crucial bottle-neck of Gibraltar. Third, communications facilities on the Azores form an important link in the US Defense Communications System.[4]

US aid to Portugal

1951 to 1974

In exchange for base rights, the United States has provided economic and military aid to Portugal since the early 1950s. Aid levels have fluctuated over the years but Portugal has always claimed that compensation by the United States has never been adequate, given the strategic importance of the bases to the United States on the one hand, and the defence and economic needs of Portugal on the other.

Up to 1961 Portugal received $302.6 million in US aid including grants from surplus stocks, such as F-84 and F-86 fighter aircraft, Lockheed PV-2 Harpoon light bombers, C-47 transport and T-6 Texan trainer aircraft, some of the latter supplied via the United Kingdom. The Portuguese Navy received mine sweepers, patrol boats and frigates, and the Army received M-41 and M-47 tanks, M-16 armoured personnel carriers and 155-mm howitzers. This aid was controversial in the United States. Congressional distaste for the right-wing dictatorship and colonial policy of the Salazar regime found expression in the 1951 agreement, which specified that US equipment sent to Portugal could not be used for purposes other than to defend the North Atlantic area, i.e., it could not be used in Portugal's colonial wars in Africa. In spite of this injunction, however, US equipment was sighted in Portuguese Angola in the late 1950s.[5]

In the early 1960s, the Kennedy Administration endorsed United Nations resolutions calling for an end to Portuguese colonial repression and for Angolan independence. To put pressure on the Portuguese Government, US aid to Portugal was in 1961 reduced from a planned $25 million to $3 million.[6] The Kennedy Administration began to look more favourably on Portugal in 1962, however, as the 1957 agreement on Lajes came up for renewal. As the agreement was being renegotiated, the United States provided a $55 million Export-Import Bank (Eximbank) loan that increased total economic and military aid from the United States to Portugal from $5.6 million in fiscal year 1961 to $72 million in fiscal year 1962. A new agreement on the Azores facilities was concluded in January 1963, and for the remainder of the 1960s the United States abstained from voting against Portugal in the United Nations. Military aid was reduced in the late 1960s, however, and Portuguese requests for bombers were denied by the State Department in 1965, although some B-26 bombers were later smuggled into Portugal.[7]

Military aid to Portugal continued to be an embarrassment for Western governments until the overthrow of the post-Salazar dictatorship of Marcelino Caetano in 1974. Some aid was given through the Military Assistance Program in the early 1970s, designed to support NATO-committed forces, primarily in air defence and anti-submarine warfare, and to enhance Portugal's NATO

Table 11.1. US public aid to Portugal 1975–1986

Type of aid	Type of financing[a]	Commitment US $m.	%	Actual expenditure US $m.	%
Programmmes and projects	c	150.0	14.6	137.0	13.5
Technical co-operation	g	22.7	2.2	17.2	1.7
Balance of payments support	c	300.0	28.8	300.0	29.7
Lajes base agreement	g	299.0	28.7	299.0	29.5
Azores	g	199.0	19.1	199.0	19.5
FLAD[b]	g	100.0	9.6	100.0	10.0
Agricultural goods import (PL 480)	c	215.0	20.6	215.0	21.3
Emergency aid	g	53.5	5.1	53.5	5.3
Total		**1 042.2**	*100.0*	**1 011.6**	*100.0*

[a] c: credits; g: grants.
[b] Fundação Luso-Americano para o Desenvolvimento.

Source: Oppenheimer, J., 'A ajuda pública dos EAU a Portugal', *Economia e Socialismo*, vol. 11, no. 72/73 (Dec. 1987).

capability through training and material. Until 1971 there was no direct compensation for the use of facilities at Lajes, but in 1971, when the agreement was renewed, the United States agreed to pay Portugal $400 million in compensation in the form of another Eximbank credit for military equipment to be purchased from the United States.

After 1974

After the fall of the Caetano Government, aid continued to fluctuate as many senior officials in the Nixon Administration worried that Portugal might fall under the control of the Communist Party. Table 11.1 shows US economic aid (excluding military aid) to Portugal for the period 1975–86. Military aid has been estimated at $630 million for the period 1962–85, with $213 received since the new aid package was negotiated in 1983. During this period the structure of military aid changed. Consisting mainly of grants (worth $215.2 million) in the 1962–81 period, it has since 1982 consisted of a combination of grants and credits (worth $414.5 million).[8]

Economic aid has been distributed unevenly. During 1975–77 funds were divided between a number of co-operative programmes for economic and social development, while during 1978–81 most US aid went to meet Portuguese balance of payments. Between 1979 and 1983 total US aid amounted to $140 million, of which $60 million went towards the reorganization of a NATO-committed air transport brigade and $80 million to finance various civil programmes.

A new aid package was negotiated in 1983 in conjunction with the extension of the United States access to Lajes through 1991. In this agreement the United

Table 11.2 Activities of Fundação Luso-Americano para o Desenvolvimento (FLAD), 1985-86

Figures are as of 31 December 1986.

Area	No. of projects	Allocations Esc. m.	US $m.
Private economic sector development	24	3 154	470
Science and technology	36	2 617	375
Education	46	3 315	494
Public administration and regional development	12	2 107	314
Culture	18	349	52
Others	2	22	3
Total	**138**	**11 463**	**1 708**

Figures may not add up to totals due to rounding.

Source: FLAD, *Relatório e Contas 1985–86* (Fundação Luso-Americana para o Desenvolvimento: Lisbon, 1987)

States promised to assist in the economic development of Portugal, and in 1985 the Fundação Luso-Americana Para o Desenvolvimento (the Portuguese-American Development Fund, FLAD) was established to channel aid to areas of technical, economic and cultural activity. Table 11.2 shows how FLAD funds were distributed in 1985–86.

By 1986, FLAD had received approximately 500 applications for funds. Distribution was difficult, however, as basic economic needs competed with efforts to catch up in science and technology. A $100 million construction programme began in 1987 to improve hotel and other tourist facilities in the Azores. While this defused some local opposition to the US facilities, it proved especially controversial in certain mainland regions, where many felt that the aid was more urgently needed.[9]

In September 1987 Portuguese Prime Minister Cavaco Silva asked for a renegotiation of the terms of the 1983 Agreement, complaining that the United States had failed to honour its pledge of annual increments in economic and military aid. Aid and trade had both fallen between 1985 and 1987. For its part, the Reagan Administration argued that the United States had not made firm commitments to increase aid but only to 'use its best efforts' with the Congress. In a visit to Washington, President Mario Soares argued that US compensation for its use of bases in the Azores was inadequate and that further military aid was necessary to modernize the armed forces. US Secretary of Defense Frank Carlucci, a former Ambassador to Portugal, persuaded Silva to renegotiate increased compensation without a formal review of the 1983 agreement. A new aid package was agreed on 24 February 1988. Consultations continued until January 1989, resulting in new US loans of $100 million per annum up to $600 million.[10]

III. Portuguese commitments to NATO

As noted in chapter 1, the US commitment to NATO is not wholly altruistic but is designed in part to generate reciprocal commitments from the West European governments. Portugal contributes 3 per cent of its gross national product (GNP) to defence, but as the country is one of the two poorest NATO allies (only Turkey has a lower per capita income) this is quite inadequate to provide even for the defence of the national territory. Portugal is thus heavily dependent on military aid from its allies.[11]

If Portugal contributes little to the alliance financially, it nevertheless scores well in Washington, Paris and London both in providing vital strategic real estate and in not raising objections to US, French and British out-of-area military adventures. Portugal was the only NATO ally to allow the United States to use its facilities in support of Israel in the 1973 Yom Kippur War, and was one of the few to support US political and economic sanctions against the Soviet Union after the invasion of Afghanistan, for example by declining the Soviet invitation to the 1980 Olympic Games. In April 1977 France was able to use the airfield of Porto Santo on Madeira to project military power southwards during the Shaba conflict in Zaire. The same facilities were used by the United Kingdom during the Falklands/Malvinas War in the early 1980s.

Given Portugal's own colonial past, this acquiescence was not surprising. Nevertheless, the governments that have ruled the country since 1974 were clearly uncomfortable with some of these decisions. In 1980 it was announced that foreign use of facilities in Portugal for out-of-area activity would henceforth be subject to case-by-case review.[12] Since then, Portuguese support for out-of-area activity has become less automatic and more pragmatic. In 1984, for example, Foreign Minister Jaime Gama made it clear that facilities in Portugal would never again be used for actions against Arab countries. In 1986 his successor Pedro Pires de Miranda deplored the US bombing raid over Libya.[13]

Portugal is seen by the USA as a more reliable and robust ally than Spain on the issue of nuclear weapons, but has not really been tested on its willingness to share the risk of nuclear basing. There are no US nuclear weapons based on Portuguese territory, although contingency plans exist to store 32 nuclear depth bombs at Lajes in wartime, and presumably US vessels visiting Lisbon and the Azores facilities are sometimes armed with nuclear weapons.[14] Moreover, when Spain and Portugal joined the newly revived Western European Union (WEU) in November 1988, they signed the 1987 Platform Document of the WEU, thereby accepting an unambiguous nuclear deterrent policy in the defence of Western Europe. Unlike Spain, Portugal did not request not to be asked to accept nuclear weapons on its territory for the time being.[15]

Although Portugal was a founding member of NATO in 1949, it was peripheral to West European politics and defence until the end of the Caetano regime. There was no obvious WTO threat to Portuguese national security, and Portugal's meagre defence resources were focused on equipping the Army to fight its colonial wars. In 1963 the United Nations placed an embargo on sales to Portugal of armaments for use in Angola, Mozambique and Portuguese Guinea.[16] From 1963 until 1974 NATO suppliers of arms to Portugal therefore insisted that the equipment was to enhance Portugal's contribution to NATO and that arms were only for use in 'metropolitan' Portugal. US and West European arms and equipment nevertheless showed up regularly in Angola and elsewhere in Africa where Portuguese forces were engaged.[17]

Although an Atlantic rather than a Central European or Mediterranean power, at the 1952 NATO summit in Lisbon Portugal earmarked troops for contingencies in the FRG under SACEUR (Supreme Allied Commander Europe) command. During the 1950s these Portuguese troops participated in a number of major NATO exercises on the Central Front, such as Lion Noir and Sidestep. During the 1960s and early 1970s, the brunt of Portugal's forces were tied down in Africa, but in the second half of the 1970s successive democratic governments again began to seek ways to demonstrate a commitment to Portugal's NATO allies in Western Europe. In 1976 the Portuguese Army set up the 1st Composite Brigade as a contribution to the alliance's forward defence strategy. This brigade was recategorized to NATO standards in 1987, and assigned to SACEUR under the operational command of COMLAND-SOUTH (Commander Allied Land Forces Southern Europe), earmarked for service in northern Italy with the III and V Italian Army Corps.[18]

Relations with Spain

Spanish membership in NATO has complicated Portugal's contribution to the integrated military command of the alliance. As mentioned, the two dominant objectives of Portugal's security policy until the mid-1970s were to maintain control over the colonies and to uphold the country's national identity on the Iberian peninsula. The 1980s saw big changes in both areas. The colonial era had definitely come to an end, and Portugal now shared membership with Spain in NATO, the WEU and the Independent European Programme Group (IEPG), the latter an inter-governmental body set up to promote better co-operation in the West European defence industry.

In 1949 some Spaniards objected to Portugal joining NATO, arguing that it would undermine the bilateral Iberian Pact, signed in March 1939 with the aim of keeping both countries out of the coming war.[19] In November 1977, when both countries had replaced their right-wing dictatorships with democratic governments, they signed a new Treaty of Friendship and Co-operation. Spain's accession to NATO in 1982 nevertheless presented Portugal with several potential problems. Like France, Spain opted to remain outside the integrated military command of the alliance. Should Spain decide to integrate its forces more fully with NATO in the future, however, Portugal would prefer that Spain be subordinate to SACEUR rather than SACLANT. In 1989 the Commander-in-Chief Iberian Atlantic Area (IBERLANT) was a Portuguese admiral. This command is responsible for an area of the Atlantic that includes the Canary

Islands and Gibraltar; clearly an area of Spanish interest. Some Portuguese analysts fear that NATO might assign Spain, with its more numerous and better-equipped forces, security tasks within the IBERLANT command for which Portuguese forces are not adequately equipped.[20] Portugal thus insisted that the Azores and Madeira, formerly under the command of WESTLANT (Western Atlantic Area), with headquarters in Norfolk, Virginia, be reassigned to IBERLANT.[21]

IV. Modernizing the defence-industrial base

Once the colonial wars were over, the Portuguese forces returned home. They were in dire need of modernization. Military equipment produced in Portugal was more suitable for counter-insurgency in Africa than for the defence of Western Europe, although orders placed by other countries kept production lines open in artillery, mortar and small arms munitions.[22]

Finding the will and the funds to modernize the Army was difficult, however, with a succession of unstable governments, an unhealthy economy, inefficient defence industries and a popular yearning for an end to militarism. Portugal had to contend with 14 different governments between 1974 and 1983, when a relatively stable coalition emerged, led by Socialist Party leader Mario Soares. All the post-colonial governments had to wrestle with enormous economic problems, including pent-up demands for jobs and higher wages. In the first years following the *coup*, policy was dominated by the Portuguese Communist Party. They nationalized the main industries and the banking system, creating large public corporations that ran up crippling debts.

The Army played a central role in bringing democracy to Portugal but leaned, unlike the Spanish armed forces, to the left. This generated alarm in the USA, with Henry Kissinger and others in the Ford Administration questioning whether continued membership in NATO was still appropriate for Portugal.[23] Portugal was excluded from NATO's Nuclear Planning Group (NPG) as long as the Marxists were in power; after their defeat in 1975, however, Portugal was again represented in the NPG.[24] Anxiety about the coalition governments that followed was effectively countered by support for the moderate Socialist Party from the US Ambassador in Lisbon, Frank Carlucci, and from West European Social Democrats, notably Helmut Schmidt in the FRG. As the Socialist Party gained ascendancy, the Government did not denationalize the state enterprises established by the Communists, however, adopting instead increasingly austere measures. When inflation reached 25 per cent per annum, for example, the ruling coalition put the economy under control of the International Monetary Fund. This led to unpopular cuts in public spending and wage controls.[25] With the austerity at home, however, came a reputation in the international community of growing stability. This was reflected in military assistance from the allies, notably from the FRG, and in successful application for membership in the European Community (EC) and the WEU. Portugal and Spain applied for admission to the EC in 1984 and were admitted in 1986. Both countries were admitted to the WEU in November 1988.[26]

West European economic assistance

Portugal clearly regards membership in the EC as important to its future prosperity and security as membership in NATO. When asked in 1984 what Portugal wanted from membership in the EC, Antonio Marta, President of the governmental committee for integrating Portugal into the EC, said that 'it should consolidate our democracy and be the motor of our modernization'.[27] In May 1988 *The Economist* reported that the mood in Portugal about EC membership was still euphoric. In the first two years EC development grants for a variety of construction projects, including roads, railways and expansion of telephone services, totalled 137 billion Escudos ($1 billion). Trade with other EC countries had doubled, foreign investment was flooding in, and the only anxieties were whether Portugal could design projects capable of absorbing this capital and grow sufficiently competitive to benefit from the post-1992 single EC market. Some dissident voices wondered whether the country would not be better off with another decade or so of protective barriers.[28]

Nevertheless, by December 1988, at its 12th party congress, even the Portuguese Communist Party had acknowledged the advantages of the EC, despite its earlier predictions that membership would bring economic disaster to Portugal, including unemployment and exploitation by the bureaucracy in Brussels. In early 1989 *New Times* in Moscow reported that Portugal was receiving some $500 million per year from various EC funds and reported that without integration in the EC, 'Portugal is unlikely to have remained a developed capitalist economy and would almost certainly have regressed into economic, scientific and technological backwardness'.[29]

West European military assistance

Like the United States, both France and the FRG provided aid to Portugal during the colonial period in exchange for access to military facilities. As a solution to the problem of its own crowded airspace, the FRG made arrangements in 1960 to use an air base in Alcochete, near Lisbon, and in 1962 established an air base and pilot training facility in Beja. In exchange the FRG provided military aid to Portugal through the 1960s. This included surplus Luftwaffe stocks of G-91 fighter aircraft. FRG spokesmen said these aircraft were to be used exclusively for the defence of NATO, but Portuguese Foreign Ministry spokesmen said they could be used anywhere in Portuguese territory, which extended to Angola, Mozambique and Portuguese Guinea.

France supplied most of the counter-insurgency weapons used by Portugal in Africa during the 1960s, in exchange for the use of a missile-tracking station on the island of Flores in the Azores. Britain also supplied arms to Portugal in this period, largely naval equipment: minesweepers, frigates, submarines and patrol boats. These arms sales were more controversial in Britain than in France, and in an interview in the *Daily Telegraph* President Salazar reported that a 1964 order for four frigates and four submarines, originally to be placed with a UK firm, was instead given to France because of British restrictions on use in Africa.[30]

Since the fall of the Caetano Government, West European military assistance to Portugal has been regarded in a more positive light. In exchange for continued use of airfields and training facilities, the FRG provides substantial economic and military aid to Portugal, and has done more than any other ally to bolster the military capabilities of those southern countries that constitute what many Western strategists see as the 'soft underbelly of NATO'. A 1985 FRG Defence White Paper claimed that 80 per cent of such aid consisted of delivery of new material free of charge and 20 per cent of serviceable Bundeswehr surplus. This is not politically controversial in the same sense as it was in the 1950s and 1960s, and many in the FRG believe that other NATO allies should do more to help.[31]

In 1987 Rainer Rupp, member of NATO's Economic Directorate, acknowledged that aid from the wealthier allies was sporadic at best and consisted mostly of transferring obsolete equipment that only complicated modernization problems in the recipient countries.[32] NATO's Secretary General, Dr Manfred Wörner, a former FRG Minister of Defence, has called for expansion of 'the very exclusive Southern Region Supporters Club', but the response has been modest at best; Canada and the Netherlands provide some help but not on the scale that satisfies the Southern Flank's yearning to modernize.[33]

In December 1985 the North Atlantic Council agreed upon an 'Armaments Cooperation Improvement Strategy' designed to help the countries with less developed defence industries (the LDDIs).[34]

At the beginning of the 1990s, however, Portuguese planners can look for help to several all-European institutions designed to promote better co-operation among defence industries. Of these the most important are the WEU, the Eurogroup, and the IEPG. The Western European Union is the oldest of these but has been moribund for most of its existence. The WEU was revived in 1984, largely at the initiative of former French Premier Jacques Chirac, as part of the wider effort to strengthen the European pillar in NATO and bind the FRG closer to the Western allies in the aftermath of intra-alliance differences over the double track decision on intermediate-range nuclear forces.[35]

Portugal was also a founding member of the Eurogroup (which excludes France) in 1968, and the IEPG (which includes France) in 1976. Under Dutch chairmanship and with the enthusiastic support of British Defence Minister Michael Heseltine, the IEPG also underwent a revival of sorts in the early 1980s and commissioned a report (The Vredeling Report) on how to improve the competitiveness of Europe's defence equipment industry.[36] The report was then translated into a new IEPG Action Plan on the development of a European

Armaments Market, which envisaged special assistance and protection for the LDDIs and established in 1988 a new secretariat in Estoril, Portugal.[37]

Unlike Michael Heseltine, some US and British officials tend to view all-European bodies such as the WEU and the IEPG as working against transatlantic co-operation and undermining alliance cohesion.[38] In 1985 Richard Burt, then serving as an Under Secretary of State for European Affairs in Washington, chastised members of the WEU for co-ordinating a European position on East–West arms control.[39] In early 1989, Mack Mattingley, NATO Assistant Secretary for defence support, criticized renewed efforts towards a common defence market in Western Europe, suggesting that the IEPG was beginning to appear as an 'inward looking cosy Europe-only club, not an outward looking force and contributor to wider Atlantic cohesion'.[40] Mattingley complained that the IEPG duplicated the work of the Conference of National Armaments Directors and generally slowed down NATO defence co-operation. Still not well understood is how the IEPG Action Plan will relate to the Single European Act of the European Community. Defence production and trade is excluded from the Treaty of Rome, but Article 30 of the Single European Act widens the area of interest of EC governments to 'maintain the technological and industrial conditions necessary for security'.[41]

Through 1988 and 1989 European officials discussed the possibility of a joint defence research agency, modelled on the Eureka project for research and development in civilian technologies set up by EC and neutral countries in Europe.[42] This was launched as The European Long Term Initiative in Defence (EUCLID) under the auspices of the IEPG rather than the EC in June 1989. Former British Defence Minister George Younger warned that the project must be careful not to trigger North American anxieties about being shut out of a European arms market, while the Portuguese Foreign Minister, Enrico de Melo, emphasized the importance of forging European co-operation so as to better meet the competition from Japan and North America.[43]

Thus, in the 1990s Portugal and the other Southern Flank countries may come to see participation in these all-European institutions as a more viable option than relying on US aid to modernize their outdated defence industries. This poses some domestic difficulty in Portugal, with the centre-left arguing for more intra-European co-operation and the centre-right anxious not to upset traditional Atlantic ties.[44]

V. Conclusion

What, then, are the costs and benefits to the United States of its facilities in Portugal, the costs and benefits to Portugal of the US military presence, and the likely impact on these costs and benefits to both countries under the two withdrawal options?

Given the small number of US personnel involved, there has been little direct political or economic cost to Portugal. US aid in developing tourist facilities successfully defused a nascent independence movement in the Azores, and there has been no hint of unhappiness with US naval facilities in Lisbon. Portugal has not claimed that the costs of the US presence are too high, rather that the benefits are not high enough. Specifically, Portuguese ministers complain that US payments are inadequate given the manifest strategic benefits that Portuguese facilities provide for the alliance in general and for the United States in particular. Clearly Portugal would rather have the bases than not, but is frustrated by its inability to bargain for a better deal. Withdrawal option B assumes retention of a skeleton US presence in Europe, which would probably include all the facilities in Portugal, but under option A, the 'gone for good' withdrawal, all US facilities on Portuguese territory would close down. This would entail the departure of approximately 1700 Air Force and 500 Navy personnel, not a sufficient presence to have a direct effect on the economy, except in communities immediately surrounding the bases. The tourist facilities already built on the Azores represent sunk costs, and would not be affected should the US military facilities disappear.

For Portugal, a more serious issue would be if withdrawal meant the loss of all US assistance to help modernize the Portuguese armed forces and the defence industrial base. Although the Portuguese Government has always been dissatisfied with the level of compensation from the United States, it probably regards something as better than nothing and would prefer not to have to rely completely on its West European allies for assistance. On the other hand, in the 1990s the European Community, the IEPG and the WEU will probably be more important motors for the modernization of Portugal and for its economic security than will bilateral relations with the United States. Jaime Gama, Socialist Party deputy and chairman of the parliamentary Defence Committee, noted that as a member of the EC Portugal can no longer be treated as a Third World country by the United States.[45] Thus Portugal's increasing links with its European partners may increase its bargaining power with Washington.

Notes and references

[1] Daguzan, J. F., 'The defence industries and armament policies of Portugal: the revival', *Defense & Armament Héraclès International*, no. 67 (Nov. 1987), pp. 57–59.

[2] Vasconcelos, A., 'Portuguese defence policy: internal politics and defence commitments', ed. J. Chipman, *NATO's Southern Allies: Internal and External Challenges* (Routledge: London, 1988), pp. 86–139.

[3] Vasconcelos (note 2), p. 2.

[4] For details of US facilities on the Azores and on mainland Portugal, see Duke, S., SIPRI, *United States Military Forces and Installations in Europe* (Oxford University Press: Oxford, 1989), pp. 236–50.

[5] SIPRI, *The Arms Trade with the Third World* (Almqvist & Wiksell: Stockholm, 1971), p. 670.

[6] SIPRI (note 5), pp. 668–74.

[7] SIPRI (note 5), pp. 668–74.

[8] Oppenheimer, J., 'A ajuda publica dos EAU a Portugal', *Economia e Socialismo*, vol. 11, no. 72/73 (Dec. 1987), p. 23.

⁹ Oppenheimer (note 8). See also: *Diario de Noticias*, 22 Feb. 1988. For a discussion of the importance of US grants to the Azores economy, see Monteiro da Silva, J. M. 'Topics for an analysis of the recent economic evolution of the autonomous region of Açores', *Recent Evolution and Transformation Perspectives of the Portuguese Economy* (CIESP, Instituto Superior de Economia, Universidade Tecnica de Lisboa: Lisbon, 1983).

¹⁰ Vasconcelos, A., 'The impact of democratization on the security of Portugal', ed. K. Maxwell, *Defence and Foreign Policy in Portugal* (Duke University Press: Durham, N. C., 1980)

¹¹ For NATO Defence Expenditures as percentage of GDP in both current and constant prices from 1978–88, see NATO's Defence Planning Committee, *Enhancing Alliance Collective Security: Shared Roles, Risks and Responsibilities in the Alliance* (NATO: Brussels, Dec. 1988), pp. 12–13.

¹² Vasconcelos (note 2), p. 118.

¹³ Vasconcelos (note 2), p. 118.

¹⁴ Arkin, W. M. and Fieldhouse, R. W., *Nuclear Battlefields: Global Links in the Arms Race* (Ballinger: Cambridge, Mass., 1985), p. 229.

¹⁵ McEwen, A., 'Revitalized WEU set to become voice of Europe on joint security policy', *The Times*, 15 Nov. 1988; and unsigned leader, 'Closer Western Union', *The Times*, 15 Nov. 1988; 'Iberian nations join defense group', *International Herald Tribune*, 15 Nov. 1988.

¹⁶ Landgren, S., SIPRI, *Embargo Disimplemented* (Oxford University Press: Oxford, 1989), p. 5.

¹⁷ SIPRI (note 5), pp. 668–74.

¹⁸ Miguel, M. F., 'Army support of the Southern Flank', *NATO's Sixteen Nations*, Oct. 1988, pp. 67–68.

¹⁹ Vasconcelos (note 2), pp. 94.

²⁰ Vasconcelos (note 2), pp. 117–18.

²¹ Ausland, J., 'Atlantic islands with military roles', *International Herald Tribune*, 10 Mar. 1989.

²² Barata, F. T., 'The defence industry of Portugal', *NATO's Sixteen Nations*, Feb. 1983, pp. 70–75.

²³ On Kissinger's anxiety about Portugal in the mid 1970s, see R. L. Garthoff, *Detente and Confrontation* (Brookings Institution: Washington, DC, 1985), pp. 486–87.

²⁴ Bell, C., *The Diplomacy of Detente: The Kissinger Era* (Martin Robertson: London, 1977), p. 164, cited in Chipman (note 2), p. 373.

²⁵ Lewis, P., 'Socialists forces to use economics of the right', *New York Times*, 1 Dec. 1983; Darnton, J., 'Ideology yields to pragmatism for socialists', *New York Times*, 30 Nov. 1983; Vinocur, J., 'Socialist governments in Western Europe find intellectual's ardor for left cooling', *New York Times*, 2 Dec. 1983.

²⁶ Clarke, M., 'Evaluating the new Western European Union: the implications for Spain and Portugal', ed. Maxwell (note 10).

²⁷ *Le Figaro*, 16 May 1984, cited in Tugendhat, C., *Making Sense of Europe* (Pelican Books: Harmondsworth, 1987), p. 115.

²⁸ 'A survey of Portugal', *The Economist*, 28 May 1988, pp. 15–17.

²⁹ Shavrova, T., 'Portugal: Common Market integration and dogmas: the surprises Common Market membership brought to the nation', *New Times* (Moscow), no. 9 (1989), p. 19.

³⁰ *Daily Telegraph*, 16 Nov. 1965, cited in SIPRI (note 5), p. 674.

³¹ Flume, W., 'Military assistance within the alliance', *NATO's Sixteen Nations*, Sep./Oct. 1984, pp. 54–63; Rupp, R., 'Assistance for Greece, Portugal and Turkey: a form of burden-sharing', *NATO's Sixteen Nations*, Dec. 1985, pp. 49–58.

³² Rupp, R., 'Defence-industrial assistance to Greece, Portugal and Turkey', *NATO's Sixteen Nations*, Oct. 1987, pp. 81–88; and Rupp (note 31).

³³ Rupp (note 32).

³⁴ 'Press statement on Armaments Cooperation issued by Foreign Ministers at the North Atlantic Council meeting on 12–13 December 1985', reprinted in Beard, R., 'New NATO approaches to improved armaments cooperation', *NATO's Sixteen Nations*, Nov. 1986, pp. 26–32.

[35] On the revitalized West European Union, see *inter alia*: Taylor, T., 'Alternative structures for European Defence Cooperation', eds K. Kaiser and J. Roper, *British–German Defence Cooperation* (Jane's: London, 1988), pp. 170–183.

[36] Vredeling, H., *Towards a Stronger Europe*, Independent European Programme Group (North Atlantic Treaty Organization: Brussels, Dec. 1986), vols 1 and 2.

[37] IEPG Communique, 22 June 1987, reprinted in *NATO's Sixteen Nations*, July 1987, p. 31; The IEPG Action Plan and its accompanying communiqué were published in *Atlantic News*, no. 2065, annexes 1 and 2, 15 Nov. 1988.

[38] On British attitudes towards the WEU, see Clarke (note 26).

[39] Bloom, B., 'US warns Europe on independent defence stance', *Financial Times*, 2 Apr. 1985.

[40] 'NATO official criticises "inward looking" IEPG', *Jane's Defence Weekly*, 4 Mar. 1989.

[41] Cited in Walker, W. and Gummett, P., 'Britain and the European arms market', *International Affairs*, vol. 65, no. 3 (Summer 1989), pp. 419–42, p. 432.

[42] Dickson, D., 'EUREKA', *Technology Review*, Aug./Sep. 1988, pp. 27–33.

[43] Fouquet, D., 'IEPG launches $120 million joint research program', *Jane's NATO and Europe Today*, vol. 4, no. 1 (4 July 1989), pp. 1–2.

[44] For an example of the Atlanticist view in Lisbon, see: Cutileiro, J., 'A Portuguese view: Keep the Atlantic Alliance', *International Herald Tribune*, 14 Apr. 1989.

[45] Vasconcelos (note 10).

12. Costs and benefits of the US military bases in Greece

Athanassios G. Platias
Panteios University, Athens, Greece

I. Introduction

This paper evaluates the costs and benefits of the US military presence in Greece. As the purely economic dimension of this presence is at times overshadowed by political considerations that are part and parcel of the arrangement between the two countries, however, the aim here is not only to provide an economic evaluation but also to explore the political costs and benefits involved.

The bilateral Greek–US relationship can fully be understood only in a larger context that also includes Turkey. Since the Truman Doctrine was first formulated in 1947, Greece and Turkey have become locked into a triangular relationship with the United States. From a US perspective, Greece and Turkey are vital allies in the global confrontation with the Soviet Union. The strategic location of the two countries, close both to the Soviet Union and to the Middle East, makes them indispensable to Western defence. Developments in the Middle East and the Persian Gulf in recent years have served only to enhance the importance of both allies to the United States. Greece and Turkey for their part, however, are less concerned with such global issues and more concerned with the ongoing struggle between them. This complicates both the strategic and bilateral relations of each of the two countries with the United States.

This chapter evaluates the political and economic aspects of Greek–US relations, as seen from a Greek perspective. The thesis presented here is that the Greek–US relationship is highly asymmetrical, to the United States' favour. It follows that if the United States were to withdraw from Greece, of the two countries it would stand to lose the most. For Greece, the economic consequences of a US withdrawal would be minimal.

This argument is subject to one important qualification. While Greece has little to lose from a US withdrawal, economically as well as politically, this applies only if it is undertaken as part of an *overall* reduction of the US military presence in Europe. If the bases were to be transferred to Turkey, and if the United States were to upgrade Turkey's capabilities, then the military and political implications would far outweigh the relatively negligible economic consequences.

II. Political aspects of Greek–US relations

The origins of the US military presence in Greece

Greece's position in the post-war order was decided in the famous 'half sheet of paper', exchanged between Stalin and Churchill on the night of 9–10 October 1944,[1] by which Greece was placed within the British sphere of influence. Although Britain's post-war dominance in Greece proved short-lived—the UK's exhaustion at the end of World War II contributed to its decision to relinquish its hegemonic responsibilities—it served to establish the West's dominance over the country in the subsequent cold war period. The declaration of the Truman Doctrine marked the beginning of an era of US dominance in the eastern Mediterranean. The US military arrived in the name of democracy, but also with a clearly stated perception of Greece's strategic importance. Since 1947 the USA has maintained an active interest in the country.

In practical terms, the Truman Doctrine meant that the United States undertook the United Kingdom's role of patron. Greece emerged in ruins from World War II and its bloody civil war and was therefore from the outset totally dependent on US support. As Geoffrey Chandler observed in the 1950s, 'Greece is being carried on the wave of American aid'.[2] Former Prime Minister Alexandros Papagos' aphorism that 'we exist not only thanks to our own decision, but because the Americans exist'[3] aptly expresses Greek dependence on the United States at the time.

In return for the aid received in a time of desperate need, the Greek political élite provided services to the United States. Greek foreign policy in the late 1940s, 1950s and early 1960s became little more than a reflection of US views. Given such an accommodating attitude, it came as no surprise when Greece signed a neo-colonial type of military agreement with the United States in 1953. The unclassified section of the agreement stipulated that 'the United States could use Greek roads, railways, and territories and construct the military bases necessary for its purposes'. US activities were exempt from all taxation, charge or duty; the United States could claim all kinds of services at rates 'not higher than those charged to the Greek armed forces', that is to say usually nothing. The United States could freely remove all installations and equipment and the Greek Government was obliged to compensate the United States for facilities not removed from Greece. In addition, US personnel and their families were granted extraterritorial rights, with immunity from the jurisdiction of Greek courts.[4] In addition to the open agreement, several secret protocols were signed covering operational aspects pertaining to the US presence.

The main US bases established during the 1952–63 period include (with first year of operation in parenthesis): Hellenikon Air Base (1953), Iraklion Air Base (1954), the Souda Bay naval complex in Crete (1959) and the Nea Makri naval communications station (1963).[5]

In addition to these major bases, several other US and NATO installations were constructed in Greece. These include five US Defense Communication System (DCS) relay stations, several early-warning radar stations, eight NATO ACE HIGH troposcatter radio communications system relay stations, a NATO missile firing range in Crete (used by six NATO countries) and the Forces Range Accuracy Control Station (FORACS) for the testing and evaluation of detection equipment for NATO naval units. More recently a forward operating base for the staging and support of NATO Airborne Early Warning Aircraft (AWACS) has been established in Preveza in western Greece.

In 1959 the Greek Government also accepted that US nuclear weapons be stored on Greek soil in several Special Ammunition Storage Sites (SAS), located in various parts of the country.[6] Furthermore, Greece provided facilities for the development and logistics support of external reinforcements and other facilities for storage of US material and supplies.

Security costs

The extended US involvement in Greece backfired in the mid-1970s. Since then a majority of Greeks, and with it the political élite, seems to believe that Greece's association with the United States has been detrimental. The main points of dissatisfaction are:

1. The United States has frequently intervened in Greek domestic politics.
2. The United States supported Britain and then Turkey at the expense of Greece in the Cyprus dispute.
3. The United States has failed to meet Greek security concerns in dealing with the perceived Turkish threat.

US interference in Greek domestic politics.

An alliance with a great power is, according to small state theorists, inherently dangerous for the small state.[7] Asymmetry in power enables the great power to coerce a smaller ally into conforming to its choice of domestic and foreign policies. The Greek–US relation offers typical examples of such interference.

A good illustration of the scope and the depth of this patron–client relationship is offered by a US State Department 1947 directive indicating matters of 'high policy' in which the US ambassador was to have the 'last word':

a) Any action by the United States representatives in connection with a change in the Greek cabinet; b) Any action by United States representatives to bring about or prevent a change in the high command of the Greek armed forces; c) Any substantial increase or decrease in the size of the Greek armed forces; d) Any disagreement arising with the Greek or British authorities [remaining in Greece] which, regardless of its source, may impair cooperation between American officials in Greece and Greek and British officials; e) Any major question involving the relations of Greece with the United Nations or any foreign nation other than the United States; f) Any major question involving the policies of the Greek Government toward Greek political parties, trade unions, subversive elements, rebel armed forces, etc., including questions of punishment, amnesties, and the like; g) Any question involving the holding of elections in Greece.[8]

The extent and intensity of US involvement in the late 1940s and early 1950s effectively blocked most Greek initiatives. As one analyst observed: 'One does not gain the impression that members of the Greek authoritative society participated on an equal or significant footing with Americans in creating policies'.[9]

As former Prime Minister Andreas Papandreou in a 1984 interview said of his father, George Papandreou, 'as Vice President of the Government and as Minister of Coordination, he had to get the signature of the Chief of Economic Commission—the American Economic Commission—in order to validate his signature, his own signature'.[10]

Another commonly cited example of domestic interference is the US support of the military junta that ruled Greece between 1967 and 1974. Leaving aside the controversial question of whether US officials had knowledge of or even were implicated in the military takeover,[11] a brief analysis of some of the ensuing events is presented below.

The colonels' regime evoked an immediate response in most of Europe, and the NATO alliance divided into two camps over the issue. One, consisting mainly of smaller European members such as Norway and Denmark, argued that Greece should be ousted from the alliance and arms supplies should be halted immediately. The other, led by leading NATO countries such as the United States and the United Kingdom, argued that Greece's strategic location and importance made it necessary to tolerate the regime.[12] As Theodore Couloumbis puts it: 'The American policy between 1967 and 1974 was presented as one of a "painful dilemma". The choice was between upholding democratic principles and maintaining strategic bases on Greek territory. The bases, however, were quickly and painlessly chosen'.[13]

Indeed, new agreements for continued use of the bases were negotiated. The most important among them was the 'home port' agreement, which provided permanent port facilities for the 6th Fleet in Greece.[14] By sending 6th Fleet families to Greece, the US Navy was able to reduce the separation of Navy personnel from their families and thus reduce the time and expense of rotating ships across the Atlantic to the Mediterranean. Inevitably, however, the massive arrival of thousands of US servicemen in Greece at this time further reinforced the idea in the minds of Greeks from left to right that the United States had somehow been involved in the 1967 coup.

Conflict of interests over Cyprus

It is widely believed in Greece that the United States supported the UK and then Turkey at the expense of Greece in the Cyprus question. It is also widely believed that the United States operates on the notion that the security of the alliance has priority over all other local issues. As a result, the perception has held sway that Greece, as the weaker part in the Cyprus conflict, has had to sacrifice its interests for the sake of Western security and unity.

Here is not the place to go into a historical account of the Cyprus conflict;[15] suffice to say that it stems from, on the one hand, the desire of an overwhelming majority of Greek-Cypriots (80 per cent of the island's population) for unification (*enosis*) with Greece, and, on the other hand, the opposition of the UK and later Turkey to such a unification.

As the process of decolonization grew around the world, the demands of the Greek-Cypriot majority for an end to British rule, which they hoped would result in union with Greece, became increasingly insistent. In response, the Turkish-Cypriots, encouraged by the UK (and Turkey), raised demands for a division of Cyprus between Turkey and Greece. For the UK, Turkish fears and suspicion of the Greeks offered themselves as means of enforcing British domination of Cyprus. Britain could justify its rejection of the Greek demands for *enosis* on the ground that the Turkish minority needed protection from future Greek domination.

The Cyprus crisis became a serious problem for NATO when the United States insisted on mediating and settling the crisis quietly within the NATO framework. A political compromise within NATO, however, was precisely what the Greeks and the Greek-Cypriot community did not want. They thought that if external powers had to be involved it was better to place the dispute in the broader context of the United Nations, which offered a forum of newly emerged countries sensitive to demands for self-determination. The issue of internalization versus NATO mediation brought Greece's relations with the United States to a crisis point.[16] In order to maintain Western access and control, the United States and NATO put strong pressure to bear on all parties to find a workable arrangement.

In mid-1957 the British Government, as part of a new defence policy, decided that the security of the UK no longer required British control in Cyprus. It announced that it would retain military bases on the island, but would otherwise abandon it. In a political compromise reflecting the balance of power within NATO, the Zürich and London agreements were signed in 1959 proclaiming Cyprus an independent state.

The parties least satisfied with these agreements were Greece and the Greek majority on Cyprus. Greece was disappointed with the United States for failing to support the Greek position. There was widespread feeling in Greece at this time that on the Cyprus issue the United States and NATO paid more heed to British and Turkish than to Greek feelings.

Against this background, co-operation between the two communities in Cyprus in implementing the provisions of the constitution was practically impossible, and disputes arose in almost every sphere of government. This precipitated a series of crises, in 1963–65, 1967 and 1974, of which the last proved decisive. In the summer of 1974 the Greek military junta overthrew Greek-Cypriot President Makarios of Cyprus. Turkey used this event as a pretext to invade the island. The response of the United States and NATO was to do nothing, which Greece viewed as favoritism towards Turkey.[17] Greek politicians from all political parties argued that the United States and NATO could have acted in several ways, before and after the invasion, which might have prevented the situation from escalating. For a start, the United States could have discouraged the leaders of the Greek junta in 1974 from staging the *coup d'état* against Makarios. Second, joint pressure on Turkey by the United States and the United Kingdom might have stopped or limited the extent of Turkey's intervention. No such pressure was applied. Third, the United States could have pressured Turkey to continue the Geneva cease-fire talks in 1974. Instead, Turkey mounted a second campaign—apparently without US objections—to secure its territorial objectives and to accomplish the long-desired partition while cease-fire talks were taking place in Geneva between Greece and Turkey.

In short, the events surrounding the Cyprus issue demonstrate a conflict of interests between Greece and a large segment of NATO, especially the United States and Turkey.

The issue of US guarantees

The viewpoint from Athens is that Greece's NATO ally Turkey has adopted a revisionist policy aimed at changing the status quo established in the treaties of Lausanne (1923), Montreux (1936) and Paris (1947). Greek strategic analysts, political leaders and the public all believe that the ultimate Turkish objective is westward expansion. This perception cuts across party lines; there are no significant party differences on the evaluation of the Turkish threat. Threatening signals universally interpreted by Greek strategic analysts as indicators of impending danger include statements by leading Turkish politicians,[18] Turkish diplomatic initiatives[19] and military preparations.[20]

Greece's unique situation in NATO is that it faces a threat from one of its own allies. After the 1974 Turkish invasion of Cyprus, the Greek Government asked the United States (and NATO) to issue formal guarantees to defend Greece in case of attack from Turkey. The United States (and NATO) refused to give such guarantees, however,[21] on the grounds that it could be seen by Turkey as yet another humiliating act similar to the 1964 Johnson letter and the 1974 arms embargo.[22] In practical terms this means that Greece will be protected from the less likely threat from the north but cannot be protected from the more likely threat from the east.

The issue of US guarantees lies at the heart of Greek–US relations and is directly related to the presence of US bases in Greece. Usually small states make trade-offs between independence and security. In order to improve security and obtain protection, a small state may trade away some of its independence by joining an alliance. Greece's policy makers feel that Greece paid the cost of alignment with the USA (i. e., dependence, risk of involvement in wars, risk of becoming a target, etc.) without enjoying the benefits that alignment with the USA was supposed to provide (i. e., protection). Greece did not move from insecurity to security. It is both insecure and dependent.

Security benefits

Along with the costs associated with the alignment with the United States and the presence of US bases in Greece come some intended and unintended benefits.

Above all, the US military presence in Greece has increased the costs and risks associated with an attack against the host country. Any attack on Greece from Turkey and the resulting Greek–Turkish war would on both sides inevitably result in the destruction of military installations and weapons that are partially or even exclusively financed by NATO[23] and the United States for the purpose of common defence. Such a regional war would place the security of the US bases in Greece and Turkey at risk. Even a short (but intensive) war could completely wipe out NATO's infrastructure in the region. Hence, NATO and the United States have every reason to try to deter any conflict between Greece and Turkey and, if deterrence fails, to intervene to stop it.

The presence of US bases in Greece also gives the host country some reverse leverage *vis-à-vis* its superpower ally. Greece used its leverage during the March 1987 crisis with Turkey, when the two NATO allies were on the brink of war.[24] Turkey had announced its intention to explore the Greek continental shelf in the Aegean. In response, Greek armed forces rushed to the eastern border in a state of full alert. For Greece, such territorial infringement constituted a cause for war. During this crisis the Greek Government ordered the US naval communications station at Nea Makri to cease operations. The order was lifted only when Turkey backed down. Intensive US pressure has contributed to Turkey's decision.

Greek reverse influence has been used above all to promote US impartiality in Greek–Turkish conflicts. This applies in particular to issues pertaining to US military aid to Greece and Turkey. By and large, the US Congress has been sensitive to Greece's desire to maintain a quantitative balance in the levels of aid to Greece and Turkey. Since 1978 that balance has been set as a ratio of 7 : 10 in Turkey's favour.[25]

The 7 : 10 ratio was the central bargain in the 1983 negotiation of the new base agreement between Greece and United States, the so-called Agreement on Defense and Economic Cooperation (DECA). According to the Greek interpretation of the agreement, the United States has 'a contractual obligation to preserve the balance of military power' in the region, with the implication that Greece has the 'right to abrogate the agreement if in [the Greek Government's] judgement, there is an attempt to upset the balance in favor of Turkey'.[26]

Alliance versus non-alliance

All in all, it has been widely perceived in Athens that Greece's association with the USA has had a negative impact on the country's security. Greece has had to accept interference in its domestic politics, it has had to sacrifice its regional interests in the name of NATO unity and it has had to accept decisions that went against its perceived national interests. Given this evaluation of Greece's political ties with the USA, the question arises why Greece remains an ally.

International relations theory suggests that for small countries that like Greece are included in the clearly marked sphere of influence of a great power, a shift to non-alignment is extremely difficult as the dominant power may intervene to secure alignment and prevent defection. Of course, in the case of a Greek withdrawal from NATO, US intervention is not expected to take the form of direct military intervention. It is expected to take the form of support for Greece's rival Turkey.

Withdrawal from NATO and closure of the US bases would leave Greece exposed to Turkey. In the words of Stephen Larrabee, a former US National Security Council official responsible for the region, 'once Greece was outside NATO, the U.S. would probably tilt more heavily toward Turkey by committing even more resources to modernizing Ankara's armed forces'.[27] This, of course, is not a pleasant prospect given the intensity and extent of the Turkish threat. The severe repercussions for the security of Greece of a genuine US tilt in favour of Turkey are readily admitted in Greece:

If we leave the Western Defense System we shall lose the military aid we are now securing from the Alliance. NATO will continue to aid only Turkey. It might even give to Turkey the share that NATO today offers to us. The balance in the Aegean will be disturbed at our expense. If, therefore, we find ourselves in such a condition of weakness and isolation, one cannot exclude the possibility that Turkey will consider the moment opportune to attack us and take whatever it claims from us.[28]

Indeed, the 'Turkish card' has very frequently been played by the United States in order to keep Greece in place. For example, in the middle of a crisis in Greek–US relations, the Reagan Administration decided to give to Turkey surplus F-5 aircraft originally intended for Greece. In another instance, during a crisis that took place in a delicate phase of the DECA negotiations, the Reagan Administration announced its intention to boost military aid to Turkey from $402 million to $758 million and to keep aid to Greece to $280 million. Greek officials had no difficulty in reading these messages.

In short, the prospect of outright interference in the form of a decisive tilt in favour of Turkey has forced the Greek Government to maintain the US military presence in Greece. As Prime Minister Papandreou admitted, Greece's freedom of action is limited by the 'structure of power that exists today'.[29]

Table 12.1. Military construction appropriations for US facilities in Greece, FYs 1977–88
Figures are in US $m.

Year	1977–85	1986	1987	1988	Total 1977–88
Appropriations	18.22	0.91	–	2.14	21.27

Source: GAO, *Security Assistance: Update of Programs and Related Activities*, report no. GAO/NSIAD-89-78FS (US General Accounting Office: Washington, DC, 1988), p. 54.

III. Economic aspects of Greek–US relations

It is estimated that US military bases pump $100 million into the Greek economy every year.[30] This includes expenditures for military construction (see table 12.1), supplies, salaries to Greek employees, social security benefits and so on. Some 2000 Greek nationals are directly employed and 8000 are indirectly involved in projects related to the US bases.

The impact of the US military presence on local economies is hard to measure. Available data indicate that at least 40 per cent of the money is directed into the economy of Athens from the nearby Hellenikon Air Base. Table 12.2 shows that Hellinikon Air Base brings at least $39 million into the local economy. Furthermore, it is reasonable to assume that part of the salaries of the US military personnel (in total approximately $15 million) is also spent in Athens. The remaining 60 per cent of the US spending is split between facilities in greater Athens (the Nea Makri communications station) and Crete (the Souda Bay naval complex and Heraklion Air Base).

Greece has also received some funds from the United States through NATO channels. Greece has received approximately $17 million per year (1980–86) from NATO's Infrastructure Programme (NIP) and $1.2 million per year (1984–87) from the 'Science for Stability Program'. The US contribution to these programmes is 27.4 per cent of the total. This means that Greece receives approximately $5 million per year from the United States through NATO.

As a supplement to the 1983 DECA agreement, Greece concluded a Defense Industrial Cooperation Agreement (DICA) with the USA in 1986. In theory the purpose of this agreement is to set up a two-way flow in military procurement between the USA and Greece and to bring about, in the long term, a state of balance in defence production transactions. In practice, the impact of DICA is negligible since it does not promise offsets or award contracts to the Greek defence industry. It only assumes that Greek firms will be allowed to compete for US Department of Defense (DOD) contracts. Contract opportunities for the Greek defence industry are slim, however, with only part of a $10 million US Navy emergency repair project in the Mediterranean in the offering.

Greek policy makers tend to discount the importance of the impact of the US bases on the Greek economy. They claim that while the benefits are marginal, the US military presence in Greece entails high political, psychological and even economic costs. Couloumbis has described Greek perceptions as follows:

Table 12.2. Hellenikon AB expenditure in the economy of Athens

Expenditure	US $m.
Salaries of Greek personnel (aprox. 400)	7.00
Construction	5.25
Supplies (food, etc.)	19.00
Social security benefits (pensions, etc.)	6.00
Other expenses	2.65
Total	**39.90**
Salaries of US personnel (aprox. 1500)	15.10

Source: *Ta Nea* (Athens), 23 Mar. 1988, p. 2.

Most Americans tended to aggregate in comfortable, suburban enclaves with their own PX facilities and entertainment centers. Their relative affluence, their ability to rent luxury homes, their unwillingness to learn the languages of host countries, their normal proclivity to employ native people as drivers and domestics all contributed to increased psychological distances between visitors and hosts. Exacerbating matters further were occasional excesses in customs privileges, black-market activities, smuggling of antiquities, and peddling hard currency. With time, a reservoir of resentment was accumulated.[31]

What are then the economic costs associated with the US presence in Greece? First, the real estate utilized by the US troops free of charge has an estimated value of $437 million (see table 12.3). Assuming an annual user value of 5 per cent (a very low estimate) this means a loss of income for the Greek Government of at least $22 million. In addition, Greece bears the costs for all infrastructure (roads, ports, airports, railroads, etc.), even where US forces are the main users.

Greece also bears the cost of the Greek military personnel assigned to the bases as well as the cost of policemen responsible for the security of the base area. An estimated 100 policemen are assigned to maintain order outside the bases. This number is multiplied in case of demonstrations or when an emergency situation has been declared.

Finally, the United States is exempt from payment of aircraft landing fees (about 10 landings take place every day), tolls, levies, custom services, etc. Adding up all these costs, the US military presence costs the Greek Govern-

Table 12.3. Value of real estate utilized by major US bases in Greece

Location	Area (m^2)	Price (US $) m^2	Total
Hellenikon	1 076 888	140	150 764 320
Other	2 605 185	110	286 570 350
Total	**3 682 073**		**437 334 670**

Source: Author's interviews with Greek government officials.

Table 12.4. Total NATO defence spending by country as percentage of GDP, 1985–86

Country	1985 %	Rank	1986 %	Rank
Belgium	3.0	10	3.0	10
Canada	2.2	13	2.2	13
Denmark	2.2	14	2.0	14
France	4.1	5	3.9	5
FRG	3.2	7	3.1	8
Greece	7.1	1	6.1	2
Italy	2.7	12	2.2	12
Luxembourg	1.1	15	1.1	15
Netherlands	3.1	8	3.1	9
Norway	3.3	6	3.1	7
Portugal	3.1	8	3.2	6
Spain	2.9	11	2.6	11
Turkey	4.5	4	4.8	4
UK	5.2	3	5.1	3
USA	6.9	2	6.8	1
Average non-US NATO	*3.5*		*3.3*	

Sources: Carlucci, F., *Report on Allied Contributions to the Common Defense*, Report to the United States Congress (US Department of Defense: Washington, DC, 1988), p. 90; Weinberger, C. W., *Report on Allied Contribution to the Common Defense*, Report to the United States Congress (US Department of Defense: Washington, DC, 1987).

ment an estimated $45 million annually.

All in all, then, the analysis shows that the annual benefit of Greece from the presence of the US bases is of the order of $105 million, while the annual cost is of the order of $45 million. In other words, the current net benefit to the Greek economy amounts to $60 million.

Given the above-mentioned perceptions of the political and economic costs and benefits associated with the US presence, it came as no surprise that the Greek Government, consisting of a grand coalition of the three major political parties (New Democracy, the Pan-Hellenic Socialist Movement and the Communist Party), in January 1990 greeted with enthusiasm the US Government decision to close down Hellenikon Air Base and the naval communications station at Nea Makri.[32]

'Free riding' and the Greek defence budget

It has been argued that there are substantive financial advantages for small states to align themselves with great powers. Through joining an alliance, a small state may enjoy a 'free ride'[33] by obtaining protection while at the same time reducing its own defence expenditures. According to to Bruce Russett,

[A]s long as the smaller state is neither coerced by the big one nor offered special incentives, and unless the threat to the small state is very grave indeed—as in actual

Table 12.5. Greece's defence expenditure as percentage of GDP, 1975–85

Year	1975	1976	1977	1978	1979	1980	1981	1982	1983	1984	1985
Percentage of GDP	6.8	6.7	6.8	6.7	6.3	5.7	7.0	6.9	6.3	7.2	7.1
Rank among NATO countries	1	1	1	1	1	2	1	1	2	1	1

Source: IISS, *The Military Balance, 1987–1988* (International Institute for Strategic Studies: London, 1987), p. 216.

wartime—the small nation is likely to regard the big country's armed forces as a substitute for its own. The small country will feel able to relax its own efforts because it has obtained great power protection.[34]

It has repeatedly been claimed that the smaller members of NATO have profited disproportionately from the common defence effort. For Greece, however, alignment with NATO has not resulted in a 'free ride'. Rather the opposite is true. Relative to what it has received, Greece has committed a large amount of its available resources to the alliance. In return, NATO has failed to address Greek security concerns in dealing with the threat of the north (in the 1950s) and the threat from the east (in the 1960s, 1970s and 1980s).

It is apparent from US Secretary of Defense Caspar Weinberger's 1987 Report to the US Congress on *Allied Contributions to the Common Defense*[35] that Greece spends the highest percentage of its national product on defence of all NATO countries. In fact, Greece spends more than double the average for European NATO countries (see table 12.4). Greece's defence spending has consistently been high. Even in the 1950s Greece spent more than five per cent of its grossd domestic product (GDP) on defence. More recently, as table 12.5 shows, Greece consistently ranks first or second among NATO countries.

Another factor which must be taken into consideration is the share of manpower devoted to defence. As tables 12.6, 12.7 and 12.8 clearly show, Greece devotes more manpower to the common defence than any other NATO country. All in all, Greece's burden is the highest in NATO in terms of economic and manpower resources devoted to defence.

What is the explanation, then, for the consistent magnitude of Greece's defence effort? Small states that like Greece live in a state of insecurity have an incentive to increase their own strengths, to counter power with power. This is what in international relations theory is called balancing behaviour.[36] Balancing can be external and/or internal. External balancing means increasing power through alliances; internal balancing means increasing one's own strength.

Greece has traditionally failed to draw support from NATO to deal with its security threats. In the 1950s, NATO authorities placed Greece's first line of defence at a distance south of the Bulgarian border. This was the Strimón River line, north of Salonika, which abandoned to the enemy nearly a million Greeks

Table 12.6. Total NATO active duty military and civilian manpower as percentage of total population, by country, 1985–86

Country	1985 %	1985 Rank	1986 %	1986 Rank
Belgium	1.15	6	1.15	7
Canada	0.48	14	0.49	14
Denmark	0.76	13	0.73	13
France	1.28	5	1.26	5
FRG	1.10	9	1.10	9
Greece	2.37	1	2.31	1
Italy	1.03	10	1.09	10
Luxembourg	0.39	15	0.37	15
Netherlands	0.90	12	0.92	12
Norway	1.14	7	1.12	8
Portugal	1.14	7	1.12	8
Spain	1.36	4	1.34	4
Turkey	1.74	2	1.81	2
UK	0.95	11	0.93	11
USA	1.41	3	1.40	3
NATO average	*3.5*		*1.27*	

Sources: Carlucci, F., *Report on Allied Contributions to the Common Defense*, Report to Congress (US Department of Defense: Washington, DC, 1988); Weinberger, C. W., *Report on Allied Contributions to the Common Defense*, Report to Congress (US Department of Defense: Washington, DC, 1987).

and miles of rich agricultural land. With no room to trade for time in the north (at several points the Bulgarian border is about twenty miles from the Aegean sea), Greece had no option but to accept this line of defence, since NATO authorities refused to adopt a forward defence strategy for the Balkans similar to the one for the Central Front. Iatrides summarizes the Greek military situation in the 1950s as follows: 'Upon the plans developed in the 1950s, the role of Greece in NATO remained that of a trip-wire whose destruction would set into motion the alliances powerful components. However heroic, such a mission held little comfort for those entrusted with the national defense'.[37]

Since the 1960s the threat to Greece comes from the east. Again, for reasons that have been explained, the USA and NATO have failed to address the Greek security concerns. It comes therefore as no surprise that Greece has adopted a strategy of internal mustering of power, which involves devoting a comparatively high portion of its economic and manpower resources to defence.

Ironically, in terms of burden-sharing, NATO is the primary beneficiary of the conflict in its south-eastern flank, since the conflict forces Greece and Turkey to devote more resources towards defence than they otherwise would.

Table 12.7. NATO military manpower as percentage of population, by country, 1985–86

Country	Unmobilized (% of pop.)	Rank	Mobilized[a] (% of pop.)	Rank	Length of military service Months	Rank
Belgium	0.7	3	2.7	6	8–10	10
Canada	0.3	14	0.4	14	voluntary	14
Denmark	0.6	13	3.7	3	9	11
France	0.8	6	1.5	11	12	9
FRG	0.8	8	2.0	9	15	7
Greece	2.0	1	5.9	2	22–26	1
Italy	0.7	11	2.0	8	12–18	4
Luxembourg	0.2	15	0.2	15	voluntary	15
Netherlands	0.7	9	1.9	10	14–17	5
Norway	0.9	5	7.9	1	12–15	8
Portugal	0.7	10	2.3	7	16–24	2
Spain	0.8	7	3.5	4	15	6
Turkey	1.9	2	3.1	9	18	3
UK	0.6	12	1.1	11	voluntary	13
USA	0.9	4	1.4	12	voluntary	12

[a] Combined total of active forces and reserves, excluding paramilitary forces.

Source: Computed by the author from IISS, *The Military Balance, 1985–1986* (International Institute for Strategic Studies: London, 1985).

Problems with US aid

The four major security assistance programmes through which the USA provides weapons, training and other defence-related services to foreign countries are the Foreign Military Sales (FMS) programme, the Economic Support Fund (ESF), the International Military Education and Training Program (IMET) and the International Military Assistance Program (MAP).[38] Since 1946 Greece has received $7.7 billion in military and economic aid. This aid proved to be very important for the survival of the Greek political élites during the late 1940s and 1950s. While the benefits associated with foreign aid are apparent, however, the US aid programmes entail high costs which are less obvious.

Small countries whose arms procurement programmes by and large are financed through US military aid programmess such as FMS often discover that: (*a*) they have neglected their own defence industry; (*b*) they have created a one-sided dependence on the USA; (*c*) they suffer from foreign exchange loss; (*d*) they purchase weapon systems that are not designed to meet their own specific military needs; (*e*) they have opened a path giving the USA direct access to their own armed forces; and (*f*) they are losing a significant force for economic and technological development.[39] Greece has faced all of these problems. Specifically, US military aid to Greece has: (*a*) helped develop a dangerous dependence on a single supplier; (*b*) given the USA a mechanism by which it has been able to influence the force structure of the Greek armed

Table 12.8. Total military and civilian defence personnel of the NATO countries as percentage of the labour force, by country, 1980–85

Country[b]	1985 %	1985 Rank	1986 %	1986 Rank
Belgium	2.8	5	2.9	4
Canada	1.9	13	1.0	13
Denmark	1.6	12	1.4	12
France	3.0	4	3.0	3
FRG	2.4	9	2.4	8
Greece	6.1	1	6.1	1
Italy	2.4	8	2.5	7
Luxembourg	0.8	14	0.8	14
Netherlands	2.5	7	2.3	9
Norway	2.6	6	2.3	10
Portugal	2.3	10	2.8	6
Turkey	4.4	2	4.6	2
UK	2.2	11	2.0	11
USA	3.3	3	2.9	5

[b] Spain is not included.

Source: Financial and economic data relating to NATO defense, press release D-DPC-2(86)39, (NATO Press Service: Brussels, 4 Dec. 1986), p. 7.

forces; (*c*) complicated and thereby made more difficult the political and economic supervision of the armed forces; and (*d*) given rise to the myth that the USA finances Greek defence procurement.

Greece emerged from the civil war completely dependent on the United States for military supplies. Indeed, up till 1974 more than 80 per cent of the total weaponry in the Greek armed forces was of US origin (see table 12.9). After the Turkish invasion of Cyprus in 1974, Greek governments have tried to minimize the dependence on US arms supplies by adopting a strategy of diversification. Within the past 10 years, the degree of dependence upon a single supplier has diminished from 81 per cent to 42 per cent. Furthermore, 20 per cent of the defence needs are currently satisfied by domestic production.[40]

The policy of reducing dependence on a single source of supply and the concomitant development of the Greek defence industry is undoubtedly a step in the right direction. A word of caution is in order, however. While the value of purchases from the United States comprises only 42 per cent of the total value of weaponry purchased from foreign suppliers, this relates only to new acquisitions, the peak of the defence procurement pyramid. If the base of the pyramid is examined, that is the value of the total stock of existing weapons at the disposal of the armed forces, it will be seen that it is almost exclusively of US origin. This has serious implications for the country's security. In the event of war with Turkey, a US embargo on spare parts to Greece would paralyse Greece's defence capability in a relatively short period of time.

Table 12.9. Changes in the distribution of countries supplying arms to Greece, 1965–85

Figures are in US $m.

Country	1965–74	%	1974–78	%	1981–85	%
France	130	9	380	22	50	2
FRG	97	7	110	7	250	13
Italy	-	-	50	3	100	5
United Kingdom	-	-	20	1	5	-
USA	1 094	81	1 100	65	925	42
Others	36	3	40	2	695	36
Total	**1 357**	*100*	**1 700**	*100*	**1 925**	*100*

Source: Prepared by the author from US Arms Control and Disarmament Agency, *World Military Expenditures and Arms Transfers*, reports for 1965–74, 1969–78 and 1986 (US Government Printing Office: Washington, DC, 1975, 1980 and 1987, respectively).

A second problem is that the military aid programmes give the United States a means by which it can influence the force structure of its allies. The United States will understandably seek to influence its allies to develop a force structure directed towards meeting wider US strategic needs, beyond those of the allies themselves. A few examples from the Greek experience are in order. During the Greek–Turkish crisis of 1964–65, the Government of George Papandreou tried to reinforce the Greek Navy and Air Force, in order to increase the country's readiness for war. It was clear that in case of war with Turkey, the Navy and the Air Force would play the dominant role. The Papandreou Government's preferences brought it into conflict with the US military mission in Greece (the Joint US Military Advisory Group, JUSMAG) and the DOD. As the pro-US Greek Minister of Defence at the time, Petros Garoufalias, explains:

The American mission in Greece [in charge of administrating US military aid] was, in co-operating with Greek military authorities, responsible for determining the structure of the Greek armed forces . . . The evaluation of the Greek authorities was that a different and stronger structure was necessary for the defence of the country, while the American military mission thought that a weaker structure was sufficient. A different quantity and quality of aid was requested by the Greek military than that which was proposed by the American military mission and approved by the Pentagon. In the end, *the American military mission had the final and decisive word, supplying the Greek armed forces with military equipment of such quantity and quality as it saw fit.*[41]

The US objective was to maintain the traditional emphasis on the Army and to reduce the roles of the Navy and Air Force. In effect, the US plans meant that Greece would not have adequate forces to fight a war by itself.[42] The USA, it seems, believed that the lack of symmetry between the three services acted as a deterrent against war between Greece and Turkey, since Greece would never engage in a war that was doomed to failure from the start. In this way, the USA sought to preserve the integrity of the south-eastern wing of NATO.

234 EUROPE AFTER AN AMERICAN WITHDRAWAL

Figure 12.1. Planned US military assistance to Greece under the 1988 FMS Foreign Assistance Program; proposed distribution

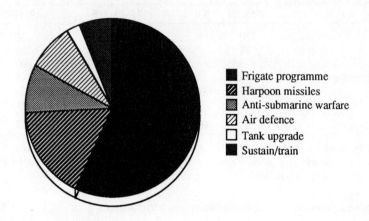

Total programme: $436.3 m.

Source: Congressional Presentation for Security Assistance Program, Fiscal Year 1988, USGPO publication no. 174-295-1987 (US Government Printing Office: Washington, DC, 1987), p. 135.

The Greek Government reacted against the US proposals. Since it could not use US funds to reinforce the Navy and the Air Force, it could only satisfy these objectives by raising the defence budget. Following the fall of the Papandreou Government and the subsequent military take-over (1967–74), however, the US preference for a strengthening of the Army prevailed. This proved fatal in 1974. The ill-equipped Navy and Air Force were in no position to deter the Turkish aggression against Cyprus.[43]

Even today the United States attempts, through more indirect means of course, to influence the Greek force structure. To give an example, figure 12.1 illustrates the direction of FMS money for 1988, showing how the United States directs FMS funds towards the construction of frigates, a weapons system which satisfies allied rather than national needs.

With these direct and indirect means at its disposal, the United States attempts to influence the Greek force structure. The problem does not stop here. Further programmes make even political and economic supervision of the armed forces difficult. Specifically, through IMET, which runs parallel to the FMS programme, 52 494 Greek officers were trained in the United States during the years 1950–85. These close contacts on an operational and technical level have created a strong and direct communication link between the US and Greek armed forces. During their training, the Greek officers become familiar with US military doctrine and tactics, and as a consequence feel more comfortable with the force structure envisioned by their trans-Atlantic allies. This serves to intensify the strong US orientation of the Greek armed forces.

Furthermore, military aid programmes such as FMS weaken the political control over defence spending. The Government's budgetary restraints are overcome and bypassed through the FMS programme. To take a hypothetical example, suppose the military demands another $10 million to purchase weapons for a given fiscal year. It is quite possible that the request will be turned down by the political leadership. On the other hand, if the military leadership uses FMS money to satisfy its requirements, this would cause no stir in the political leadership, since the FMS money does not appear in the current budget. In fact, the FMS repayments would only surface in the budget a few years later, under the general title of 'unfulfilled obligations of Greece', and very likely the respective Ministers would not be aware of just exactly where the money went in the first place. This manner of payment allows the military to sidestep political control and supervision in budgetary matters.

Finally, a more detailed scrutiny of the terms and provisions of the FMS financing programme, the major military aid programme, is in order. Many are erroneously under the impression that this money is free US aid to Greece. FMS money is no more than a loan,[44] which the US Government provides to foreign governments in order that they may purchase US material and services. FMS loans cannot be used for the purchase of weapons from non-US sources.

There are two categories of FMS credits: those which are lent on free market terms, the Treasury loans, and those which are lent at a reduced interest rate, the so-called concessionary loans. A typical Treasury FMS loan repayment scheme consists of a 30-year repayment schedule with a 10-year grace period, with a market interest rate. These credits must be absorbed within three years but can easily be extended by two more years. The conditions of the concessionary loans are different. Concessionary loans must be repaid within 12 years, with a 5-year grace period. The interest rate on this loan is usually half the going market rate, but it cannot fall below 5 per cent. Like the Treasury loans, the absorption period of three years can easily be extended to five.

Greece contracted most of its FMS loans at the high Treasury interest rates. Some of these interest rates are higher than 11–12 per cent, whereas today the free market lending rates are around 6 per cent (see table 12.10). These high-interest rate loans place a heavy burden on Greece's defence budget.[45] Every year 30–80 per cent of the new FMS loans are indirectly devoted to the repayment of the old loans (see table 12.11). As the situation stands today, Greece will be spending more on interest repayments than on principal right up until the millenium. Table 12.12 shows that for the next 10 years, 60 per cent of all new loans will go towards interest repayments of old ones.

If Greece had not resorted to this relatively 'easy' FMS money, it could have borrowed at lower rates from other countries. For example, a recently signed contract with the FRG concerning the procurement of MILAN anti-tank weapons was financed at an interest rate no higher than 4 per cent.

Table 12.10. US aid (grants and loans) to Greece, 1974–88
Figures are in US $m.

	Grants and loans						
	FMS financing (loans)						
	Treasury		Concessionary		MAP[a]	IMET[b]	
Year	Amount	% Interest	Amount	% Interest	(grants)	(grants)	Total
1974	50.5	7.54	–	–	–	–	50.5
1975	133	8.17	–	–	–	–	133
1976	86	7.58	–	–	36.6	–	122.6
1977	122	7.33	–	–	33	0.9	155.9
1978	140	8.42	–	–	33	0.2	175
1979	140	10.92	–	–	32.5	–	172.5
1980	145.1	12.15	–	–	–	1.2	146.3
1981	176.5	11.41	–	–	–	1.1	177.6
1982	280	9.00	–	–	–	1.2	281.2
1983	280	7.50	–	–	–	1.3	281.3
1984	500	7.50	–	–	–	1.4	501.4
1985	242	10.75	258	5.31	–	1.4	501.4
1986	83.5	7.50	347.1	5.00	–	1.2	431.9
Total	2 378.6		605.1		135.1	11.8	3 130.6

Figures may not add up to totals due to rounding.
[a] Military Assistance Program.
[b] International Military Education and Training Program.

Source: Greek Ministry of Defence.

The FMS loans appear to have long grace periods, at first sight. As mentioned above, grace periods are five and ten years for concessionary and Treasury loans respectively. In practice, however, Greece's delay in credit absorption, which often reaches five years, uses up completely the grace period on concessionary loans, and reduces by half that on Treasury loans.

Worse, in 1986 Greece started using the high-interest rate loans contracted in 1981 without having made the purchases in 1981. In fact, Greece rarely manages to take advantage of the grace period by purchasing now and paying later. Most of the time it purchases equipment and begins paying immediately in pre-specified installments.

The complexities of the FMS programme do not stop here. If one examines carefully the legal arrangements of the Letter of Offer and Acceptance (LOA) signed between the purchaser (the Greek Defence Ministry) and the DOD, one discovers a number of costs that must be shouldered by the purchaser.

First, acting as a middleman between contractors and foreign customers, the DOD charges a fee for administrative oversight, often 3 per cent of the purchase value plus a surcharge of up to 1.5 per cent. These administrative charges go towards payment of all personnel associated with or involved in the FMS loans. For example, FMS loans were used to pay $13.38 million towards

Table 12.11. Percentage of new FMS loans indirectly devoted to the repayment of old FMS loans, 1981–85
Figures are in US $m.

Year	Old FMS loans (principal and interest)	New FMS loans	Percentage
1981	140.6	175.5	80
1982	146.8	280	52
1983	147.8	280	53
1984	146.78	500	30
1985	138.6	500	28

Source: Greek Ministry of Defence data.

payment of American Military Mission (JUSMAG) personnel responsible for handling FMS loans in Greece during the period 1968–74.

Second, every client must shoulder a series of extra costs. These are: (*a*) non-recurring research, development, testing and evaluation costs; (*b*) recurring support costs (repairs, spare parts, etc.); and (*c*) an asset user cost. All of these are added to the basic price as a percentage of the total cost. A perhaps more serious problem is that there is no guarantee in the legal agreement (LOA) that the material will be new. The LOA says that material can be either 'new' or 'like new'. As one high ranking officer said, it is not uncommon to find aircraft procured through FMS with many hours of flight already logged.

A third problem is that in signing the LOA there is no price guarantee for the purchase of equipment, since all prices are considered estimates. Form 1513, which all FMS recipients must sign, says that 'the client will pay the U.S. Government the full cost of the purchase, even when the cost exceeds the estimated cost specified in the LOA'.

A fourth difficulty which arises with the costing of FMS-related purchases is that there is no guarantee that the purchases will be delivered on time. The contract does not refer to a specified but to an estimated date of delivery. This gives the US Government increased leverage over client governments, since purposeful delays in the delivery of major weapon systems or spare parts can create a brief but decisive 'window of vulnerability' for the armed forces of the purchaser, providing a potential aggressor with a 'window of opportunity'.

A fifth problem is that there are no legal sanctions which would require the United States to fulfill its part of the bargain to provide the material in question. It is not uncommon for the United States to deny delivery of weapons which have already been paid for. A weapons and spare parts embargo on a small country such as Greece, which depends to a large degree on the United States for its arms supplies, could prove to be its undoing in a time of war.

In sum, a close examination of the US aid programmes to Greece shows that the political, military and economic costs of the US aid are substantial.

Table 12.12. Greek obligations for FMS repayment, 1987–96
Figures are in US $ thou.

Year	Principal	Interest	Total
1987	52 250	162 356	214 606
1988	41 309	179 669	220 978
1989	24 484	183 715	208 199
1990	30 354	182 988	213 342
1991	63 889	179 930	243 819
1992	98 168	175 046	273 214
1993	111 826	169 531	281 174
1994	130 850	161 531	292 381
1995	148 949	152 080	301 029
1996	156 888	141 641	298 529
Total	**1 683 812**	**2 189 205**	**3 873 017**

Source: *Congressional Presentation for Security Assistance Program, Fiscal Year 1988* USGPO publication no. 174-295-1987 (US Government Printing Office: Washington, DC, 1987), p. 137.

IV. Implications for Greece of a US withdrawal

The US bases in Greece are designed to promote US global interests. They provide peace- as well as wartime support to the US forces. Because the bases are not oriented for the defence of Greece, the Government of Andreas Papandreou argued that the US bases in Greece exclusively serve US rather than Greek interests. Given this interpretation of the role of the US bases, Greece has no incentive to assume the operation of the bases if the United States decides to withdraw. The bases would therefore have to be closed.

If the United States withdrew from Europe and Greece but maintained a commitment to return in the case of a crisis, the costs to Greece of keeping the bases functional would be insignificant. The US forces in Greece are in terms of numbers equivalent to approximately 1.5 per cent of Greek defence personnel. Greece need only make a marginal rearrangement in the allocation of its military personnel, at no additional cost to the defence budget, in order to replace the departing US servicemen.

The impact on Greece from a US withdrawal is summarized in table 12.13. As has been shown, the current net benefit of the US military presence in Greece amounts to a total of $60 million. For the Greek economy, this loss of approximately 0.011 per cent of the GDP is insignificant.

Potential loss of the FMS financing for military procurement will have minimal economic implications on Greece. As shown above, the terms of the FMS loans are highly unfavourable. Greece will only have to manage the short-term difficulties associated with the shift from US to European financing. As a result of such a shift the US share of Greek arms procurement will decline further while the share of European suppliers will increase. Loss of the FMS financing will also encourage Greek defence industries to undertake joint

Table 12.13. Impact for Greece of a US military withdrawal

Defence budget	↔
Economy	↔
Indigenous defence industry	↑
European defence collaboration	↑
Military procurement from US sources	↓
Military procurement from European sources	↑

ventures with their European counterparts and to participate in multinational collaborative research and development projects within the framework of NATO's Conference for National Armaments Directors as well as the Independent European Program Group.[46] Last, loss of the FMS financing will stimulate the further development of the Greek arms industry. This will raise the level of procurement by the Greek armed forces from their current 20 per cent rate to 30–40 per cent of the entire military procurement.

All in all, the costs to Greece from a US withdrawal from Europe would be negligible. Given the asymmetries in Greek–US relations, the United States currently gains more than Greece from this relationship.

Notes and references

[1] See Churchill, W. S., *The Second World War: Triumph and Tragedy* (Houghton Mifflin: Boston, Mass., 1953), p. 227.

[2] Chandler, G., 'Greece: relapse or recovery', *International Affairs*, Apr. 1950, p. 191.

[3] Interview with General Pagagos, *To Vema* (Athens), 27 Apr. 1952, p. 1 (in Greek).

[4] US Department of State, *United States Treaties and Other International Agreements, 1953* (US Government Printing Office: Washington, 1955), vol. 4, pp. 2189–98. For a discussion of the 1953 agreement, see Couloumbis, T., *Greek Political Reactions to American and NATO Influences* (Yale University Press: New Haven, Conn., 1966), pp. 77–79, 222–27. For a recent discussion and re-evaluation of this agreement, see *Practica Voulis* (Journal of Parliamentary Debates) (Government Printer: Athens, 31 Oct. 1983), pp. 814 (in Greek).

[5] For a discussion of the military missions served by installations in Greece, see Grimmett, R. F., *United States Military Installations in Greece*, Congressional Research Service Report no. 84-24 F (US Library of Congress: Washington, DC, 16 Feb. 1984), pp. 8–10.

[6] For a discussion of the nuclear weapons agreements, see Christos Sazanides, *Foreigners, Bases and Nuclear Weapons in Greece* (published by the author: Thessaloniki, 1985), pp. 412–24 (in Greek). It is interesting to note that even President Eisenhower considered the US nuclear weapons in Greece as provocative to the Soviet Union. Brigadier General A. J. Goodpaster summarized the President's views as follows: '[Eisenhower] said that he could see reason for putting IRBM's into such areas as Britain, Germany and France. However, when it comes to the "flank" or advanced areas such as Greece, the matter seems very questionable. He reverted to his analogy—if Cuba or Mexico were to become Communist inclined, and the Soviets were to send arms and equipment, what would we feel we had to do then? He thought we would feel we had to intervene, militarily if necessary'. 'Memorandum of Conference with the President' (17 June 1959). See also Eisenhower, D. D., 'Memorandum for the Secretary of Defense' (3 June 1959); The Secretary of Defense, 'Memorandum to the President' (12 June 1959), Dwight D. Eisenhower Library, Abilene, Kans.

[7] Niccolo Machiavelli's classic treatise *The Prince* (The Modern Library: New Yory, 1950) contains an analysis of the issue that remains valid, p. 34. For discussion of the problem in the literature on small states see Platias, A., 'High politics in small countries: an inquiry into the security policies of Greece, Israel and Sweden', Ph.D. dissertation, Cornell University, 1986, chapters 1 and 2; Handel, M., *Weak States in the International System* (Frank Cass: London, 1981), pp. 131–39.

[8] See US Department of State, *Foreign Relations of the United States, 1947* (US Government Printing Office: Washington, DC, 1971), p. 394. For an analysis of the scope and depth of the US involvement in Greek affairs see Iatrides, J. O., 'The Truman doctrine: the beginning of United States penetration in Greece', eds T. A. Couloumbis and S. M. Hicks, *U.S. Foreign Policy Toward Greece and Cyprus: The Clash of Principle and Pragmatism* (Center for Mediterranean Studies: Washington, DC, 1975), pp. 11–18.

[9] Amen, M., A., 'American institutional penetration into Greek military and political policy-making structures: June 1947–October 1949', *Journal of Hellenic Diaspora*, vol. 4 (autumn 1979), p. 112.

[10] BBC, *Face the Press*, British Broadcasting Commision, Channel 4, London, 12 Mar. 1984 (television broadcast).

[11] It has been established that the US State Department knew that a 'rightist Greek military conspiratorial group' was ready to stage a coup. See 'Greece in political crisis', *Administrative History: The Department of State during the Administration of President Lyndon B. Johnson, November 1963–January 1969*, ch. 4, section J, Lyndon B. Johnson Library, Austin, Tex. It does not follow, however, that the United States engineered the *coup d'état*. This conspiratorial explanation is superficial and unconvincing. For an excellent analysis of the domestic determinants of the 1967 coup, see Mouzelis, N. P., *Politics in the Semi-Periphery: Early Parliamentarism and Late Industrialization in the Balkans and Latin America* (Macmillan: London, 1986), pp. 134–45.

[12] See for example the position taken by Assistant Secretary of State for European Affairs Martin J. Hillenbrand, in *International Herald Tribune*, 15 May 1971. See also *The Decision to Homeport in Greece*, Report by the Subcommittee on the Near East, Comittee on Foreign Affairs, US House of Representatives, 92nd Congress, 2nd Session (US Government Printing Office: Washington, DC, 31 Dec. 1972).

[13] Couloumbis, T. A., 'A new model for Greek–American Relations: From Dependence to Interdependence', eds T. A. Couloumbis and J. O. Iatrides, *Greek-American Relations: A Critical Review* (Pella: New York, 1980), p. 189.

[14] These facilities were to serve an estimated 6500 military personnel attached to the 6th Fleet and an estimated 3350 dependents. For an analysis of the homeporting decision, see Keagy, T. and Roubatis, Y., 'Homeporting with the Greek Junta: something new and more of the same in the U.S. foreign policy', eds Couloumbis and Hicks (note 8), pp. 49–65.

[15] For historical accounts, see Alastos, D., *Cyprus in History* (Zeno Publishers: London, 1955); Crawshaw, N., *The Cyprus Revolt: An Account of the Struggle for Union with Greece* (George Allen & Unwin: London, 1978); Poliviou, P., *Cyprus: The Tragedy and the Challenge* (John Swain & Son: London, 1975); Xydis, S. G., *Cyprus: Conflict and Conciliation, 1954–1958* (Ohio State University Press: Columbus, Ohio, 1967).

[16] See Coufoudakis-Petroussis, E., 'International organizations and small state conflicts: the Greek experience', Ph.D. dissertation, University of Michigan, Ann Arbour, Mich., 1972, pp. 245–329.

[17] See Stavrou, N. A., 'Kissinger's tilt on Cyprus: the new style of crisis diplomacy', eds Couloumbis and Hicks (note 8), pp. 98–105.

[18] These statements have been published in two semi-official booklets: *Turkish Officials Speak on Turkey's Aims* (Institute for Political Studies: Athens, c. 1985) and *Threat in the Aegean* (Journalists Union of the Athens Daily Newspapers: Athens, c. 1984).

[19] For a review of the major Turkish diplomatic initiatives, see Wilson, A., *The Aegean Dispute*, Adelphi Paper no.155 (International Institute for Strategic Studies: London, 1979–80).

[20] 'Threat to ... West!', *Spotlight* (Athens), (15 Feb. 1985), p. 6. See also *Greece's Security Problems* (Institute for Political Studies: Athens, c. 1985) and *Practica Voulis* (Journal of Parliamentary Debates) (Government Printer: Athens, 23 Jan. 1987) (in Greek).

[21] The United States has even refused to reaffirm a generally worded pledge contained in a 1976 letter by Secretary of State Henry A. Kissinger. In it Kissinger said that the USA 'would actively and unequivocally oppose either side's seeking a military solution' in the Aegean dispute. Bitsios, D., *Pera Apo ta Synora, 1974–1977* (Hestia: Athens, 1983), pp. 253–54.

[22] During the 1963–64 Cyprus crisis President Johnson sent a letter to Turkish President Inonu warning that the United States might not come to Turkey's defence if a Turkish invasion of Cyprus provoked a Soviet intervention. This letter shocked many Turks. See Rustow, D. A., *Turkey: America's Forgotten Ally* (Council on Foreign Relations, New York, 1987), p. 95.

[23] NATO infrastructure cost-sharing percentages for the 1980 to 1984 period are as follows:

Member Nation	Percentage	Member Nation	Percentage
United States	27.82	Denmark	3.77
FRG	26.70	Norway	3.70
France	13.34	Turkey	0.81
United Kingdom	12.18	Greece	0.79
Italy	8.07	Luxembourg	0.23
Canada	6.43	Portugal	0.20
Netherlands	5.19	Iceland	..
Belgium	4.59	Spain	..

France's participation in the NIP is limited to programs involving early-warning installations. See GAO, *Security Assistance: An Update of Programs and Related Activities*, Report no. GAO/NSIAD-89-79FS (US General Accounting Office: Washington, DC, 1988), p. 55.

[24] The March 1987 crisis showed how easily tensions between Greece and Turkey could lead to a rapid and uncontrollable escalation and to all-out war. This crisis had a sobering effect in both countries and brought about a *rapprochement* between Greece and Turkey. This began in January 1988 at the international economic forum held annually in Davos, Switzerland where the Greek and Turkish Prime Ministers met for the first time. After the talks in Davos and a subsequent meeting in Brussels in March, the two Prime Ministers met again in Athens in June 1988. During these meetings the two leaders have agreed upon 'confidence-building measures' and the establishment of a crisis-management mechanism. These meetings represent a step along the road towards peaceful co-existence. However, little further progress appears to have been made so far.

[25] For a discussion of this issue see Laipson, E. , *A Congressional Tradition: 7/10 Ratio in Military Aid to Greece and Turkey*, Congressional Research Service Report no. 83-524 F (US Library of Congress: Washington, DC, 15 June 1983). During FY 1988, Greece received $343 million and Turkey $480 million in US military aid. To be sure, the 7 : 10 ratio does not serve exclusively the Greek interests. As Monteagle Stearns, a US diplomat and former Ambassador in Athens, has persuasively argued, 'the ratio is an all-purpose beureaucratic device that serves everyone's interests, even those of the Turks... It provides the American Congress with a reassuring rationale for aid to Turkey and Greece, which would otherwise provoke sharp domestic controversy, and at the same time a rationale for limiting overall levels of aid. It enables the Aministration to blame Congress when its 'best efforts' fail to obtain projected amounts of Turkish aid, and for the Turks to justify tight control of the American military activities by blaming the Administration when aid falls short of levels that Congress would have been unlikely to approve anyway. Everyone, in other words, has a politically expedient excuse for doing the politically expedient thing. It is hard to avoid the conclusion that if the Greeks had not invented 7 : 10, someone else would have had to do so.' Stearns, M., 'Updating the Truman Doctrine: The U.S. and NATO's Southern Flank', working paper no. 86 (International Security Studies Program, Wilson Center: Washington, DC, 1988).

[26] See *Praktic Voulis* (note 4). The maintenance of the 7 : 10 ratio does not satisfy all the Greek demands. Greece wants the United States to maintain its impartiality not only as to the quantity but also as to the quality of the aid to Greece and Turkey. For this reason Greece has asked for more military aid on a grant basis and the allocation of economic aid as well.

[27] Larrabee, F. S., 'Papandreou: national interests are the Key', *The Atlantic*, Mar. 1983, p. 24.

[28] *Acropolis* (Athens), 28 July 1981 (in Greek).

[29] BBC (note 10).

[30] Most of the data used in this section have been obtained through interviews with Greek Government officials.

[31] Couloumbis, T. A., *The United States, Greece and Turkey: The Trouble Triangle* (Praeger: New York, 1983), p. 16.

[32] The US decision pre-empted a Greek Government decision to close at least the Hellenikon Air Base. The existing DECA expired on 31 Dec. 1989; negotiations for a new DECA covering the remaining bases and facilities will resume following the Greek general elections, to be held on 8 Apr. 1990.

[33] See Olson, M. and Zeckhauser, R., 'An economic theory of alliances', ed. B. M. Russett, *Economic Theories of International Politics* (Markham Publishing: Chicago, 1968), pp. 25–49.

[34] Russett, B. M., *What Price Vigilance?* (Yale University Press: New Haven, Conn., 1970), p. 94.

[35] Weinberger, C. W., *Allied Contributions to the Common Defense*, Report to Congress (US Department of Defense: Washington, DC, 1987).

[36] See Waltz, K., *Theory of International Politics* (Addison–Wesley: Reading, Mass., 1979).

[37] Iatrides, J. O., 'Greece and the United States: the strained partnership', ed. Richard Clogg, *Greece in the 1980's* (Macmillan: London, 1983), p. 163.

[38] The FMS program helps to finance foreign governments' purchase of US defence articles and services. Through the ESF, the US provides economic assistance to countries of special military, political and economic significance. The IMET program was established to provide instruction in US doctrine to foreign military personnel on a grant basis. MAP provides grant funding for defence articles and services.

[39] For a discussion of these issues, see Levite, A. and Platias A., *Evaluating Small States' Dependence on Arms Imports: An Alternative Perspective* (Cornell University Peace Studies Program: Ithaca, N.Y., 1983), p. 35–48.

[40] In particular, Greece has recently reached self-sufficiency in the production of ammunition, small arms, military trucks and armored personnel carriers. Greece now also undertakes all its own aircraft and helicopter maintenance. See *Practica Voulis* (note 20).

[41] Garoufalias, P. G., *Greece and Cyprus: Tragic Mistakes and Lost Opportunities* (Bergadis: Athens, 1982), pp. 37–38 (in Greek); emphasis added.

[42] Roubatis, Y., 'The U.S. involvement in the Army and politics of Greece', Ph.D dissertation, School of Advanced International Studies, Johns Hopkins University, Washington, DC, 1980, p. 298.

[43] Roubatis (note 42).

[44] For the first time in recent years during FY 1988 a small portion of the US military aid to Greece—$30 million out of $343 million—was awarded in the form of grant aid.

[45] For an analysis of the impact of high interest rates on FMS recipients see US General Accounting Office, *Unrealistic Use of Loans to Support Foreign Military Sales*, Report to Congress (US Government Printing Office: Washington, DC, 1983).

[46] Currently Greece participates only in two R&D projects within the framework of IEPG: the Stinger anti-aircraft missile and the Trigat anti-tank missile. Greece has also signed a number of Memoranda of Understanding (MOU) with the following European countries: France (1982), the UK (1983), Italy (1984) and Spain (1985). The purpose of these MOU was, among other things, to facilitate bilateral co-operation in research, development, production and procurement of defence materials.

13. The economic costs and benefits of US–Turkish military relations

Saadet Deger
SIPRI

I. Introduction and background

The military, economic and political relations between Turkey and the United States are complex, both at the bilateral level and within the context of NATO, not least because Turkey plays a pivotal role in US strategic and foreign policy from the Atlantic to the Gulf. Geography is a key factor. The country bridges two continents; borders on the Soviet Union as well as the Middle East; guards the access to the Mediterranean from the Black Sea; participates actively in the defence of the Southern Flank; and, finally, provides a unique terrain for intelligence gathering. A viable, secure and strong Turkey is essential not only from the point of view of NATO's strategic stability but also from the viewpoint of US global interests.

Political changes in Eastern Europe, and the lessening of tension and conflict in the region, do not affect significantly the relationship between Washington and Ankara. If anything, Turkey, as the only member of NATO bordering on Middle Eastern countries, may become more important in the US global perspective. The events of 1989 have certainly helped to improve Turkish relations with the USSR and Bulgaria, although much more remains to be done. Balkan co-operation is now a more feasible option. Although this chapter was written before the dramatic events of 1989, its central tenets remain unchanged.

The primary purpose of this chapter is to analyse the economic costs and benefits of US–Turkish relations. It includes a detailed cost–benefit analysis of the basing facilities that the United States (and by proxy, NATO) enjoys in Turkey. But it goes beyond that in estimating all the other gains and expenditures that accrue to the two countries in maintaining their special relations. For a country like Turkey, the symbiotic nature of military and economic security makes it imperative that a broad perspective be taken. The economics of burden-sharing are therefore analytically discussed both in terms of the narrow dimension of the US military presence as well as a much wider dimension encompassing factors such as trade and aid.

The main conclusion seems to be that the relations are not a 'one-way street'. Rather, the USA tends to gain considerably from its special relations with Turkey. For Turkey the potential economic losses incurred from the closure of the bases would have negligible impact on the macro-economy; the regional effects are also minimal given the small number of personnel and the lack of integration with the local economy. The economics of withdrawal are easy to handle. A much more serious issue is the loss of protection and deterrence that the postulated withdrawal entails. For a non-nuclear weapon country such as Turkey, having the longest borders (within NATO) with the USSR to boot, the reduction of security would be particularly significant. However, it should be kept in mind that Turkey performs a 'bodyguard' function for its superpower ally, as well as providing the USA with many other types of benefit that are discussed in this chapter. From a security point of view, therefore, there seems to be an even balance in their relationship.

As the primary concern here is to discuss economic and security aspects of US–Turkish relations, other security issues, peripheral in this context, are given scant attention. One such issue is the Cyprus problem. The Turkish position, as one of the three signatories (with Great Britain and Greece) of the 1959 London Agreement, of coming to the aid of Turkish Cypriots when it become necessary to protect their lives from terrorism and possible genocide, is well known; thus it is not discussed here.

The chapter necessarily concentrates on economic issues. But these cannot be properly understood without an appreciation of the complex framework within which the interconnections are embedded. The next section discusses Turkey's own security perceptions while the third section clarifies the web of US–Turkish relations. The last two sections provide economic data and estimates as well as an analysis of the implications of the hypothesized withdrawal.

II. Three questions about Turkish security

Is Turkey unique?

There are many reasons why the Turkish situation can be considered unique, not only from the point of view of NATO but, more important, from the global perspective of the United States. The first is of course geostrategic and, understandably, much attention has been paid to this factor. During a war the country may need to play a pivotal role, at least in the initial stages, in holding the lines of defence before support can arrive. As the gateway to the Mediterranean its navy and the armed forces will have to make a major effort to hold back the increasingly sophisticated Soviet fleet. In addition, it may have to take on the role of the Western first line of defence against Soviet pincer movements towards the Persian Gulf and the Middle East. With increasing instability in this region, and the failure of a middle-ranking power to emerge (as was expected of pre-revolution Iran), Turkey may be required to defend a number of fronts. For a technologically rather weak armed force such as Turkey's the requirements are formidable.

In a sense the transition away from the doctrine of Mutual Assured Destruction (MAD) to the more rational doctrine of flexible response may paradoxically have worked against the best security interests of Turkey, at least in a scenario involving a limited nuclear attack. Robert McNamara said in 1969:

I am convinced that our forces must be sufficiently large to possess an 'Assured Destruction' capability—an ability to inflict at all times and under all foreseeable conditions an unacceptable degree of damage upon any single aggressor or any combination of aggressors—the fundamental principle involved is simply this: it is the clear and present ability to destroy the attacker as a viable Twentieth Century nation and an unwavering will to use these forces in retaliation to a nuclear attack upon ourselves *or our allies* that provides the deterrent, and not the ability partially to limit damage to ourselves.[1]

It is a moot question whether MAD would actually have been used.[2] In terms of probability calculations, however, the anticipated costs of a calculated attack would be impossibly high. Hence smaller allies such as Turkey would be protected by the symbolic presence of the superpower's nuclear might. With a flexible response doctrine, however, it is not inconceivable that Turkey would face a conventional attack without massive interference by the major powers. As one scholar states:

Given the divided theatres of possible battle, it is clear that initial defence in the Southern Region must be national. The forces of each country must be able to resist an aggression at least long enough for reinforcements to arrive or for NATO authorities to warn an aggressor that continued action could lead to the use of nuclear weapons. Clearly, the priority must be on improving local defence capacities and the ability to integrate reinforcements efficiently.[3]

Within this scenario, the vulnerability and insecurity of a country with a defence system technologically far less advanced than that of the Soviet Union are substantially greater than for most other NATO states. The difference with other countries is not only one of degree but also one of kind.

But even during times of peace, Turkish relations with the Soviet Union have been of a special kind. There are long-term historical reasons for mutual distrust between the two countries, dating back to rivalries between the Ottoman and Czarist empires.[4] The intervening years have seen a curious interplay of suspicion and cordiality; strategic relations have been dominated by the former; economic relations have been tempered with the latter. Russia's invasion of the eastern Ottoman empire in World War I was stopped by Lenin, thus allowing the beleaguered country a welcome respite. Atatürk was wary of Soviet external relations; however, his economic philosophy of state-sponsored development, etatism, was influenced by early socialist thinking in the USSR.

The geographical position of Turkey, with its long sea and land borders with the Soviet Union, makes it fearful of the latter not just as a superpower but as a very powerful neighbour. Even if Turkey was not a member of NATO it would still be involved in an arms race simply because of the nearby threat of the USSR. Thus global *détente*, or entente between the USA and USSR, although desirable is not sufficient to allay all Turkish fears about expansionism. One is

Table 13.1. Land area and exposed borders of European NATO allies
Figures are percentages.

Country	Area[a]	Exposed[b]
Belgium	0.8	0.2
Denmark	1.6	4.2
France	19.8	8.4
FRG	8.7	5.1
Greece	4.8	11.0
Italy	11.1	17.3
Netherlands	1.6	1.7
Norway	11.9	10.8
Portugal	3.2	2.9
Turkey	27.8	20.3
UK	8.7	17.7

[a] Area: share of NATO Europe's total land mass.
[b] Exposed: share of NATO Europe's total kilometres of exposed borders.

Source: Sandler, T., 'NATO burden-sharing: rules or reality?', eds C. Schmidt and F. Blackaby, *Peace, Defence and Economic Analysis* (Macmillan Press: London, 1987), p. 37. Data for Spain not available.

reminded of the old Sri Lankan proverb: when two elephants fight—or make love—the grass is always crushed. Table 13.1 shows the land areas of European NATO countries as a percentage of total area; it also gives each ally's share of total exposed border miles. Turkey ranks first in terms of exposure (having a greater area to defend against attacks); unlike the United Kingdom, for example, its exposed borders are mostly with the Soviet Union. Hence, at the prospect of limited war, the element of insecurity is all the more obvious.

From the Soviet Union's point of view as well, Turkey merits unique consideration. Despite its superpower status the USSR has always worried about securing its borders and has been extremely conscious of the military strength of its neighbours. In addition, the USSR is sensitive about its access to the Mediterranean, as its ships must pass through the Turkish-controlled waters of Bosporus. As the Soviet Mediterranean Squadron (Sovmedron, the Fifth Eskadra) has grown in stature so has the importance of Turkey as an adversary. The 1936 Montreaux Convention, which limits access by Soviet ships from the Black Sea, is increasingly difficult to uphold; in war situations it would obviously be useless.

At the same time, the USSR realizes that an optimal strategy would be to have a neutral or at least a non-belligerent Turkey on its borders. Trade prospects are bright. Geographical proximity makes goods cheaper to transport. Turkey also has high growth rates and offers an expanding market for Soviet products. The increasingly restive nationalities of the south contain numerous people of Turkish races; good neighbourliness would benefit domestic politics, and a friendly Turkey would be a bonus. Secular Turkey is also a good role model to counter Islamic fundamentalist aspirations within the Soviet Union

Table 13.2. Soviet economic aid to less developed countries (LDCs), 1954–79

Recipient	Value (US $m.)	Rank
Africa		
Algeria	715	9
Morocco	2 100	3
Middle East and Asia		
Afganistan	1 290	5
Bangladesh	3 05	11
Egypt	1 440	4
India	2 280	2
Iran	1 165	6
Iraq	7 05	10
Pakistan	9 20	7
Syria	7 70	8
Turkey	3 370	1
Others	3 170	
Total	**18 190**	

Source: Desai, P., *The Soviet Economy* (Basil Blackwell: Oxford, 1987).

itself. It is interesting to note that given Turkey's long history of military and political antagonism with the Soviet Union, economic relations tend to be good. Table 13.2 gives Soviet aid disbursements to non-socialist countries during 1954–80, which includes the cold war period (when Turkey was a member of both the Central Treaty Organization, CENTO, and NATO). In spite of the generally held belief that the USSR only gives aid to its allies in the Third World, Turkey appears to be the largest recipient of Soviet foreign economic aid.[5]

The political developments in Europe in the autumn/winter of 1989/90 raise hopes that security relations between Moscow and Ankara will be characterized by greater *rapprochement*. With the lowering of tensions between NATO and the Warsaw Treaty Organization (WTO), and the success of arms control negotiations, economic and political aspects of security will tend to become increasingly important. Given the opening of Black Sea trade and the prospects of greater technological co-operation between the two countries, Turkey and the Soviet Union may be entering in their bilateral relations a period of cordiality, eclipsing the mutual suspicion of the past. Fears of instability remain, however, due to military factors as well as to ethnic tensions within the USSR, including the outburst in 1989 of nationalist conflict in the Caucasus region.

From a US point of view, Turkey is special and probably unique within the NATO framework. The geostrategic factor is of course the most crucial. In addition to the points mentioned above, Turkey's proximity to the USSR means that it is well located in terms of communications and intelligence gathering. There are 14 early-warning sites in the country, useful for alerting the 6th Fleet of impending attacks, monitoring military activity and weapons production in the south-western areas of the Soviet Union, as well as for data collection for the verification of strategic arms limitation agreements.

Equally important, however, is the socio-economic dimension, an area rarely emphasized by analysts. Turkey is the only Third World country within NATO, in terms of overall economic development. In this sense it can be regarded as a test case of US global policy towards its allies in the rest of the world. If Turkey 'fails', either in terms of security or in terms of development, there is the risk of a psychological domino effect among US global allies. For this reason, military security cannot define all the parameters of the relationship between the two countries. Developmental goals are central to Turkish–US relations. The emphasis that is placed on economic matters in the Defence and Economic Cooperation Agreements (DECA) is therefore not surprising. Turkey has insisted on close links between economic and military assistance; related the negotiations of the bases with trade concessions; and made defence modernization dependent on technology transfer and offset agreements. In short, a close interrelationship between military and economic affairs is a *sine qua non* of Turkish security policy. As the Turkish Under Secretary of Foreign Affairs said in 1985:

Turkey's ability to ensure an effective defense in the southeastern flank of NATO ... is closely connected with the rapid development of her economic and military capabilities. Turkey is spending great efforts in these fields. But trade restrictions imposed by some of our friends and allies on major Turkish export items affect negatively Turkey's economic development and limit her efforts in the field of defense *since defense and economy are closely related.*[6]

Finally, while the United States shares the defence concerns of its European allies, given its global interests, it is unable to share their Eurocentric perspective. Turkey, however, plays a role in both US European and 'out of area' planning. While the country belongs to Europe, it is equally part of the Middle East and of South-West Asia—a geographical, cultural, economic, social, political and military link between two areas of strategic concern. The Soviet Union, also an Asian as well as a European power, has also to consider this duality when appraising its strategic needs.

Is Turkey a 'free-rider'?

An alliance, such as NATO, is essentially a club in which members have certain rights and obligations. The former allows members to enjoy a good called 'security'; the latter requires them to pay a cost in terms of defence spending—an obligation often labelled 'burden-sharing'. However, NATO security is a public good; a member can enjoy its benefit (say that of nuclear deterrence against Soviet threat) without others getting less of it. But more important, it may be difficult to exclude a member from enjoying the benefit, once it is provided, even if obligations of payment are not met. This characteristic of non-excludability[7] gives rise to the 'free-rider' problem, used to describe a country that shoulders a disproportionately low burden and yet enjoys the advantages of the alliance.

It is sometimes thought that smaller (poorer, weaker) countries tend to free ride because their military expenditure is such a small proportion of the aggregate total. As has been claimed: 'Researchers have found strong evidence of free riding in NATO throughout the 1950s and during the first half of the 1960s. In particular, the wealthy allies shouldered the defence burden of the smaller, poorer allies who rode free'.[8] However, the implicit supposition that a smaller country necessarily free-rides simply because it pays little is wrong. Unless one stipulates the method by which the spending is allocated among the alliance members it is not possible to come to a definite conclusion purely on the basis of low contribution.

Why is the issue of free-rider important? On a general level, it has implications for the burden-sharing debate now prevalent in the USA. A common misconception is the claim that smaller NATO countries, such as Greece and Turkey, should contribute a greater proportion of total alliance costs since they tend to benefit from membership of the security club. As shown below, however, in terms of incremental spending Turkey has kept up its commitments; it has certainly exceeded the 3 per cent 'rule' that NATO has set itself for this decade. In terms of its ability to pay, given a relatively poor economy, it has done more than its fair share.

A more important point relates to the specific concern of this chapter. If Turkey is indeed a free-rider, then a US withdrawal—and concomitant reductions in military effort—would force Turkey to spend even more than the calculations would suggest. In a sense there would have to be a 'catching-up' effect since the present level of military effort could be relatively low (within the free-rider scenario). An evaluation of the 'true' cost would then require an adjustment to compensate for the alleged underspending. On the basis of the analysis, however, it is clear that this is not a problem when conducting the cost–benefit analysis.

There are two normative ways in which contribution to a public good may be calculated. The benefit approach claims that the participants should pay according to the benefits received. The ability-to-pay approach, on the other hand, stipulates that members' contribution to aggregate cost must be related to their ability to pay. Both these criteria imply therefore that major countries *should pay more*; they receive more security benefits since their political, strategic and economic stakes are higher; they also have greater resources and hence can afford to pay proportionately more.

A good index, at least from the point of view of the ability-to-pay approach, is to look at the defence burden, that is, the ratio of military expenditure to the total amount of the gross domestic product (the GDP). If this is high then clearly the relevant country is allocating a relatively large proportion of its (possibly meagre) output towards its own and, *ceteris paribus*, the alliance's

Table 13.3. Estimated defence burden of a Turkish professional armed force, 1985–87

	1985	1986	1987
Actual military expenditure (TL b.)	1234.6	1868	2467.9
Military expenditure with a professional force (TL b.)	1566.6	2340.8	3139.3
Increase (%)	27	25	27
Actual defence burden (%)	4.9	5.2	4.7
New defence burden (%)	6.2	6.5	6.0
US defence burden (%)	6.6	6.7	6.4

Source: Author's calculations using data from Turkish sources and the SIPRI data base.

security. Turkey, as the member of NATO with the lowest per capita income, tends to contribute a relatively high share of the common defence effort; its military burden is the fourth highest in NATO.

An additional, and crucial, point to remember is that the Turkish defence burden is kept at an artificially low level through the policy of keeping a conscript army and also paying regular officers at rates far lower than those which the market would dictate. For example, a pilot in the Turkish Air Force is paid a *quarter* of what he would receive as a commercial pilot in the state airlines. Not only does Turkey have one of the largest standing armies in NATO, but the proportion of conscripts (90 per cent of the total armed forces) is the highest in the alliance. If all these conscripts were paid a market wage, the defence expenditure figures would be exceedingly high.

Table 13.3 shows estimates of the military burden under a hypothesized scenario in which all conscripts are paid a very modest wage rather than, as now, not at all. The payment attributed to the soldiers is assumed to be the per capita GDP, which is much lower than the average industrial wage. In addition, no allowance is made for the fact that the regulars and officers are paid much less than their civilian counterparts. Thus the figures are clearly underestimates. On average the defence budget would have to increase by over 25 per cent per year. Even as an underestimate the ratio of military spending to GDP would cross the 6 per cent mark every year. In effect, on a comparable basis with the United States, which has the largest defence burden within NATO and also relies on a 100 per cent fully paid volunteer force, Turkey has a very high burden indeed.

Potentially any country, large or small, rich or poor, powerful or weak, can free-ride. The fundamental criterion for checking this is the following: if the relevant country decreases its expenditure at a time when the rest of the alliance is increasing its defence spending then that country is using the services of the public good without paying a commensurate share.

Table 13.4 provides data for the growth rates of real defence spending for the United Sates (as the largest member of NATO), the European NATO countries excluding Turkey (which come closest to sharing similar geostrategic threats) and Turkey itself. An empirical method for detecting the free-rider status of Turkey is to compare these three cases. Throughout the 1980s as US military spending has reached new heights under the Reagan Administration, European

Table 13.4. Growth rate of real military expenditure for the United States, European NATO and Turkey, 1980–87
Figures are percentages.

	1980	1981	1982	1983	1984	1985	1986	1987	Average
USA	2	8.4	5.6	7.8	4.7	7.1	5.1	–2.7	4.8
European NATO excl. Turkey	2.5	1.1	3.2	1.8	1.4	0.6	1.1	1.1	1.6
Turkey	–0.9	16.8	16.1	–6.5	–2.8	6.1	12.4	–2.8	4.8

Source: SIPRI data base.

NATO's defence effort has not matched this high growth rate. The average year-to-year growth rate for the United States has been about 4.8 per cent; the corresponding figure for European NATO is of the order of 1.6 per cent. On the other hand, the Turkish military sector has also grown at an average of 4.8 per cent during these eight years. Comparing only 1980 and 1987 the annual compounded growth rate of real US defence has been 5.1 per cent; the figure for Turkey is 5.2 per cent; and that for Europe without Turkey is 1.6 per cent. Clearly aggregate data do not reveal wide inter-country variations. But purely from the Turkish point of view, it is clear that there has been no attempt at free-riding. In a period of unprecedented peacetime budgetary increases of the military on the part of the largest partner of NATO, it would have been easy to forgo defence spending for more developmental expenditure, leaving legitimate security concerns to be covered by the US 'umbrella'. This has not been the case, however; Turkey has tended to spend increasing amounts to match incrementally the US defence effort.

Is Turkish security viable?

Defining the concept of 'security' for a country like Turkey is extremely problematic. Not only is it multidimensional but the various dimensions are also closely related. In addition, the various objectives are to a certain extent in conflict. Turkey defines its security problems in three contexts, which all interact with each other: global, regional and national. Within each context further sub-divisions can be made: military security (NATO, WTO, Greece, Middle-East conflicts); internal threat (terrorism, legitimacy of the Government although *not* of the state); and economic security (rapid growth to satisfy the aspirations of an increasing and restive population).

At the global military level, the difficulties that Turkey faces are similar to those of NATO in general. The southern region has five theatres of military operation: northern Italy; northern Greece/Greek Thrace; western Turkey/Turkish Thrace; the Mediterranean Sea; and eastern Turkey (described in detail by George Thibault in chapter 17). The problem of *thinning*, in terms of long frontiers, incompatible and backward communications systems as well as the

Table 13.5. NATO and WTO force comparison, 1987

	NATO	WTO
Air forces southern region		
Fighter ground attack aircraft	615	695
Interceptor aircraft	295	1 560
Reconnaissance aircraft	90	195
Land forces eastern Turkey		
Divisions	12	20
Tanks	1 000	2 435
Artillery/mortar	1 800	4 800
Balkan Front		
Divisions	25	34
Tanks	3 000	6 570
Artillery/mortar	2 800	6 400

Source: AFSOUTH Headquarters Italy (unclassified), quoted in Chipman, J., 'NATO and the security problems of the southern region: from the Azores to Ardahan', ed. J. Chipman, *NATO's Southern Allies: Internal and External Challenges* (Routledge: London, 1988), p. 39.

numerical inferiority of NATO forces, means that the costs of war will be high. Most important for Turkey, however, is that it will need to be involved in three of the four potential theatres. If primary responsibility at the initial stages of a holding operation is national, then the strain on the technologically inferior Turkish forces may become intolerable. Table 13.5 provides some information on comparative force capabilities; the discrepancy between NATO and WTO is rather wide. It is generally agreed that in terms of naval forces NATO is superior. The military problem arises in the other two areas.

Turkey also aspires, given its wealth, size and population, to be a benevolent regional power. Its good relations with both Iran and Iraq, through substantial trade, could have been helpful in initiating the process which led to the cease-fire. It is also an important country within the Islamic world, where it has a potential influential role in resolving strategic/economic problems. But all of this may or may not be compatible with its global relations with the United States. For example, during the Arab–Israeli war of 1973 Turkey refused to grant out-of-area facilities to US aircraft in support of Israel, although it had supported the creation of the state of Israel in spite of its Islamic background; on the other hand, Soviet overflights of Turkish airspace were accepted.

A vital problem for the Turkish armed forces is the lack of sophisticated weaponry and equipment. A common complaint is that armaments of Korean War vintage are still being used in the absence of an overdue massive overhaul. In spite of large-scale imports and military assistance programmes, the Turkish Army is essentially under-armed, particularly when compared to its formidable neighbour. There seems to be little protection against nuclear, biological and chemical weapons; air defence systems are weak; and, more generally, C^3I equipment is obsolete.[9] In the past, military aid has essentially consisted of dumping vintage weapons. For a viable national security policy defence modernization must be given foremost priority.

The Air Force is particularly vulnerable, with obsolescence threatening a large proportion of its combat aircraft. It has been claimed that there are about two accidents a month involving training flights. The F-100 has been named the 'widow machine' after its lethal failures; even the more modern F-104 is sarcastically nicknamed the 'flying coffin'. The recent modernization programme, embarked upon in the 1980s, is central to the efforts being made by the defence authorities to combat threats to the country's military security. This topic is discussed in the context of policy options in the last section.

A major aspect of Turkish security perception is the importance it attaches to non-military threats to security. This is mainly a reflection of the low level of economic development as evidenced by Turkey's per capita income, which is the lowest among NATO and OECD (Organization for Economic Cooperation and Development) countries. Average growth rates for the past two decades have been high. At the same time, this has created an aspirational crisis, since the rapidly increasing population has come to expect more, with the responsibility of raising the standard of living placed squarely on the Government. While the legitimacy of the state remains firm, there exists a legitimacy problem for governments. The three military take-overs that have occurred in the past three decades are evidence of this problem.

Cross-section evidence for large samples of developing countries shows that military burdens tend to reduce growth rates for less developed countries (LDCs).[10] No such comparable econometrics is available for Turkey. But a common supposition is that defence expenditure takes away resources from investment and government spending on infrastructure and human capital formation. This long-term crowding out, therefore, has adverse effects on the economy. Thus, a balance needs to be struck for developing countries such as Turkey between the competing needs of enhanced security and greater growth. The conflict between the security dilemma and the development trap is a real one for the Turkish economy.

This linkage, and concomitant problems, has not been sufficiently understood. For example, the 1988 Report of the Defense Burdensharing Panel of the US House of Representatives is sceptical of the 'linking' between basing rights and foreign aid: 'Although the Panel believes that Portugal, Turkey, as well as certain other non-industrial allies, are deserving of substantial economic and security assistance, the linkage between foreign aid and base rights and the "purchase" of basing agreements subverts the mutual security arrangements to which the United States and its allies are parties'.[11]

This is clearly based on a profound misunderstanding of the nature of security in the so-called non-industrial allies. For them security implies having both tanks and tractors, and the absence of either is perceived as a threat. Hence in the negotiations that occur, the demands for both feature prominently. The multifarious and often conflicting objectives in the complex web of Turkish security have been aptly summarized by one scholar as:

> playing an active and constructive role within the Alliance and contributing to the strengthening of allied deterrence capacity without challenging vital Soviet security interests and without reawakening suspicions in Moscow; improving relations with the Soviet Union without raising American and West European fears of a neutralist foreign policy shift; playing a more effective role in the Middle East without becoming entangled in regional conflicts or provoking Western anxieties about Islamic resurgence; and, finally, defending its vital interests in Cyprus and the Aegean without antagonizing an ally and complicating its own position within the Alliance.[12]

To this can be added: increasing investment and growth without allocating too high a share to defence yet maintaining a credible and independent domestic military capability; and bolstering internal security without destroying the legitimacy of democratic governments and the secular state.

The central tenets of Turkish security policy, and its implications for Turko-US relations including basing rights, can be summarized in the following way.

1. Security should not be confined to military aspects alone nor at all times expressed in terms of defence and armaments.

2. Although military security is crucial and essential, in the absence of economic development it will be ineffective.

3. Turkey is simultaneously a Balkan, Middle Eastern, Mediterranean and European country. Hence its security interests as well as its foreign policy must be multidimensional and complex.

4. Turkey is a staunch US ally and an important NATO member. At the same time, its military preparedness must not antagonize the superpower that also happens to be its largest neighbour.

5. Many of the country's military capability problems can only be solved through force modernization. Given Turkey's relative poverty, this can only be achieved through substantial economic support from other NATO allies. In addition, such defence industrialization is crucial both for effective defence in times of crisis but also as a engine of growth for other industries within the economy.

6. Turkey is neither 'selfish' nor a 'free-rider'. Its needs are great; but its budgetary constraints are, relatively speaking, tighter. For this reason, its demand on common resources is correspondingly high.

III. The tangled web: Turkish–US relations

Turkish relations with the USA have been both volatile and stable; economic, political and military factors are all entwined; the country's geostrategic importance (proximity to Europe, the Middle East and the Soviet Union) has been utilized and exploited by both players; and finally, the 'leader–follower' framework has been tempered by outbursts of nationalism and growing independence.

An interesting feature, from a historical point of view, has been the long periods of stable relations punctured by short periods of strong turbulence. Turkey was neutral during World War II—a natural reaction to the disastrous consequences of World War I. In 1945 relations with the Soviet Union deteriorated significantly after the territorial demands on certain parts of northeastern Turkey; the tension was exacerbated by Soviet claims on the Dardanelles/Bosporus. By the late 1940s, worried by its awesome strategic responsibility in the Straits (linking the Black Sea and the Mediterranean) and encouraged by the assurances of the Truman Doctrine, Turkey was wholeheartedly in favour of the Western security alliances initiated by the United States. Its commitment was further demonstrated by active support in the Korean War. In spite of the initial reluctance to include it, and Greece, in the alliance, Turkey formally joined NATO in 1952, ending this period of turbulence. Although US–Turkish relations were fostered within a multilateral framework, the emphasis right from the beginning was on the bilateral strands between the two countries.

In 1954 the secret Military Facilities Agreement was signed which was to govern stable strategic relationships for almost two decades.[13] A series of bilateral accords in the intervening years were formalized by the next major treaty in 1969—a Defense Cooperation Agreement. Although the contents were classified, Prime Minister Süleyman Demirel in a press conference in 1970 spelt out some of the implications.[14] His tone implied the beginning of Turkish belligerence and heralded a period of relative instability in bilateral relations. The Cyprus War of 1974, and the US arms embargo until 1978, were of course the nadir of Turkish–US relations and have been well documented. During this period Prime Minister Bülent Ecevit even went so far as to claim that the historical role of the USSR as the nation's principal threat was no longer valid.

In 1980, the Defense and Economic Cooperation Agreement (DECA) was signed between Ankara and Washington to herald a new era of stable relations. Although the military aspects of DECA worked relatively well, the economic— in particular the trade-related—areas had become a point of discontent by the mid-1980s. In 1987, a new DECA was signed and ratified.[15] Turkey remained concerned about a number of bilateral issues, however: the limited nature of assistance for the vital arms modernization programme; the 7:10 'balance' between security assistance to Greece and Turkey, which Congress stipulates, but which is considered asymmetric given Turkey's unquestioned greater military needs; the difficulties of subsidized technology transfer; and, possibly most important, the restriction of Turkish exports to US markets, particularly the quota on textiles. It is not fanciful to claim that Turkey is once again entering a turbulent stage with the USA unless steps are taken on both sides to actively change the situation.

In a sense Turkish authorities would like to see the country treated, in terms of trade and economic relations, as Japan, South Korea and Taiwan were in the 1950s and the 1960s. Their economic miracles were successful because of the favourable trading regimes within which they operated; a paternalistic United States allowed some form of 'most favoured nation' treatment to these countries, which helped the tremendous drive towards export promotion. It is doubtful whether the current atmosphere of protectionism could ever foster such an active outward-looking growth strategy. Perceived US security interests explain, at least partly, why such a benevolent policy was followed by the United States. As similar advantages exist in the security relations with Turkey as well, it is difficult to understand why such free-trade relations are not allowed. Multilateral institutions, such as the World Bank, believe that Turkey is poised for high growth and an economic miracle is possible. Much will depend on trade and the ability to export without trade barriers, however. Washington has to understand that if the current Turkish economic reforms fail to perform the claimed 'miracles', then national security will be jeopardized. This may also trigger off the domino effect discussed above, whereby Third World countries fail to reconcile collective security commitments with the demands of economic growth. Equally important, if Turkey is economically successful then it will be more able to share a higher proportion of the defence burden of NATO as a whole; this must be beneficial to all the allies. It will then follow the path taken by Western Europe with economic prosperity aiding military preparedness.

It is interesting to note that even the names of the agreements governing relations between Ankara and Washington have changed over the years; this certainly reflects the structural and systemic shifts in what they consider to be vital to their respective welfares. What began as a regime governing military facilities changed to one pertaining to defence co-operation and finally transformed to one encompassing defence and economic co-operation. The changing nature of substantive issues, centred on basing and military facilities but steadily going much further than that, is of crucial importance in the two countries' bilateral perspective. It is this 'coupling' or 'linkage' between the military and economic dimensions of security that has to be accepted as a cornerstone of the current relationship. The numerous facets of the economic cost–benefit analysis below cannot be appreciated unless this point is clear.

There is now disquiet, and even dissatisfaction, in the US Congress at these types of linkage between basing rights and foreign assistance, that the southern region countries (Greece, Portugal and Turkey) insist upon. An argument often made in Congress is that if these bases protect the country in question as well as NATO, then it is unfair for the USA to 'pay' the overwhelming share; in addition, since these facilities have a narrow military function they should not be linked with other economic (aid/trade), military (arms sale/modernization programmes) or political issues.

Although the arguments seem reasonable there are a number of good reasons why such linkages have occurred. Historically, both the United States and the host country preferred to deal with basing rights and other related issues at a bilateral level. It is important to note that these are called US bases and not NATO bases. Even though the preamble to the (Turkish) DECAs mention NATO, it is quite obvious to whom they belong. If they are matters of bilateral concern then the accompanying (implicit) assumption is that such installations are only one of the whole mosaic of factors that constitute bilateralism; hence the so-called linkages. If the US Government had made it clear, right from the beginning, that the bases are for NATO purposes alone and that the USA had no unique or specific interest in them, then of course no such problems would be arising today.

However, the USA could not do so since the bases do constitute an important ingredient in the US global security strategy, as distinct from that of NATO. The US concern about the possible denial of access for 'out of area' operations is evidence that these bases could serve such purposes as well. From the Eisenhower Administration days on, it has been clear that Turkey has played a much wider role than simply being a part of NATO's Southern Flank. The formation of CENTO and Turkey's participation in that organization was an indication of the country's role in US non-NATO strategic policy. CENTO was certainly 'out of area', from NATO's point of view, but slotted neatly into the encirclement policy pursued at that time by the US Administration.

From an economic point of view as well Turkey was part of international US policy. Since the Marshall Plan days alliance members all round the world (NATO, CENTO, SEATO) were the chief beneficiaries of US economic assistance. There was an implicit recognition that if a country was not prosperous it would not be able to defend itself from external threats simply by military power. In addition, the 1950s saw a great opening up of the Turkish economy, reverting the trend of self-reliant development promoted by Atatürk. Multinational corporations, mostly from the United States, benefited from this 'open door' policy.

Turkish–US relations have been characterized by a complicated mix of military, political and economic relationships. The issue of bases and concomitant military facilities is only one of many aspects of this whole mosaic of interdependencies. Today, the environment, both in terms of the US global security perspective and Turkish economic development, is not substantially different from that of three decades ago. Hence, attitudes and bargaining postures remain very much the same. The fundamental assumption is that all these factors and elements are linked. To call now for a 'de-linkage' or a 'decoupling' of the bases from other issues is impossible. The technical analysis in section IV is based on the supposition that all these linkages are crucial.

A major aspect of the wider definition of Turkish–US relations pertains to security assistance, particularly those arising from Foreign Military Sales (FMS). These also demonstrate that the bilateral relations are not all a 'one-way

Table 13.6. Turkey's debt repayment schedule under US Foreign Military Sales, fiscal years 1985–90
Figures are in US $m.

Fiscal year	Interest	Principal	Total	Actual FMS
1985				485.0
1986	267.4	152.9	420.3	409.5
1987	394.5	248.0	642.5	177.9
1988	263.1	143.6	406.7	235.0
1989	261.0	109.7	370.7	..
1990	251.4	126.4	377.8	..

Source: *Congressional Presentation for Security Assistance Programs*, volumes for fiscal years 1986–88, US House of Representatives (US Government Printing Office: Washington DC, 1985, 1986 and 1987, respectively).

street', with the USA simply 'paying' exorbitant amounts to defend what is after all the Turkish homeland. Turkish sources show that between 1972 and 1986 the country received $4.7 billion of military assistance, both through FMS and through the US Military Assistance Program (MAP) (FMS 74.2 per cent, MAP 25.8 per cent).[16] The repayments, to be made in 1987–2015, will amount to $2.9 billion as principal and $6.5 billion as interest. In other words, total repayments are twice that of the original amount. Although the payments profile varies considerably from year to year, these estimates show that in 1987 Turkey paid back $560 million; this was higher than the total security assistance received, which was of the order of $490 million. Currently, for market (interest) rate FMS credits, the amount repaid over 20 years amounts to $4 for each dollar received; for concessional rate credits this becomes $1.92.

In the absence of radical changes, Turkey will start suffering from the so-called debt trap: repayments exceed the new amounts received so that the current loans cannot be used for any equipment purchases but are simply used up for debt retirement and interest payment.

Table 13.6 shows year-to-year repayments for FMS loans received on defence sales and transfers. These are from US sources. For recent years, the amounts received under FMS do not cover the interest and principal to be paid back. Unless there is debt retirement through forgiveness or repudiation the situation will become absurd. The problems of military debt servicing are not fundamentally different from the overall debt crisis. Its burden has equally explosive potential. The point to stress is that Turkey is paying exorbitant amounts and the benefits accruing to the United States from this transaction are significant. With the drop in inflation in the 1980s, both for civilian and military products, the real rate of interest that Turkey pays and the USA receives is exceedingly high.

IV. Turkey with and without the US forces: economic costs and benefits

As discussed above the required cost–benefit analysis cannot, and should not, be conducted within the narrow framework of the bases and military facilities alone. A comprehensive overview is called for, since the multifarious aspects of the relationship are all equally important. Naturally, given the brief of this chapter, the focus will always be on the economic aspects, wherefore the standard strong *ceterus paribus* assumption is made throughout.

In addition to the paucity of data, the purely analytical problems of computing the costs and benefits of the current US presence, as well as the hypothesized withdrawal, are complex for a variety of interdependent reasons. First, it is not clear at what level one needs to do the costing: government (budgetary) or the economy; national or local. If the costs were mutually exclusive, exhaustive and additive there would be no theoretical problems (although practical difficulties remain). But they are not. Consider a US loan to the recipient country to buy arms. This is a budgetary cost since the US Government has to pay. However, since the money is used to purchase arms produced by US companies it is a gain for them and hence to the US economy. If the recipient country uses a loan for construction, and pays local contractors to build the installations, it is a benefit to the economy but a possible loss to the government when it pays back the loan.

Second, when evaluating the costs and benefits, should both countries be considered separately or jointly? It is not the case that one country's cost (benefit) is the other country's benefit (cost). These gains and losses are asymmetric. I is not a zero-sum game where computation of incremental costs on one side will imply an equivalent amount for the other side. It is therefore not possible simply to add up US costs for, say, its overseas bases, and argue that the host country needs to bear the same amount of burden. The actual burden may be more—or it may be less.

Third, there are problems of using factual data to analyse counterfactual situations. In other words, it is not always possible to use information regarding trends to get plausible answers to 'what if' situations. Another way of putting this is to say that a discussion of Europe with the USA may not be the best way of forecasting Europe without the USA. Two examples will clarify.

Consider the payments received through the NATO Infrastructure Programme (NIP), although this is not purely a US issue. It is possible to look at the current payment and receipt of a specific country and calculate the net gains. But the numbers will be different in the withdrawal scenario since the United States may not contribute and each country will have to pay more. On the other hand the total expenditures from the NIP will diminish since the US reimbursements are not there either. Hence, there may well be a net gain from the NIP if the United States withdraws.

Table 13.7. Number of personnel at US military installations in Turkey, 1987

US military	US dependants	US civilians	Turkish civilians	Total
4 823	3 927	576	4 450	13 776

Sources: *Manpower Requirements Report, Fiscal Year 1988* (US Department of Defense: Washington, DC, Feb. 1987); *List of All Military Installations Including Authorized Full Time Personnel* (US Department of Defense: Washington, DC, 1985); Turkish sources.

Consider also the income and expenditures of US servicemen in Europe. The latter is a good indication of the beneficial effects to the local economy of the bases. But after withdrawal they will need to be replaced by local armed forces. Their salaries will be different, hence the incomes earned; more important, the expenditure patterns will be totally different. In general, lower incomes should be expected, but greater propensity towards local consumption. The comparative effects could be substantially different.

Finally, there is a distinction to be made between stocks and flows. The US bases have large amounts of weapons, particularly nuclear warheads, as well as other types of strategic commodity reserves such as oil. There are also the installations, buildings and other sophisticated equipment. The Turkish DECA stipulates that the buildings, equipment, and so on will be the property of the host country at termination of the Agreement, although the position is not clear about the weapons. It is impossible to calculate the monetary value of the weapons, however. The most manageable assumption is that all of these stocks will be taken away if the facilities are shut down; alternatively, if the bases continue to function under Turkish authority, they will remain. It is then possible to concentrate on the flows of recurrent expenditures, costs and benefits that the nation will face consequent to US withdrawal. A unique problem arises with nuclear weapons, for non-nuclear countries, but this issue goes beyond the economics of burden-sharing.

The implication of all this is that there may be different ways of computing costs and benefits (assuming for each option that the United States either may or may not return); these might even be incompatible with each other. Even if all the information required was available there would still be a range of numbers and values of net costs and benefits. Tables 13.7–13.10 show the various alternatives. The final evaluation of the implications for Turkey (in section V) uses normative judgements to balance these approaches.

One point of interest, for the whole of NATO, is the employment effects of US servicemen stationed in Europe. For Turkey the total is small and almost insignificant in terms of overall NATO, its own population and numbers in armed forces. David Greenwood[17] and the US Department of Defense (DOD)[18] show that in 1987 there were 4823 active-duty US military personnel in Turkey out of a European NATO total of 332 516. Given a domestic population of over 50 million and total armed force of over 650 000 the presence of US

Table 13.8. US gross annual expenditure for outlays associated with defence arangements with Turkey, 1986–87 average
Figures are in US $m.

Outlay	Expenditure
1. Consumption and investment (outlays for for goods and services purchased abroad by DOD; includes pay of local civilian employees)	59.5
2. NATO Infrastructure Programme (NIP) payments	21.0
3. US personnel salaries minus local consuption expenditure by forces	110.0
4. Military grants (MAP)	259.0
5. Military loans (FMS credits)	293.7
6. Economic aid (grants and loans from ESF[a])	109.8
Total	**853.0**

[a] Economic Support Fund

Source: Author's calculations.

servicemen *per se* has a negligible impact on the country. Table 13.7 gives a breakdown of personnel figures for US bases in Turkey.

More so than in other countries where they are stationed, US servicemen and families stationed in Turkey are almost totally cut off from the local economy. Their only interest seems to be an occasional weekend holiday and buying souvenirs when returning home. The upper limit of the indirect employment effect for the local economy would be about 5000 outside jobs attributable to the bases. The impact, even locally, would be insignificant.

As for the overall economic and financial impact, since there are significant year-to-year fluctuations, both in the amounts involved and the exchange-rates, all monetary values are specifically calculated as current price averages for 1986 and 1987 in million US dollars. The averaging helps in smoothing out the erratic elements.[19]

With the advent of a floating exchange-rate system, and the extreme fluctuations observed in the 1980s, it is difficult to have a stable measure of converting local currency values to dollars. In the Turkish case, since the Turkish lira (TL) has been rapidly depreciating against the dollar, the United States has benefited from the strength of its currency, particularly when local wages have not always kept up with general inflation and currency depreciation. It has therefore been possible to spend 'fewer' dollars to get the same amount of goods and services. For the purposes of this chapter, conversion from/to Turkish liras, where relevant, are done on the basis of a weighted average of the dollar/TL exchange-rates for 1986 and 1987. The appendix to this chapter provides the precise method of calculating this conversion factor.

Table 13.8 presents information on expenditures (costs) incurred annually by the US Government both for the bases as well as through its wider military and assistance relations with Turkey. It includes direct DOD spending, economic and military grants and loans, salaries (incomes) of US forces in Turkey as well an estimate of the amount that the home country receives from US contribution to the NIP. All these figures are gross; thus for example new loans do not include adjustments for repayments and interest payments.

Survey of Current Business provides information for total direct defence expenditure abroad (payment for all goods and services purchased abroad by the DOD).[20] In addition to local expenditure by US military and civilian personnel, it also includes payments of local personnel hired in the country. Further, all construction, operation and maintenance of US bases are included. The category is comprehensive except for the inclusion of NIP.

To arrive at the amount that Turkey specifically receives from the US contribution to NIP a *pro rata* distribution is assumed. Currently it receives about 10 per cent of the total fund. Thus the figure in table 13.8, row 2, is 10 per cent of US contribution to NIP in 1986–87.

The other estimate was for salaries of US forces and civilian employees stationed in Turkey. From table 13.7 this would be around 5500 people whose salaries are to be paid by the US Government. It is assumed that the per head average salary is around $25 000 per annum giving a total of $137.5 million. But some of this is covered in the values given in table 13.8, row 1, since the latter includes local expenditures by the servicemen. Thus there could be a double counting unless the consumption element of this income is netted out. Assuming the propensity of the forces to consume to be around 0.2 (since their integration with the local economy is small) this leaves the value in table 13.8, row 3. These three rows account for the actual cost of the bases. The rest cover items under economic and military assistance.

The total is of the order of $850 million. It should be noted that this number is gross; the gains to the United States, from which net figures may be obtained, are considered later. Overall, therefore, the US Government needs to spend around $850 million gross for its presence in Turkey, including all military installations, security assistance and economic support funding.

Consider now the benefits that the Turkish economy is currently getting through its special relations with the USA within the framework of NATO. Table 13.9 gives the net value of Turkish financial gains. In the Benefits column, rows 1a and 1b are derived from table 13.8. These expenditures, on consumption and investment goods, obviously have some multiplier effects on the local economy. The income multiplier value is assumed to be 1.25, which is on the high side for a developing country like Turkey, where the problem is one of supply constraints rather than the lack of Keynesian effective demand. It is similar to that suggested by David Greenwood in chapter 5. The multiplier value, and the corresponding net figure in table 13.9, row 2, is definitely an overestimate.[21] The other three rows are similar to table 13.8.

Now consider the cost. The loans need to be considered net of interest and principal payments. For FMS credits the gains over these two years are negative; Turkey is caught in a debt trap whereby the actual amount received,

Table 13.9. Annual financial costs and benefits to Turkey relating to the US military presence, 1986–87 average
Figures are in US $m.

Benefits		Costs	
1. US expenditure		1. FMS debt repayment	531.4
(a) consumption and investment	59.5		
(b) NIP	21.0		
2. Multiplier effect on local economy[a]	20.0	2. Debt service on economic aid	71.3
3. Military grants (MAP)	259.0	3. Government expenditure on bases	18.8
4. Military loans (FMS)	293.7	4. User value of property	..
5. Economic aid	109.8	5. Environmental and social cost	..
Total	763.0		621.5
Net benefit	141.5		

[a] Income multiplier = 1.25.

Source: Author's calculations.

and more, must be given away to service the old debt outstanding. As mentioned above, the situation needs to be reversed, and recent authorizations are advocating concessional and waivered loans rather than the old form which charged market rates of interest. However, this is not the position for 1986–87 as depicted in row 1 of the Costs column. For economic loans the historical values of interest and principal payments on past debt have been used to arrive at the figure in row 2 (debt service on economic aid). Row 3 gives the Turkish costs for local and central government expenditure on roads, infrastructure and so on, to maintain the bases. Data on user value of property as well as other costs are not available. The figure for total costs is therefore an underestimate.

The current net benefit to the Turkish economy of the US military presence amounts to a total of about $140 million. The actual net earnings (for the country) from the bases alone are about $100 million. The rest comes from foreign assistance and transfers directly connected with the military relationship. The amount is small, which is understandable. The US presence in Turkey is quite insignificant compared to that in the FRG or the UK. Direct DOD expenditures in these countries were $6834 million and $1105 million dollars, respectively, in 1987; for Turkey it was only $68 million. Overall, this table sums up the financial statement for Turkey *with* the US military presence.

Consider now the financial benefits that the US economy has in its total dealing with Turkey. Table 13.10 gives the relevant figures again as 1986–87 averages. Turkish purchases of military goods, given in row 1, amount to around $385 million. This is the total value of transfers under the Military Sales Agency Contract. It is among the highest in NATO. The next two rows give the interest and principal (debt service) that the country pays back to the USA for military and economic loans taken earlier. Finally there is the US trade surplus. This is of the order of $435 million. One of the recent bones of contention in

Table 13.10. Annual financial benefit to the United States of payments on military sales and loans to Turkey, 1986–87 average

Figures are in US $m.

Payment by Turkey	Benefit to the USA
1. Turkish purchases of arms (includes US commercial arms exports, FMS and transfers under US military agency sales contracts)	384.5
2. Repayments on FMS credit	531.4
3. Debt service on economic aid	71.3
4. US trade surplus with Turkey (favourable US trade balance)	435.0
Total	1 422.2

Source: Author's calculations.

DECA re-negotiations has been this high trade surplus that the US runs; Turkey would certainly like a reduction in US protectionism and growth of its own exports to correct the imbalance. The overall gross gain for the USA is around $1400 million—not an insignificant sum. This table provides the basis of the thesis that the USA derives large economic benefits from Turkey.

Finally, a central issue regarding Turkey *without* the US military presence is what the running costs of maintaining the bases when the United States leaves would be. The results are summarized in table 13.11 as the *incremental* costs, both to the Government and the economy, after withdrawal. It may first be assumed that US forces are replaced by local servicemen who are already available to the domestic army; hence there are no extra costs to the defence budget for replacement of military personnel. However, the salaries of Turkish civilian employees, currently paid by the DOD, need to be accounted for. The average monthly income of such employees is about TL 250 000. This is used to calculate the value of row 1 in table 13.11.

Table 13.9 shows the estimated consumption expenditure of US personnel at the bases (around $27.5 million). This consumption expenditure of US servicemen, as well as the salaries of Turkish civilian employees, is substracted from total DOD spending (row 1 of table 13.8) to get a residual figure of around $20.6 million as investment. This is row 2 of table 13.11. If Turkey wishes to keep the bases running it will have to bear this investment spending cost in full.

One major technical problem remains. The US servicemen are spending around $59.5 million at the bases, which has an indirect multiplier effect on local incomes. Assuming that everything else remains the same, would the replacement of US servicemen (whose consumption expenditure is $27.5 million) by Turkish serving officers have a differential impact on the local economy? Clearly the income structure will be radically different. The average net income of an officer in the Air Force is currently around TL 750 000 per month. Making inflationary adjustments, converting into dollars and estimating total pay, it comes out that such personnel will be paid about $36.95 million. Of

COSTS AND BENEFITS OF US–TURKISH MILITARY RELATIONS   265

Table 13.11. Incremental or additional running costs per year to Turkey of maintaining bases after a US withdrawal (option A)
Figures are in US $m.

	Cost
1. Salaries of Turkish civilian replacement personnel	11.4
2. Investment (construction, etc.)	20.6
3. NATO Infrastructure Programme	– 2.7
4. Incremental loss to the local economy (multiplier effect)	..
Total	**29.3**

Source: Author's calculations.

course only a part of this will be spent; the rest will be saved. Assuming that 75 per cent is spent on consumption, the expenditure of replacement Turkish servicemen is about $27.7 million. Thus from an expenditure point of view the effects of the replacement are precisely the same; US forces earn more and spend less; the balance is about the same.

The conclusion of this exercise is that compared to total DOD spending now, run by Turkish authorities the bases will generate exactly the same amount of income to the economy; hence the additional multiplier effects will not be significant.

In addition, the net payments by the Government for NIP have been recalculated (see row 3 of table 13.11). The assumption is that although the USA withdraws its contribution (of 27 per cent of the total) there is also a reduction in the total disbursements since the US was the major beneficiary of NIP. There is evidence that the USA could potentially receive almost 40 per cent of the total funding.[22] Hence there will be a net gain to the European participants after withdrawal of the order of 13 per cent. Distributed *pro rata* gives the relevant savings for Turkey (as a minus figure). Clearly total NIP expenditures will remain unchanged since Europe benefits in this respect.

Overall, table 13.11 shows that Turkey will need to pay around $30 million per year to maintain the bases in the near future. Clearly this cannot continue forever since the stocks will run down. As the analysis below shows, however, it would be in the interest of the USA to supply the technological hardware, particularly for intelligence gathering, since the bases predominantly serve US security interests.

As regards aspects of the relationships between the two countries outside the narrow realm of the basing, no radical changes are expected. The main reason is that Turkey will still be a vital element in US global strategy quite independent of NATO. The discussions in section II suggest that Turkey's relatively unique position as the link between the Atlantic and the Gulf gives it a crucial role in US security planning. In the period 1985–87 countries such as Egypt, Israel and Pakistan received more US arms (FMS) than Turkey; there is no reason to believe that their security is any more important than that of Turkey in US global perceptions. So the basic assumption is that as a Middle Eastern/Balkan/Islamic country it will still have a relatively special position independent of the postulated changes to NATO.

Turkey stands to incur extra expenditures of about $30 million in the case of a US withdrawal, assuming that the bases need to be maintained as before. The other effects are also relatively insignificant. Since there are only around 5400 US servicemen at the bases, the local economy would not be much affected by a withdrawal. As discussed above, employment multipliers would be small. Section V looks at the implications of these figures.

V. Implications of a US withdrawal

The two options

The predominant functions of the US bases in Turkey are geared to peacetime operations, of surveillance and warning to the United States, rather than to the wartime defence of Turkey itself. Most of these are connected with intelligence gathering whereby Soviet weapons and signals are monitored *vis-à-vis* their offensive capability with respect to the USA. This is the general state of affairs although there are some exceptions such as the Incirlik base, which houses two squadrons of the 401st Tactical Fighter Wing on rotation from Europe. Overall, Turkish sources believe that the bases serve predominantly US security interests. It is therefore most likely that these facilities would be shut down in case of a withdrawal.

With regard to the two alternative withdrawal options described in the introductory chapter, under option A, which assumes a US commitment to return in times of crisis, the implicit price for this security guarantee could be the demand that the bases be kept functional. It is therefore necessary to calculate the costs of maintaining the bases. The land and buildings, according to DECA, belong to the host country; so there is no extra cost of acquisition. However, given their vital role in intelligence gathering for the United States, it is assumed that the sophisticated equipment would be donated to the Turkish authorities. Under option B, which assumes that the United States will not return under any circumstances, clearly the bases would be shut down, and the opportunity cost needs to be accounted for.

As shown above, if the Turkish Government has to operate these bases the total financial cost will amount to around $60 million for both investment on the bases and expenditure on personnel (both military and civilian). However, the pay of military personnel will have already been accounted for in the defence budget since Turkish forces will be deployed to replace the departing US forces. Hence the additional cost to replace the US presence would be for investment and for wages to civilian Turkish workers currently paid by the DOD. Even here, recurrent capital formation expenditure would be lower than the figures given earlier since these are once and for all spending which will not

Table 13.12. Cost comparison of a US withdrawal under options A and B

	Cost (US $m.)	Proportion of defence expenditure (%)	Proportion of GDP (%)
Option A	29.3	1	..
Option B	59.5	..	.08

Source: Author's calculations.

be maintained at previous levels in the absence of any expansion of activities. Thus the investment figure of $20 million is an overestimate. Be that as it may, the incremental cost to the defence budget of preserving the bases would be around $32 million: investment expenditure is around $20.6 million; civilian salaries are around $11.4 million. If the infrastructure programme works out as assumed then savings of $2.7 million will result.

Option A therefore requires the Turkish defence budget to be increased by around $29.3 million. This is approximately 1 per cent of the total military expenditure of the country. Thus the sum is by no means substantial and well within the reach of the defence authorities. In the near future, the cost to Turkey of keeping the facilities operational is minor and can be borne within the existing financial framework.

If the bases are shut down around $60 million of external spending is lost immediately. In addition, the beneficial indirect effects, although low, are further reduced; this loss comes to around $15 million (note that the multiplier is 1.25). The latter is offset by the benefits of closure, however. On the benefit side, the land and buildings are recovered for other purposes. In addition, government spending on maintenance falls to zero. Therefore, the overall loss to the economy cannot be more than $60 million—which is equal to the original spending. For the Turkish macro-economy the loss is negligible. As table 13.12 shows it corresponds to about .08 per cent of the GDP.

What about local consequences? Once again the effects are barely noticeable since the US servicemen are totally resident in an 'enclave' economy and are not integrated with local and regional economic structures. Their numbers are in any case small and would even under favourable scenarios following a withdrawal hardly cause any ripple effects. Taken together, Turkish civilian employees for all the bases amount to less than 5000 people. Thus at any one station the numbers are insignificant; even at the large Incirlik base the total local employment is around 1800—a small fraction of the population in the thriving local economy of the Adana region. Compared to a total population of over 50 million people the the number of base employees is insignificant, even if local variations are taken into account.

The conclusion is clear. Neither at the local, regional or national level would the alternative options present any serious economic and financial problems to Turkey. The Turkish defence agencies can accommodate the running costs of the bases. Alternatively, closure poses no economic burden. The reason is that the relative size of the operations is small compared to the country's resources.

Response of the United States

The overwhelming importance of the Turkish bases to the USA lies in intelligence gathering. Sinop, on the Black Sea, hosts facilities used for electromagnetic monitoring; Pirinclik, in south-eastern Turkey, provides access for long-range radar penetration into the Soviet Union and is also the site for space monitoring facilities; Belbasi hosts a seismic array for monitoring Soviet nuclear tests; others locations host related intelligence facilities. Geography provides the vital link. Thanks to its location and high mountainous terrain, Turkey is well situated for gathering information not only from the Soviet Union but also from Bulgaria, Iran, Iraq and Syria, all potential areas of superpower confrontation. In addition, recent political turmoil has contributed to the increasing importance of the Turkish bases. Since the fall of the Shah, and thereby the loss of basing access in Iran, Turkey has become central to US intelligence gathering in the region. With instability in Afghanistan and political difficulties in Pakistan, the only country that can be relied upon to monitor the southern regions of the USSR is Turkey. In effect, the US presence is determined by a range of unique factors that sets Turkey apart from other NATO host nations.

The benefits that accrue to the USA from these operations are great and in some respects incalculable. There are at least four areas where intelligence gathered from facilities in Turkey can be of crucial value: the state of Soviet technology and its impact on US research and development (R&D); verification of Soviet compliance with recent treaties; testing of new equipment and training; and early warning.[23]

The first area, pertaining to efforts to gauge Soviet technological advances in certain fields, has considerable economic potential. Although not directly measurable, knowledge gleaned from available intelligence data can help reduce R&D expenditures and thereby boost efficiency. The costs of acquiring sophisticated military technology whenever the Soviet Union has a comparative technological advantage can be very high; intelligence data from the facilities in Turkey are therefore of great economic interest.

If progress continues to be made in arms control, then access to facilities in Turkey for verification of compliance will become increasingly important. In similar fashion, training and testing of new equipment serve to enhance technological progress and in the process reduce unit costs. In addition, the environmental dangers inherent in such testing are borne by Turkey rather than by the USA.

In conclusion, it seems that in addition to serving security interests, the US bases in Turkey provide considerable technical and economic benefits to the USA. All of these factors do not enter into the type of cost–benefit analysis conducted so far as sufficient economic data are not available to make reasoned judgements about the magnitude of the financial benefits made. Nevertheless, it is important to emphasize that this is a dimension of the problem that has been left out of the calculation, and which does not necessarily exist for other NATO countries. From the US point of view, both in terms of its narrow security interests and in terms of the wider technological and economic issues, invaluable benefits are accrued through its special relationship with Turkey and through the established base access. It is not surprising that in recent years the host country has become increasingly reticent about keeping these facilities while the USA has tried its best to maintain them.

Future policy considerations for Turkey: to arm or not to arm

The discussion has until now focused on one important conclusion. US–Turkish relations are currently asymmetric, with most of the economic benefits flowing towards the richer ally. In terms of trade relations, military assistance and loan repayment, as well as the benefits from intelligence data, the United States tends to gain substantial amounts while Turkish returns are commensurately low. The burden-sharing debate takes on a new relevance in this framework. In the final analysis the question is therefore why such an asymmetric relation should continue and what the alternatives are.

The basic reason for the perpetuation of *status quo* is that the USA and its military might constitute an 'insurance' for Turkey in the event of a conflict with the WTO and the USSR. Even though Ankara works on the basis of *rapprochement*, the long-standing history of suspicion means that it will remain wary, in the near future, of Moscow's intentions. Its vulnerability is all the more acute given its location and strategic value in a war situation. From the early post-war period it has always relied on Washington to provide deterrence in peacetime and military protection in war. Since it believes that the provision of security is the ultimate function of the state, no economic costs have been considered to great to deviate from the policy of extreme dependence.

Even though the Turkish military burden is high, and would have been much higher if Turkey had a volunteer army paid at market rates (see table 13.3 for the magnitudes involved), one consequence of this dependence has been an exclusive reliance on imported arms, particularly from the USA. In recent years the policy of arming national forces with low-quality second-hand equipment financed by heavy borrowing has become untenable. In this context the presence or absence of the US bases is an irrelevant issue both from an economic as well as a security point of view. More central to Turkish security are the questions of arms modernization and indigenous arms production.

It has been estimated that a sensible armaments programme aimed at re-equipping Turkey's forces with some modern weapons would cost a minimum of $1 billion per year for over a decade. This amounts to approximately 36 per cent of the current defence budget. Compared with the figures for basing costs given in table 13.11, the figures for arms procurement are astronomical.

How this massive rearmament is to be financed is a difficult question. The Government has shown its serious commitment by establishing, in 1985, a Defence Fund, which through special taxation and imaginative budgetary savings and transfers is expected to generate large amounts of domestic resources. Estimates claim that around $600 million were generated in 1986.[24] Clearly it will be impossible to keep up this pace in light of Turkey's serious budget deficits and concomitant inflationary pressures. In addition, the burden-sharing concept has to be seen not only from the point of view of how much a country is spending but also how much defence resources it is expected to allocate in the event of a conflict. Turkish military planners believe that they will have to bear a disproportionate burden and cost in the case of a war on the Southern Flank, or even in a more general one. Thus the rest of NATO also has a responsibility to support—as well as an interest in—Turkey's arms modernization process, not only for reasons of Turkish regional security but also for the alliance's own long-term security.

Equally important is the issue of whether Turkey once again will opt for imported equipment or encourage nascent domestic arms industrialization. All the present indications are that indigenous production will be encouraged, particularly to take advantage of the linkages provided by the booming civilian industrial sectors. The F-16 co-production project, which has already started domestic production for the Turkish Air Force, is a good example of the new resolve. Another encouraging factor is the imaginative way some of these projects are being financed. The use of offset arrangements helps not only in technology transfer but also in reducing the economic burdens implicit in the traditional credit system.

Ultimately, the only viable option for Turkey is to have a well-equipped army, a strong economy, an advanced industrial structure, a sustainable domestic arms production base and less reliance on the United States during 'normal' times. This applies not only to its own security needs but also to the security of Europe itself. The value of the USA as an ally should be in its role as a final deterrent. Turkey should look to countries such as Norway or even neutral Sweden for models: countries that are relatively independent in terms of arms supply and that pursue policies favouring a strong defence. The alternative of a Turkey overly dependent on one superpower and overly vulnerable to another is an untenable situation. Furthermore, the current situation does not serve the best interests of NATO, in particular not the European members of the alliance. In the long term, given the excellent morale and fighting spirit of the Turkish armed forces, a well-equipped and relatively independent Turkish defence structure would no doubt be to the benefit of the whole of Western Europe.

VI. Concluding remarks

There are no straightforward estimates which can encapsulate the costs and benefits of the US presence in, and hypothesized withdrawal from, Turkey. In addition to the problem of the bases there are a whole host of supplementary factors that complicate the issue. The only conclusion seems to be that the gains and losses are finely balanced; both countries gain from the association and could suffer potential losses if the *status quo* was changed drastically.

From the point of view of the US facilities, Turkey's direct economic costs are small under both options and can be easily incorporated within the defence budget and the national economy. It is also understandable that the United States in its own global interests would prefer to continue the association in the present form. US assistance should therefore not be considered 'payment' for services rendered; given the quality of aid and the negative resource transfers involved it is not worth much anyway. What Turkey really needs is the ability to stand by itself through *free trade, technology transfer, defence industrialization and modernization, as well as reduction of import dependence.* It is in these areas that it needs support.

The basis of a successful bilateral relation is that policy co-ordination at the different levels should make both countries better off compared to a situation in which such co-operation does not exists. This paper argues that US–Turkish relations are mutually beneficial. Since Turkey occupies a unique position within NATO, however, it has a right to expect more understanding for its position. The case is clearly stated in US Secretary of State George Shultz's letter of 16 March 1987, renewing the 1980 DECA:

In furtherance of our common interests and in recognition of Turkey's important and *unique contribution* to our mutual security, and acting on the belief that the DECA is a solemn commitment on the part of the United States to assist in strengthening the *armed forces as well as the economy* of the Turkish Republic, the Administration is resolved . . . to propose . . . a high level of support—commensurate with Turkey's important contribution to the common defense.[25]

The positive results of these declared intentions have yet to materialize. Turkey's current requirements are clear: to modernize armaments through domestic industrialization; to become more independent of arms imports; to give substance to the concept 'trade, not aid'; to become more self-reliant and hence able to contribute more effectively to the common defence. However, none of this is possible under the debt trap, with restrictions on technology transfer, implicit pressure to import more armaments than it produces and a high degree of protectionism towards products for which the country has a comparative advantage. These are the central issues and they are far from being resolved. The negotiations and discussions about the bases can act as a catalyst to this process; alternatively, the basing problem may become irrelevant.

Notes and references

[1] Statement by Secretary of Defense Robert S. McNamara on the 1969-73 Defense Program and the 1969 Defense Budget; cited in McGuire, M. C., 'Economic considerations in the comparison between assured destruction and assured survival', eds C. Schmidt and F. Blackaby, *Peace, Defence and Economic Analysis* (Macmillan: London, 1987), p. 123; emphasis added.

[2] Although the allies were informed first in the late 1960s, Turkish officials have in private communications with the author claimed that the US Government was thinking of abandoning MAD as early as 1962.

[3] Chipman, J., 'NATO and the security problems of the southern region: from the Azores to Ardahan', ed. J. Chipman, *NATO's Southern Allies: Internal and External Challenges* (Routledge: London, 1988), p. 39.

[4] The two empires fought 14 wars in 200 years.

[5] Ankara's own version of *Ostpolitik* has been reasonably successful, although a fundamental distrust of Moscow remains.

[6] Tezel, N., 'Opening speech to the Seminar on Eastern Mediterranean Security NATO Perspectives', reprinted in *Dis Politika*, vol. 13, nos 1–2 (1986), p. 6; emphasis added.

[7] Sandler, T., 'NATO burden-sharing: rules or reality?', in Schmidt and Blackaby (note 1), pp. 363–83.

[8] Sandler (note 7), p. 365.

[9] See Copley, G , 'Turkey's bold new strategic initiative', *Defence and Foreign Affairs*, Nov. 1984, pp. 9–17.

[10] Deger, S., *Military Expenditure in Third World Countries: The Economic Effects* (Routledge & Kegan Paul: London, 1986).

[11] *Report of the Defense Burdensharing Panel of the Committee on Armed Services*, US House of Representatives (US Government Printing Office: Washington, DC, 1988), p. 53.

[12] Karaosmanoglu, A., 'Turkey and the Southern Flank: domestic and external contexts', Chipman (note 3), p. 288.

[13] Congressional Research Service, *U.S. Military Installations in NATO's Southern Region*, Report prepared for the Subcommitte on Europe and the Middle East of the Committee on Foreign Affairs, US House of Representatives, 99th Congress, 2nd Session (US Government Printing Office: Washington, DC, 7 Oct. 1986), p. 50.

[14] Details are provided in Congressional Resarch Service (note 13). Demirel is to have said '[u]s yoktur tesis vardir' ('there are no bases but there are facilities').

[15] Agreement for Cooperation on Defense and the Economy between the Governments of the Republic of Turkey and of the United States of America in accordance with Articles II and III of the North Atlantic Treaty (DECA), *Treaties and Other International Acts Series 9901* (US Department of State: Washington, DC, 1980).

[16] Turkish Ministry of Defence, 'Defence policies of Turkey and Turkish armed forces', supplement to *NATO's Sixteen Nations* , vol. 32, no. 8 (Dec./Jan. 1987–88), p. 65. See also *Turkiyenin Savunma Politikalari ve Silahi Kuvvetlerinin Yapisi* (Turkish Defence Policies and the Structure of the Armed Forces) (Turkish Ministry of Defence: Ankara, 1987).

[17] See chapter 5, table 5.2.

[18] *Manpower Requirements Report, Fiscal Year 1988* (US Department of Defence: Washington, DC, Feb. 1987), chapter 9.

[19] In similar fashion it is necessary to construct an average exchange-rate which reflects the purchasing power parity of the currencies of two countries. This may be called a conversion factor converting Turkish lira (TL) to dollars and vice versa; see the World Bank, *World Development Report* (Oxford University Press: Oxford, 1987), p. 271. The conversion factor is the arithmetic mean (average) of the 1987 exchange rate and the adjusted 1986 exchange-rate. The latter is the actual 1986 rate multiplied by the relative rates of inflation (or the ratios of the price indices) of Turkey and the USA between 1986 and 1987. The specific conversion factor or deflated exchange rate (e) used in chapter 13 for the period 1986 and 1987 is:

$$e = 1/2\ [674.5\{(932.7/687.7)/(137.9/133.1)\} + 857.2]$$

(674.5 is the lira/dollar rate for 1986; 932.7 and 687.7 are the price indices for the years 1987 and 1986 for Turkey; 137.9 and 133.1 are the appropriate price indices for the United States;

857.2 is the exchange rate for 1987). The actual value of 'e' used, where necessary, comes out to be TL 870.1/US $1.

[20] McCormick, W., 'Selected military transactions in the US international accounts', *Survey of Current Business*, vol. 68, no. 6 (June 1988), pp. 56–63.

[21] Turkish sources claim that the *additional indirect* effects are almost zero. Any positive effects are cancelled out by inflation consequences on the local economy, which is a serious problem in Turkey. Furthermore, most of the income generated goes to high savers such as construction companies and owners of luxurious apartments, etc., and hence generates little ripple effect. Finally, some of the beneficiaries tend to save their earnings in foreign exchange, thus there are no benefits to the local economy.

[22] Duke, S., 'The US military presence in Europe: economic aspects', *RUSI and Brassey's Defence Yearbook 1988* (Brassey's Defence Publishers: London, 1988), pp. 81–97.

[23] In his statement to a subcommitte of the House Committe on Appropriations, US Under Secretary of Defence Edward Derwinski said in 1988: 'Turkey provides the Western Alliance with irreplaceable defense facilities and intelligence sites, due to its obvious geographical location'. See *Foreign Operations, Export Financing, and Related Program Appropriations for 1989*, Hearings before a subcommittee of the Committee on Appropriations, US House of Representatives, 100th Congress, 2nd Session (US Government Printing Office: Washington, DC, 1988), p. 247. According to James Phillips, Senior Policy Analyst at the Heritage Foundation, 'NATO gets 25% of its hard intelligence on Soviet military activities from Turkish facilities', *Defense News*, 19 Oct. 1987, p. 21.

[24] For a general discussion of the current financing of the Fund see *1989 Yili Butce ve Finansman Programi* [1989 Annual Budget and Financial Programm]) (Government Printer: Ankara, 1988). The actual 1987 figures in Turkish Lira given in the text are from *Mili Savunma Bakanligi 1989 Mali Yili Butcesi ve Savunma Sanayii Destekleme Fonu'na Ait Raportorier Raporu* [Report for 1989 of Defence Ministries Annual Defence Budget and Defence Industries Support Fund] (Turkish Ministry of Defence: Ankara, 1988).

[25] *New York Times*, 17 Mar. 1987, p. 5; emphasis added.

14. Economic costs and benefits of US military bases and facilities in Italy

Jennifer Sims
University of Maryland, College Park, Maryland

I. Introduction

Italy is awakening to a new activism in security affairs in response to a perception of a more complex threat emanating from NATO's 'grey areas'. Although Rome has long identified its security interests as coincident with NATO's, NATO has first and foremost represented the bilateral relationship with the United States.[1] Washington has defined its interests more broadly than NATO's charter and has tacitly supported Italy's growing 'out of area' interests in the Mediterranean with the presence of the 6th Fleet. Italy has tried but never entirely succeeded in getting its northern neighbours to take Mediterranean security and Italy's pivotal role in it as seriously as many Italians deem necessary.[2]

It is thus not surprising that talk of a significantly reduced US presence puts Italy particularly on edge. Although it is conceivable that a reduced US profile in Italy might reduce tensions with certain Arab or North African states, demographic, ideological and economic trends suggest that growing tensions in the Mediterranean are structural.[3] At the same time there is little precedent for believing that other European NATO countries will be willing to spend as much as the USA has spent to bolster Mediterranean security. The Southern Flank is demonstrably tangential to their primary concern: the problem of a divided or even possibly united Germany and the military balance in Central Europe. Besides, the Mediterranean offers a Pandora's Box of politically volatile issues related to the Middle East and the Persian Gulf area, issues which if addressed would not only pose more demands on already tight defence budgets but also fracture domestic constituencies along politically sensitive lines.

To summarize, the security effects for Italy of a US military withdrawal might have to be borne largely by Italy alone. Such a withdrawal would at the same time lead to increased costs for US security commitments to non-NATO states in the Mediterranean littoral. Since no NATO bases, or, in the extreme case, not even a communications infrastructure, would be available to provide support for the 6th Fleet, a removal of bases from the southern region could end up costing the USA more than keeping them.

These particular twists to the issue of US withdrawal from Italy provide a critically important lesson, not the least for the USA. There is an unhealthy tendency in the United States to consider a whole range of NATO issues in the binary terms of us or them; indeed reference to a 'European view' or 'Europe's interest' is probably more often fallacious than not. Moreover, the Italian case serves to remind us that the costs and benefits of NATO do not find their definition in the Central Front alone.

Any analysis of the economic impact of US bases in Italy is constrained by a paucity of data, a consequence of classification requirements and accounting methods. Nevertheless, a rough sketch of the economic impact of a potential US withdrawal is possible if it is well understood that the analysis will be impressionistic at best. The following discussion is divided into two sections. First, the size and general distribution of the US presence are outlined, linking these with the structure of NATO commands in the southern region. This description of force distribution is followed by rough data on the total dollar flows to Italy resulting from the employment of Italian nationals, construction contracts, housing (rents/leases) and the purchase of supplies and raw materials (e.g., petroleum). More detailed information on dollar flows is provided in a case study of the Comiso Air Station, for which a breakdown of data for certain sectors is available.

Estimates on the economic costs of hosting US bases have proven difficult because the amount of dedicated acreage and thus forgone taxes are either unknown or considered sensitive—as are all figures pertaining to the size of the Carabinieri forces assigned to perimeter security on the bases. Such figures are undoubtedly substantial. Nevertheless, at the end of this section an attempt is been made to draw broad conclusions regarding costs and benefits on the basis of available information.

Section III of the chapter addresses the logical follow-on question: If the United States were to withdraw, could the Italian economy support an enhanced military effort to replace the lost US presence?

II. The US military presence: status and economic impact

The status of forces

On 19 June 1951 parties to the North Atlantic Treaty signed status of forces agreements covering the legal obligations of receiving and sending states with regard to exchanges of forces pursuant to the adoption of Alliance obligations. At this time, Italy and France were the only NATO members of the Mediterranean littoral to adopt this agreement. With the French decision to withdraw from NATO's military structure in 1966, Italy remained the only Mediterranean state in NATO obligated by an ongoing accord for which regular renegotiation is not required. This has had the two-fold effect of diminishing the political volatility of the issue of US bases in the country and of obscuring over time their economic impact since this is not normally quantified for the purpose of negotiations or political debate.[4]

US bases in Italy exist to support NATO objectives, particularly with regard to the defence of the southern region, which comprises over 10 million square kilometres. Allied Forces Southern Europe (AFSOUTH), with headquarters in Naples, is responsible for the defence of Italy, Greece and Turkey plus the sea lines of communication (SLOC) in the Mediterranean and Black Seas. Allied Air Forces Southern Europe (AIRSOUTH) is also located in Naples and is commanded by a US Air Force General who is responsible for all land-based air operations along the 3600-kilometre border between northern Italy and eastern Turkey. In support of this mission, Italy hosts the 5th Allied Tactical Air Force (ATAF) at Vicenza.[5] The Commander Allied Land Forces Southern Europe (COMLANDSOUTH), an Italian general whose headquarters are in Verona, is responsible for the forward defence of the Veneto-Friuli Plain and thus Italy's industrial heartland. The Commander Allied Naval Forces Southern Europe (COMNAVSOUTH), with headquarters in Naples, is responsible for ensuring continuous surveillance, security and freedom of navigation through all areas of the Mediterranean.

The Naval Striking and Support Forces for Southern Europe (STRIKEFOR-SOUTH, the US 6th Fleet) works with COMNAVSOUTH to deter aggression in the region and to project power in support of land operations in time of war. Because of the disconnected theatres for land operations in the southern region, the Mediterranean provides essential mobility and depth for integrated NATO defence. Of critical importance in this regard is the Channel of Sicily which stretches 145 km from Italy to Tunisia, separating the eastern and western basins.

The US bases in Italy

The US forces in Italy support NATO's land, sea and air missions and in time of war come under the operational control of NATO commanders in Verona and Naples.[6] All US troops use US hardware and equipment which would, in the event of withdrawal, be removed with them or sold to the Italian Government at prices determined by residual values. Such assets include communications equipment such as long-range navigational aids (Loran), satellite links for naval communications, radio relays and, at San Vito, an optical space-tracking system.

The principal US air bases are: Comiso (Sicily), which had the 487th Tactical Missile Wing with ground-launched cruise missiles (GLCMS), now removed as a result of the Treaty between the USA and the USSR on the elimination of their intermediate-range and shorter-range missiles (the INF Treaty); San Vito dei Normanni, which provides communications, reconnaissance and electronic intelligence for support of Ramstein Air Base (in the Federal Republic of Germany); and Aviano, which serves as the rotational base for the 401st Tactical Fighter Wing (nuclear capable F-16s) assigned to US Air Force Europe (USAFE) in time of war. In addition, the Italian Government recently approved NATO's request to accept those F-16s whose removal from Torrejón was conceded to Spain during US base negotiations in 1988. They may now be based near Crotone in Calabria, although munitions will probably continue to be stored at Aviano.[7]

Naval bases include the Sigonella–Catania complex in Sicily for support of anti-submarine warfare (ASW) operations in the Mediterranean; La Maddalena (Sardinia), which services US submarines, and Naples–Gaeta for support of the US 6th Fleet and its communications requirements. Headquarters for STRIKEFORSOUTH is at Gaeta.

The main US Army bases are around Vicenza and at Camp Darby. The 8th Logistics Command at Camp Darby, near Leghorn, serves as a major command and storage post for the US Army Europe (USAREUR). Vicenza's Camp Ederle has the 4th battalion of the 325th Airborne team, part of the 5th Allied Task Force. Airlift for this team is provided partly out of Aviano. In addition, the 559th US Field Artillery Group is spread out in detachments located in nine nearby towns. It serves in a custodial role for deployed nuclear-capable artillery and surface-to-air missiles.[8]

In addition to these bases, the United States supports defence of the southern region through participation in Allied Command Europe's Allied Mobile Force (AMF) and the Naval On-Call Force Mediterranean (NAVOCFORMED). The former, established in 1960 under the command of the Supreme Allied Commander Europe (SACEUR), whose headquarters are in Heidelberg, FRG, is designed for rapid reinforcement to NATO's Northern and Southern flanks—in particular to Norway, Denmark, Greece and Turkey. The AMF regularly conducts exercises simulating collaborative wartime combat support with ground and air forces from Belgium, Canada, the FRG, Italy, the USA, the Netherlands, Luxembourg and the UK.[9] Italy has incorporated the AMF into its general defence plans and contributes an alpine battalion combat group to its land component (AMF-L).[10] The AMF-L is, however, very small (7000 men on average) and of limited readiness and mobility; only the air component (AMF-A) could go to both flanks at one time. The 4th Battalion (Airborne) 325th Infantry at Vicenza constitutes the forward-deployed US contribution to the AMF in Italy; it is also a forward based element of the 82d Airborne Division, which is part of the US Rapid Deployment Force.[11]

NAVOCFORMED, a quick-reaction force composed of mostly frigates and destroyers, exercises twice a year for 30 days with combatant warships provided by Italy (usually one frigate), Turkey, the United States and the UK. It is under the command of COMNAVSOUTH during these periods. NAVOCFORMED does not rely on land-basing of US assets or troops in Italy.

NATO's only other instrument for flank defence in a crisis, the Rapid Reinforcement Plan (RRP), is completely dependent on US contributions: 2000 US combat aircraft with up to 700 available for the southern region. As of March 1990, pre-positioning of supplies in southern region countries has not been done, despite recommendations for this from NATO staff since the plan was adopted in 1982. As a result, many analysts consider this only a 'paper plan'.[12]

Financial implications for Italy of the US bases[13]

Costs

As already mentioned, perimeter security for the US bases is partly the responsibility of the Italian Government which provides, according to one source, 'many hundreds of Carabinieri' or state police for this purpose.[14] About 20 per cent of the Italian defence budget is allotted to the Carabinieri, generally responsible for internal security, although the precise number assigned to US bases remains classified.

The amount of land purchased by the Italian Government for US military bases is also classified. According to the Office for Defense Cooperation of the US Embassy in Rome, this land is not leased but granted to the US Government for its indefinite use. In the event of a US withdrawal, the land would remain the property of the Italian Government and, in principle, be available for resale to private entities. In the meantime, however, receipt of taxes on the appreciating land is forgone by the Italian Government.

US military authorities currently require more land for the enlargement of several bases and the relocation of some facilities—particularly in the Naples area. The costs of land acquisition have sky-rocketed in recent years, however, as a result of limited supply aggravated by strict Italian laws and regulations regarding historic sites.

The issue of land acquisition raises an important point: support of US bases brings additional indirect costs to the host country which are extremely difficult to quantify. Not least of these are legal fees. In the case of the base at Agnano, for example, relocation of facilities has become desirable owing to the poor local infrastructure and to earthquake risks. The owner of the desired land in Capua, however, has repeatedly raised his asking price, using the legal argument regarding historic monuments. The process of acquiring the land has therefore become enmeshed in the Italian legal system at some cost to the Italian Government.

Other social and legal costs are involved with processing criminal charges (a small percentage of which are adjudicated in Italy) and supplying welfare or social services to Italian nationals related by blood or marriage to US servicemen. Manœuvre damages related to the operation of US equipment on Italian soil are unknown but probably minimal.

Although infrastructure and capital improvements on US bases are generally paid for by the US Government, maintenance and construction of infrastructure outside of base perimeters (access roads, utilities) are paid for by the Italian localities concerned. This has posed problems for some of the smaller facilities, especially mountain-top communications sites which are highly dependent on regular road maintenance by local authorities. This is less of a problem for the larger facilities which have the advantage of being collocated with Italian bases.

Benefits

Italian nationals employed on US bases number approximately 2000. On the average they receive from the US Government approximately $17 857 per capita annually, which amounts to approximately $35 714 285 in total annual pay for all. These employees, while hired by the US Government, are in many cases covered by special Italian laws which guarantee their continued employment in the event of a US withdrawal. According to law number 98 of 9 March 1971, all Italian citizens who were employed on 30 June 1969 in military entities of states belonging to NATO are guaranteed employment by the Italian Government in the event of job termination. On 23 November 1979, law number 596 extended this benefit to cover those who were employed as of 30 June 1979. Proposals now exist to extend this provision to cover those currently employed.

The total number of US military personnel in Italy is approximately 14 900. Their dependents number approximately 16 000. Off-base housing for these US nationals costs the US Government up to 200 billion lire (L), or approximately $143 million, in rents per year.[15]

Construction contracts for on-base improvements are in most instances allotted to local firms. They total approximately $12.9 million annually. Support services provided by Italy at US expense (petroleum, minor construction and subsistence support) amounted to approximately $241.4 million in fiscal year (FY) 1986.[16]

Italy receives no military assistance from the US Government and no special financial or commercial benefits for hosting US bases. However, the USA has committed itself, to the goal of balanced trade in the defence sector with its ally. In April 1978 the USA and Italy signed a Memorandum of Understanding (MOU) to improve the two-way street in defence purchases. At the time the USA enjoyed a trade balance in its favour of 9:1. As of 1987 it had been reduced to 2:1. The Office of Defense Cooperation of the US Embassy in Rome notes that if all support services provided by the Italian Government were included in the calculation (i.e., the $241.4 million figure noted above for FY 1986), the balance would be slightly in Italy's favour (see table 14.1).

Besides this special bilateral effort, significant economic advantages to Italy are not gained by the hosting of US forces on Italian soil. All bidding for US military contracts is by law done competitively and without 'sole source' arrangements. Therefore, despite the Italian Government's generally co-operative attitude as a receiving state, the bidding process results in few gains for Italian firms since their labour costs are relatively high.[17] The USA has purchased some military hardware from Italy, including a recent contract for radar for marine communications from Selenia (in co-operation with Sperry) worth

Table 14.1. USA–Italy defence balance of trade summary, as of 30 September 1986 Figures are in US $m.

	1982	1983	1984	1985	1986
Italy					
FMS[a]	86.9	18.3	50.8	175.3	44.4
Exports[b]	135.0	150.0	47.5	62.2	73.1
Total	**221.9**	**168.3**	**98.3**	**237.5**	**117.5**
DOD					
Prime contracts[c]	25.9	30.2	55.0	164.5	65.3
Sub-contract	3.0	15.1	10.0	19.2	24.2
Total	**28.9**	**45.3**	**65.0**	**183.7**	**89.5**
DOD computed ratio	7.69 : 1	3.71 : 1	1.51 : 1	1.29 : 1	1.31 : 1
Negotiated ratio[d]	..	4.7 : 1	4.64 : 1	1.94 : 1	..

[a] Foreign Military Sales.
[b] Commercial Exports: estimated total of commercial exports licensed under the Arms Control Act.
[c] Prime Contract Awards: do not include contracts for subsistence, petroleum, construction and support services.
[d] The negotiated ratio results from a comparison of DOD's data with Italian Government data.

Source: Office of Defense Cooperation, US Embassy, Rome.

$15 million, the acquisition of Beretta hand-guns[18] and a contract for helicopter maintenance with Agusta. These are all relatively small contracts, however.

The only major US–Italian co-production agreement concerns the Patriot, a modern, conventionally armed air-defence missile to replace the aging Nike-Hercules. This 5 billion defence package, accepted in principle by both parties in 1988, awaits assurances from Rome that the Italian Government will cover its share of the agreement. If such assurances are not received by 31 March 1990, the US Department Of Defence (DOD) will be unable to sign the first of two contracts, and the USA and Italy will have to reopen negotiations. The most immediate impact of such a delay would be felt by Rayethon, which is providing 20 sets of radars and fire control stations. If Italy fails to pass the requisite special legislation, 1200 Rayethon workers will have to be laid off.[19]

Case study: the economic impact of closing the Comiso Air Station

In January 1990 US Defense Secretary Richard Cheney announced the closure of the USAF base at Comiso. According to an unclassified 1986 report prepared for the US Embassy in Rome, 202 Italian nationals were employed at the base during FY 1986, each earning an average annual salary of $10 106.[20] Of these employees roughly 57 per cent were from Comiso. US military and civilian personnel numbered approximately 1513 and 122 respectively. The total pay-roll for Italian and US personnel was $30 271 022 in FY 1986. Of this $3 446 412 was exchanged for lire through the base's Cashier's window and an

Table 14.2. Value of ROICC-controlled construction in Italy, 1982–86

Year	Value in lire	Value in US $
1982	3 313 065 60	2 347 705
1983	31 885 288 117	21 070 101
1984	25 491 405 056	14 815 905
1985	21 924 611 726	11 702 451
1986[a]	8 320 177 051	5 521 278
Total	**90 934 547 550**	**555 457 140**

[a] As of August 1986.

Source: US Embassy, Rome.

estimated 80 to 90 per cent was spent on the local economy. The report estimates that the total monthly impact of the US personnel's pay-roll alone was approximately $287 201 for that fiscal year.

Personnel apartments leased from the Italian Government numbered 157 as of October 1986, at a total cost of L631 807 500, or approximately $451 291.[21] In addition, utilities paid by the US Government amounted to L210 million or $150 000 annually. Other property leased from the Italian Government (warehouses, office space, cold storage, etc.) cost approximately L264 819 150 or about $114 954 annually.[22] Contract quarters (hotels) in FY 1986 amounted to $491 935.[23] Thus the combined yearly total of off-base leased property (excluding contract hotels and privately leased apartments) and utilities was approximately $716 245. Privately leased apartments by base employees totalled $229 440 per year.[24]

Construction and services

Construction and services for Comiso are purchased in any of four ways: through the civil engineering office on the base; through the Resident Officer in Charge of Construction (ROICC), in this case a naval officer; through NATO offices for infrastructure support; or through Italian-sponsored construction. The computed ROICC-controlled construction, including material and labour, completed between 1982 and 1986 was distributed as shown in table 14.2.

ROICC Comiso construction under contract but not completed as of 31 August 1986 was valued at $45 583 417.[25] Thus total construction completed or under contract totalled $101 040 557. The ROICC's prime contractors are all Italian firms and all the facilities built under his supervision are US-owned. They include, *inter alia*, such diverse facilities as tennis courts, an auto hobby shop, an addition to a sewage treatment facility and a small training range.

The Office of Civil Engineering planned to design and award 38 projects worth $2.89 million (at L1680 to the dollar) during FY 1987. Projects done out of this office include small constructions, facilities upgrades and facilities designs. The total value of the civil engineer's construction projects as of FY 1985 was approximately $17.9 million. Projects completed in FY 1986 equalled

$150 294. The total dollar value of all civil engineering projects under way (as of first quarter FY 1987) was L54.6 billion, or $24.14 million at the USAFE Headquarters rate of 2241.93 lire to the dollar. Italian construction costs for the base were projected in November 1986 to equal approximately $32 million. NATO funded construction as of 1 October 1986 totalled roughly $30.5 and included such facilities as a missile storage building, a warhead igloo, a warehouse, six shelters, various roads and a utility and electrical sub-station. All prime contracts were awarded to Italian firms.

With the withdrawal of the GLCMs under the provisions of the INF Treaty, the future of the Comiso base remains uncertain. There has been some discussion of maintaining the base as an Italian, US or NATO facility. However, since many of the facilities located on the base are US-owned, any new, non-US owner would have to reimburse the US Government for the residual value of the facilities. An alternative suggestion offered by Italian Foreign Minister Giulio Andreotti is to turn the base into a 'peace park'.

With the US decision to close the base at Comiso, the impact on the local economy will be appreciable. Of the Italian nationals employed on the base, about 57 per cent reside in the immediate vicinity and receive 55 per cent of the total salary earned by all Italian nationals. Few of these employees would be eligible for guaranteed re-employment rights with the Italian Government. Local business would be affected by the loss of approximately $287 201 in purchases by US personnel and dependents each month, and by the loss of public contracts, 57 per cent of the value of which goes to local firms. Six out of the ten prime contractors used by the ROICC are Sicilian, and 91 per cent of those employed for US-funded construction have come from Sicily.

Nevertheless, it should also be noted that the peak years for base construction are now past. Since a considerable part of the maintenance and small facilities improvements can be done by the civil engineering office on base, the lost income to Italian firms in the event of base closure would not be as substantial as would first appear. Indeed, given the strategic location of Comiso, it would be likely that the departure of the US forces would be followed by conversion of the base to other uses in the national or NATO context, leading to continued and perhaps even greater opportunities for national construction firms. The issue of whether the Italian Government could afford such expenditures is discussed briefly in the second part of this study.

Summary

In the event of an abrupt US military withdrawal from Italy, the Italian economy would lose up to L200 billion or $142.85 million in rents per year: $35.7 million in annual compensation to Italian base employees—a large percentage of which would have to be assumed by the Italian Government as a result of employment guarantees; approximately $241.4 million per year in service contracts (supplies of petroleum, minor construction, etc.); and up to

Table 14.3. NATO and WTO land forces in north-eastern Italy, 1988

	NATO	WTO
Divisions	8	17
Tanks	1 250	4 340
Artillery/Mortar	1 400	2 860

Source: AFSOUTH Headquarters, derived from Chipman, J. (ed.), *NATO's Southern Allies: Internal and External Challenges* (Routledge: London, 1988), p. 39.

$13.2 million in annual construction contracts. Moreover, the economic shock of the withdrawal would be felt disproportionately in the south, where elimination of the naval bases and Comiso Air Station would exacerbate the persistent economic problems of relatively slow growth and unemployment.

Trade in the defence sector would probably initially move in the United States' favour as contracts for the maintenance of US military assets were terminated and the Italian Government purchased fixed assets on the former US bases for their residual value. However, purchasing such fixed assets 'second hand' might be advantageous should the Italian Government seek to expand its national military capabilities in substitution for those lost through the withdrawal. In this sense a US departure could be considered an economic advantage from the Italian Government's and military's point of view. Still, unless other Europeans assumed part of the responsibility for base conversion and maintenance, these costs would fall on Italy alone.

The capacity of the Italian Government to absorb the costs of sustaining adequate defence capabilities after a US withdrawal and of responding to the opportunities that might emerge in the internal European weapon market is the subject of the following section. In order to establish the parameters of the discussion—what the Italian Government perceives as its defence requirements—it is necessary to begin with a general discussion of the Italian 'defence model' and the practical contribution that the USA makes towards meeting what are believed to be the central threats to Italian security.

III. Implications of a US withdrawal for Italy's defence costs

The Italian defence model[26]

Italy has based its defence requirements on a threat perceived as having three central elements. The first has been the prospect of an air–ground offensive on the north-eastern border where several mountain passes could provide invading forces access to the Po Valley—probably in conjunction with other territorial attacks to prevent reinforcement of forward-deployed troops. Although past comparisons of forces suggested a heavy advantage to the Warsaw Treaty Organization (WTO) in such a contingency (see table 14.3), the inhospitable terrain in neutral Austria and Yugoslavia and the likely dedication of a substantial portion of these WTO troops to other areas in the event of conflict have led strategists, including George Thibault in this volume (chapter 17), to downplay the threat from the north-east. Others, such as Brigadier-General Luigi Caligaris and General Carlo Jean, both highly respected Italian strategists, have taken a less sanguine view.[27] In 1990, obviously, the changing balance of forces in Europe is altering calculations of threats in the north-east. However, until the European structure stabilizes it is safe to assume that Rome will continue to be concerned with ways of addressing the vulnerabilities in this sector.

The second category of threat is to the south and involves the sea lines of communication. Air–naval forces, if concentrated or co-ordinated, could jeopardize NATO's capabilities in the Mediterranean, particularly its re-supply efforts and communications.

However, Italy has not only been concerned about large-scale East–West contingencies in this context. Italy's vulnerability in the south, especially to harassment from Libya's growing conventional capabilities, were evident with the 1986 missile attack on Lampedusa.[28] Although it is not possible to discuss fully all the sources of instability in the Mediterranean that concern Italian security, the case of Libya's threat to NATO's south is illustrative of the kind of 'grey area' problem that would become more salient after a reduction in the US military presence in the region.

Libya has hosted military personnel from Syria, Cuba, North Korea, Eastern Europe and the Soviet Union and had, as of 1982, spent $12 billion in purchases of Soviet arms. These include MiG-23 interceptor/bombers, Tu-22 medium bombers, short-range FROG-7 and Scud B surface-to-surface missiles (SSMs), T-62 and T-72 main battle tanks, SA-2, SA-3 and SA-6 surface-to-air missiles (SAMs), missile patrol boats and submarines.[29]

It is true that the most sophisticated Soviet equipment is often not operated at its peak by the Libyan military, leading a number of analysts to conclude that Libyan forces threaten only vulnerable neighbours (such as Chad or potentially Tunisia).[30] Nevertheless, Libya must also be recognized as possessing a significant capacity to harass—even some distance from home—with limited air and naval strikes and to contribute to efforts in more apocalyptic world war scenarios.

Indeed, the third category of threat identified by the Italian Government concerns the country's vulnerability to attack from the air. The WTO has had the potential to establish air superiority over the north-eastern sector, the Mediterranean and the country's territory more generally (see table 14.4). Since Italy perceives its security as critically linked to the security of the Mediterranean region as a whole, threats to the eastern Mediterranean basin are considered part of the general threat to which Italian forces must be able to respond.

The Italian Government has specified five 'joint operational missions' for addressing these threats. The first, defence in the north-east, would employ air forces and long-range ground weapons for early reconnaissance and interdiction; mechanized units and alpine units at intrinsically strong defensive positions for containment of the offensive; armoured forces and fire-power to

Table 14.4. NATO and WTO air forces in the southern region, 1988

Type of aircraft	NATO	WTO
Fighter/bomber/ground attack	615	695
Interceptors	295	1 560
Reconnaissance	90	195

Source: AFSOUTH HQ, derived from Chipman, J. (ed.), *NATO's Southern Allies: Internal and External Challenges* (Routledge: London, 1988), p. 39.

reinforce the defence; and naval operations to control the sea-front and rear areas to avoid encirclement.

Given the small number of US and allied troops deployed in the northeastern sector and the relative isolation of the area with respect to other NATO theatres of operation, the Government has concluded that any initial defence will rest primarily with national forces. It will depend for success in part on national capabilities to conduct a forward defence in mountainous areas guarding the industrial plain and in part on the extent to which neutral countries engage in blunting an enemy advance. Available ground forces include 20 brigades divided into three Army Corps, one of which is held in reserve.[31] Naval assets include missile hydrofoils and gun-boats for patrol, coastal radars covering much of the upper Adriatic, anti-ship helicopters and sea-mines. The air forces include four squadrons of fighter-bombers, two squadrons of fighter-reconnaissance aircraft and three squadrons of light fighter-bombers. These air assets are deployed on northern airfields and could be reinforced with four air defence squadrons.

The second joint mission, defence in the south and protection of the SLOC, primarily involves Italian air and naval forces. The SLOC, shipping (upon which Italy must rely for almost all its raw materials) and an 8000-km coastline must be protected by means of naval coastal and amphibious forces, two ocean-going air-naval groups on either end of the Channel of Sicily, and air- and sea-borne reconnaissance forces which, in many instances, would have to work in conjunction with the 6th Fleet.[32]

The air forces available for this mission are based in the south and on the islands and include one squadron of fighter-bombers/interceptors, two squadrons of fighter-bombers, two squadrons of interceptors, two squadrons of light fighter-bombers, two squadrons of anti-submarine warfare (ASW) maritime patrol aircraft, one electronic warfare squadron and one helicopter squadron. In addition, some of those air assets deployed in north-central Italy could be redeployed south if operational conditions permitted it.

The ground forces are mechanized and motorized infantry units; artillery; engineer, signal and light aviation units; plus reserve units to be mobilized.

The Italian Government aspires to a more independent capability to achieve the second joint mission—that is, with less reliance on the 6th Fleet which could, in time of crisis, be employed in the eastern basin or withdrawn from the Mediterranean completely.[33] This orientation partially accounts for Italy's recent concentration on the development of ocean-going task groups based on light carriers, the relative weight given to the Navy in recent defence budgets, and the Franco-Italian naval *entente*, which has included joint military exercises. There has also been bilateral co-operation between Italy and France in the areas of undersea research, rescue and amphibious warfare.[34]

The third joint mission, air defence of the territory, addresses the enhanced threat to Italian territory arising from the increased range of enemy aircraft in the region, particularly the deployment of Soviet Tu-265 Backfire and Tu-225 Blinder bombers in the Crimea and thus the omni-directional nature of any potential air offensive. Italy relies on F-104 interceptors and Tornado fighter-bombers for air defence and counter-air tasks. Defensive missile systems include the Air Force's aged Nike-Hercules (for medium-high altitudes), the Army's Hawk (for low altitudes) and ship-borne missiles of the standard NATO type. Point defence includes the Army's 40/70 guns and 12.7 mm four-barrel guns; the Navy's Sea Sparrow and Albatross short-range missile systems, guns (127/54, 76/62) and the Dardo anti-aircraft system; and the Air Force's four-barrel guns.

Command, control, communications and intelligence (C^3I) is provided by a complex of land, sea and air-based radar systems, integrated in part with the NATO Air Defense Ground Environment (NADGE) system. Italy also hosts two of NATO's Airborne Early Warning (NAEW) craft based at Trapani. An upgrade of C^3I is planned with the acquisition of the CATRIN field communications system, which is designed to be partially integrated with a larger NATO network.

The fourth joint mission, operational defence of the territory, includes the defence of the Italian peninsula and insular territory exclusive of the north-east sector. It is anticipated that offensives may range from major air and naval bombing campaigns to smaller-scale attacks against communication sites or key strategic points. For the purposes of planning and organization, the country is divided into seven military regions. Each region is commanded by a Military Region Commander (MRC) responsible for the co-ordination of all forces in his area and accountable to the Army Chief of Staff. The forces are in turn divided into security companies (static defence), infantry battalions (reserve force for protection of key areas), artillery battalions, engineer battalions, motorized or mechanized infantry brigades and the paratroop brigade (for reinforcement, as co-operatively organized by the three armed services). They are supported by air and naval assets mentioned above and, in particular, the Air Force's C-130 and G-222 airlift aircraft which supplement the Army Light Aviation's CH-47 medium helicopters and AB-212 multi-role helicopters. Italy has also created a rapid deployment force designed to be quickly available as needed for reinforcement of the territorial defence effort.

The fifth joint mission, peace-keeping, security and civil protection operations, has been emphasized ever since Italy's deployment of peace-keeping troops to Lebanon in 1982–84. It is divided into two sets of sub-tasks:

international security and support for civil protection. Although at first both tasks were to be accomplished by a single rapid deployment force, the decision was later made to split the force in two: the Forza di Pronto Intervento (FOPI), dedicated to civilian contingencies, and the Forza di Intervento Rapido (FIR), assigned to overseas crises or emergency defence of the territory. The FIR consists of about 10 000 men assigned (and seconded as the need arises) from the three services. They are under the command of an Army general who reports directly to the Chief of Defence Staff. Apart from the inter-service rivalry these arrangements generate, the force suffers from deficient air transport capabilities, logistical support arrangements and specialized weaponry.[35]

In addition to this rapid intervention capability, the Italian Government has identified a humanitarian and security interest in maintaining a peace-keeping capability. In this regard it has agreed to assist Malta in assuring its security, to protect shipping in the Persian Gulf and to contribute national forces to the United Nations Interim Force in Lebanon (UNIFIL) and the Multinational Force and Observers (MFO) in the Sinai.[36]

In support of the joint missions defined by the defence model, Italy maintains a sizeable military force: 265 000 in the Army, 100 000 in the Carabinieri, 51 800 in the Navy (including its air arm) and 73 000 in the Air Force.[37] The Army is equipped with 1720 tanks, 1110 artillery pieces, 230 air defence guns, 60 Improved Hawk plus Stinger missiles and 429 aircraft and helicopters. The Navy maintains in addition to a variety of SAMs and SSMs, 9 submarines, 2 helicopter carriers, 2 cruisers, 4 destroyers, 14 frigates, 11 corvettes, 7 hydrofoils, 23 mine counter-measure vessels, 23 amphibious vessels, 8 support ships (tankers and coastal transport) and 83 armed helicopters. The Air Force has 460 combat aircraft, 76 helicopters and a variety of missiles including 96 Nike-Hercules.

Yet despite the apparent quantitative adequacy of the Italian armed forces, there continue to be problems with ageing weapon systems and the uncertain operational effectiveness of forces only recently required to combine in joint missions. This chapter is directly concerned with only the first problem. It is linked to the defence budget and the availability of resources more generally. The resource problem and continued dependence on US capabilities in certain sectors for achieving mission goals is discussed below.

The effect of a US withdrawal on Italy's defence posture

The first and fourth joint missions require significant modernization of the Army's equipment—particularly the replacement of M-47 tanks and the acquisition of self-propelled artillery, multiple missile launchers, short-range and portable SSMs and combat helicopters.[38] The ground forces are rather severely undercapitalized: while the Army's share of the defence budget has shrunk, its operational forces have not been cut back proportionately.[39] Investment programmes have repeatedly been delayed and although expenditures for these increased at one stage (from 29 per cent of the Army's discretionary expenditures in 1984 to 36 per cent in 1987), they dropped in the 1988 budget request to 32.2 per cent. In any case, these outlays still amount to barely one-fifth of the Army's total when fixed expenditures (including personnel costs) are taken into account. According to some estimates, in order to fulfil the missions described in the defence model, the Army should have double or triple the resources that it currently claims to have.[40]

Part of the problem with modernizing the Army has been the need to ensure adequate supplies and maintenance for ageing systems while still allocating enough resources to systems modernization. Since the Army has been updating across the board, without preference to any one mission, progress has been slow and perhaps more expensive than necessary.

With respect to air defence, Italy does not have an adequate airborne early warning capability or an independent satellite reconnaissance capability. After 'Operazione Girasole' in 1986, the Air Force requested 2–4 AWACS and the Navy argued for co-operation with France in satellite surveillance.[41] Since then Italy has agreed to take a 14 per cent share in the French Helios military satellite programme, for which it is building a national receiver station. France also suggested in December of 1987 that it would consider creating an airborne early warning force for the Mediterranean in co-operation with Spain and Italy. This would be done by using part of the Boeing E-3 AWACS force ordered for the French Air Force earlier that year. The two additional AWACS on which options have been placed might also be used in this role.[42]

In addition, and particularly given the increased threat from Soviet bombers deployed in the Crimea, Italy has sought improved air combat, penetration and defence capabilities for maintaining the effectiveness of the second, third and fifth joint missions. The Ministry of Defence in its Additional Note to the 1988 Defence Budget identified the replacement of the F-104S Starfighter interceptors and obsolete Nike and Spada missile defence systems as the most critical.[43]

Also important are tanker aircraft for in-flight refuelling of Tornado fighter-bombers and transport aircraft. Italy is participating in the production of the European Fighter Aircraft (EFA) and plans to buy 100–150 of them, thus accepting a 21 per cent share of the programme in which the FRG, the UK and Spain are also involved. Delays in the programme, partly attributable to the uncertain commitment of participating governments, have contributed to its rising costs. The Government is likely to have to find replacements for the F-104 before the EFA enters into service.[44] Italy is also replacing its obsolete F-104Gs with 187 AM-X close air support aircraft.

The gaps in naval capabilities are perhaps the most problematic. The growth in the Italian 'high seas' capability during the 1960s and 1970s was a response to the increasing Soviet presence in the region and later, in the 1980s, to the uncertainties regarding the employment of the 6th Fleet in a Mediterranean crisis. Yet since an independent defence of even the central Mediterranean has been rejected as unfeasible, the the aim of the building programme has been to improve Italy's ability to contribute to 6th Fleet operations in the eastern Mediterranean. Unfortunately, the single most glaring gap in Italy's naval capabilities in the Mediterranean is the lack of adequate air cover—a gap that renders the increasing number of capital ships as much an added vulnerability as a new benefit. Even if inter-service rivalry stays quelled, and short take-off and landing aircraft are purchased for Italy's light carriers, the situation may not be much improved.[45]

It thus seems that Italy's naval forces are unbalanced in favour of vulnerable surface ships whose value, apart from 6th Fleet support, lie primarily in the radar picket role. Italy needs improved ASW capabilities and coastal patrol and defence units—especially if, as suggested by one source, the Italian Navy Staff expects that French and Spanish navies may concentrate on defence of the SLOC in the Atlantic and leave the western Mediterranean and the Gibraltar–Suez routes to Italian forces.[46] Recent reports indicate that the Italian Navy recognizes some of these weaknesses and intends to order two Sauro Class conventional submarines in the summer of 1988. In addition, improvements are being made on ASW capabilities for the aircraft-carrier *Vittorio Veneto*.[47]

From the foregoing brief analysis several general conclusions can be made about Italy's dependence on US military forces for national defence. First, defence of the north-eastern region, as for territorial defence as a whole, are not dependent on the US military presence for success in the initial phase of any war. Although there are significant weaknesses in Italy's independent capabilities for these missions, they relate to the adequacy of current equipment, the under-capitalization of the armed forces and the lack of mobility—not to reliance on foreign troops. However, in any sustained conflict, Italian forces would need to be resupplied and reinforced and this could introduce significant problems insofar as such an effort would have to involve moving supplies through the Mediterranean. For this, and for the defence of the southern region more generally, the presence and function of the US 6th Fleet are, while perhaps not adequate, certainly indispensable. Although a withdrawal of US forces from Europe would not necessarily mean that the US 6th Fleet could no longer function in the Mediterranean—the USSR has long made use of off-shore anchorages for its naval presence there—it could well make the logistical requirements prohibitive or require radical changes in deployment strategies.

Furthermore, it is not clear how these costs and logistical problems would be eased by option B, the second hypothetical scenario—a withdrawal 'with promise to return'. The USA could not promise to return to resupply Italy by sea in the event of a crisis without a well-established and smoothly functioning base structure there. The Italian military would have to develop costly substitute capabilities to guard against the possibility that the US forces would not return when needed, while suffering the risk of having the US contingent basing infrastructure on Italian soil targeted all the same. If the decision were made to proceed with a US naval withdrawal, and this was not accompanied by a withdrawal of the Soviet Mediterranean Squadron (SOVMEDRON, or 5th Eskadra), Italy would likely face a sizeable Soviet presence: 7 combatants, 6 submarines and 31 auxiliary ships on any given day. These ships use anchorages perilously close to home: the Gulf of Hammamet (off Tunisia), the Gulf of Sollum (off Libya) and one south of Cape Passero (Sicily) during exercises.

Apart from increasing naval expenditures to improve Italy's ASW, anti-ship, air-defence and reconnaissance capabilities, improved Italian capabilities could perhaps be achieved through increased co-operation with France and Spain in reconnaissance and a commitment from Italy's European partners to increase the NAVOCFORMED to a more robust standing force. Such options would indeed have to be explored in the event of a US withdrawal from the Mediterranean, given the limited capacity of the Italian economy to support a rapidly growing defence budget.

To summarize, if one changes the hypothetical situation from a permanent US withdrawal to a temporary withdrawal with a promise to return in the event of crisis, it would be more likely that the 6th Fleet would retain a contingent base network and infrastructure in the Mediterranean region. However, given that Italy also regards protection of the SLOC as an ongoing security requirement in the face of local, non-NATO threats, such a US promise would be unlikely to relieve pressure for more independent deterrent and war-fighting capabilities in the Mediterranean; indeed, it would likely increase the cost of achieving current levels of security for both countries. For reasons discussed below, such increased costs could not easily be shouldered by Italy alone.

The Italian defence budget and the economy

After undergoing a recession during 1981–83, the Italian economy slowly began to recover between 1984 and 1986, with increases in its gross national product that averaged 2.6 per cent annually.[48] Prospects for growth in the economy are uncertain and are partly dependent on growth in the economies of Italy's main trading partners, especially the FRG. Italian economic well-being is also highly dependent on trade with Arab and North African states. The relief for the balance of payments that came with reductions in oil prices was in part counteracted by the effects of reduced exports to oil-producing nations.

The rate of inflation in Italy, although falling, is still relatively high compared with other NATO nations. At 20 per cent in 1980 it subsequently fell to approximately 5.3 per cent in 1988.[49] Unemployment has been high—11.25 per cent—and disproportionately high in the south and among the young.[50] Moreover, industrial production, which had improved substantially during the period of stability provided by the relatively long-lived centrist Pentapartito Government, has been affected by the re-emergence of instability in the workplace as the number of strikes have increased during late 1987 and 1988. Positive signs do exist: large companies are reorganizing, expanding into foreign markets, improving the technological level of Italian products and

Table 14.5. The Italian defence budget, 1982–84
Figures are in billion lire.

Allocations	1982	1983	1984
Force programmes			
Officers	179.7	206.3	273.8
Non-commissioned officers	293.4	334.3	393.4
Troops	658.5	879.4	983.9
Force programmes total	**1 131.6**	**1 420.0**	**1 651.1**
Operation			
Training	382.6	495.8	570.1
Technical/logistic support	862.4	1 147.9	1 499.2
Infrastructure	275.3	344.2	376.0
Requirements of commands and other bodies	373.0	458.2	555.6
Welfare	58.9	67.6	90.4
Operation total	**1 952.2**	**2 513.7**	**3 091.3**
Investment			
Special programmes (special laws)	1 201.1	1 596.7	1 895.4
Means (materials and equipment)	438.8	632.1	857.6
Infrastructure	179.5	174.8	339.3
Stocks	2.0	26.1	35.7
Research and development	40.7	64.3	105.6
Investment total	**1 862.1**	**2 494.0**	**3 233.6**

Source: *Defence White Paper 1985* (Italian Ministry of Defence: Rome, 1985).

helping to propel the Italian economy forward. Indeed, in 1987 Italy claimed it had surpassed the UK in GNP.

Nevertheless, to sustain growth into the 1990s, Italian financial authorities have determined that current governmental expenditures should not grow in real terms. Of major concern is the size of public debt which is greater than 100 per cent of GNP. Of the total, one-third is renewed every year. Governmental borrowing equalled 14.3 per cent of GNP in 1986 which has continued upward pressure on interest rates. Although the Finance Ministry wants a reduction in government borrowing to 8.7 per cent of GNP in the period 1987–89 and cancellation of the debt by 1991, prospects for achieving these goals do not seem good. Success depends in part on keeping a tight lid on defence expenditures. Last year's planning figures called for zero growth in fixed (primarily personnel) expenditures and increases in spending on goods and services linked to the GNP.

However, this plan to increase GNP and reduce inflation and Government borrowing has run up against the military's expressed need to improve the structure and sophistication of Italian military forces, while meeting relatively high fixed costs (personnel and infrastructure). The Italian defence community has argued strongly for increased and rationalized military spending. A process of re-structuring was begun in 1975 and included special laws passed outside the budgetary process for rapid improvement of all three military services.[51]

Table 14.6. The Italian defence budget as percentage of GDP and total state budget, 1975–84

Year	A Defence budget	B GDP	C A:B (%)	D State budget	E A:D (%)
1975	2 451.3	125 378	1.95	30 373.9	8.07
1976	2 956.7	156 657	1.88	38 071.7	7.77
1977	3 530.6	190 083	1.86	47 083.5	7.49
1978	4 313.8	222 254	1.94	64 444.0	6.69
1979	5 119.1	270 198	1.89	119 376.0	4.28
1980	5 780.0	338 743	1.71	150 249.0	3.85
1981	7 500.7	401 579	1.87	189 606.0	3.96
1982	9 918.0	471 390	2.11	235 366.0	4.21
1083	11 648.7	535 904	2.17	273 227.0	4.26
1984	13 820.0	603 964	2.29	345 586.7	4.00

Source: Defence White Paper 1985 (Italian Ministry of Defence: Rome, 1985).

Unfortunately, inflation soon drove up the costs of these special programmes and delayed their completion. The 10-year military renewal plan was therefore changed in 1980 and a new 10-year plan for 1982–91 was announced. Included were additional requests for special funding of joint research and development (R&D) programmes (the AM-X close-support aircraft, the EH-101 helicopter, the CATRIN C^{3}I system) and dual-capable (civil–military) systems. However, finances during the 1981–83 period continued to fall short of requirements and in 1983 the completion of the programme was again postponed until 1992–93. As mentioned above, these delays affected the Army most of all. Horizontal movement of funds as a result of special financing provisions brought the Army's budget share down from 50 per cent to 42 per cent while the shares of the Air Force and Navy grew. As the Army's share decreased, savings were made in training, infrastructure and readiness.[52]

In 1986 the share of defence spending to GNP was 2.7 per cent (compared to 3.1 per cent in the FRG and 3.2 per cent in Belgium). According to the DOD 1986 *Report on Allied Contributions to the Common Defense*, Italy ranked only 12th among NATO allies in the ratio of its share of total NATO defence spending to its share of total NATO gross domestic product (59 per cent).[53] Military spending per inhabitant is low—$188 in 1986—and growing relatively slowly (see table 14.7), leaving Italy outdone in this measure by all NATO allies except Luxembourg, Portugal and Turkey. Expenditure per serviceman is also very low at about one-half to one-third that of leading NATO allies.[54] Indeed, as mentioned above, the under-capitalization of the Italian armed forces is one of the key problems Italy faces in the military sector. There is, in general, a significant gap between the size of the forces and the resources available for their maintenance and modernization despite the cut-backs of 1975. These reduced by one-third the number of operational units in the Army and cut certain Navy and Air Force programmes.

Table 14.7. Italian defence expenditure per capita: selected years, 1970–86
Figures are in US 230$, at constant (1980) prices.

Year	1970	1975	1980	1983	1984	1985	1986
Total	141	152	170	177	182	187	188

Source: *Jane's Defence Weekly*, vol. 8, no. 23 (12 Dec. 1987), p. 1342.

Given the restrictive plans of Italy's financial authorities and the armed forces' need to increase expenditures both on personnel and weapons modernization, it is difficult to predict how much additional flexibility there is for an expansion of the defence budget. The budget did grow from 2.4 per cent of GNP in 1977 to 2.7 per cent in 1986 and the military's interest in up-dating equipment has been reflected in increases in investment (R&D plus procurement)—although this measure dipped slightly in 1987. Moreover, approximately L200 billion (1 per cent of the defence budget total) goes to defence purposes from the budgets of other ministries (the Ministries of Industry, of Scientific Research and Technology and of State Industry).[55] It is also true that the financial authorities' restrictive guide-lines for defence spending have not been accurate predictors of defence spending since they tend to be abstractly derived and therefore somewhat unrealistic. Indeed, accounting and defence planning instruments are weak in Italy, making unpredictability a key problem for effective resource management.

Three additional factors also suggest that Italy will continue its efforts to sustain the upward trend in defence expenditures: first, there is evident political support for increasing conventional as opposed to nuclear capabilities within Europe generally; second, Italy is demonstrating an increased commitment to assuring the security of the Mediterranean area—as evidenced by its commitments to Malta, UNIFIL, the MFO and its mine-sweeping operations in the Persian Gulf; and third, the state of Italy's defence-related industries is improving both in terms of their productivity and organizational links to government. They now satisfy 80 per cent of national armament needs. However, their future prospects depend in part on the Government's procurement policies and in part on continued productive involvement, with adequate governmental support, in collaborative projects with non-Italian firms. Apart from the collaborative European projects mentioned earlier (the A-129 light attack helicopter, the EFA, the Helios satellite and the EH-101 and NH-90 helicopters), Italian industries have joined a number of other collaborative projects as part of the Nunn initiative.[56] These include the Ada Project Support Environment, the Multifunctional Information Distribution System, the 155-mm autonomous precision guided missile, the NATO Identification System, the Modular Stand-off Weapon, the Battlefield Information Collection and Exploitation System and the NATO Frigate Replacement (NFR-90) programmes.

Prospects for industrial collaboration in the defence sector

If a US withdrawal were to take place in a political context of hostility and non-cooperation, it could be expected that collaborative projects, reciprocal procurement, side-by side testing, liberalized technology flows and agreements for balanced trade might also be curtailed, suspended or terminated.

Moreover, if the US Government were persuaded to demobilize troops and cut back on weapons purchases, it would not be surprising to find US defence industries marketing their products more aggressively abroad. Such developments might threaten European industries or raise procurement costs to European governments no longer benefiting from US economies of scale in equipment production and indirect 'standardization'. On the other hand, to the extent that the US withdrawal were to remove a political pressure to purchase highly advanced, expensive or otherwise inappropriate technologies for the Italian theatre of operations, it could prove beneficial in the long run to Italian industries and the economy.

If withdrawal were to bring with it a firmer European commitment to an open market in the defence sector, Europe's defence industries might find their competitiveness boosted as the market in the defence sector would become rationalized, governments would be forced to collaborate on R&D and production costs would fall.

Although such restructuring is meant to be a positive result of the European Community's decision to constitute a free and open market by 1992, European governments do have the option to exclude those industries considered vital to national defence. Whether they would be interested in taking advantage of this caveat may depend to some extent on the costs of subsidizing inefficient producers—a cost that would increase appreciably if US equipment were to become scarce and expensive. The issue is not a minor one, especially for Italian industries where, according to Western European Union data, there were approximately 50 000 people employed exclusively on the production of military equipment in 1983 and a further 83 000 'directly employed part-time on armaments production'.[57]

Just as unification of the Italian national market led to a growing gap between the development of the north and the poorer south, so some Italians fear that the impressive advantages of market rationalization will mostly flow to central and northern Europe. In the defence sector such economic 'thinning' could affect some of those industries developing cutting edge technologies with important offsets to national, civilian production. If compensatory mechanisms such as assurance of 'fair returns' regarding the economic off-sets of collaborative projects (production shares, employment, etc.) are not provided, countries in NATO's southern tier may find themselves without industrial mobilization capabilities in time of crisis.[58] Since this would parallel their growing concern about their particular exposure to 'out-of-area' threats, such a development could heighten tension significantly within the alliance.

Of course, others fear that any compensatory arrangements such as 'fair shares' would be excessively protectionist, raise costs and delay those benefits which Europe could otherwise expect to obtain from a rationalized internal market. The savings that could be obtained by pursuit of maximum efficiency might be directed toward financing a common R&D programme—particularly in Europe's lagging electronics sector—that would spill over through liberalized information flows to civilian economies in all of NATO Europe, thus spreading the positive benefits of reduced barriers to collaborative economic activity. Of course such rationalization will not necessarily help those national defence industries which depend for their survival on the export of *whole* systems to third parties.[59]

Of the southern tier states faced with the challenge of the 1992 open market, Italy probably has the most advantageous position. A number of Italian firms are involved in European-wide projects such as the NH-90, Tonal and EH-101 collaborative helicopter projects, the Future International Military Airlifter (FIMA) and the EFA.[60] Nevertheless, the fear exists in Italy that, robust and dynamic as some selected firms are, the heavy burden of poor national infrastructure and an over-inflated, unresponsive public sector will prove major obstacles to Italian industry's ability to compete for lucrative contracts. This has already had the effect of driving away top scientific and engineering talent from Italian firms despite relatively abundant national funds for R&D in recent years.

This being said, if one assumes that the status quo is maintained with respect to Europe's international market and that a US withdrawal would be accompanied by an end to military sales to Italy, the economic effects would probably be mixed and difficult to quantify in the abstract.[61] Although weapon-related industries might gain certain advantages, investment and consumption in other sectors of the economy would be squeezed, with potentially negative effects for the economy as a whole.[62] Moreover, even if the DOD's Foreign Weapons Evaluation Program and the NATO Comparative Test Program were to continue, Italian access to these programmes, already limited by bureaucratic obstacles, would be further impaired by the reduced circulation of US military personnel on Italian bases (one-way knowledge of important foreign technological capabilities is funnelled back to these testing programmes in the USA).[63]

Italian defence industries have exerted considerable pressure on the Government to assist in marketing and in improvement of the investment environment.[64] Yet at the same time most firms—especially the larger ones such as FIAT and Olivetti—have welcomed the 1992 deadline as a national challenge to revitalize and trim the public sector. The health of firms such as these, which may turn on the extent to which this challenge is met, will help determine not only the quality of today's military output but that of tomorrow as well. Military R&D is performed by specialized firms which are largely state-owned (Aeritalia, Italcantieri, Agusta, Otomelara, Selenia, etc.), as well as by certain military centres, the National Research Centre and universities which conduct scientific research.

The extent to which Italian industry would suffer from any reduced US role in collaborative R&D would depend on the willingness of European governments to co-ordinate the planning, design and R&D of advanced systems either through a military variant of the Eureka project or preferably from the perspective of southern European states, by means of a more top-down effort, such as a European variant of the US Defense Advanced Research Projects Agency, with adequate financial assistance for those countries which cannot allocate adequate funds for independent R&D.

IV. Conclusion

In order to sustain Italy's defence capabilities, given what the Government has perceived to be its requirements, Italy's defence budget would have to grow considerably in the coming years. Even with a reduced WTO threat, Italy's security depends on an ability to address 'grey area' third-party contingencies that may continue to increase in number and gravity. Devoting additional resources to the defence effort will be difficult both because of the changing East–West political environment and because of the financial pressures which already exist as a result of the size of public debt, the relatively slow pace of growth expected for GNP and the heavy tax burden which the country bears.

If the United States were to eliminate its bases in Italy, the economic burden would increase as the Italian Government would be required to absorb the costs of acquiring fixed assets on the former US bases for their residual value. As a result the defence trade balance would probably move more in the US favour. In addition, several of Italy's major joint missions, particularly air defence of the territory and defence of the south and SLOC would demand significant increases in Italy's military capabilities in these areas. To some extent such improvements are already built into Italian defence planning insofar as the role of short-range tactical nuclear weapons has been reduced (since 1975), the share of the Navy in the total defence budget has recently grown, and the Government is pursuing new co-operative arrangements with France regarding Helios and probably AWACS. However, sufficient expenditures to make up for the loss of US assets, particularly the 6th Fleet and the protection it could potentially provide a resupply effort in time of war, are simply not available.

In this context it might be noted that the economic model of military alliances offered by Keith Hartley and Nicholas Hooper in chapter 8 of this volume neglects a key fact, namely, that the 'free ride' enjoyed by smaller allies has a hidden cost. This is the opportunity cost of having planned, developed and standardized one's military capital stocks around a larger ally's capabilities. If these capabilities are suddenly withdrawn, the smaller state might be left with an unbalanced and vulnerable military structure for a long time to come. Such imbalances, inevitably reflected in the drawn-out procurement process, are costly and time-consuming to correct.

A US withdrawal with a promise to return would not save much for either side: the Italian armed forces would not have the benefit of acquiring second-hand, low-cost military facilities for a rapid buildup of national capabilities, since the United States would have to retain a substantial basing infrastructure for a possible resupply effort. At the same time they would have to plan both for the possibility that the US pre-positioned assets would be targeted and for the possibility that the US might not show when needed—a much more likely contingency when the 6th Fleet is not actually in the Mediterranean at the start of a conflict.

As the above analysis suggests, the economic or strategic effects of a US withdrawal from NATO's southern region would depend on the extent to which such a withdrawal were accompanied by a reduction in the threats to Italy either through formal agreement or as a by-product of the process. Italy is already orienting its defence requirements to perceptions of a threat that is more complex than that posed by the hypothetical NATO contingency of an outbreak of war on the Central Front. Italy perceives its interests as directly connected to the security of the Mediterranean basin—an area only partially covered by NATO security arrangements and involving potential instabilities that could concern Italy alone.

One may also infer from this analysis that even the continued hosting of US military assets in the region does not guarantee to Italy that US forces, particularly the 6th Fleet but also reinforcements to the north-eastern sector, will be where Italy wants them to be at any given moment in a crisis. Western security would seem to require a more robust air defence and defence of SLOC in the Mediterranean. Although these improvements may not be achievable in the current NATO context, they also most certainly cannot be accomplished by Italy alone.

Notes and references

[1] For general background on Italian defence policy, see Cremasco, M., 'Italy: a new role in the Mediterranean?', ed. J. Chipman, *Nato's Southern Allies: Internal and External Challenges* (Routledge: London, 1988), pp. 195–236. This volume offers an excellent short course on southern flank security issues.

[2] See Ruggiero, R., 'The Atlantic Alliance and challenges to security in the Mediterranean: problems and policy choices', *Prospects for Security in the Mediterranean, Part 3*, Adelphi Paper no. 231 (International Institute for Strategic Studies: London, spring 1988), pp. 3–13.

[3] After the US bombing of Tripoli and Libya's failed counter-attack on the Italian island of Lampedusa, Libya announced that Italy had become its 'number one enemy' because it was 'a base for official American terrorism'. Italy later conducted military exercises on Sicily and redeployed naval forces to the Channel of Sicily, probably to demonstrate a capability to defend against an invasion from the south. See Burger, W. and Stanger, T. 'Kaddafi: Italy is our enemy', *Newsweek*, 27 Oct. 1986. On the structural nature of Mediterranean tensions see Chipman (note 1), *passim*; and *Sources of Change in the Future Security Environment*, Report

of the Future Security Environment Working Group, submitted to the Commission on Integrated Long-Term Strategy (US Department of Defense: Washington, DC, Apr. 1988), pp. 2–7.

[4] All Italian parties have supported NATO since the Socialist and Communist parties accepted capitalism in the 1960s and 1970s respectively.

[5] The Sixth ATAF is located at Izmir, Turkey.

[6] Operational control is here distinct from 'full command'. Non-US military commanders cannot subdivide forces and attach, for instance, a brigade to a foreign battalion.

[7] The US Government has a policy of neither confirming nor denying the presence of nuclear munitions at any base. A DOD official has confirmed, however, that Aviano would not be reduced in size as a result of the development of a new F-16 base in Italy.

[8] These include 6 MGM-52C Lance short-range ballistic missiles, 23 M-110 203-mm SP howitzers, 256 M-109 155mm SP howitzers and 96 M1M-14B Nike-Hercules surface-to-air missiles. Most of these systems are quite old and not all are nuclear-equipped. See *The Military balance 1987–1988* (International Institute for Strategic Studies: London, 1988), pp. 202–209.

[9] The United Kingdom and Luxembourg provide only land forces; the Netherlands provides only air support.

[10] See, for example, Italy's *Defence White Paper 1985* (Italian Ministry of Defence: Rome, 1985), p. 56. However, the AMF and NAVOCFORMED have not been mentioned in any of the following Additional Notes to the Defence Budget.

[11] Chipman, J., 'Allies in the Mediterranean: legacy of fragmentation', Chipman (note 1), p. 60.

[12] For more on rapid reinforcement plans and a critical view of the RRP, see Chipman, J. 'NATO and the security problems of the southern region: from the Azores to Ardahan', Chipman (note 1), pp. 40–44.

[13] Data for this section were obtained primarily through private interviews at the Sending State Office, the Office for Defense Cooperation and the Political/Military Affairs Bureau of the US Embassy in Rome; at the Center for Advanced Defense Studies in Rome; and at the US Department of Defense, Washington, DC.

[14] Napoletano, A. P. (Capt.) and Jones, M. (TSgt), 'Local economic impact study: Comiso Air Station, Italy', unpublished report prepared for the US Embassy, Rome, Nov. 1986.

[15] A 1986 exchange-rate of L1400 to the dollar is used throughout this analysis unless stated otherwise.

[16] Data provided by the Office for Defense Cooperation, US Embassy, Rome.

[17] Exceptions occasionally occur. In what has been termed by one source as a 'political favour', Industria Aeronautica Meridionale recently received a contract for the maintenance of C-130 transports in Brindisi worth approximately $3 million a year.

[18] The contract for Beretta hand-guns is under review and may well not be renewed.

[19] *Defense Week*, vol 11, no. 9, 20 Feb. 1990. An Italian consortium would build the air defence missile if the deal goes through. Analysts are speculating that the delay reflects the political changes that have swept Europe in the autumn/winter of 1989/90.

[20] Napoletano and Jones (note 14). The salary is calculated using an exchange-rate of L1600 to the dollar.

[21] Using an exchange-rate of L1400 to the dollar.

[22] At the FY 1986 Headquarters, US Air Force Europe/ACB (HQUSAFE/ACB) rate of L2261.93 to the dollar.

[23] At the same HQUSAFE/ACB budget rate as above (note 22).

[24] Exchange-rate of L1400 to the dollar.

[25] This figure presumes that the contracts would be completed within the next fiscal year (exchange-rate of L1400 to the dollar).

[26] Data for this section were provided by the Italian 1985 Defence *White Paper* (note 10); and the *Nota Aggiuntiva allo Stato di Previsione per la Difesa* for 1985, 1986, 1987 and 1988, presented to Parliament in October of each year by the Minister of Defence.

[27] These views were expressed during interviews in the fall of 1987 and spring 1988. See also Caligaris, L., 'Italy's Strategic Dilemma', *NATO's Sixteen Nations*, vol. 32, no. 4 (July/Aug. 1988).

[28] Evaluations of a potential threat to the south are based on the growing resources, military capabilities (e.g. diesel electric submarines) and nuclear aspirations of littoral states, taken together with the rise of Islamic fundamentalism.

[29] See McCormick, G., 'Soviet strategic aims and capabilities in the Mediterranean: part 2', *Prospects for Security in the Mediterranean, Part 1*, Adelphi Paper no. 229 (International Institute for Strategic Studies: London, spring 1988), p. 43; and Cremasco, M., 'Two uncertain futures: Tunisia and Libya', *Prospects for Security in the Mediterranean, Part 3*, Adelphi Paper no. 231 (International Institute for Strategic Studies: London, spring 1988), p. 50.

[30] Cremasco (note 29), pp. 50–51.

[31] The 4th and 5th Army Corps are 1st echelon forces in the mountains and on the Veneto-Friuli Plain. The 3rd Army Corps is held in reserve.

[32] The air naval groups consist of at least one cruiser, 2–3 destroyers and 5–6 frigates. The Italian Government's 1985 *White Paper* (note 10) refers to the 6th Fleet as playing a decisive role in the protection of the sea lines of communication.

[33] See Snyder, J. C., *Defending the Fringe: NATO, the Mediterranean and the Persian Gulf*, SAIS Papers in International Affairs (Westview Press/Foreign Policy Institute: Boulder, Colo., 1987), pp. 16–27.

[34] Exercises have included, for example, the 'Olive Noir' and the 'Trident' series. Naval co-operation agreements are also underway with Spain. Cremasco (note 1), p. 216.

[35] See Cremasco (note 1), pp. 228–29.

[36] See *Nota Aggiuntiva allo Stato di Previsione per la Difesa 1988* (note 26), p. 11.

[37] See *The Military Balance 1987–1988* (note 8).

[38] AGUSTA is participating (38%) in the development of the Tonal A-129 MK2 light attack helicopter with Westland (38%), Fokker (19%) and Casa (5%). Feasibility and cost studies are due by the end of 1988. Italy will probably purchase 90 of these.

[39] Jean, C., 'Economic and budget restraints on Italian defense', *International Spectator*, vol. 22, no. 2 (Apr.–June 1987), p. 72.

[40] Ilari, V., 'The new model of Italian defense, doctrinal options, issues and trends', *International Spectator*, vol. 22, no. 2 (Apr.–June 1987), p. 83.

[41] Ilari (note 40).

[42] *Jane's Defence Weekly*, 19 Dec. 1987, p. 1405.

[43] *Jane's Defence Weekly*, 14 May 1988, p. 921.

[44] Assembly of the WEU, Document 1119 (Western European Union: Paris, 5 Nov. 1987), p. 11.

[45] An old Italian law prohibits the Navy from operating fixed-wing aircraft. Parliament has considered ending this restriction despite strong objections from the Air Force, which maintains that it can perform the defensive missions required, provided Italy's objectives in the Mediterranean remain appropriately limited.

[46] Ilari (note 40), p. 24.

[47] *Jane's Defence Weekly*, 2 Apr. 1988, p. 639.

[48] Jean (note 39), p. 73.

[49] Jean (note 39).

[50] Jean (note 39), p. 73.

[51] See Italy's *Defence White Paper* (note 10), pp. 63–69.

[52] Ilari (note 40), p. 82.

[53] US Department of Defense, *Report on Allied Contributions to the Common Defense*, a report to the US Congress (US Department of Defense: Washington, DC, Mar. 1986), p. 20.

[54] Jean (note 39), p. 72.

[55] However, it should be noted that 20% of the Italian defence budget goes to what are essentially non-military expenditures.

[56] In 1985 Senator Nunn initiated legislation designed to enhance co-operative weapon procurement among allies. Included in the Nunn Amendment were two funding programmes. One provided the DOD with venture capital to encourage co-operative research and development efforts with allies; the other provided funding for the testing of weapons produced in other countries, essentially building on the foreign weapon evaluation programme which already existed at that time.

[57] WEU document no. CPA (85) IA-D/32 (Western European Union: Paris, 1985), p. 21.

[58] 'Fair Return' is the concept 'whereby national industries are compensated by being attributed a fair share of each programme on a case by case basis, the division of the industrial workload reflecting each individual nation's investment in a collaborative programme'. 'European Cooperation in Armaments Research and Development—Guidelines Drawn from the Colloquy', report submitted on behalf of the Committee on Scientific, Technological and Aerospace Questions, Assembly of the WEU, 34th Ordinary Session, 10 May 1988, 1st Part, Document 1141 (Western European Union: Paris, 1988), p. 5.

[59] The Italian naval consortium Melara Club experienced a major downturn in 1987 after a 50% drop in exports. Foreign sales were L300 billion that year as compared to an annual L600 billion during the previous two. After an annual turnover of L1200 billion between 1984 and 1986, total turnover in 1987 was L850 billion. A further decline, perhaps as much as 20%, is expected in 1988. The companies involved include Ansaldo, Breda MB, Elettronica, Elmer, Elsag, Fiat Aviazione, Fincantieri, OTO Melara, Riva Calzoni, Selenia and Whitehead.

[60] Several of these collaborative programmes, such as the NH-90 and the EFA, are delayed and unlikely to survive as a consequence of uncertain threat scenarios, limited resources and increased competition in the defence sector. See *National Defense*, vol. 74, no. 455 (Feb. 1990), pp. 22–25.

[61] This is a radical assumption but of interest in clarifying the argument. The US military sales to Italy totalled approximately $117.5 million (see table 14.1).

[62] For some recent research regarding the relationship between military spending and national economic growth, see Rasley, K. and Thompson, W. R., 'Defense burdens, capital formation and economic growth', *Journal of Conflict Resolution*, vol. 32, no. 7 (Mar. 1988), pp. 61–86.

[63] See Roos, J. G., 'Allies success in US programs hinges on equipment requirements', *Armed Forces Journal International*, Sep. 1988, pp. 36, 38.

[64] See comments by Enrico Gimelli, Head of the Italian Aerospace Industry (AIA) in *Jane's Defence Weekly*, 12 Dec. 1987, p. 1385.

Part III
Military implications of a US military withdrawal from Europe

15. US withdrawal and NATO's Northern Flank: impact and implications

Steven E. Miller

SIPRI

I. Introduction

Of the two withdrawal options posited for this study, option B, the skeletal US presence with a commitment to reinforce in a crisis, is the status quo in the northern region. The analysis in this chapter therefore deals exclusively with option A, the 'gone for good' withdrawal. It will discuss the implications for northern Europe of the termination of all military collaboration between the United States and its NATO allies. The implications of a US military retrenchment for NATO's Northern Flank are less obvious and less immediate than for the Central Front, but they are no less significant and no less worthy of study. This chapter will focus on the region encompassed by Allied Forces North (AFNORTH).[1] The discussion naturally subdivides into two parts: the first has to do with the defence of Norway and the military situation in adjacent sea areas; it is not possible to consider problems of security in the Norwegian components of AFNORTH without examining also the military situation in the waters of the North Atlantic and Norwegian Seas (which encompasses Iceland as well). The second component of AFNORTH has to do with the defence of Denmark and the Danish Straits, in NATO terminology the Baltic Approaches (BALTAP), and with the military situation in the Baltic Sea; while not entirely unrelated to security concerns further north, BALTAP raises a distinctly different set of issues and is therefore discussed separately.

It is important to begin by noting the stark contrast between the Northern Flank/BALTAP region and the more familiar Central Front. NATO's AFNORTH differs from the Central Front in three significant ways. First, to a very large extent, the Northern Flank is, in peacetime, already free of US forces and bases. Indeed, many of the plans and collaborative efforts that have been made between the USA and its allies in the north (most notably Norway, but also Denmark and Iceland) are essentially attempts to compensate for the peacetime absence of US (as well as other allied) military power. Reinforcement from outside the theatre is, of course, important in many scenarios for the Central Front, but it is the very essence of NATO strategy for the north; this heavy reliance on reinforcement has resulted in a number of arrangements intended to provide support of US forces in the region in the event of war.

Second, while the defence of Norway is obviously important, the Northern Flank is, especially from the US point of view, primarily a maritime theatre, one in which several important naval missions must be executed. These include defence of the North Atlantic sea lines of communications (SLOCs), strategic counterforce operations against Soviet strategic submarines (SSBNs), and naval land attack against targets in the Soviet Union. The strategic importance of Norway and Iceland derives from the influence they could have on these operations, and many of the security arrangements that exist between the USA and these two Northern Flank allies have to do with the provision of various sorts of supporting infrastructure for the US *naval* forces that might be introduced into the region in wartime. Thus, the collaboration between the USA and the northern allies takes a different form than that which characterizes the Central Front; and, while far from insubstantial or unimportant, it is far less visible in peacetime than is the case in the Central Front, since it does not involve the regular stationing of US forces. It is these *wartime* arrangements that would be disrupted in the event of a severance of military relations between the USA and its allies, and their implications are felt at least as much at sea as on land, if not more.

Third, with respect to BALTAP, this command is already one in which European NATO allies have overwhelmingly assumed responsibility. Although, as is shown below, some air reinforcements are expected, the defence of Denmark and the Danish Straits will fall primarily on Denmark and the Federal Republic of Germany.

As a consequence of these considerations, analysis of the Northern Flank after a US withdrawal from NATO's military arrangements must be quite different from that for the Central Front. It is not simply a question of how many US forces will be withdrawn, producing what gaps in NATO standing forces, imposing what demands on NATO Europe to compensate. Rather, on the Northern Flank, the important consequences of a US withdrawal will be the forces that *will not arrive* and the naval power that *cannot be counted upon* in the event of war. What happens to a strategy that calls for holding out until the US forces arrive if they are not going to arrive? Closely related is the matter of the arrangements for peacetime intelligence collaboration and for the wartime support of US forces in the Northern Flank that *would not exist* if the USA were to extricate itself from all military entanglements with NATO allies. What happens to naval power currently reliant on support ashore if the support ashore is not available?

II. The USA and the Northern Flank: bases and other arrangements

Although the importance of the Northern Flank[2] is often invoked, and knowledgeable observers have suggested that NATO could actually lose a large-scale East–West war by allowing the USSR to triumph in the region, what is most remarkable about the security environment there is how small the US peacetime presence is. There are very few US bases and forces in the region, and indeed, Norway observes a policy of not permitting foreign bases on its territory in peacetime. Moreover, even with respect to exercises the US presence has been surprisingly (and, some believe, distressingly) limited. Ground forces exercise in Norway once or twice a year, but there have been chronic complaints that this is insufficient to provide the skill in mountain and arctic warfare that is necessary to function effectively in the inhospitable conditions often found in the far north.[3] US surface naval forces[4] have been notably absent from these allegedly critical waters, although efforts are now under way to raise the exercise frequency for US carriers in these waters to a week or two every other year.[5] Indeed, virtually every other sea receives more regular attention from US Navy surface vessels than does the Norwegian Sea. This is not a region in which large deployments of US military power are found in peacetime. There are, nevertheless, some facilities and forces to be taken into account, and although minuscule by the standard of the Central Front, they are not inconsequential in the context of NATO and US strategy in the north.

Keflavík, Iceland

Although a tiny nation with a population of only 240 000, Iceland occupies a crucial geostrategic location astride the waters that connect the Norwegian and Barents Seas to the open ocean of the North Atlantic. Iceland is, it is sometimes said, the cork that keeps the Soviet Navy bottled up north of the Greenland–Iceland–United Kingdom (GIUK) gap.[6] Iceland's importance in any campaign for the control of the North Atlantic is indicated by the fact that it was occupied by British forces during World War II, primarily to deny Germany the use of Iceland in its U-boat campaign against Atlantic shipping. The immutable facts of geography have preserved Iceland's strategic importance in connection with a NATO–Warsaw Pact conflict, particularly because NATO is dependent on sea-borne reinforcements in any war that lasts more than a short time.

Iceland itself possesses no military forces.[7] It is nevertheless a member of NATO and, equally importantly, party to a 1951 defence agreement with the United States in which the United States acquired base rights in Iceland and in return pledged to defend the island nation.[8] The Icelandic Defense Force (IDF) is comprised of US forces. The main consequence of the security arrangement between Iceland and the United States is a single US base at Keflavík, in the south-west corner of the island. The size of this base is strictly limited by the Icelandic Government, owing to sensitivity about the social and cultural repercussions of having large numbers of US personnel living in a country whose entire population is no more than that of a small US city. As a result, there are only 3000 US servicemen stationed at Keflavík; 1700 are US Navy personnel, and 1300 are from the US Air Force Tactical Air Command.[9]

Although small, however, the US capabilities deployed on Iceland perform several important functions. First, of course, they provide some defence of

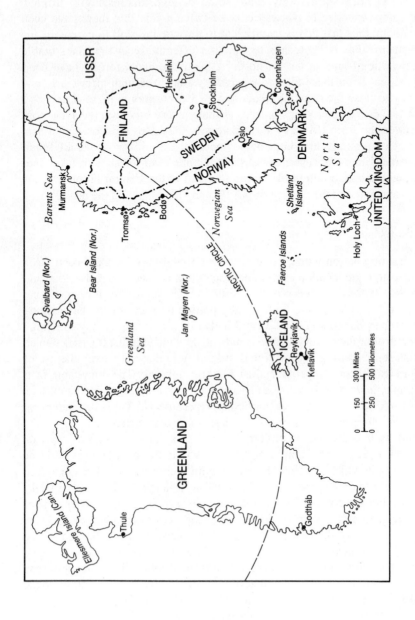

Figure 15.1. The Nordic Flank

Source: Bjøl, E., *Nordic Security*, Adelphi Papers no. 181, (International Institute for Strategic Studies: London, 1983), map 'The Nordic Countries', p. 3.

Iceland itself (in particular, the base and other facilities on Iceland that are the militarily significant targets of any potential attack), denying the Soviet Union a 'free ride' with respect to a potential Soviet occupation of this strategically valuable piece of real estate. There can be no question that Soviet control of Iceland would completely alter the geostrategic picture in the North Atlantic and Norwegian Seas. Norwegian security expert John Kristen Skogan of the Norwegian Institute of International Affairs has written, for example, 'If the Americans left Iceland, and even more if the Soviets arrived there themselves, Soviet control of the Norwegian Sea would be much easier to achieve and maintain . . .'[10] Or, as Major General D. J. Murphy of the US Marine Corps has explained, if Iceland is lost 'success will be difficult, if not impossible, in the Norwegian Sea. It will take much valuable time to regain Iceland if it is lost in the early stages. That will cause further delay—at a most critical point in time—before the seaborne reinforcement of NATO could commence'.[11] Fortunately, Iceland is isolated by distance from Soviet military capability, it is an inhospitable environment for any attacker, and there is interposed between Iceland and the relevant Soviet bases a substantial quantity of Western air and naval power in Norway and the Norwegian Sea. In conjunction with these other considerations, the modest forces deployed in Iceland can make the island difficult to attack (particularly if those forces were augmented in crisis or war).

Second, Iceland plays a substantial role in NATO's anti-submarine warfare (ASW) capability in the GIUK gap, in two ways. One is that nine US Navy P-3 Orion maritime patrol aircraft are based at Keflavík; they carry not only onboard sensors of various types but also anti-submarine weapons.[12] Obviously, these provide ASW coverage of the waters surrounding Iceland;[13] the range of these aircraft (the P-3 has a maximum range of nearly 9000 kilometres) is such that it can cover enormous ocean areas in the Norwegian Sea from the base at Keflavík, particularly when P-3s from Iceland can be rotated through air bases in north Norway.[14] In addition, Iceland sits in the middle of a Sound Surveillance System (known as SOSUS), which provides hydrophonic surveillance of the waters between Iceland and Greenland in the one direction, and the UK in the other. An ASW command centre, which collects data from these arrays and can play a role in controlling ASW operations, is located on Iceland and is being hardened.[15] This facility may be particularly important because ASW assets appear to be much more effective when information from SOSUS is able to be exploited by maritime patrol aircraft and attack submarines.[16]

Third, capabilities deployed in Iceland provide air defence of the island and the adjacent waters of the GIUK gap. The US Air Force operates 18 F-15 fighter aircraft (for which hardened shelters are being provided) and 1–3 E-3 Airborne Warning and Control System (AWACS) early-warning aircraft out of Keflavík.[17] In addition, two radar stations on Iceland function as part of NATO's North Atlantic Defense System; these radars are being upgraded. Overall, this capability has been improved in recent years, reflecting growing concern about the increased threat to the area around Iceland from long-range

Soviet bombers and cruise missiles (both air- and sea-launched).[18] Furthermore, now under consideration is the construction by NATO (through the NATO infrastructure fund) of a second airport in Iceland that would be be available in the event of war.[19] This Iceland-based air defence capability is intended to provide protection to Iceland itself, to maritime patrol aircraft (which have no ability to defend themselves against aerial attack), to US naval forces that might be operating in the vicinity, and to merchant shipping in the North Atlantic that might otherwise be exposed to attack from the air.[20]

In a severe crisis or at the outbreak of war, the modest but important US forces on Iceland would be considerably augmented.[21] After reinforcement, one squadron of 18 F-15 aircraft will have grown to three squadrons and 42 F-15 aircraft. These will be joined by an Air National Guard squadron and an Air Force Reserve squadron, each comprising 24 F-4s, for a total of 48 additional combat aircraft.[22] Thus, the number of *combat* aircraft would swell from 18 to 90 in the event of the reinforcement of Iceland.[23] In addition, the number of P-3s would be doubled to 18, an additional two or three AWACS aircraft would probably be deployed (for a total of five), and an additional four aerial tankers would join the one KC-135R that supports peacetime air operations. Taken together, this reinforcement amounts to a fourfold increase (from 30 to 118) in the number of fixed-wing aircraft that would be deployed at Keflavík, and this augmented force represents a significant increment of airborne early-warning, air defence and ASW patrol capabilities. While these air reinforcements are the most significant consideration in the broader context of the Northern Flank, Iceland might also receive ground reinforcements as well. One brigade of the US Army's 7th Light Infantry Division, stationed at Fort Ord, California, has been designated for the mission of reinforcing Iceland. However, it has other mission designations too, and so it is not absolutely certain that it would arrive in Iceland in an emergency. Further, the 187th Brigade of the US Army Reserve has been designated for Iceland; this unit, however, would not be available unless there had been presidential authorization of US national mobilization. The purpose of these ground forces would be to protect the Keflavík base and other military installations on Iceland from any sort of attack by Soviet amphibious or special forces.

Because of its geographic location, Iceland occupies an important, even pivotal, role in maritime scenarios involving the North Atlantic and the Norwegian Sea. Even in peacetime, its strategic importance is great. After reinforcement it obviously represents a considerably larger increment of military capability, which would be sorely missed if it were withdrawn by the USA and not replaced, at least to some extent, by European forces. Without forces on Iceland, one of the West's great advantages in the northern waters would be lost. It should be noted, however, that the arrangements for US forces on Iceland are the result of a bilateral arrangement—the 1951 defence agreement—that need not necessarily be rescinded even if the USA is eliminating its military involvement with NATO Europe. The USA might wish

NATO'S NORTHERN FLANK: IMPACT AND IMPLICATIONS 309

to preserve its defense relationship with Iceland because of its permanent concern with the Soviet Navy (and, in particular, because of its enduring interest in pursuing Soviet ballistic missile submarines in northern waters); Iceland might wish to preserve this relationship because without US forces it is defenceless, and it might well prefer its long-standing involvement with the USA to an uncertain and unproven commitment from a NATO Europe that would probably be struggling to cope with the effects of a US withdrawal. Whatever one thinks of this speculation, however, there can be no doubt that Iceland is a major piece of the military puzzle in the north, and the picture will look entirely different if its status is altered.

Holy Loch, Scotland

There are a large number of US bases in the United Kingdom, but most of them are oriented towards the Central Front. However, one other important US base relevant to, if perhaps not quite in, the Northern Flank is the naval base at Holy Loch, near Glasgow. It is home to 10 US nuclear-powered ballistic missile submarines (SSBNs) and to some 2000 US naval personnel.[24] Information about Holy Loch is scarce,[25] but it clearly must have considerable capability to service US submarines, and while its main function is to support the SSBNs that are based there, it would be surprising if it did not have at least some capacity to service and support US attack submarines[26] (and perhaps even surface vessels). In addition, it seems likely that US nuclear-powered attack submarines (SSNs) could make use of facilities of the Royal Navy, which also possesses nuclear-powered submarines.[27] Obviously, without some source of support in the United Kingdom, US SSNs would have to return to Norfolk, Virginia, from deployments in northern waters whenever they needed supplies or repairs, making support facilities in the UK very attractive and useful. Hints that such support is available can be found in the Norwegian press, in the context of discussions between the USA and Norway over the possibility of creating advanced logistics support sites that would enhance Norway's ability to support US naval forces in wartime. One report, based on a Pentagon interview with US Navy officials, comments: *'Today service and repair work is done from Great Britain,* but Norway lies closer to the operational area and it is consequently desirable that in a crisis or during wartime service and supply functions can be carried out from there'.[28] What such statements imply, of course, is that *in peacetime* US forces operating in the north (presumably SSNs) would use facilities in the United Kingdom for essential support functions, but that these could or might be assumed (or supplemented) *in wartime* by facilities in Norway, organized via Host Nation Support agreements. This makes US capabilities in Scotland a significant factor in the military equation on the Northern Flank, although it rarely figures in discussions of the naval balance. While information is not available that gives an indication of what role Holy Loch plays in supporting US naval (and especially attack submarine) operations in the Norwegian and Barents seas, it can be safely said that the larger the role it plays, the greater is the US capability to deploy and sustain forward SSN operations in northern waters.

Thus, while on the Central Front a US withdrawal involves dozens, if not hundreds of significant bases as well as several hundred thousand men with associated equipment, the picture on the Northern Flank is quite different. In peacetime, only two prominent bases, involving a total of 5000 US servicemen, constitute the basic core of the infrastructure supporting US combat capabilities in the north.

There are, of course, a number of other ways in which the US military co-operates with its Northern Flank allies, and these manifest themselves in facilities, installations and military collaboration as well. Generally, however, these are small, often devoted to a single function, frequently have no combat capability associated with them, and in many cases involve no US personnel. The functions performed by such installations and collaborations are significant, nevertheless.

Intelligence, warning, logistics and reinforcement

The United States is involved in a fairly extensive web of arrangements in the Northern Flank that have to do with the provision of intelligence and warning in peace and war, and with the provision of logistical support for US forces and reinforcements for allied forces in the region in times of war.

The intelligence and warning function has two key elements: radars for warning of aerial attack and various sensors for the provision of ASW information. There are, in addition, an array of communications, meteorological and space-tracking stations strewn across locations in the northern waters, stretching from Greenland through Iceland, the Faeroes, the Shetlands and including Norway on the eastern side of the Norwegian Sea.[29] Thus, Greenland (which enjoys an autonomous status in the Danish realm) is host to two prominent US bases (Thule and Søndre Strømfjord) which are primarily intended for the provision of early warning against strategic nuclear attack, but which also may contribute to the monitoring of Soviet aerial activities over northern waters.[30] Furthermore, P-3 ASW patrol aircraft can, and occasionally do, operate from the airstrips at these bases, contributing to US ASW capability in the Greenland and northern Norwegian seas. However, the utility of these bases for either air defence or ASW patrol in the Norwegian Sea (particularly the Greenland Sea and the Denmark Strait) is limited by their location on Greenland's west coast, and consequently there has for some years been interest in developing an air base in on the east coast of Greenland.[31] Hence, it is not surprising that discussions are under way regarding the possibility of adding an additional US air field on the east-central coast of Greenland, at Mestersvig; while not the north of Greenland, this is nevertheless a very northerly location (between 70 and 75 degrees latitude, for example, and further north than the

North Cape of Norway) and is well situated to provide air coverage of the Greenland Sea. According to the publicly available information, this project, which would be funded by the NATO Infrastructure Fund, would serve primarily as an alternative to Keflavík, Iceland, during bad weather.[32] While at present it is uncertain whether this project will go forward and public information is sketchy, it would be very surprising if such a base were not intended to support, and perhaps serve as the home base in crisis or war for, air defence interceptors and maritime patrol aircraft.[33] In addition, the eastern portion of the Distant Early Warning (DEW) line of radars for the detection of aerial attack against North America stretches across southern Greenland. Four such installations (known as DYE stations) are located in Greenland, from Holsteinsborg on the west coast to Kulusuk on the east coast; a fifth DYE station is found in Keflavík, Iceland.[34] All told, these facilities provide considerable radar coverage of northern airspace, and in the event of war could provide warning of Soviet bomber or cruise missile attack in the northern region, whether against targets on land or against maritime assets at sea. Because of technological advances and modernizations, however, the radars in Greenland are declining in importance, and in the summer of 1989 the US announced that it would leave the base at Søndre Strømfjord and close down the two early-warning radar stations it supported by 1992.[35] Facilities on Greenland also provide meteorological and communications support that can be relevant in northern waters.

Similarly, the Faeroe Islands, which occupy a prime strategic location in the middle of the ocean passage between Iceland and Norway, are home to the Faeroes Command, subordinate to the Supreme Allied Commander Atlantic (SACLANT). The islands accommodate a small airfield, a NATO Early Warning Station, a North Atlantic Relay System (NARS) Forward Scatter communications station as well as a Loran-C navigation facility.[36] The Shetland Islands, which lie between the Faeroes and Norway, host similar facilities, including a Loran-C facility, a shore station for ASW sensors, as well as communications installations.[37] Taken together, these facilities across the GIUK gap, while small and not very visible, constitute an integral part of the fabric of NATO ASW, air defence and communications activities in the Northern Flank.

To these must be added the more extensive intelligence and communications facilities in Norway.[38] Because of its strategic location, in proximity to the Soviet bases on the Kola Peninsula and adjacent to waters in which the Soviet Northern Fleet must operate, Norway is especially well situated to provide critical intelligence on Soviet naval activities (including submarine movements) and warning of Soviet aerial operations. As a result of Norway's base policy (which prohibits foreign bases in peacetime) these intelligence functions are performed by Norway, and the information is shared through the NATO command structure and through bilateral consultation. The nature of the arrangement is suggested in the following statement to the Storting by Defence Minister Johan Jørgen Holst:

Norway of course co-operates with the United States, which is our foremost pillar of support. Norway has an exposed and important strategic location . . . We are dependent upon a good intelligence service for timely warning of a potential attack on our country. The monitoring of submarine traffic in our local waters is of special significance for us in this connection. Co-operation takes place through NATO's political and military organs and through direct contact with member countries . . . This co-ordination also includes surveillance, warning and command and control. All surveillance is conducted by Norwegian personnel under full Norwegian control . . . The information which is gathered in the Nordic region and shared with our allies is evaluated and cleared by Norwegian authorities. This of course also applies to information gathered by underseas listening devices . . . A *long series* of installations are financed and operated in co-operation with NATO member states.[39]

Some believe that this official public description overstates Norway's control and minimizes the extent of US involvement for the sake of avoiding controversy over Norway's base policy. Nevertheless, it does convey something of the nature and importance of the arrangements that exist. These arrangements, one must surmise, provide valuable information to US forces operating in the Northern Flank, in particular with respect to ASW operations. As detection and localization are essential and difficult aspects of that mission, US forces will be more effective the more information they have.

In addition to these intelligence and communications arrangements, Norway is also party to a series of preparations to receive US reinforcements in the event of war. It is consistent with Norway's base policy to make such preparations as are necessary to expedite and enhance reinforcement capabilities. In effect, there has evolved over the years a fairly extensive set of arrangements between the USA and Norway designed to allow US forces to operate effectively on the Northern Flank *despite* the absence of permanent peacetime bases.[40] Three types of arrangements are notable. First, in 1974, Norway reached a Collocated Operating Bases (COB) Agreement with the US Air Force, which calls for preparations in Norway (including the stockpiling of ammunition, fuel and spare parts and the provision of shelters for aircraft) to receive wartime deployments of USAF fighter squadrons.[41] Second, there is the 1980 Invictus Arrangement between Norway and the US Navy, in which Norway agrees to provide logistical and operational support to the US Navy and to allow carrier aircraft to operate from airfields in Norway if necessary in wartime.[42] Finally, in 1981, Norway reached an agreement with the US Marine Corps on the stationing of equipment for one Marine Amphibious Brigade—since renamed Marine Expeditionary Brigade (MEB)—in central Norway. As a consequence of this agreement, heavy equipment (including artillery, trucks, towing vehicles and so on), ammunition and fuel are being prepositioned in storage facilities carved into rocky mountains in Trøndelag, allowing the MEB to be flown rapidly rather than arriving by sea; this project is scheduled for completion in 1989.[43] In addition to these existing arrangements, discussions are continuing between Norway and the USA about extending the logistical support arrangements for the US Navy in Norway. This might involve another

Host Nation Support agreement, perhaps leading to the creation of an advanced logistical support site in Norway. Such preparations would significantly augment the US Navy's ability to operate in the northern Norwegian Sea in the event of war.[44]

To summarize, while the United States has no forces deployed in Norway in peacetime, preparations have been made to facilitate land, sea and air reinforcement. Without these preparations, it would be much more difficult for US forces to arrive quickly and operate effectively. When these preparations are added to the intelligence and other forms of co-operation between the United States and its Northern Flank allies, it is apparent that a fairly substantial network of military collaboration exists, despite (indeed, to a certain extent because of) the quite limited infrastructure of US forces and bases in the region. These arrangements would, of course, be disrupted should the USA cease to participate in co-operative military endeavours with its European allies.

III. Defending the Northern Flank without US involvement

The issue that this analysis seeks to address is what the security environment in the Northern Flank would be if all forms of US direct military participation in Europe's defence were dismantled. What would the Northern Flank look like after a US withdrawal? What would be the key changes, and what deficiencies would arise?[45]

It should be obvious from the previous discussion that on the Northern Flank a US withdrawal would involve very little in the way of standing forces. From the perspective of deployed military capability—especially in Norway—there would be very little visible difference before and after a withdrawal.[46] What would be affected, however, is *potential* military power. Since the NATO concept for the Northern Flank depends heavily on US ground, sea and air reinforcements in time of crisis or war, it is the elimination of arrangements to receive and properly support these reinforcements that would be the primary *military* consequence of US military disentanglement. From a *political* perspective, it is probably the elimination of plans for and expectations of US reinforcements that would be of consequence, as this would undermine the *deterrent* effects such arrangements are thought to provide by guaranteeing that a Soviet attack on the Northern Flank would bring it into direct military confrontation with the USA as well as with the NATO states within the region.

In assessing the military consequences for the Northern Flank of a US withdrawal, it is the situation in northern Norway that commands the most attention, for it is the lynchpin of the entire northern campaign. If the Soviet conquest of north Norway can be prevented, and if the Soviet effort to achieve command of the air in the region is defeated, then most likely the Soviet Union can be prevented from dominating the entire flank. The focus here is therefore primarily on the northern Norway scenario. For purposes of analytical clarity, land, sea and air campaigns are examined separately, but it should not be forgotten that they affect one another; in particular, the air campaign can have an enormous impact on the ground campaign (because Soviet forces will be vulnerable to interdiction if not well protected), and the ground campaign in turn will greatly influence the war at sea, since the side that controls northern Norway will also have an advantage in adjacent waters.

The ground campaign: defending northern Norway[47]

There are several inherent disadvantages that the Soviet Union must contend with should it decide to attack northern Norway. The distances are great; it is roughly 1000 kilometres along the coastal road from the Soviet border to the important Norwegian bases just north of the Arctic Circle (see figure 15.2). Much of the terrain is difficult, dominated by rough, rocky, mountainous areas and cut by fjords; this limits mobility, prevents operational flexibility and channels attacking forces along predictable lines of advance. The transportation network north of the Arctic Circle is very sparse; there is no railroad line north of Bodø, and only one major road along the coast from the Soviet border to the militarily significant portions of northern Norway.[48] Finally, for a good portion of the year, climate is a significant constraint. It is not simply a matter of cold temperatures, but even more of dramatic variations of temperature from place to place and from time to time, of cycles of freezing and thawing, and of high winds, all of which take their toll on both men and machines.[49] In short, in order to capture the most important Norwegian bases in the north, Soviet forces would have to move great distances in a very unfavourable environment.[50] Those who assume that the Soviet Union could easily accomplish a 'northern Norway grab' have not taken these considerations into account.[51]

The northernmost county in Norway, Finnmark, is thinly populated (the total population is approximately 75 000),[52] very lightly defended, and contains little in the way of militarily significant infrastructure. It is the next county to the south, Troms, that is the critical military target, for it contains almost all the important air bases and command facilities for Norway and NATO in the far north. It is surely these that would be the main target of any Soviet invasion;[53] advancing Soviet forces would wish to neutralize NATO air power and perhaps seize the air bases for their own use. The aim of Norway's strategy is to prevent this from happening. To accomplish this, it seeks to exploit some of the advantages of terrain and geography to provide time for the reinforcement of this remote area. In essence, Norway is willing to trade space for time: as the Soviet Union attempts to traverse Finnmark, NATO forces will be moving into Troms to create in the region a kind of land bastion encompassing the main bases. In effect, in this scenario, Soviet and NATO forces race to see which can first reach Troms in force.[54]

The central question here is what role US ground forces play in this race. It appears that they are a minor factor, for they are relatively few in number and

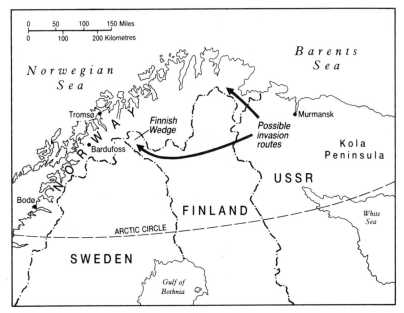

Figure 15.2. Northern Norway

arrive relatively late in the game. To support this conclusion, the overall NATO capabilities that could be mobilized to defend northern Norway are examined.

In terms of ground forces, the defence of Norway depends on four increments of capability.[55] First, there are Norway's own standing forces, consisting of a 19 000-man conscript Army (of whom only 6000 are professional military). Of these, approximately 1500 are deployed in Finnmark. The 5000-man Brigade North is deployed in Troms.

Second, Norway possesses a rapidly mobilizable Home Guard of 85 000–90 000 men. Locally based, annually trained and armed in peacetime, the primary mission of the Home Guard is to provide time for the mobilization of Army reserves.[56] This is to be accomplished not only through guerrilla-type harassment operations, but also through demolition efforts (undertaken by specialized demolition platoons) intended to exacerbate the logistical problems described above: 'The Home Guard is capable of demolishing communications, bridges and railway lines'.[57] Thus, Soviet forces attempting to invade along Norway's lone coastal highway are likely to find mountain passes blocked, ferries sunk and bridges down; this is not likely to hasten their advance or to increase their chances of winning the race for Troms.

While these demolition operations are likely to be critically important in slowing the rate of Soviet advance and allowing time for mobilization and reinforcements, the Home Guard is more than just a demolition crew. Indeed, it also has the important mission of helping to protect Norway's north-south logistics lines of communication (sometimes called land-LOCs).[58] Its organization includes sniper platoons, light infantry platoons and support platoons.

While these units are generally lightly armed, they do include some heavier weapons, including heavy machine-guns and 60-mm mortars, as well as some anti-armour capability, including 106-mm recoilless guns and the recently acquired 84-mm Carl Gustav anti-tank gun. The Home Guard also plays a role in air defence (particularly of crucial facilities) and coastal defence (for which purpose it possesses 400 small vessels and 8 motor torpedo boats); specialized units are organized to serve these purposes as well. These capabilities are not intended to be adequate to thwart an attack, but are meant to prevent a *fait accompli* attack, to guarantee resistance from the outset of an attack, to buy time for the larger mobilization, and to provide protection for key installations and transport routes. Hence, for the Home Guard to have any meaning, it must be able to mobilize quickly. In unannounced exercises, Home Guard units have been found to be 70 per cent manned after three hours and operational after four. This constitutes the first stage in the defence of Norway.[59]

Third, Norway's defence strategy relies heavily on an internal mobilization and reinforcement programme, involving the rapid mobilization of forces in southern Norway and their transport to the north.[60] To facilitate this operation extensive preparations have been made, including pre-stocking of heavy equipment in the north and detailed planning of logistics.[61] Within a week after the beginning of mobilization, according to one source, Norway's total strength in the north will exceed 80 000.[62] All told, the Army can draw upon 165 000 first-line and 41 000 second-line reserves, while Norway's total strength, on full national mobilization, is reported to be 320 000.[63] Moreover, Norway mobilizes not only manpower but other civilian assets that would be useful in wartime. As one impressed journalist reported after being exposed to Norway's mobilization preparations, 'Norway mobilizes . . . virtually all its civilian assets usable in war—ships, ferries, equipment, the off-shore oil industry's helicopter fleets, etc. All this is pre-planned, contracted for and periodically tested.'[64] Norway's own internal mobilization constitutes the largest increment of capability that will be available for the defence of the north (and of the rest of the country for that matter). The extent to which Norway would attempt to draw on internal resources in the event of war is indicated by the fact that its internal mobilization involves the highest share of the civilian population of any country in NATO (approximately four times the rate of the Federal Republic of Germany or the United States), its fully mobilized force is only slightly smaller than that of the United Kingdom, whose population is twelve times that of Norway, and its per capita defence expenditures are the highest in Europe.[65] Because of its policy prohibiting foreign bases and nuclear weapons on its soil, Norway is often viewed as half-hearted about matters pertaining to NATO and defence. The facts prove otherwise.

Fourth and finally, Norway hopes to receive external reinforcements from its NATO allies. Until recently, there were four main potential sources of such reinforcements: the United States (in the form of the US Marines), NATO's Allied Mobile Force (AMF), the United Kingdom and Canada. In 1987, however, Canada withdrew from its commitment to allocate the Canadian Air/Sea Transportable (CAST) brigade to northern Norway in a crisis or war.[66] This was disturbing to the Norwegians, who value the deterrent effect of multiple NATO flags committed to the defence of northern Norway, and after a brief search for a replacement NATO announced the creation of a new multinational force intended to compensate for the withdrawal of the Canadian brigade. The new force of 2500 men will consist of a Canadian infantry battalion, a US artillery battalion and a West German artillery battalion.[67]

The AMF, based in Seckenheim, FRG, and numbering some 5000 men, is available for deployment to northern Norway but is probably considered more a deterrent force than a true reinforcement capability. As one analysis explains: 'Essentially a light brigade, the AMF is a show-of-force unit designed to deter Soviet escalation of a crisis by early insertion of mixed NATO forces, who then serve as "trip-wire" hostages should deterrence fail'.[68] Or as another analyst describes it, the AMF 'is designed to deploy rapidly to crisis areas, to demonstrate NATO resolve and solidarity'.[69] As applied to Norway, the idea is that the presence of the AMF in northern Norway might deter a Soviet attack by ensuring wide NATO involvement in any conflict in the region; if deterrence should fail the AMF would involve Norway's allies from the outset of a conflict, would give them a stake (in the sense of both presence and casualties) in any battle for the north, and thereby would facilitate decisions for further reinforcements.[70] The AMF is not dedicated solely to the northern Norway mission, however, but is available for other contingencies as well, at the discretion of NATO's Supreme Allied Commander Europe (SACEUR).

As for Britain, the United Kingdom/Netherlands Landing Force (the 3rd Commando Brigade) has been assigned northern Norway as a priority mission (although it, too, could be used elsewhere at the discretion of SACLANT) and would be airlifted to Norway in the event of a reinforcement decision.[71] This is a highly regarded force, much respected by the Norwegians, that trains in Norway three months each year and is considered well-prepared to operate effectively in the difficult conditions found in northern Norway.[72]

The US Marines earmarked for Norway by virtue of the 1981 agreement between Norway and the United States represent the final increment of external reinforcement for Norway. As noted above, the prepositioning arrangement involves the storage of equipment for one Marine Expeditionary Brigade, allowing these forces to be flown in quickly rather than coming by sea.[73] The MEB dedicated to Norway, however, represents a ground force capability of only 5000 men.[74] A second MEB could be deployed to Norway, but it would come by sea and would not be available for 30–45 days after it received its orders. In terms of the race for Troms, this is probably too slow for these forces to be dramatically significant—unless, of course, they were put out to sea in the midst of a crisis and arrived before or soon after the fighting began. So at most 10 000 and more likely 5000 US Marines would contribute to the ground defence of northern Norway. Numerically, this is a relatively small increment compared to the numbers of reinforcement generated (quickly and certainly) by the Norwegians themselves. Even compared to the number of forces the Norwegians expect to have deployed north after a week of mobilization (80 000), the 5000 Marines represent only somewhat more than 5 per cent.

For several reasons, however, this purely numerical evaluation almost surely undervalues the contribution of the US Marines to the defence of northern Norway. First, the MEB is about equal in size to the entire professional cadre of the Norwegian Army. While conscripts and reservists should not be excessively discounted, it does not seem unreasonable to assume that units of professional soldiers are likely to generate more combat power on the battlefield. Second, the MEB is rather heavily armed, certainly compared to many of the Norwegian forces (such as the Home Guard) that count equally in a numerical comparison; consequently, the MEB should weigh somewhat more heavily in assessing the balance than its size would suggest.[75] Third, and most important, the MEB comes equipped with about 75 heavy-lift and troop-carrying helicopters, which means that it has more tactical mobility and operational flexibility than many of the other forces available for the defence of northern Norway;[76] this can be a significant factor in the rugged terrain that characterizes Troms county. (One might add a fourth point: only the MEB provides the deterrent effect of guaranteeing that an attack on Norway will involve combat with US forces. This is not a matter of military effectiveness, but a political consideration that is considered important by Norway and perhaps by the Soviet Union.) To summarize, the MEB is more professional, more powerful and more mobile than many of the forces available in the north.

On the other hand, the advantages that the MEB possesses in power and mobility may be offset by its relative lack of experience, compared with Norwegian forces, in operating in mountain and Arctic conditions. Training is essential to effectiveness in this harsh environment, and it is unlikely that the Marines can acquire, in a few weeks of annual exercises in Norway, the skill of indigenous forces.[77]

All things considered, then, it appears that the MEB is a relatively small component of the forces available for the defence of northern Norway. Even if one credits the MEB with a disproportionately large capacity because of its power and mobility, far the largest share of ground capability in the north is that generated by Norway itself. Furthermore, as a ground unit, the MEB is not so large or so capable that it would be especially difficult or onerous to replace it with European forces in the event of a US withdrawal (assuming, of course, that this would be possible in such a situation, given the additional demand for European forces in Central Europe). It is primarily as a deterrent force that the MEB is irreplaceable, but the deterrent effect of US military power would be forgone by all of Europe, not only Norway, if one assumes that US military involvement is terminated.

The air war

The picture is very different with respect to air power. The air balance in the north is extremely important for three reasons. First, the outcome of the air war will determine which infrastructure of air bases, NATO's in Troms County or the USSR's on the Kola Peninsula, is most exposed to aerial attack. Once NATO's defences against air attack are seriously weakened, relentless assaults can be expected and further deterioration of the air balance is likely. This, in turn, would inhibit and could prevent ground and naval reinforcement of northern Norway, because the reinforcement efforts would be exposed to air attack.[78] Second, if NATO can dominate the air, Soviet ground forces, strung out as they would be on the narrow roads of northern Norway, would be vulnerable to aerial interdiction. Conversely, if the Soviets can provide these forces with meaningful air cover, their task will be considerably eased. And third, the side that wins the air battle will be in a position to use land-based air to extend coverage to the maritime theatre in northern waters. The air war thus has important implications for both ground and naval operations in the region. Indeed, it is likely to be the *decisive* consideration in the campaign for northern Norway and for control of the nearby ocean areas.

In the conduct of the air war, Norway is much more reliant on outside reinforcements than is the case with ground forces. As Defence Minister Holst has explained, 'For Norway air reinforcements are of critical importance. Whereas ground force reinforcements of Norway constitute but a fraction of the forces Norway will mobilize on her own, the allied air reinforcements will provide the bulk of air power in Norway in the event of war.'[79] The heart of the Norwegian Air Force consists of 67 F-16s, supplemented with one squadron of 16 operational F-5s. The *potential* Soviet threat against which this force might have to operate is very much larger (although uneven in quality). Three Soviet forces are of relevance.[80] The land-based naval aviation associated with the Soviet Northern Fleet consists of some 430 aircraft. A considerable number of these are reconnaissance or ASW aircraft that do not represent a direct threat to the Norwegian Air Force, but many dozens are fighter-bombers or strike aircraft that could be brought to bear against Norway. The Archangelsk Air Defence District, of which the Kola Peninsula is a part, has about 300 aircraft, most of them air defence fighters; 100 of these are deployed on bases on the Kola Peninsula, and include some of the most advanced Soviet fighters, such as the MiG-31 Foxhound and the Su-27 Flanker. In addition, there are another 450 aircraft and helicopters in the Leningrad Military District; included among these are three regiments of ground attack aircraft (a total of 150 planes). No fighter-bombers are permanently deployed on the Kola Peninsula, but they do regularly exercise there, and these assets could be employed in the northern campaign.[81]

An additional consideration is that the Kola Peninsula possess an extensive infrastructure of air bases capable of supporting many more aircraft than are normally deployed there in peacetime. There is the potential for substantial Soviet air reinforcement of the Kola Peninsula; the region could easily sustain a doubling of peacetime deployment levels. While there is reason to doubt the priority the Soviet Union would give to the Kola Peninsula if war was raging on the Central Front, it is evident that it possesses considerable air capability in the region and an even greater *potential* capability should it choose to reinforce.[82] A last consideration that should not be forgotten is that many Soviet submarines and surface vessels of the Northern Fleet possess a cruise missile land attack capability, and any such capability deployed in northern waters could of course be used against the air bases in northern Norway.[83] Defence against a cruise missile attack from off-shore is thus another aerial threat against which Norway must be defended, adding further to the demands on NATO air defences.[84] While this threat is rather circumscribed at present and certainly secondary to the other components of the air threat, it could grow more challenging in the future.[85]

Mere recitation of numbers, of course, does not constitute an analysis of the air balance, particularly since the extent of Soviet air capability in the north depends so heavily on the choices it would make about the actual disposition of aircraft in the event of war; but outlining the major assets and deployment options can serve as a crude indicator of the magnitude of the air threat to Norway.[86] It should be apparent that the challenge to Norway's Air Force is substantial. On the other hand, with allied air reinforcements in the picture, the air balance is far from being disastrously in favour of the Soviet Union.[87]

It is this circumstance that has led to the fairly extensive preparations that exist for the air reinforcement of the north. Norway's reliance on these reinforcements is indicated by the fact that the air power that could arrive is very much greater than the capability of the Norwegian Air Force itself. As Holst has emphasized:

In the field of air reinforcements the allied contribution will dominate. Norway is unable to maintain an air force large enough to provide sustained protection to the large national territory. All studies show that if we should lose air superiority in Norwegian airspace we would be likely to lose a fight for military control in Norway. Allied air reinforcements could arrive rapidly provided proper preparations have been made for reception facilities, host nation support, prestocking of ammunition, spare parts, etc.[88]

As noted, efforts to make such preparations have been undertaken.

The potential air reinforcements come in four separate increments:[89]

1. Four squadrons of fighters would accompany SACEUR's Allied Mobile Force.
2. The air power indigenous to the US MEB consists of two squadrons of fighters and two of ground attack aircraft. All told, the MEB brings with it 64 combat aircraft. This is roughly the equivalent of the Norwegian Air Force. In addition, the MEB includes two batteries of Improved Hawk (I-Hawk) air

defence missiles, augmenting Norway's improving but still limited point defences.⁹⁰ Finally, the MEB provides air-battle management capability (in the form of command, control, communications and intelligence (C³I) assets) that may be crucial to the air campaign in north Norway. One analyst has written, for example, that the Marine's Tactical Operations Centers (TAOC) 'may well be the single most valuable system in the theatre during the Norwegian Sea campaign, providing linkage between Navy ships and NADGE systems ashore, and between NATO or USAF AWACS and the ship'.⁹¹ It is evident that the MEB's contribution to the air balance is considerably more important than its contribution to the ground balance.

3. In addition, the NATO Rapid Reinforcement Program (RRP) allocates an additional 6–8 squadrons of aircraft to the Northern Flank.⁹²

The sum of all this potential air reinforcement amounts to 15 squadrons, or a maximum of 300 aircraft. While it is not certain that all this capability would actually be made available to Norway in the event of war,⁹³ the scale of the potential air reinforcements is nevertheless such that it can be measured in multiples of the Norwegian Air Force. What matters most for this analysis is that 11 of the squadrons that constitute potential air reinforcement for Norway come from the USA. Removing them, the equation changes significantly. Furthermore, high-capability combat aircraft (F-16s, F-18s, A-4s and so on) in these numbers would be extremely expensive to replace, and not easily within reach of Norway or its European allies.

4. The final increment of potential air reinforcement is that represented by the US aircraft-carriers. Depending on how they are loaded, each carrier will have 80–100 aircraft aboard.⁹⁴ That is to say, like the air complement that would accompany the MEB, each carrier is roughly the equivalent of the Norwegian Air Force. There is no firm plan for the operation of carriers in wartime, but conceivably as many as four or five could operate in the Norwegian Sea. There are indications that Striking Fleet Atlantic's Carrier Striking Force would include at least three carrier battle groups, and this would represent a substantial contribution of capability to the air battle over north Norway:

A three carrier force could be configured to provide more than 70 fighters and over 100 attack aircraft. This would include the F-14 Tomcat all-weather air superiority interceptor/fighter with its Phoenix, Sparrow, and Sidewinder missiles and guns; the A-6 all-weather night-capable medium attack bomber and either the A-7 light attack bomber or the new F/A-18 combination fighter-bomber . . . Not only does this provide a credible force with considerable power to wield in theater, but it also provides sufficient power that the force itself should be able to fight its way into the theater if an enemy should attempt to deny access to the force.⁹⁵

Critics of the carrier are wont to point out that much of its capability goes to defending itself; while true, it is nevertheless also the case that when operating close to shore the carrier helps to defend the airspace over northern Norway by defending itself. This is reflected in exercises that have taken place in the past several years in which carriers have practised operating in fjords in northern

Norway.[96] From such locations, the carrier's air defence perimeter (including its air-battle management assets) will certainly contribute to the defence of northern Norway's air bases. It could well be, however, that in the event of war the disadvantages of operating in fjords (confined waters, proximity to potential shore-based threats, potential vulnerability to mines and so on) would outweigh the advantages. Nevertheless, the range of carrier-based aircraft is such that the carriers do not have to operate in the fjords in order to be relevant to the air campaign in the north. Indeed, a carrier battle group would routinely extend its air defence perimeter out to distances of 250–500 kilometres and more.[97] Carriers off-shore could help defend the airspace over northern Norway without ever venturing near a fjord.

From the Norwegian perspective, the carriers are significant, even essential, for several additional reasons—apart from the sheer increment of capability they represent. First, the high-quality air-to-air/air defence capacity provided by the carriers (notably in the form of the F-14) is particularly valuable because many of the air reinforcements for Norway are ground-attack aircraft for close air support and battlefield air interdiction missions.[98] This will strengthen NATO's position in the battle for air superiority over northern Norway.[99] Second, the limited infrastructure of air bases in northern Norway is a major constraint on both reinforcements and operations. Beyond some point, northern Norway simply cannot usefully absorb any more aircraft, and congestion may be a particularly acute problem at early stages of the war when ground forces are being airlifted into the region.[100] Carriers obviously represent a means of circumventing this problem to at least some extent and of augmenting NATO's ability to support and operate aircraft in the north.[101] The potential utility of employing carriers to enlarge NATO's air infrastructure in northern Norway is clear; some analysts even believe it is essential. One analyst concludes, for example, 'Carrier aircraft will clearly be vital given . . . the limited number of aircraft that could operate from northern Norwegian airbases'.[102] Third, Norway's air bases in the north will presumably be high-priority targets for Soviet air power, and the limited number of such bases means that they can be subjected to more concentrated attack. Each carrier operating in the north represents an additional high-value target, and may dilute the intensity of attacks on air bases in northern Norway by attracting air attack upon itself; in this argument, the carriers serve as a 'honeypot' to lure Soviet air power away from Norway's air bases. Fourth, Norwegian authorities expect that a major component of any Soviet attack on Norway would be sea-borne—largely because of the evident difficulties, noted above, associated with movement on land in northern Norway. Hence, quickly establishing NATO naval dominance in these waters is viewed as critical. According to Holst, for example:

Any attack on northern Norway would probably have to be mounted as a combined ground-sea-air operation. Possible axes of attack over land are limited, and would involve long and vulnerable lines of supply, thus Norway would expect an attacker to transport troops, materiel and supplies mainly by sea. Consequently, the rapid establishment of control by NATO of the ocean areas off northern Norway constitutes a key component of allied forward defence arrangements . . .[103]

It is arguable that Soviet amphibious and transport vessels would in any case be quite vulnerable in the defended inner waters along the Norwegian coast, but from the Norwegian point of view this is yet another argument for early forward deployment of the carriers.

Obviously, each carrier that arrives on the scene will represent a significant augmentation of NATO/Norwegian air capability in the north. Whether any carriers will arrive in a timely fashion cannot be known in advance.[104] Further, whether the carrier-based air capacity will be necessary if Norway receives substantial air reinforcements from other sources is an open question, answerable only by detailed analysis of how one thinks the air war will unfold; obviously, those who conclude that land-based air reinforcements will be sufficient to allow NATO to hold its own in the air battle over northern Norway will not believe that the carriers are urgently required even if they can be helpful.[105] However, while it seems at least plausible that the carriers are supplementary rather than essential, the official view is nevertheless unambiguous: it is considered imperative that the carriers arrive, and arrive early. Major General Murphy, for example, has written: 'There is really no strategic alternative to the early deployment of the Striking Fleet into the Norwegian Sea'.[106] Or, as AFNORTH Commander General Howlett rather more vividly puts it, the carriers 'are a number one reinforcement and I pray that they would come and come early'.[107] Holst has suggested that the carriers could be a decisive factor: 'It is questionable whether allied forces could hold out very long in northern Norway in the absence of forward naval reinforcements'.[108] However debatable one considers these views, there can be no question that the carriers represent a large amount of capability that could augment NATO's side of the air balance in northern Norway.

If one subtracts these as well as the ground-based US squadrons from the equation, then one has eliminated far the largest share of the air reinforcements that might have gone to the aid of Norway in the event of war. If one measures by reference to the maximum possible US reinforcement outlined here, attempting to defend the Northern Flank without the United States involves the subtraction of some 500 or more combat aircraft. Little wonder, then, as Sverre Lodgaard puts it, that 'all studies show that with the defence concept presently applied, the defence of Norway depends very much on allied air support'.[109]

The war at sea

The severance of all military collaboration between the United States and its European allies would have a substantial impact on the naval balance in the North Atlantic and the Norwegian Sea. NATO naval capability is substantially diminished if the US Atlantic Fleet is not a participant. Conversely, the ability of the US Navy to operate in northern waters is at least somewhat diminished by the loss of bases and intelligence. If one assumes that NATO Europe will really be left to fend for itself against the Soviet Union, even in the event of war, this will matter relatively little since the SLOCs from North America to Europe will cease to be pipelines for US reinforcements. On the other hand, if one assumes that the United States might still, if necessary, come to the rescue of Europe, then the SLOCs would become, if anything, even more important because *all* US capability would have to be transported across the Atlantic. Indeed, even the prospect of US reductions in Europe issuing from the Conventional Armed Forces in Europe (CFE) Negotiation in Vienna has led to expressions of concern that the North Atlantic sea lanes, the Norwegian Sea, and hence Norway, would increase in importance because NATO's dependence on sea-borne reinforcements would grow.[110] Thus, the implications of a US withdrawal for the maritime environment depend very much on one's assumptions, but that there is a significant diminishment of capability is unquestioned.

One basic proposition provides the foundation for analysing the situation at sea in the aftermath of a US withdrawal: it would almost certainly be impossible for NATO Europe to replicate the US Atlantic Fleet. The US Navy provides virtually all the surface combatants of cruiser size and larger (with the notable exception of ASW carriers) that are available to the Alliance, including seven aircraft-carriers.[111] It accounts for approximately 75 per cent of the nuclear-powered attack submarines in the NATO region. It possesses a large fraction of the naval aircraft operating in the North Atlantic region. Even with respect to the smaller classes of vessel, which have traditionally been the speciality of the navies of the European NATO countries, the US Atlantic Fleet provides nearly half the destroyers and more than one-fourth the frigates available to NATO in the waters of the North Atlantic. No European navy comes even close to approximating the size and strength of this force. The United Kingdom, for example, the second most formidable NATO naval power, possesses 59 principal surface combatants, of which 35 are frigates and 15 nuclear-powered submarines (SSNs); the US Atlantic Fleet possesses 7 carriers, 93 principle surface combatants (of which 53 are frigates) and 51 SSNs. If US naval forces were to sit out the war, the Soviet Union would face a dramatically reduced naval power in northern waters.

This picture is improved by several important caveats, however. One is that the US Navy contemplates missions, such as attacks on Soviet SSBNs and air strikes against bases on the Kola Peninsula, that are very demanding in terms of forces. If NATO Europe did not feel it necessary to emulate the US Navy's mission structure, its requirements for naval forces could be considerably less than is suggested by looking at the US Atlantic Fleet. (This point is buttressed by the fact that US naval force requirements are driven at least in part by US 'global interests', particularly in the Third World,[112] which a NATO European navy is unlikely to pursue.) A less demanding role allows for a smaller and less capable navy.

Another consideration is that in protecting the sea lanes, the US Navy has a very strong preference for an offensive doctrine that involves seeking out and destroying (or at least threatening) Soviet naval forces in ocean areas adjacent to the USSR, including Soviet home waters. As the Soviet Navy is, for several reasons (including high concentrations of forces, the presence of land-based air support and reduced logistical demands), most formidable in these waters, this again results in a requirement for large and very capable naval forces.

If NATO Europe, standing without US military support, wished to have a reasonable capacity to protect the sea lines of communication (SLOC), in order to secure intra-theatre sea lanes or to preserve a US reinforcement option, it could choose a different, more defensive strategy, one that avoids Soviet maritime strongholds and instead focuses on convoy activities in the lower reaches of the North Atlantic, where the Soviet Navy has more difficulty operating.[113] Such a strategy, which was effective in both world wars, is much more compatible with the naval infrastructure possessed by NATO Europe (ASW carriers, escort vessels and mostly diesel-powered submarines) than one that involves challenging the teeth of Soviet naval capability with nuclear-powered attack submarines and Nimitz Class aircraft-carriers. Once again, this strategy is likely to result in smaller force requirements.

Moreover, the capability required to protect the sea lanes is obviously dependent on the extent of the Soviet threat. There is a tendency to assume a vast and menacing Soviet threat to the sea lanes, but careful examination reveals that the Soviet potential for interrupting the SLOC in the North Atlantic, while not negligible, is quite circumscribed; available merchant tonnage is simply too great, and modern Soviet SSNs are simply too few.[114] This means that if NATO European navies wished to have the capability to provide some protection to the SLOC, they would not be setting themselves an impossible task. If they chose to do this defensively, they would probably confront naval requirements that are not impossible to fulfil.

A final caveat to the rather straightforward observation that the loss of the US Atlantic Fleet would represent a significant reduction in naval power for NATO Europe is the fact that the Soviets would still have to worry about the US Navy even if the European allies could no longer count on it. Even without participating in NATO, the US Navy could and probably would still have an incentive to operate in northern waters to carry out strategic counterforce operations against Soviet strategic submarines; the US instinct to pursue counterforce is only partially related to its NATO commitment and would be unlikely to disappear even if NATO's military arrangements were altered. In addition, it could still wish to keep the Soviet Northern Fleet bottled up north of the GIUK gap in order to prevent it from assuming a larger power projection role in the Third World. Hence, unless the Soviet SSNs were to leave their home waters and the SSBNs patrolling there largely unprotected from the US naval threat, they would be unable to unleash the full brunt of their capability against the NATO Europe effort to protect the SLOC. In other words, the US

Navy would still provide some *de facto* assistance to the NATO European allies even after a US withdrawal.

To summarize, the subtraction of the US Atlantic Fleet would represent the loss of a very significant increment of naval power. NATO Europe almost surely could not compensate for this lost capability. However, it might neither need nor want to. If it were to assign its navies a less demanding set of missions and a less demanding strategy, oriented towards a finite Soviet threat, and with some *de facto* support from the United States, it could well end up with attainable naval force requirements. Simply thinking in terms of replacing the US Atlantic Fleet overdramatizes the problem.

Nevertheless, there are two factors to consider that cut the other way. First, a defensive sea control strategy, reliant on barriers in the GIUK gap and escort forces for convoys in the North Atlantic, will cede to Soviet naval forces control of northern waters in any war that does not involve the USA. This might jeopardize northern Norway, and if the latter were lost, this in turn could have repercussions for the naval war elsewhere, for reasons described above.

Second, the effectiveness of the US Navy in northern waters would almost certainly be reduced if military collaboration with the allies were abandoned. The disruption of intelligence co-operation (including loss of SOSUS data and of maritime patrol aircraft operating out of Iceland and Norway) would certainly hamper those activities—notably ASW—that are highly sensitive to the availability of information. The loss of logistical support, both that available in peacetime and that arranged for wartime, would impose virtual attrition on the US force; much time and capability would be lost as US forces steam back to Norfolk for resupply and maintenance, rather than head for an advanced logistical support site in Norway or for Holy Loch. In short, the US Navy might still be in the vicinity, but it would be a less capable force.

Thus, it is not easy to arrive at a single tidy assessment of the effect of US withdrawal on the naval environment. Much depends on assumptions made about US, Soviet and NATO European behaviour under the circumstances. One can say that while the loss in capability would be sizeable, it is possible to envision NATO Europe finding an adequate solution to its naval needs in the North Atlantic. The protection of northern Norway, however, seems more problematic.

IV. The United States and the defence of the Baltic Approaches

While the problems and issues associated with the defence of Norway, Iceland and the Norwegian Sea tend to dominate discussions of security issues in northern Europe, NATO's northern command, Allied Forces North, is also responsible for the defence of Denmark, the Danish Straits and the adjacent waters of the Baltic Sea—these being the domain of the subordinate BALTAP command. Consequently, BALTAP deserves some attention here, although the issues it raises are for the most part distinct from those of the rest of

AFNORTH. It can be treated more briefly, however, as the matters relevant to this discussion are both circumscribed and straightforward.

In terms of national territory, BALTAP consists of Denmark and a small slice of the Federal Republic of Germany on the Jutland Peninsula; for all practical purposes, the defence of BALTAP boils down to the defence of Denmark (including, of course, the naval dimension in the Baltic Sea). BALTAP is an awkward command, wedged between the Northern Flank and the Central Front, relevant to both but not fully a part of either. In truth, it is more directly connected, both physically and operationally, to the latter,[115] but was included in AFNORTH primarily because the legacy of World War II left political sensitivities in Denmark about being bound too closely to the FRG and because Denmark preferred to emphasize its Nordic connection.[116]

The geostrategic importance of BALTAP derives from several considerations. First, just as Iceland serves as the cork that blocks the passage from the Kola Peninsula to the open waters of the North Atlantic, so does Denmark serve—even more dramatically—as the stopper that prevents easy access between the Baltic and the North Sea; for most vessels of any consequence, passage is possible only through a small number of straits through or around Danish territory (see figure 15.3). This is militarily significant because the Soviet Union has incentives to get both out of and into the Baltic. It may wish to exit through the Danish Straits so that its Baltic Fleet can co-ordinate with the powerful Northern Fleet and/or disrupt NATO sea-borne reinforcements in the North Sea;[117] it may wish to be able to enter the Baltic because fully half of the Soviet shipbuilding and repair infrastructure is located on the Baltic.[118] Whether NATO or the USSR can control the Baltic approaches will determine whether Soviet maritime assets are constrained or unfettered in these regards.

Second, Denmark also serves as a forward defence of southern Norway, the United Kingdom, and even of the crucial channel ports through which NATO's reinforcements and supplies must flow.[119] This is particularly true in terms of air power, for the loss of Denmark would open up an attack corridor for Warsaw Pact air power and perhaps also provide forward basing options for the Soviet Union in conducting such air operations. In addition, it would be very difficult for the Soviet Union to mount a direct attack on southern Norway without first gaining control of the Baltic approaches; indeed, this would be impossible if it wished to avoid violating the territory of neutral Sweden.[120] Comments by RAF Vice Marshal Michael Graydon illustrate these concerns:

The loss of the Baltic Approaches to the Warsaw Pact would open up the North Sea for the Soviet Baltic Fleet. In effect, the Warsaw Pact would have driven a wedge through Allied Command Europe and, in so doing, breached our air defences. The consequences would be greatly improved WP capabilities for their maritime and air campaigns, in particular against the UK, and amphibious operations against adjacent territory.[121]

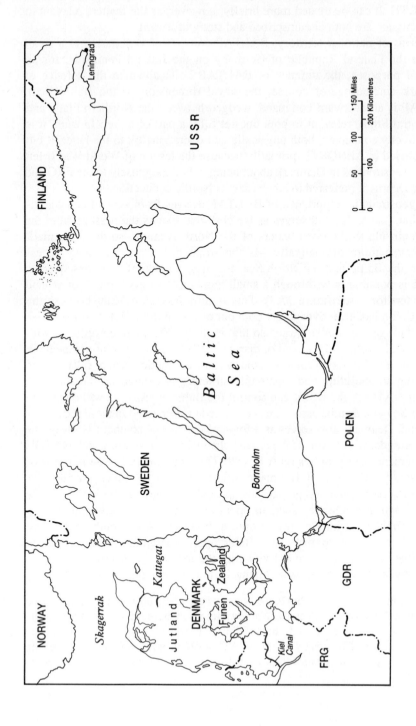

Figure 15.3. The Baltic region

In a sense, then, Denmark serves as a kind of buffer between the Warsaw Pact and several potentially significant and attractive targets.[122]

Third, in relation to the Central Front, BALTAP serves as an immediate tactical flank. As such, it constitutes an inherent threat of flank attack against forces operating on an East–West axis; whoever controls Denmark has the potential to launch a perpendicular southerly attack against forces in the central region. As von zur Gathen puts it: 'As long as NATO controls Jutland, Schleswig-Holstein and the Danish Isles, then a flank threat to any Warsaw Pact aggression is retained. If the Warsaw Pact were to occupy this area, then its forces would be able to project a flank threat against NATO's central region'.[123] In short, allowing the opponent to control BALTAP increases NATO's vulnerability on the critical Central Front.

Finally, control of BALTAP may be strategically significant because the Baltic can serve as an avenue for the reinforcement and resupply of Soviet forces in Central Europe. A Baltic SLOC could be desirable or necessary for the Soviet Union for several reasons. It could supplement supply lines on land, which will inevitably be limited by the finite availability of railroad tracks and cars, and of trucks and roads. Further, logistic lines on land are vulnerable to attack, and NATO, with such notions as Follow-on-Forces Attack, has explicitly expressed its intention to disrupt these lines as much as is possible. A SLOC, while certainly not invulnerable, can be more difficult to disrupt or interrupt than supply lines over land, provided adequate protection can be afforded to port facilities (a condition that may not, however, be easy to fill). In addition, it is very efficient to carry large amounts of cargo by sea.[124] While it is unlikely that the USSR would want to depend entirely or even heavily on the Baltic SLOC, it is quite plausible that, in the event of war, the Baltic could come to play an important, perhaps even critical role in the support and reinforcement of Soviet forces on the Central Front, especially with NATO's growing emphasis on attacking land logistics lines. This naturally makes control of the Baltic a major strategic consideration for the Soviet Union.

In short, although understandably overshadowed by both the Central Front and the Northern Flank, BALTAP is not lacking in strategic importance and, in the event of war, could have significant implications for one or both of the more prominent theatres. It affects both access to and use of the Baltic. It influences the potential vulnerability of forces and assets in Central Europe, on the North Sea and in southern Norway. Thus, while BALTAP will not be at the centre of a major East–West war, nor will it directly determine the outcome of such a war, developments within BALTAP can have a significant impact on the primary theatres. Indeed, a study of the Soviet perspective on BALTAP notes the great importance the USSR appears to attribute to this area:

Militarily, Denmark stands at the crossroads of naval and air communications links between central and northern Europe and between the Baltic Sea and Atlantic. Consequently, Denmark becomes the keystone to a successful and short Soviet conventional war in Europe and, thereby, perhaps the fulcrum of peace and security in Europe ... In the Soviet view, NATO's success in a European war will depend to a large extent on keeping control over the western part of the Baltic and the Danish straits.[125]

This formulation seems to overstate the importance of BALTAP, but still it is reasonable to conclude that a Europe standing alone after US withdrawal could not afford to leave this area undefended.

Similarly, it is widely assumed that as a consequence of such considerations the Soviet Union will have a strong incentive to control the Baltic and to open up the exits from the Baltic by defeating NATO's BALTAP Command.[126] Indeed, there is evidence to suggest that the Soviet Union has given relatively high priority to these tasks and has organized its forces in the Baltic accordingly. As retired Vice Admiral Helmut Kampe of the West German Navy puts it, 'Soviet military doctrine, the composition of the Warsaw Pact navies in the Baltic Sea, and their exercises indicate that seizing the BALTAP area is a primary Soviet military objective'.[127]

The exact extent of the Soviet threat to BALTAP is extremely difficult to gauge, however, for the simple reason that Denmark is situated on the edge of continental Europe and is within reach of Soviet and Warsaw Pact land and air power deployed in the centre region. Hence, the *potential* capability that could be brought to bear against Denmark is substantial, but will likely be circumscribed by the enormous competing demands for forces on the Central Front. As one recent analysis explains:

There is little doubt about the capability of the Warsaw Pact to effect a number of different attack alternatives. One important restraining factor is how large forces the Warsaw Pact will be able to reserve for operations against the Baltic exits and possibly South Norway when operations against NATO's main forces in the central region will probably be given first priority. It is estimated, however, that 12–14 divisions stationed in the northern part of the GDR and Poland will be able to form a northern front for an attack on NATO's BALTAP Command.[128]

In addition, the Soviet Union has approximately 700 tactical combat aircraft deployed in the German Democratic Republic, and some portion of these could be used to support an attack on Denmark—subject of course to the same constraint that most of these aircraft will be allocated to the Central Front. In short, Denmark is, in effect, an adjunct to the Central Front and exists in proximity to the large military capabilities deployed there.[129] Consequently, it faces a considerable potential threat from the south, but one that is difficult to specify with any precision.

Much easier to account for are Soviet forces directly assigned to the Baltic. These involve three forces, all connected with the Soviet Navy. There is, first, the deployment of naval forces themselves. The Soviet Baltic Fleet, while not remotely equal to the powerful Northern Fleet, is nevertheless not a trivial force. It includes more than 30 diesel-powered submarines, 45 major surface combatants (of which 30 are frigates and corvettes) and more than 100 minor surface combatants.[130] In addition, the Soviet Baltic Fleet includes 123 minelayers and mine-sweepers, which is significant because of the obvious utility of mines in the very confined waters of the Danish Straits. Finally, the Baltic Fleet possesses 49 amphibious ships, of consequence because they raise the threat of amphibious attack against Denmark.

Operating in conjunction with these amphibious ships is the second component of Soviet Baltic capability: naval infantry.[131] The Soviets have deployed one naval infantry brigade of 5000 men in the Baltic, and it is supplemented by a small Polish force of about 3000 men. Although very small compared to the large ground forces that could be sent against Denmark from the GDR, this specialized force could be useful on the Danish islands or elsewhere in conjunction with a much larger attack from the south.

Finally, Soviet naval aviation has about 135 combat aircraft deployed in the Baltic, along with a like number of reconnaissance, ASW and transport aircraft and helicopters. These presumably would be devoted to Baltic tasks and hence constitute a kind of minimum air threat to BALTAP that could be enlarged by contributions from the centre region if that were thought feasible or necessary.

Because of the highly contingent nature of the threat to Denmark, it is difficult to assess what all this adds up to. Soviet Baltic forces alone, while not insignificant, are quite limited—dramatically so in terms of ground forces. On the other hand, the threat from the south, from Warsaw Pact forces in the GDR, is quite substantial. To the extent that the Soviet Union credits arguments about the strategic importance of BALTAP and hence accords it some priority, it certainly can muster the military power to put Denmark in jeopardy.[132]

Denmark's capacity to defend itself depends on three increments of capability.[133] First, there are the Danish forces themselves. As in the case of Norway, these consist primarily of a small active Army (numbering 17 000 in peacetime), supplemented by approximately 55 000 reserves, as well as a Home Guard numbering 75 000. Beyond this, Denmark has a small, specialized Navy and a small Air Force, the heart of which is 48 F-16 aircraft.[134]

BALTAP is a combined Danish–FRG command, and the second increment of capability for the defence of Denmark is comprised of the Baltic-oriented forces of the Federal Republic of Germany, which contribute on land, at sea and in the air.[135] The FRG devotes an armoured infantry division of 20 000, a Home Defence Brigade, as well as more than a brigade of territorial forces to BALTAP. The FRG has primary responsibility for defending the Jutland Peninsula. The West German Navy is also primarily responsible for NATO naval missions in the Baltic. Apart from the modest Danish capabilities intended for use mostly in the Danish Straits, the FRG is the only NATO power that allocates naval capability to the Baltic, including two dozen diesel submarines, nearly 50 medium surface combatants and roughly 70 light surface combatants (mostly mine-sweepers), as well as a Naval Air Arm that provides approximately 110 combat aircraft.[136] In addition, the West German Air Force has assigned roughly 60 combat aircraft to the BALTAP region. Obviously, the contribution of the FRG more than doubles the overall capability available to NATO in defending the Baltic Approaches.

The third increment of capability for the defence of Denmark and BALTAP is reinforcements from outside. Because Denmark, like Norway, does not allow foreign bases in peacetime, these reinforcements would have to be brought into the region in the event of crisis or war. It is in this context that US forces are seen as playing an important role, but as will be shown, they do not come into play in any dramatic way. The defence of BALTAP is primarily a European show.

In terms of ground forces, BALTAP could compete with Norway for some of the reinforcements (such as the AMF) available for AFNORTH contingencies. Reference has also been made to the possibility that Denmark might under some circumstances receive a US Marine Expeditionary Brigade, but it is generally assumed that BALTAP will have lower priority than other commands that have a claim on these forces.[137] The one reinforcement especially designated for Denmark is the UK Mobile Force, totalling some 13 000 men.[138]

It is only with respect to air reinforcements that US forces will figure in the defence of BALTAP. Like Norway, Denmark in 1976 entered into a Collocated Operating Bases (COB) Agreement with the United States that arranged for the reception of US air reinforcements.[139] This agreement calls for the deployment of five US Air Force squadrons (approximately 75 aircraft) to four airfields in Denmark. It is this arrangement that would be disrupted if the United States were to terminate its military co-operation with NATO Europe. So the problem for Europe after US withdrawal is again one of replacing a fairly significant number of expensive, high-capability aircraft. As is the case with respect to northern Norway, the number of aircraft involved is not great compared to the central region, but the replacement needs in AFNORTH would be taking place in the context of huge demands for replacements in Central Europe. If, on the other hand, the United States were to withdraw from Europe but retain a commitment to reinforce, then something like the present arrangement with Denmark could be preserved; since there currently are no US forces in Denmark, a US withdrawal would not signify any dramatic change.

V. Conclusion

Of the two withdrawal options posited by this study, option B, the skeletal peacetime presence with a commitment to reinforce in a crisis, is the status quo in the Northern Flank. The analysis in this chapter has thus focused exclusively on the impact of option A, the 'gone for good' withdrawal. With respect to BALTAP the problem exposed by this analysis is quite circumscribed, and amounts to the question of whether NATO Europe would be in a position, in case of a total US withdrawal, to provide BALTAP with a few squadrons of combat aircraft.

The situation further north in AFNORTH is more complex. It has been shown that the United States possesses only a very modest infrastructure of peacetime bases and forces on NATO's Northern Flank. At the same time, however, it also collaborates extensively with its Northern Flank allies in a variety of arrangements relating to reinforcement, intelligence and logistics. The aim here has been to assess what would happen if all these bases, forces and arrangements were withdrawn.

The answer has been different in each of the three domains of warfare. With respect to ground forces, the situation after a US withdrawal looks manageable. The Marine Expeditionary Brigade, while a potent force, is not a large ground force, and represents only a small fraction of the capability available to defend northern Norway. It would not seem to be that difficult to make up for the loss of 5000 Marines (although it should not be forgotten that NATO Europe will simultaneously be attempting to compensate for US withdrawal on the Central Front, and the demand for ground forces may be great).

The situation in the air is much more worrisome. Without the United States, the majority of the air reinforcements on which Norway relies will no longer be available. Just how great Norway's predicament will be at that point depends on one's view of the magnitude of the Soviet air threat. Without the US units the available air reinforcements are dramatically smaller and perhaps subject to greater competition for use on the Central Front. Alone, the Norwegian Air Force is likely to be overmatched.

The effects at sea are ambiguous, since the SLOC are not critical if one assumes that there will be no US reinforcements and that some of the lost US naval capability is devoted to missions that the European allies probably will not wish to pursue. It is plausible that NATO Europe alone (with some tacit assistance from the US Navy) could preserve its minimum maritime interests in the North Atlantic, but adverse consequences in the northern Norwegian Sea may be unavoidable, with troubling implications for the defence of Norway.

Finally, one must guard against the fallacy of evaluating the land, sea and air contingencies separately. In reality, of course, they interact with one another in ways that would probably exacerbate the difficulties for Europe of a US withdrawal. The greater likelihood that the Soviet Union would fare well in the air battle and the possibility that it would be able to operate its naval forces more freely in the Norwegian Sea increases the chance that it would succeed in securing effective control of northern Norway. This in turn would enhance its ability to operate still further south in the Norwegian Sea, further compounding the difficulties that would face a Europe lacking US military support in securing its interests in northern Europe. The most important component in this sequence is the air balance. Much would depend on the extent to which Europe is able to replace the US air reinforcements that no longer would be available.

Notes and references

[1] AFNORTH is subdivided into three subordinate commands: North Norway (COMNON), South Norway (COMSONOR) and the Baltic Approaches (COMBALTAP), the latter tasked with the defence of Denmark and the securing of NATO interests in the Baltic. For a concise description of NATO Command arrangements in Northern Europe, see von zur Gathen, H., 'The Federal Republic of Germany's contribution to the defense of Northern Europe', eds P. Cole and D. Hart, *Northern Europe: Security Issues for the 1990s* (Westview Press: Boulder, Colo., 1986), pp. 62–67.

[2] For the sake of convenience, the term 'Northern Flank' shall from here on refer only to the region encompassing Norway, Iceland, Greenland and the interjacent seas. A separate discussion of BALTAP is found below.

[3] See, for example, Schepe, G., *Mountain Warfare in Europe*, National Security Series no. 2/83 (Centre for International Relations, Queen's University: Kingston, Ont., 1983), pp. 6–10, 42–49. Schepe concludes: 'Only troops with an adequate training and equipment for mountain warfare will cope successfully with the numerous environmental situations—and the enemy' (p. 9). He complains that the majority of NATO reinforcements intended for Norway 'will not have sufficient experiences to cope with the environment in northern Norway' (p. 48). According to studies undertaken in the Canadian Arctic, troops operating in Arctic conditions may expend 80 per cent of their energy merely to survive. This is reported in Honderich, J., *Artcic Imperative: Is Canada Losing the North?* (University of Toronto Press: Toronto, 1987), p. 89. As Honderich sensibly concludes: 'That leaves little [energy] with which to do battle'.

[4] Submarine deployments are a different matter altogether. It is generally assumed that US attack submarines (SSNs) operate in the Norwegian Sea and waters further north as part of the West's effort to block Soviet access to the North Atlantic and to keep the Soviet Northern Fleet on the defensive in its home waters. However, public indications of this are rare, and official data on the size, frequency and duration of US SSN operations in northern waters are, to my knowledge, unavailable. General indications of US SSN operations in northern waters may be found in Ball, D., 'Nuclear war at sea', and Posen, B. R., 'Inadvertent nuclear war? Escalation and NATO's Northern Flank', eds S. E. Miller and S. Van Evera, *Naval Strategy and National Security* (Princeton University Press: Princeton, N. J., 1988). Børresen, J. (Capt.), *USA-Marinens Operasjoner i Nord-Atlanteren og Norskehavet*, NUPI report no. 89 (Norsk Utenrikspolitisk Institutt: Oslo, May 1985), provides an analysis of US attack submarine options and suggests that a typical *peacetime* deployment would (or might) include roughly five submarines operating in the northern Norwegian Sea or further north, with another five on patrol in the North Atlantic. (The numbers, of course, could be substantially augmented in the event of crisis or war.) See especially pp. 95–100.

[5] One source reports that US carriers spent only 41 days in the Norwegian Sea during the decade ending in 1987. See Petersen, P. A., 'Iceland: is a red storm rising?', *International Defense Review*, vol. 20, no. 8 (1987), p. 1011.

[6] According to one account, the Soviet Union views Iceland as controlling 'the Arctic gates to the Atlantic'. Peterson (note 5), pp. 1008–1009.

[7] Iceland has, however, adopted what has been described as 'a long-term policy of playing a more active role in selective aspects of its own defence.' See 'East Iceland radar facility now manned by Icelanders', *News from Iceland*, Nov. 1988, p. 1.

[8] For the full details on the US role in Iceland, including the location and function of all US installations, see Duke, S., SIPRI, *United States Military Forces and Installations in Europe*, (Oxford University Press: Oxford, 1989), pp. 181–94. For an overview of recent improvements in US forces and facilities on Iceland, see Greeley, Jr., B. B., 'Iceland defense modernized to offset Soviet advances', *Aviation Week & Space Technology*, 14 Nov. 1988, pp. 50–54. For a brief official description, see Iceland Minister for Foreign Affairs Steingrímur Hermannsson's *Foreign Affairs: A Report to the Althing*, (Reykjavík, Feb. 1988), pp. 47–54. For general background on Iceland's security policies, see de Lee, N., 'Iceland: unarmed ally', eds C. Archer and D. Scrivener, *Northern Waters: Security and Resource Issues* (Croom Helm: London, 1986), pp. 190–207. See also Ausland, J., *Nordic Security and the Great Powers* (Westview Press: Boulder, Colo., 1986), pp. 170–73. Particularly instructive are the essays by

two of Iceland's leading security experts: Gunnarsson, G., 'Icelandic security policy: context and trends', *Cooperation and Conflict*, vol. 17, no. 4 (Dec. 1982), pp. 257–72; and Bjarnason, B., 'Iceland's security policy: vulnerability and responsibility', eds J. J. Holst, K. Hunt and A. Sjaastad, *Deterrence and Defense in the North* (Norwegian University Press: Oslo, 1985). Also helpful are de Lee, N., 'Bastion of the north? Icelandic security policy', *RUSI Journal*, Dec. 1981, pp. 47–53, which includes a very useful map showing the distances from Iceland to adjacent landfalls; and Gunnarson, K. (ed.), *Iceland, NATO, and Security in the Norwegian Sea* (The Icelandic Association for Western Cooperation: Reykjavík, 1987). On the sometimes contentious domestic politics of security policy in Iceland, see Arnason, R. T., *Political Parties and Defence: the Case of Iceland, 1945–1980*, National Security Series no. 4/80 (Centre for International Relations, Queen's University: Kingston, Ont., 1980).

[9] Data on US forces are from IISS, *The Military Balance, 1987–1988* (International Institute for Strategic Studies: London, 1987), pp. 24–25. Including dependants, there are a total of 5000 US personnel on Iceland. While this might seem a rather small threat to Icelandic culture, Icelanders are prone to point out that having 5000 US servicemen at Keflavík is equivalent to having five million foreigners stationed at a military base 50 kilometres from Washington, DC. See, for example, Bjarnason, B., 'Iceland and NATO', *NATO Review*, Feb. 1986, p. 10. For additional discussion of Iceland's base policy, see Gunnarsson, G.,'Icelandic security policy and the high north', paper presented at the Second Harvard Nordic Conference, Reykjavík, 7–10 Aug. 1987, pp. 8–11.

[10] Skogan, J. K.,'Scandinavia and war operations in the North Atlantic', ed. L. B.Wallin, *The Northern Flank in a Central European War* (The Swedish National Defence Research Institute (FOA): Stockholm, 1980), pp. 69–70. Of the Soviet interest in occupying Iceland, Petersen (note 5) writes: 'There can be no question but that the Soviets would like to capture or at least neutralize Iceland' (p. 1009). An attack on Iceland figures prominently in a popular novel, Clancy, T., *Red Storm Rising* (Berkeley Books: New York, 1987), which although fiction is illuminating on many operational matters. In this scenario, a Soviet air attack with air-to-surface missiles destroys much of the US capability on the island and is followed by an invasion borne by armed hovercraft. See pp. 184–226. See also Vego, M., 'The Soviet envelopment option on the Northern Flank', *Naval War College Review*, autumn 1986, pp. 32–33.

[11] Murphy, D. J. (Maj. Gen.), 'Role of the US Marine Corps', ed. E. Ellingsen, *Reinforcing the Northern Flank* (The Norwegian Atlantic Committee: Oslo, 1989), p. 50.

[12] Bjarnason (note 8), p. 9, is the source for the number of P-3s deployed on Iceland. All sources agree that there are P-3s in Iceland, but most are not specific as to their number. For discussion of the role and capability of these aircraft, see Porth, J. S., 'Submarine hunter', *Defense & Foreign Affairs*, vol. 10, no. 7 (July 1982), pp. 10–13, which specifically profiles the P-3; and, more generally on the role of maritime partrol aircraft, Cairns, G. C., 'Maritime air patrol versus the submarine', *NATO's Fifteeen Nations*, Apr./May 1980, pp. 57–59.

[13] Indications of the patrol patterns of the P-3s in Iceland may be found in Wit, J. S., 'Advances in antisubmarine warfare', *Scientific American*, Feb. 1981, p. 41; Ausland (note 8), p. 158. See also Erickson, J., 'The Soviet Northern Fleet: commitments, capabilities, constraints', paper presented at the Oslo Symposium, Oslo, Norway, 10–14 Aug. 1986. In appendix B the author provides several interesting maps, drawn from Soviet sources, that display the Soviet conception of NATO ASW measures in the GIUK gap, including ASW aircraft patrol zones. For a brief discussion of this mission, see 'Patrol aircraft based in Iceland monitor Soviet submarines, ships', *Aviation Week & Space Technology*, 14 Nov. 1988, p. 78.

[14] See Stefanick, T., *Strategic Antisubmarine Warfare and Naval Strategy* (Lexington Books: Lexington, Mass., 1987), p. 165 for data on the P-3. Stefanick lists its 'combat radius', defined as one-third of the maximum range, as 2760 km. The range of this aircraft should not be confused with its ability to patrol, which is constrained by its capacity to deploy and monitor airdropped sonobuoys. According to Shunji Taoka, a P-3 can patrol an area about 50% larger than England. Although a considerable area, it is still much smaller than the whole of the Norwegian Sea and much smaller than might be implied by the sheer range of the aircraft. The range does, however, permit it to reach distant patrol areas. See Taoka, S., 'East–West naval force comparison', ed. R. Fieldhouse and S. Taoka, SIPRI, *Superpowers at Sea: An Assessment of the Naval Arms Race* (Oxford University Press: Oxford, 1989), pp. 71–72.

¹⁵ See Gunnarsson, G., *The Keflavík Base: Plans and Projects*, occasional paper no. 3 (Icelandic Commission on Security and International Affairs: Reykjavík, 1986), p. 7. This paper is very useful in providing information on the facilities in Iceland and the improvements that have been undertaken in recent years. See also Harkavy, R. E., SIPRI, *Bases Abroad: The Global Foreign Military Presence* (Oxford University Press: Oxford, 1989), pp. 192–94, for more on the SOSUS arrays, how they function and why they are important.

¹⁶ Stefanick (note 14), p. 7: 'The US Orion P-3 antisubmarine aircraft is considered at its most effective when operating with information provided by the US fixed Sound Ocean Surveillance System (SOSUS)'. This is undoubtedly true of attack submarines as well.

¹⁷ For an account of the 57th Fighter Interceptor Squadron based on Iceland, see Ross, W. A., 'Black Knights: the GIUK gap guardians', *Jane's Defence Weekly*, 8 Aug. 1987, pp. 236–38. According to Ross, F-15s on Iceland are permanently maintained on 10-minute alert status, and they intercept more Soviet aircraft in a year than all other US Air Force units together.

¹⁸ See, for example, 'Iceland-based fighters intercept high percentage of Soviet targets', *Aviation Week & Space Technology*, 14 Nov. 1988, pp. 62–69; and 'East Iceland radar facility now manned by Icelanders' (note 7).

¹⁹ According to Iceland's Foreign Minister, Jón Baldvin Hannibalsson, 'The Infrastructure Fund is prepared to pay for such an airport in Iceland, and would probably meet the entire cost. It would be manned by Icelanders under the control of the Icelandic aviation authorities, and NATO's only condition is that it would take the airport over in the event of war'. Quoted in 'NATO prepared to fund new airport', *News from Iceland*, Dec. 1988, p. 28. There is considerable opposition in Iceland to this airport, and it remains uncertain whether it will be built.

²⁰ For discussion of the Soviet air threat to shipping and naval forces in the North Atlantic, including the argument that this has been underrated relative to the Soviet submarine threat, see Friedman, N., *The US Maritime Strategy* (Jane's Publishing Co.: London, 1988), pp. 173–75. Friedman emphasizes the potential importance of airborne early-warning capability based on Iceland, which can complement and augment carrier-based early warning assets. See p. 187.

²¹ This paragraph is based entirely on Jonsson, A., *Ísland, Atlantshafsbandalagið og Keflavíkurstöðin*, (Icelandic Commission on Security and International Affairs: Reykjavík, 1989), pp. 47, 81–83. This is the most detailed study of the question available. I am grateful to Albert Jonsson for making this material available to me.

²² For an account of an exercise involving some of these units, see 'Air Guard Fighters Join Exercises in Iceland', *Aviation Week & Space Technology*, 14 Nov. 1988, pp. 69–70.

²³ The Air National Guard and Air Force Reserve squadrons are, however, being upgraded, with F-16s replacing F-4s. This will cause the squadron size to shrink from 24 to 18, with a corresponding reduction in the anticipated reinforcement numbers for Iceland.

²⁴ IISS (note 9), p. 24, lists US Navy strength in the United Kingdom at a total of 2300 but under the heading 'Holy Loch and other'. For a brief discussion of Holy Loch's role as an SSBN forward base, see Harkavy (note 15), p. 256.

²⁵ A fair bit of searching among open sources on Holy Loch uncovered very little in the way of detailed or useful information.

²⁶ Published reports indicate that US attack submarines do visit Holy Loch. Furthermore, a large support ship, the *USS Simon Lake*, is stationed at Holy Loch and is reported to be fully capable of servicing submarines. See Cameron, D., 'Fears over security at Holy Loch base', *The Scotsman*, 22 Apr. 1988, p. 3. Cameron reports that the *Simon Lake* 'arrived for the first time last year [1987]'. If he is suggesting that this represented something new in Holy Loch (and it is unclear from the article whether this is the case), then one might speculate that the arival of this vessel had more to do with increased attack submarine activities than with the support of SSBNs.

²⁷ It is worth noting that the US base at Holy Loch is collocated with a Royal Navy nuclear submarine base at nearby Gareloch. For a map illustrating this proximity, see 'US Nuclear secrets found: Holy Loch submarine fleet documents discovered in bag on Clydeside beach', *The Scotsman*, 22 Apr. 1988, p. 1.

²⁸ Elvik, H., 'Stryket US-marine', *Arbeiderbladet*, 8 May 1987, p. 7 (in Norwegian; emphasis added). It is worth pointing out that while Scotland is much closer to northern waters than Norfolk, Virginia, northern Norway is a further 1600 km closer to the critical waters of the

northern Norwegian and Barents Seas. The ability to resupply and repair forces in northern Norway would be a substantial force multiplier by reducing transit times. (I am grateful to Shannon Kile for research assistance and help in translating material from Norwegian.)

[29] There are also two large phased-array radars, one in Greenland and one in the United Kingdom, that form part of the US Ballistic Missile Early Warning System (BMEWS). Although of obvious importance to the defence of the North American continent, they are not very relevant to Northern Flank operations.

[30] On Greenland (and the Faeroes), the most informative sources are Faurby, I. and Petersen, N., 'The far north in Danish security policy' (Institute of Political Science, University of Aarhus: Aarhus, Denmark, Apr. 1988); Archer, C., 'Greenland and the Atlantic Alliance', *Centrepiece*, no. 7 (Centre for Defence Studies, University of Aberdeen: Aberdeen, summer 1985); Archer, C., 'The United States defence areas in Greenland', paper prepared for Nordiska Statskunskapskongressen (The Nordic Political Science Conference), Copenhagen, Denmark, Aug. 1987; and Duke (note 8), pp. 43–48, 53. Dated but informative is Bach, H. C. and Taagholt, J., *Greenland and the Arctic Region in the Light of Defence Policies* (Forsvarets Oplysnings- og Velfaerdstjeneste: Copenhagen, 1977). Particularly emphasizing the strategic importance of Greenland is Vego (note 10), pp. 33–34. Also useful is Taagholt, J., 'Greenland and the Faeroes', in Archer and Scrivener (note 8) pp. 174–89; and Arnason, R. T., 'Iceland, Greenland, and North Atlantic security', *Washington Quarterly*, vol. 4, no. 2 (spring 1981), pp. 68–81. For a background on political and economic factors, see Gröndal, B., 'People and politics on the Atlantic barrier islands', *NATO Review*, vol. 29, no. 4 (Aug. 1981), pp. 14–81.

[31] See, for example, Zakheim, D. S., 'NATO's Northern Front: developments and prospects', *Cooperation and Conflict*, vol. 17, no. 4, 1982, pp. 202–203. Zakheim commented that 'the current basing structure on Greenland is not optimally configured to support interceptor operations over the Denmark Strait. US bases are in western Greenland . . . [and] both are too distant to support supersonic fighter intercepts of Soviet aircraft transiting the Denmark Strait . . It may well be necessary to expand current facilities in Greenland to permit the basing of a larger, more responsive air defense capability.' Zakheim also expressed concern that the absence of an airbase in eastern Greenland placed excessive stress on Iceland-based air assets.

[32] See 'NATO/Mestersvig Airfield', US Department of Defense, *Current News*, 8 Nov. 1988, p. 14. See also 'Enighed om at undersøge grønlandsk alternative for Keflavík', *Nyheder fra Grønlands radio* (Greenlandic Radio Press Information Service: Nuuk, 3 Oct. 1988); and 'NATO-lufthavn i Mestersvig med i NATO-program', *Nyheder fra Grønlands Radio* (Greenlandic Radio Press Information Service: Nuuk, 7 Oct. 1988). I am grateful to Clive Archer of the University of Aberdeen for drawing these latter two sources to my attention. There has been speculation that NATO's interest in Mestersvig is primarily intended to goad the Icelanders into accepting a second airport in their country. *See News from Iceland* (note 19).

[33] The US interest in a new air field in Greenland has been confirmed by political authorities from Greenland and Iceland. It should be pointed out that the waters off north-eastern Greenland—the Greenland Sea—are thought to be a significant deployment area for Soviet strategic submarines, and hence a nearby air base would be extremely useful. On Soviet SSBN deployments in these waters see, for example, Østreng, W., 'Strategic submarines in the Barents Sea: tasks and functions', *International Challenges*, vol. 7, no. 1 (1987), pp. 37–45, especially p. 44.

[34] For an extemely informative discussion of these issues, including a very useful map, see Petersen, N., 'Denmark, Greenland, and Arctic security', ed. K. Möttölä, *The Arctic Challenge: Nordic and Canadian Approaches to Security and Cooperation in an Emerging International Region* (Westview Press: Boulder, Colo., 1988), pp. 43–49 and 57–58. For background on the role of strategic airpower in the north, see George Lindsey's discussion of 'Strategic Defense and the Air-Breathing Threat', in *Strategic Stability in the Arctic*, Adelphi Papers no. 241 (International Institute for Strategic Studies: London, 1989), pp. 14–21; and Cox, D., 'Ballistic missile defences, cruise missiles, air defences', paper presented to the International Conference on Arctic Cooperation, Toronto, Canada, 26–28 Oct. 1988.

[35] See 'US to close Greenland base', *Jane's NATO and Europe Today*, 18 July 1989, p. 6. This announcement has caused an angry outcry in Greenland because Søndre Strømfjord, funded by the US, has played an important role in Greenland's transport system. This article

reports that some politicians in Greenland are suggesting that the US be required to keep open Søndre Strømfjord as a condition for continued use of the Thule radar station.

[36] Duke (note 8), p. 54.

[37] Duke (note 8), p. 322.

[38] For an overview, see the very useful paper, Gleditsch, N. P., 'Foreign funded military bases in Norway', *PRIO Inform*, no. 8/86 (International Peace Research Institute: Oslo, 1986), which includes maps identifying the location of known facilities. An extensive and very detailed analysis is Wilkes, O. and Gleditsch, N. P., PRIO, *Intelligence Installations in Norway: Their Number, Location, Function, and Legality* (International Peace Research Institute: Oslo, 1979). Duke (note 8), pp. 215–235 provides a very thorough discussion. A brief overview may be found in Richelson, J., *The US Intelligence Community* (Ballinger: Cambridge, Mass., 1985), pp. 146–49 and 212. Ausland (note 8) devotes a chapter to the arrangements for electronic warfare, pp. 23–31. On the role of Loran-C facilities, see Wilkes, O. and Gleditsch, N. P., PRIO, 'Polaris and Loran-C: The importance of electronic infrastructure' (International Peace Research Institute: Oslo, 1986). The same two authors have also addressed this subject in great detail in *Loran-C and Omega: As Study of the Military Importance of Radio Navigation Aids* (Norwegian University Press: Oslo, 1987). An extensive compilation of public references to the US SOSUS capability in Norway, including both English and Scandinavian sources, is *S.O.S. U.S. Andoya: Ikkevoldssaken og Amerikansk Antiubåtkrigføring*, FMKs fredspolitiske skrifter (Folkereising mot Krig: Oslo, 1986).

[39] From 'Spørsmal til spørretimen, Onsdag 7 Januar 1987', *FD-Informasjon*, no. 1 (Jan. 1987) (Norwegian Ministry of Defence: Oslo, 1987). The Norwegian Government has been extremely sensitive to public discussion of these facilities. In 1981, for example, two researchers were tried and convicted of violating secrecy provisions of the penal code for publishing a report on intelligence installations in Norway that was based entirely on open sources—including telephone directories. For a collection of material relating to this bizarre affair, see Gleditsch, N. P. and Wilkes, O., *The Oslo Rabbit Trial: A Record of the National Security Trial Against Owen Wilkes and Nils Petter Gleditsch in the Oslo Town Court, May 1981* (International Peace Research Institute: Oslo, Dec. 1981).

[40] For discussion of a number of the issues associated with these arrangements, including particular concern that worries about crisis stability may prevent timely reinforcement of Norway, see Lund J., *Rocking the Boat: American Reinforcements and Norwegian Security* (Rand Corp: Santa Monica, Calif., Aug. 1989); and Lund J., *Don't Rock the Boat: Reinforcing Norway in Crisis and War* (Rand Corp: Santa Monica, Calif., 1989).

[41] Ausland (note 8), pp.138–39, provides useful information on the COB agreement. For fuller discussion, see Gleditsch, N. P., 'US–Norway Collocated Operating Bases declassified', *PRIO Inform* (International Peace Research Institute: Oslo, 1983).

[42] For discussion of the Invictus Agreement, as well as the declassified text of the agreement, see Gleditsch, N. P., 'Invictus agreement declassified', *PRIO Inform*, no. 14/84 (International Peace Research Institute: Oslo, 1984).

[43] The details of the prepositioning may be found in 'Lagring av utstyr i Norge for en amerikansk brigade', *Fakta* no. 0586 (Norwegian Ministry of Defence: Oslo, June 1986).

[44] See Elvik, H., 'Ny amerikansk forhandslagring?', *Arbeiderbladet*, 11 May 1987, in which Pentagon official George Bater is quoted as saying that the USA and Norway have discussed the possibility of additional logistical preparations for the US Navy in Norway (but adding that formal negotiations had not begun). In comments before the Norwegian Parliament, Norwegian Minister of Defence Holst has acknowledged that the US Navy has expressed interest in establishing an advanced logistical support site in Norway but denied that a formal proposal had been made. See his comments in 'Spørsmal til spørretimen, Onsdag 13 Mai 1987', *FD-Informasjon*, no. 5 (May 1987) (Norwegian Ministry of Defence: Oslo, 1987).

[45] It should be emphasized that the purpose of the analysis that follows is to gauge the magnitude of the current US contribution to NATO's defence of the north and the difficulty that European NATO members might have in replacing it if it were eliminated. It is not the aim, nor is it possible within the confines of this chapter, to provide a detailed threat assessment that allows for an evaluation of the adequacy of current NATO capabilities. Obviously, those who

believe present NATO forces are inadequate would be inclined to think that Europe would need to do more than just replace withdrawn US forces.

[46] Iceland is in certain respects an obvious exception, not because the number of US forces involved is great but because Iceland would be left defenceless if they withdrew. In the Northern Flank, this would be the most dramatic visible result of US retrenchment. The small scale of this consequence, and the ability of Europe alone to compensate, is illustrated by the recent suggestion that Canada take over the defence of Iceland instead of preserving its token commitment of forces to central Europe. (There is no indication, however, that this proposal is receiving serious consideration.) See George, P., 'New NATO role for Canada', *International Perspectives*, Nov./Dec. 1987, pp. 8–10.

[47] For a thorough background on the military situation on the Northern Flank, see Furlong, R. D. M., 'The threat to northern Europe', *International Defense Review*, no. 4 (1979), pp. 517–25; Furlong, R. D. M., 'The strategic situation in northern Europe: improvements vital for NATO', *International Defense Review*, no. 6, 1979, pp. 889–910; Lellenberg, J. L., 'The North Flank military balance' (The BDM Corp., Oct. 1979); Ries, T., 'Defending the far north', *International Defense Review*, no. 7, 1984, pp. 873–80; and Lellenberg, J. L., 'The military balance', eds Holst, Hunt and Sjaastad (note 8), pp. 41–66.

[48] There is, however, a second invasion route, through the Finnish wedge (see map 15.2), but it suffers similar disadvantages.

[49] Also not to be forgotten is the the unusual cycle of night and day in the far north. In winter, it is dark nearly all the time. This makes it difficult to mount any sort of military operation. Conversely, the summer is marked by nearly 24 hours of daylight.

[50] For informative discussions of the special difficulties of operating in this environment, see for example Davis, P. O., 'The Marines: out in the cold', US Naval Institute *Proceedings*, Nov. 1981, pp. 105–108, which surveys the special equipment, shelter, medical, and mobility problems associated with military operations in Arctic conditions; Bruck, A. (Lt. Col.), 'Winter training', *NATO's Sixteen Nations*, vol. 30, no. 1 (Feb./Mar. 1985), pp. 60–61, which emphasizes the importance of training and argues that the success or failure of operations will largely be determind by relative abilities to adapt to the environment; and Hill, G. J. (Capt.), 'And some will have it cold', US Naval Institute *Proceedings*, Nov. 1983, pp. 125–27. See also Poulsson, J. A. (Col.), 'Operations on the Northern Flank of NATO', *NATO's Fifteen Nations*, Apr./May 1981, pp. 58–63. Poulsson provides a useful overview of operational difficulties in the high north but is off-target on some details. He suggests, for instance, that under Arctic conditions in rough terrain (such as that found in Troms), pack horses may be one of the few effective ways to achieve mobility. The Norwegian Army eliminated pack horses as they were not found to be very useful. Schepe (note 3), pp. 42–45, also emphasizes the problems that accompany attempts to operate military forces in this region. For analysis of the Soviet perspective on operations in difficult terrain, see Donnelly, C. N., 'Soviet mountain warfare operations', *International Defense Review*, no. 6 (1980), pp. 823–34.

[51] Regarding the idea of a 'north Norway grab', one analysis suggests that 'the Russians will most likely mount a surprise, lightening thrust across northern Norway, turn south, and sue for peace before any effective Allied retaliation can be organized'. Smaldone, R. A. (Lt. Col.), 'A case for pre-positioning', US Naval Institute *Proceedings*, Nov. 1977, p. 110. This scenario not only ignores the constraints imposed by the environment but also the fact that most Soviet forces in the region are not kept at high levels of readiness (that is to say, they are mostly Category II and III divisions) and hence would require considerable mobilization before they were combat ready. This would make surprise very difficult to achieve, rendering a 'surprise, lightening thrust' rather unlikely. On the readiness of Soviet forces on the Kola Peninsula, see *The Military Balance in Northern Europe, 1985–1986* (The Norwegian Atlantic Committee: Oslo, 1986), p. 9. The difficulty of rapid movement in this region is indicated by the experience of German forces in June 1941, when they struck towards Murmansk from northern Norway. In a campaign stretching over 10 weeks they advanced only 24 kilometres at the price of 10 000 casualties. See Lellenberg (note 47), p. 44. On pp. 43–45 Lellenberg provides a discussion of the unusual conditions for military operations found in the north. Also instructive is the Soviet attack from Murmansk against German forces in the north in 1944. While more successful than the 1941 German effort, it nevertheless illustrates the difficulties that make the 'north Norway

grab' scenario unlikely. In the first phase of the battle, for example, the Soviet Union, attacking with fresh troops and a 4 : 1 advantage in infantry forces, managed between 35 and 60 kilometers of forward movement, after which its forces were so exhausted that it found it necessary to order a three day pause in the campaign. Later in the battle, with the Germans withdrawing, Soviets forces advanced 150 kilometers in 10 days. Such rates of advance would obviously not get the Soviet forces to Troms with lightning quickness. See the very interesting detailed account in Gebhardt, J. F., 'Petsamo–Kirkenes Operation (7–30 October 1944): A Soviet joint and combined arms operation in Arctic terrain', *Journal of Soviet Military Studies*, vol. 2, no. 1 (Mar. 1989), pp. 60, 65.

[52] *Norwegian Defense Review/Status of Norwegian Defence, 1987* (Norwegian Ministry of Defence: Oslo, 1987), p. 1.

[53] One cannot preclude the possibility, however, that the Soviet Union might be motivated by defensive concerns to occupy Finnmark even if it did not wish to attempt the more demanding assault on the critical facilities in Troms. A Soviet seizure of Finnmark would have the effect of moving the border between Norway and the USSR some hundreds of kilometres to the west, providing a buffer for the Kola Peninsula and preventing the use of Finnmark as a base for NATO operations against it. This would surely be a more comfortable situation for the Soviet Union than the current one, in which the Soviet–Norwegian border sits a mere 18 kilometres from important Soviet facilities in Pechenga. This situation resembles that which existed between Finland and the Soviet Union at the outset of World War II: the Soviet Union coveted the Karelian Isthmus, then part of Finland, as a protective buffer for Leningrad. In 1938 and 1939, the Soviet Union sought to achieve this aim by negotiation. On 30 Nov. 1939 the Soviet Union attacked Finland, commencing the Winter War; while operations were not limited to the Karelian Isthmus, the brunt of the Soviet attack came there, and at the conclusion of the Winter War the entire isthmus, along with adjacent areas of south-eastern Finland, was ceded to the Soviet Union. One substantial difference between Finland in 1939 and Norway in 1989 is that the Karelian Isthmus was heavily defended, whereas Finnmark is not. The Norwegian equivalent to the Finnish Mannerheim line is found in Troms. On the other hand, a Soviet attack that stopped in Finnmark would leave intact the Norwegian infrastructure in Troms. (For discussion of Soviet motivations and behaviour in the Finnish case, see Nevakivi, J., 'The great powers and Finland's Winter War' and Vuorenma, A., 'Defensive strategy and basic operational decisions in the Finland–Soviet Winter War, 1939–1940', both in *Revue Internationale d'Histoire Militaire*, no. 62 (1985), pp. 55–96.) It is perhaps worth noting that the territory on the Soviet side of the Soviet–Norwegian border was once Finnish, and that the Soviet base at Pechenga is the site of what was once the Finnish town of Petsamo. For a detailed discussion, see Krosby, H. P., *Finland, Germany and the Soviet Union, 1940–1941: The Petsamo Dispute* (University of Wisconsin Press: Madison, Wis., 1968.)

[54] On this point, see Berg, J., 'Norway's vital defense changes', *Armed Forces Journal International*, Dec. 1980, pp. 40–50.

[55] The 1985–86 and 1987–88 volumes of the annual *The Military Balance in Northern Europe* (The Norwegian Atlantic Committee: Oslo, 1986 and 1988) provide essential data and analysis; Huitfeldt, T., 'SHAPE and the Northern Flank', paper presented at the Oslo Symposium, 10–14 Aug. 1986, provides an unusually thorough and informative treatment of the mobilization and reinforcement issue. See also Huitfeldt, T., 'Force mobilization: a new assessment of the Norway case', paper presented at the NATO–Warsaw Pact Force Mobilization Conference, National Defense University, Washington DC, 3–4 Nov. 1987, which contains a detailed account of the internal Norwegian mobilization capability (pp. 14–17). Also very helpful are Leonard, M., 'Planning reinforcements: an American perspective', and Huitfeldt, T., 'Planning reinforcements: a Norwegian perspective', both in Holst, Hunt and Sjaastad (note 8).

[56] The Inspector General of the Norwegian Home Guard, *The Norwegian Home Guard* (Norwegian Ministry of Defence: Oslo, 1973). This document claims that the entire Home Guard can be mobilized in four hours (p. 6). It explains the mission of the Home Guard as follows: 'The main part of the Army is not combat ready in peacetime. Only when war threatens will the units be mobilized and issued with weapons, equipment and ammunition. The army consequently needs time before it is ready to fight. The securing of this mobilization is the primary mission of the Home Guard, which is capable of reacting swiftly' (p. 15). For other

discussions of the Home Guard, see Berg, O. (Maj. Gen.), 'The Home Guard: deep roots in the local community', *Norwegian Defence Review* (Status of Norwegian Defence, 1987), pp. 22–23; Berg, O. (Maj. Gen.), 'Position consolidated in the Home Guard', *Norwegian Defense Review/Status of Norwegian Defence, 1986* (Norwegian Ministry of Defence: Oslo, 1986), pp. 22–23; and Howard, P., 'The Norwegian Home Guard', *Jane's Defence Weekly*, 20 Dec. 1986, pp. 1460–61.

[57] *The Norwegian Home Guard* (note 56), p. 21.

[58] See Sunde, H. I. (Lt. Gen.), 'South Norway and the reinforcement of the Northern Flank', in Ellingsen (note 11), p. 44.

[59] The details in this paragraph are drawn from Berg, 1986 and 1987 (note 56).

[60] For an overview, see Sunde (note 58).

[61] For a useful concise overview, see Breidlid, O. (Maj. Gen.), 'The Norwegian mobilisation system', *NATO's Sixteen Nations*, no. 1, 1985, especially p. 73.

[62] *The Military Balance in Northern Europe, 1985–1986* (note 51), p. 4.

[63] The Army figures are taken from IISS (note 9), p.72. IISS, however, reports total mobilized strength as 435 000. Tönne Huitfeldt offers the figure 380 000 (mobilizable in as little as 36 hours) in his 'Force mobilization in Norway', ed. J. Simon, *NATO–Warsaw Pact Force Mobilization* (National Defense University Press: Washington, DC, 1988), pp. 527–28. *Facts on Norwegian Defence* (Norwegian Ministry of Defense: Oslo, undated) p. 3, reports the total mobilized strength as being 325 000. General Sir Geoffrey Howlett (Commander in Chief Allied Forces Northern Europe) provides the figure 320 000 in his 'A summary', in Ellingsen (note 11), p. 110. I have used the lowest figure, from the most authoritative source, in the text.

[64] Pfaff, W., 'NATO: building down toward sufficient defense', *International Herald Tribune*, 23–24 Sept. 1989. On this same point, see 'Norway: NATO's Northern Flank', *Norway Times/Nordisk Tidende*, 20 July 1989, p. 5, which comments: 'All sectors of Norwegian society are obligated to take part in the defense of Norway. The concept of total defense preconditions a close cooperation between civilian and military authorities. In case of war, Norway's military can requisition supplies, transportation, materiel, maintenance services, buildings and facilities from civilian sectors and individuals'.

[65] These facts are drawn from 'Norway: NATO's Northern Flank' (note 64), p. 5; Howlett (note 63), p. 110; and Pfaff (note 64). See also Fitchett, J., 'Norway, bucking trend, shores up defenses and NATO ties', *International Herald Tribune*, 26 Sept. 1989. Fitchett comments that Norway 'surprises visitors these days by its determination to modernize its defenses and consolidate its links with the Western alliance'.

[66] See, for example, Soderlind, R., 'Canada cancels commitment to NATO's Northern Flank', *Armed Forces Journal International*, Aug. 1987, pp. 36–37; 'Norway's search for CAST replacement', *Jane's Defence Weekly*, 3 Oct. 1987, p. 719. For a detailed discussion of Canada's connection with the Northern Flank, see Jockel, J. T., *Canada and NATO's Northern Flank*, (York University Centre for International and Strategic Studies: Toronto, 1986).

[67] See 'Norway hails new unit of NATO for the north', *International Herald Tribune*, 25–26 June 1988; 'NATO composite force for Norway', *International Defense Review*, vol. 21, no. 8 (Aug. 1988), p. 882. The inclusion of the West German unit is noteworthy as it represents the first instance since the German occupation of Norway in World War II that German forces are devoted to contingencies in that country. (The West German role was announced in 'Vest-Tyske soldater til Norge', *Aftenposten*, 24 Mar. 1988.) While the new NATO force is smaller than the Canadian brigade it is replacing, the inclusion of two artillery battalions (including some 30–40 guns) makes it heavier.

[68] Alexander, J. H. (Col.), 'The role of the U.S. Marines in the defense of north Norway', US Naval Institute *Proceedings*, May 1984, p. 185.

[69] Schopfel, W. H. (Lt. Col.), 'The MAB in Norway', US Naval Institute *Proceedings*, Nov. 1986, p. 36.

[70] In other words, the AMF is meant to serve the same function as US ground forces serve in Central Europe, albeit in a small and contingent way.

[71] On Britain's role in the defence of Norway, the best source is Archer, C., *Uncertain Trust: The British–Norwegian Defence Relationship*, Forsvarsstudier 2/1989 (Institutt for Forsvarsstudier: Oslo, 1989). On the UK/NL force, see Larken, E. S. J. (Comm.), 'The role of

the UK/NL forces reinforcing the Northern Flank', and Ross, R. J., 'Organization of UK/NL landing force', both in Ellingsen, (note 11), pp. 61–70. See also Robertson, M., 'Britain's contribution to Norwegian defence, 1940–1980', *Centrepiece*, no. 11 (Centre for Defence Studies, University of Aberdeen: Aberdeen, summer 1987); and Robertson, M., 'Reinforcing north Norway: the United Kingdom's role', *Centrepiece*, no. 12 (Centre for Defence Studies, University of Aberdeen: Aberdeen, summer 1987). Also useful are Marcus, J., 'Britain and NATO's Northern Flank', *Armed Forces Journal International*, Oct. 1987, pp. 66–74; Farrar-Hockley, A., 'The United Kingdom and the Northern Flank', paper presented at the Oslo Symposium, 10–14 Aug. 1986; and Geoffrey Till (ed.), *Britain and NATO's Northern Flank* (St. Martin's Press: London, 1988).

[72] See Schopfel (note 69), p. 36. Discussion of British training in Norway can be found in Adshead, R., 'The Royal Marine Reserve in Norway', *Armed Forces*, vol. 7, no. 6 (June 1988), pp. 261–65.

[73] The advantage of prepositioning is substantial because the logistics requirements of the Marine units are large. See, for example, Stauch, Jr., V. D., 'Facing MEF logistics facts', *Marine Corps Gazette*, Aug. 1989, pp. 25–27, which comments that 'Today's MEF [Marine Expeditionary Force] requires massive tonnage to sustain it'.

[74] The overall manpower level of a MAB is 15 000, but this figure includes both the Marine Aircraft Group and the logistics support group, and hence is not indicative of ground capability. See Stephan, H. J., 'USMC to bolster Norwegian flank: 4th MAB ready to deploy in a crisis', *Armed Forces Journal International*, Aug. 1987, p. 34.

[75] Brig. General Mathew P. Caulfield, Commander of the 4th MAB, has claimed that the MAB generates more firepower than full divisions deployed in Europe. See Stephan (note 74), p. 34. Caulfield's remark, however, referred to the entire MAB, including the air power component, and not simply to the ground forces. The MAB's airpower is discussed further below. See also Rein, T. (V. Adm.), 'Reinforcing the Northern Flank: the case of north Norway', in Ellingsen (note 11), p. 34.

[76] According to IISS (note 9), the Norwegian Army possesses no helicopters in its standing forces. Its Air Force possesses 28 helicopters that are available to support the ground forces, but only two are configured especially for transport, the rest serving a utility function (p. 73). Recent Norwegian defence budgets have included funds for the purchase of 12 transport helicopters for the Army. See *The Defense Budget 1987*, Fact Sheet no. 0986 (Norwegian Ministry of Defence: Oslo, Oct. 1986), pp. 13–16. More importantly, because of the extensive offshore oil development in the Norwegian Sea, there are a great number of helicopters in Norway available for mobilization. In peacetime these are in the service of the oil industry, but they would provide the Norwegian Army with tactical mobility in the event of mobilization.

[77] The MAB has exercised annually in Norway only since 1984; Stephan (note 74), p. 34. It might be noted, however, that the US practice of rotating men and officers through units on relatively short tours of duty limits their familiarity with the special conditions in the north. AFNORTH Commander in Chief, General Howlett, has observed, for example, 'I do get worried when I see different people coming every year to give somebody else a turn at going to North Norway. That is not good enough. The people who come here must be those people who are going to return . . .' Howlett (note 63), p. 113. Also, substantial NATO exercises in Norway (the 'Express' series) occur only every other year. For an overview of NATO exercises in Norway since 1964, see 'Større øvelser i Norge med alliert deltakelse', *FD-Informasjon*, no. 0386 (Norwegian Ministry of Defence: Oslo, Apr. 1986). For a recent assessment, which argues that the US Marines are improving but are still 'not so good' at many activities associated with Arctic warfare, see Robeson, E. J. IV (Lt. Col.), 'Tactical reflections on Norway', US Naval Institute *Proceedings*, Nov. 1989, pp. 108–11. Discussions with officials in the Norwegian Defence Ministry have left me with the impression that the US Marines are rated below Norwegian and British forces in their preparations for and skill in coping with the special conditions found in northern Norway.

[78] On this point, see Rein (note 75), p. 30: 'Threat assessments leave no doubt that our ability to conduct effective air defence from the initial stages of a conflict will be decisive in executing an effective force buildup, including mobilization and reception of reinforcements as well as the effective employment of ground and naval forces'.

[79] Holst, J. J., 'The contribution of allied reinforcements to Norwegian security', in Ellingsen (note 11), p. 15.

[80] See *The Military Balance in Northern Europe, 1985–1986* (note 51) pp. 5–6; Ries, T. and Skorve, J., *Investigating Kola: A Study of Military Bases Using Satellite Photography* (Brassey's Defence Publishers: London, 1987), pp. 32–33 and 62–70. The latter provides a detailed discussion of Soviet air bases and deployments on the Kola Peninsula. Other sources include: Aalbaek-Nielsen, B., 'The Royal Norwegian Air Force', *Air Force Magazine*, Sep. 1985, pp. 120–23; 'The air defence of Norway', *NATO's Sixteen Nations*, Feb./Mar. 1987, p. 89; Aamoth, O. (Maj. Gen.), 'Changing challenges facing the air defence of Norway', *Norwegian Defense Review/Status of Norwegian Defence, 1988* (Norwegian Ministry of Defence: Oslo, 1988), pp. 24–25; Aamoth, O. (Maj. Gen.), 'Advanced material for the Air Force but difficult personnel situation', *Norwegian Defense Review/Status of Norwegian Defence, 1986* (Norwegian Ministry of Defence: Oslo, 1986), pp. 20–21; and Berg, A. R. (Maj. Gen.), 'The air threat', ed. J. A. Olsen, *The Air Situation in the North in the Year 2000 and Beyond*, (Den Norske Atlanterhavskomité: Oslo, 1989), pp. 35–41.

[81] Data in this paragraph is based on *The Military Balance in Northern Europe, 1987–1988* (note 55), pp. 10–12. A similar assessment of Soviet air strength in the north can be found in Norwegian Ministry of Defence, *Hovedretningslinjer for Forsvarets Virksomhet in Tiden 1989–93*, Stortingsmelding nr. 54 (1987–88) (Government Printer: Oslo, 1988), pp. 68–69.

[82] Alternatively, one may view the surplus Soviet air basing structure as a buffer against NATO efforts to suppress Soviet airpower by attacking bases. Air bases are in any case difficult to put out of action, but the Soviet Union has the additional luxury of 'backup' bases in the region on which they can operate if NATO attacks damage current operating bases. Thus, even if the Soviet Union does not swing substantial air reinforcements to the Kola Peninsula in the event of war, it derives benefit from the large basing capacity in the region.

[83] For an indication of the scope of Soviet deployment of sea-launched cruise missiles, see *The Military Balance in Northern Europe, 1985–1986* (note 51), pp. 16–23. The USSR has 66 or 67 cruise missile submarines, of which about half are deployed in northern waters. See also IISS (note 9), p. 37. The overall submarine force represents a cruise missile threat that numbers roughly 300–400. In addition, the Soviet Kiev and Kirov Class surface vessels, as well as many others, include cruise missiles among their force loadings. Some relevant discussion can be found in Gottemoeller, R. E., *Land Attack Cruise Missiles*, Adelphi Paper no. 226 (International Institute for Strategic Studies: London, winter 1987–88), although the focus of this chapter is long-range cruise missiles capable of homeland strikes with nuclear weapons.

[84] Clear evidence of Norwegian concern about 'the SLCM threat' can be found in Holst, J. J., 'Arms control and security on NATO's Northern Flank', *NATO Review*, Oct. 1988, pp. 12–13. Holst warns: 'Competitive deployments of SLCMs for attack on land could cast heavy shadows on the shores of littoral states like Norway . . . Norway's vulnerability to such attacks is extensive'. (p. 12).

[85] The discussion of a possible arms control agreement limiting sea-launched cruise missiles is not without implications for the Northern Flank—for this among other reasons. The idea of such a limitation has found champions in the United States. See, for example, Sigal, L. V., 'Ban this missile from the seas', *New York Times*, 9 June 1988.

[86] For more detailed discussion, see Sokolsky, J. J., 'Soviet naval aviation and the Northern Flank: its military and political implications', *Naval War College Review*, vol. 34, no. 1 (Jan./Feb. 1981), pp. 34–45; Ermarth, F. W. and Silverstein, R. L., 'Air power and strategy in the northern region', unpublished paper, Feb. 1981; and Lellenberg 1979 (note 47), pp. 18–19.

[87] Alberts, D. J. (Lt. Col.), *Deterrence in the 1980s, Part II: the Role of Conventional Air Power*, Adelphi Papers no. 193 (International Institute for Strategic Studies: London, 1983), provides one of the few publicly available, sensible attempts to allocate aircraft to theatres and to assess the resulting air balances. Assuming reinforcement, Alberts arrives at a ratio for the Northern Flank of 1.37 : 1 to the USSR's advantage (p. 16). There is no reason why a disadvantage of this magnitude need spell catastrophe for NATO, which may be able to offset the Soviet numerical superiority in several ways. It may, for instance, be able to generate more sorties per aircraft, a not unreasonable assumption given NATO's heavy investment in logistics 'tail' relative to the USSR. It may also be able to achieve favourable exchange rates in combat, a not

unreasonable possibility given NATO's widely presumed (and expensive) technological superiority. Finally, Albert's analysis appears to exclude US carrier-based aircraft; as will be shown, the arrival of one or several carriers adds considerably to NATO's air capability in north Norway. For a more pessimistic appraisal, however, see Lellenberg (note 47), pp. 56–57, which emphasizes the weakness and inadequacy of NATO air power in the northern region.

[88] Holst, J. J., 'Norway and strategic developments on NATO's Northern Flank', in Olsen (note 80), p. 20.

[89] A detailed breakdown may be found in *The Military Balance in Northern Europe, 1985–1986* (note 51), p. 10–11.

[90] See 'Lagring Av Utstyr I Norge For En Amerikansk Brigade' (note 43) for data on aircraft and I-Hawk. Norway itself has ordered six I-Hawk batteries for point defence of airfields, but even when these are deployed, the capability that accompanies the MAB will represent a significant augmentation. For discussion of Norway's difficulties with respect to air defence, see Storvik, O. T., 'Evenes får ikke Hawk', *Aftenposten*, 18 Mar. 1988, which notes Norway's inability to afford Hawk missiles for all bases that ought to have them. A brief account of Norway's plans for upgrading its air defences is found in Wilson, J.R., 'Norway updates layered air defense', *International Defense Review*, no. 3 (1989), pp. 351–52.

[91] Alexander (note 68). See also Goodman, G., 'Marines cancel MIFASS but move ahead on other C3 fronts', *Armed Forces Journal International*, Aug. 1987, pp. 110–14.

[92] For an overview of the RRP, see Welch, L. D. (Gen.), 'NATO rapid reinforcement: a key to conventional deterrence', *NATO's Sixteen Nations*, Feb./Mar. 1987, pp. 114–16. On the assignment of RRP aircraft to the Northern Flank, see Riste, O. and Tamnes, R., 'The Soviet naval threat and Norway', *FHFS Notat*, no. 3/86 (Research Center for Defence History, Norwegian National Defence College: Oslo, 1986), p. 19.

[93] For example, Huitfeldt 1986 (note 55), p. 36, expresses doubt on this score.

[94] For details on the composition of a carrier airwing, see O'Rourke, R., *The Cost of A US Navy Aircraft Carrier Battlegroup*, Congressional Research Service report no. 87-532 F (US Library of Congress: Washington, DC, 26 June 1987), p. 17. The notional carrier airwing includes 20 F-14s, 20 F/A-18s and 20 A-6s, as well as anti-submarine, electronic warfare and early-warning aircraft. For detailed discussion of present and future carrier air capabilities, see Sweetman, B., 'Carrier aviation in the 1990s', *International Defense Review*, vol. 21, no. 2 (Feb. 1988), pp. 149–52. This article notes that the move toward multi-role carriers will mean more aircraft for each mission, whether air defence or ground attack.

[95] Larson, C., 'Forward power: the aircraft carrier group in NATO's plans', *NATO's Sixteen Nations*, Feb./Mar. 1988, pp. 26, 29.

[96] For a thoughtful discussion of the advantages and disadvantages of operating carriers in fjords, see Breivik, R. (V. Adm.), 'Fjord operations', *NATO's Sixteen Nations*, Feb./Mar. 1988, p. 32–36.

[97] Friedman (note 20), pp. 80–82. Sweetman (note 94), p. 150, reports that the so-called 'outer air battle' in defence of the carrier will take place 'at least 370 km from the centre of the carrier group'. Taylor, J. W. R. (ed.), *Jane's All the World's Aircraft 1988–89* (Jane's Defence Data: Coulsdon, Surrey, 1988), p. 398, lists the maximum range of the carrier-based F-14, for example, at more than 3000 km.

[98] See, for example, Alberts (note 87), p. 56, table 3, which shows a substantial disproportion in favour of ground attack versus dedicated air-to-air assets in the northern theatre. See also Rein (note 75), p. 30, who comments: 'For several years we have recommended that reinforcing nations provide air defence aircraft to the extent possible.'

[99] One Norwegian military officer writes, 'a presence of, for instance, three CVBGs [carrier battle groups] in the Norwegian Sea may increase the air defence capacity on NATO's Northern Flank by a factor of five . . .', Børresen, J. (Capt.), 'Norway and the US Maritime Strategy', *Naval Forces*, vol. 7, no. 6 (1986), pp. 14–15.

[100] *The Military Balance in Northern Europe, 1985–1986* (note 51), p. 4, refers to the existence of five military air bases in northern Norway. In a military emergency, civilian air strips could, with some preparations, also be used. In addition, air reinforcements could be cycled into bases in central and southern Norway and deployed to the north when attrition opened space at air bases in northern Norway. On the issue of the air base constraint, see

Leonard (note 55), especially p. 161: 'The real question is how much tactical air reinforcement Norway can absorb. There is a large amount available . . . The problem is where to put them . . .' Leonard recommends (p. 166) an expansion of airfield capacity in northern Norway. See also, on this point, Huitfeldt 1987 (note 55), p. 185. Huitfeldt observes: 'Airfield capacity could be a limiting factor for allied air reinforcement squadrons for North Norway. Obviously, there could be a capacity problem given the number and types of aircraft which may be required to operate from airfields in North Norway'.

[101] Of course, this is true only so long as the carriers remain functional. If carrier aircraft have to be evacuated to Norwegian bases, as provided for by the Invictus Agreement, the congestion problem will be exacerbated, not alleviated. Holst suggests that there is a single base in central Norway, with pre-stocked supplies, allocated to meeting this contingency. Holst, J. J., 'Norwegian Security Policy: the Strategic Dimension', eds Holst, Hunt and Sjaastad (note 8), pp. 110–11.

[102] Pay, D. J., 'The US Navy and the defence of Europe, part 2', *Naval Forces,* vol. 11, no. 2 (1988), pp. 18–25. Pay's argument, however, is undermined by his own analysis, which shows NATO possessing a considerable numerical superiority in land-based fighter aircraft and overall air parity with the Soviet Union in the north. (See the table of forces for the Norwegian Sea campaign, p. 20.) Obviously, if NATO can hold its own without carrier aircraft, then the latter is not 'vital'. However, in his table Pay does not make clear what he is and is not counting on both sides, so it is hard to know what to make of his numbers. (On the Soviet side, for example, he appears to be counting primarily, if not exclusively, Soviet naval aviation, ignoring the several hundred aircraft—including 100 deployed on the Kola Peninsula—that belong to the Archangelsk Air Defence district and the Leningrad Military District. While it is unlikely that all of these aircraft would be thrown into the battle in the north, some assumption must be made about the extent of Soviet air reinforcement of the Kola Peninsula if the air threat is to be meaningfully assessed.) Ironically, Pay's line-up of forces appears to undervalue somewhat the capability of carriers; he credits NATO with as many as seven carriers in the Norwegian Sea, but suggests that these only bring to bear some 300 combat aircraft. This represents a smaller force loading than is mentioned in the discussion above. It is possible that inconsistency in Pay's analysis derives from the fact that, as an advocate of the US Maritime Strategy, he is seeking to demonstrate both the *feasibility* of forward carrier operations (hence the circumscribing of the Soviet threat) and the *necessity* of forward carrier operations (a claim which is weakened if the Soviet threat is circumscribed).

[103] Holst (note 84), p. 13.

[104] If one thinks the air war will be won or lost quickly, then it is imperative to get the carriers into position as quickly as possible. There is always the possibility, of course, that the carriers might arrive too late. Robert Wood of the US Naval War College, for example, has suggested that the air battle in north Norway might be decided in ten days (see Pay, note 102, p. 24).

[105] Conversely, one ought not to overlook the argument that the timely arrival of carriers in the northern Norwegian Sea might, if the air war were going well in the region, 'liberate' land-based air capability for use elsewhere in a NATO–WTO war.

[106] Murphy (note 11), p. 59.

[107] Howlett (note 63), p. 111. See also Breivik, R. (V. Adm.), 'SACLANT's role in the support of the NEC reinforcing the Northern Flank', in Ellingsen (note 11), pp. 93–95. Breivik comments: 'The need for forward movement into the Norwegian Sea is critical'.

[108] Holst (note 84), p. 13. For further evidence of the Norwegian concern about the carriers, see Storvik, O.T., 'Hangarskip in nord viktig NATO-bidrag', Oslo *Aftenposten,* 31 Jan. 1989.

[109] Lodgaard, S., 'The big powers and Nordic security', eds. B. Huldt and A. Lejins, *Security in the North: Nordic and Superpower Perceptions* (The Swedish Institute of International Affairs: Stockholm, 1984), p. 71.

[110] See, for example, Fitchett (note 65), p. 2: 'The importance of controlling . . . Norwegian bases is likely to grow . . . if arms control accords thin the allied forces in central Europe. That would increase Western Europe's dependence on US reinforcements crossing the Atlantic'. The same point is expressed in Graaf, P. J. (Gen), 'Why NATO in 1989? A European View', paper delivered to the conference on 'NATO and the evolving political climate in Eastern Europe', Brussels, 19–22 Sep. 1989, p. 9.

¹¹¹ See West, F. J., Davis, J. K., Dougherty, J. E., Hanks, R. J. and Perry, C. M., *Naval Forces and Western Security* (Pergamon/Brassey's: Washington, DC, 1986), p. 54, for a useful table comparing the naval forces of the United States with those of its allies. For interesting discussions of the role of West European navies in the current environment, see Hobson, S., 'Europe studies defence of Norway', *Jane's Defence Weekly*, 8 July 1989, p. 9, which reports on a NATO study that contemplates a larger European maritime role in the defence of Norway; and Bruntland, A. O. and Gjelsten, R., *European Naval Activity in the Northern Areas*, Security Policy Library No. 10, (The Norwegain Atlantic Committee: Oslo, 1987), which explores the questions related to a separate West European naval force for the Norwegian Sea. Also relevant here is Tunander, O., PRIO, *Cold Water Politics: The Maritime Strategy and Geopolitics of the Northern Front* (Sage Publications: London, 1989), pp. 144–47, which proposes that a sort of West European naval buffer zone be created in the Norwegian Sea to separate the naval forces of the superpowers.

¹¹² The US Atlantic Fleet, for example, spends more of its time in the Caribbean, in the neighbourhood of Cuba and Nicaragua, than it does in the Norwegian Sea.

¹¹³ The most detailed analysis is Posen, B. R.,'Offensive and defensive sea control: a comparative assessment', eds C. Glaser and S. E. Miller, *The Navy and Nuclear War* (Cornell University Press: Ithaca, N.Y., forthcoming). Posen concludes: 'Fighting Soviet submarines in the seas proximate to the Soviet Union is much more difficult than fighting them on ASW barriers in the north Norwegian Sea, and much more difficult than fighting them on the GIUK Gap and in the Atlantic' (p. 32).

¹¹⁴ See especially the thorough analysis of the issue in Lautenschläger, K., 'The submarine in naval warfare', *International Security*, vol. 11, no. 3 (winter 1986–87), pp. 94–140, especially pp. 109–22, 134–38. Lautenschläger notes, for example, that the Soviet Navy produces 10 or fewer submarines a year, while Germany, during the unsuccessful U-boat campaign of the World War II, were producing hundreds. (Germany lost 784 submarines in the Battle of the Atlantic.) See also Posen (note 113). For discussions of the Soviet submarine force, which constitutes the primary threat to the SLOC, see Compton-Hall, R., 'The Soviet attack submarine: power and problems', *Naval Forces*, vol. 8, no. 1 (1987), pp. 50–61 (which calls into question the Soviet ability to sustain high SSN availability rates in wartime); and Breyer, S., 'The Soviet submarine force today', *International Defense Review*, vol. 20, no. 9 (Sep. 1987), pp. 1155–59. Breyer speculates that Soviet submarine production may be as low as six per year. This would be clearly insufficient to compensate for the wartime losses which a campaign against the SLOC would generate. If war came, of course, the Soviet Union could attempt to produce more rapidly but would face two constraints: modern attack submarines are so technologically sophisticated that they are difficult to produce quickly; and the industrial infrastructure for producing submarines will likewise not be easy to expand rapidly in wartime.

¹¹⁵ As the focus here is on the north, the connection to the Central Front is here only mentioned in passing. It is discussed briefly below.

¹¹⁶ On this point, see von zur Gathen (note 1), p. 62: 'It would have been logical . . . for NATO to place the Baltic Approaches under the command of the central region of Allied Command Europe'. See also Hunt, K., 'The security of the center and the north', in Holst, Hunt and Sjaastad (note 8), especially p. 66. (The Soviet Union includes both Denmark and southern Norway in its Western Theatre of Operations, which also encompasses the Central Front.)

¹¹⁷ On the notion of BALTAP as a 'forward defence' of the North Sea, see Toyka, V. (Cmdr.), 'A submerged forward defense', US Naval Institute *Proceedings*, Mar. 1984, p. 145.

¹¹⁸ See von zur Gathen (note 1), p. 61; and Kampe, H. (V. Adm.), 'Defending the Baltic Approaches', US Naval Institute *Proceedings*, Mar. 1986, pp. 88–93.

¹¹⁹ See for example, Lind, O. K. (Gen.), 'Denmark and the Baltic: forward defence for the UK?', *RUSI Journal*, vol. 130, no. 2 (June 1985), pp. 9–13. Lind writes that 'the UK has accepted that the defence of the BALTAP area is to be considered as "forward defence for the UK"' (p. 11). Regarding the Channel ports, these naturally will be much less important if reinforcements and supplies are not going to be sent by sea from the United States. Regarding Norway, note the comment of Sunde (note 58), p. 41, 'It is better to defend Oslo in Denmark rather than [in] Norway'.

120 For a variety of reasons, however, the Soviet Union may have incentives to violate Swedish neutrality. For a thoughtful discussion of Sweden's connections to NATO and Warsaw Pact scenarios in the Baltic, see Tunander (note 111), ch. 9.

121 Graydon, M. (V. Mar.), 'SACEUR's plans and priorities: the Northern Flank', in Ellingsen (note 11), p. 79. See also Vego (note 10), p. 28, who comments, 'The Soviets consider the Baltic, the Danish Straits, and the North Sea as a strategic whole'.

122 On these points see Kampe, H. (V. Adm.), 'Amphibious objective: Baltic Approaches', US Naval Institute *Proceedings*, vol. 114, no. 3 (Mar. 1988), pp. 113–17. Kampe notes, for example, that continued NATO control of the Baltic Approaches will 'maintain a forward barrier of air defense against the aircraft of the Soviet naval and air forces . . . [and] allow Commander-in-Chief Northern Europe to concentrate his forces for the defense of Norway on the northern front' (p. 116).

123 von zur Gathen (note 1), p. 61. See also *The Military Balance in Northern Europe, 1987–1988* (note 55), p. 6, which comments as follows: 'Schleswig-Holstein and Denmark are primarily important as part of the West European continent and as the left flank to NATO's central region. For a Soviet advance against vital areas of Western Europe it would be important to secure the Schleswig-Holstein/Denmark Flank'.

124 For a short but focused discussion of the Soviet Baltic SLOC, see Jansson, C. N.-O. (Cmdr.), 'The Baltic: a sea of contention', *Naval War College Review*, vol. 41, no. 3 (summer 1988), pp. 47–61. Jansson writes, 'Sea lines of communications are more efficient in hauling cargo than are railroads and truck convoys. Recent construction of port facilities in Soviet Baltic ports and in East Germany may be a sign of the sea lines increasing importance' (p. 51). See also Tunander (note 111), especially pp. 108, 116 and 126. He comments, 'After the Polish Crisis in 1980–1981, increasing priority has been given to the Soviet SLOCs [in the Baltic]—with a new ferry line between the Soviet Union and East Germany—probably because of the vulnerable Polish lines of communication'.

125 Donnelly, C. N. and Petersen, P. A., 'Soviet strategists target Denmark', *International Defense Review*, vol. 19, no. 8 (Aug. 1986), pp. 1047–48.

126 The salience of the Baltic to the Soviet Union is perhaps indicated by the fact that in the early postwar era the Soviet Government made it clear to its Danish counterpart that it considered Denmark to lie within the Soviet sphere of influence. For discussion of this and subsequent Soviet pressure to keep Denmark out of the Western alliance, see Jensen, B., 'The Soviet Union and Denmark/Scandinavia: perception and policy, pressure and promise', *Nordic Journal of Soviet and East European Studies*, vol. 3, no. 3–4 (1986), pp. 55–59.

127 Kampe (note 118), p. 114. See also *The Military Balance in Northern Europe, 1987–1988* (note 55), p.6: 'The composition of the Soviet Baltic fleet seems to indicate that its primary task is to achieve control of the Baltic and carry out operations to open its exits'.

128 *The Military Balance in Northern Europe, 1987–1988* (note 55), p. 15.

129 This includes, of course, NATO capabilities deployed on the north German plain, which under certain circumstances presumably could afford Denmark some protection.

130 These numbers are drawn from IISS (note 9), p. 42; and *The Military Balance in Northern Europe, 1987–1988* (note 55), pp. 30–35. There are some minor discrepancies between these two sources, but none are consequential.

131 On the amphibious threat, see especially Kampe (note 118). For an alarmed view of this threat, which emphasizes the improvement of these forces and particularly their ability to operate with speed, see Cross, N. H., 'Soviet inshore capabilities: closing the grip on Europe', *Navy International*, Apr. 1989, pp. 153–57. See also Pritchard, C. G., 'Warsaw Pact amphibious forces under Gorbachev', *International Defense Review*, vol. 22, no. 4 (Apr. 1989), pp. 401–404; and Cable, J., 'The freedom of the Baltic', *Navy International*, May 1989, p. 227. Vego (note 10), p. 34, spins out a Soviet invasion scenario in which amphibious operations figure prominently.

132 On Danish perceptions of insecurity, see Heurlin, B., 'Danish security policy', *Cooperation and Conflict*, vol. 17, no. 4 (Dec. 1982), pp. 237–55. Heurlin puts the current situation in the context of a Danish tradition of feeling acutely vulnerable. A substantial minority in Denmark, he claims, generally believe that the country cannot be defended (p. 239).

¹³³ There is not an extensive literature on Danish defence policy. An extremely useful overview is Petersen, N., 'Denmark and NATO, 1949–1987', *Forsvarsstudier*, no. 2/87 (Forsvarshistorisk Forskningscenter: Copenhagen, 1987). See also Heisler, M., 'Denmark's quest for security: constraints and opportunities within the alliance', ed. G. Flynn, *NATO's Northern Allies: The National Security Policies of Belgium, Denmark, the Netherlands and Norway* (Croom Helm: London, 1985), pp. 57–112; Holbraad, C., 'Denmark: half-hearted partner', ed. N. Ørvik, *Semialignment and Western Security* (St. Martin's Press: New York, 1986), pp. 15–60; Thune, C. and Petersen, N., 'Denmark', eds W. Taylor, Jr. and P. Cole, *Nordic Defense: Comparative Decision Making* (Lexington Books: Lexington, Mass., 1985), pp. 1–36. For a sample of the recurrent criticism of Denmark's modest defence effort, see Sundaram, G. S., 'A la carte NATO membership must stop', *International Defense Review*, vol. 21, no. 6 (June 1988), p. 605.

¹³⁴ IISS (note 9), pp. 59–60; *The Military Balance in Northern Europe, 1987–1988* (note 55) p. 14.

¹³⁵ IISS (note 9), pp. 59–60; *The Military Balance in Northern Europe, 1987–1988* (note 55) p. 14. In addition, see von zur Gathen (note 1), which provides the most extensive discussion of the role of the FRG in the Baltic.

¹³⁶ The West German Navy, like many others, is struggling with mounting procurement costs and budget constraints. There is every indication that it will shrink over the next 10–15 years. See, for example, Sauerwein, B., 'Streamlining the German Navy: operational minimum by 2005?', *International Defense Review*, vol. 22, no. 8 (Aug. 1989), pp. 1071–74, which predicts a reduction from 189 to 88 vessels by 2005 if present trends continue. See also 'West German Navy faces major cutback', *Jane's Defence Weekly*, 15 July 1989, p. 74. Of course, it is the Soviet Navy above all that faces this problem because of the large number of obsolescing vessels it possesses, so the shrinking of the German Navy does not necessarily imply a worsening of NATO's relative position in the Baltic.

¹³⁷ See Faurby and Petersen (note 30), p. 16.

¹³⁸ *The Military Balance in Northern Europe, 1987–1988* (note 55), p. 18. However, it should be noted that the UK is considering the reduction or withdrawal of this force. For a discussion of the British deliberations on this issue, see Archer, C., 'The replacement of the United Kingdom's amphibious force', *Naval Forces*, vol. 8, no. 1 (1987). See also Duke (note 8), p. 40.

¹³⁹ On the Danish COB agreement, see Ausland (note 8), p. 138; Bjøl, E., *Nordic Security*, Adelphi Papers no. 181, (International Institute for Strategic Studies: London, 1983), pp. 36–37; Faurby and Petersen (note 30), p. 16; *The Military Balance in Northern Europe, 1987–1988* (note 55), p. 19; and Duke (note 8), pp. 40, 48.

16. Mission gaps in Central Europe

Hilmar Linnenkamp
Führungsakademie der Bundeswehr, Hamburg, Federal Republic of Germany

I. Introduction

The military confrontation between NATO and the Warsaw Treaty Organization (WTO) in Europe cannot adequately be described by comparing numbers of men, divisions, tanks, aircraft, ships, or the like. While such a statement has in fact become an almost trivial proposition, shared by scholars, politicians and military people alike, it is still lacking real significance in the public mind. How could it otherwise be explained that traditional force comparisons, which hardly ever go beyond the simplest form of bean counting, enjoy so much publicity, regardless of whether they are issued by governments, NATO, research institutes or, recently, the WTO.[1] It is not the purpose of this paper to discuss in detail what could be called the 'number surface' of the military game in Europe; it is necessary, however, to mention some of the main faces and facets of the existing confrontational *system* in order to put military missions, whose deficiencies in case of a US withdrawal are to be elaborated, into a wider political perspective.[2]

What are the suggested complexities of the confrontational military system in Europe? There is, first, the intermingling of conventional and nuclear forces, and—by no means less important—of the respective options, operational doctrines and strategies. Every attempt to portray the dynamics of the potential battlefield[3] or to anticipate possible combat actions is bound to suffer from the intentional uncertainty about how the attacker as well as the defender might actually use its arsenal. Second, fundamental to the system is, and will remain, the fact that the opposing parties are alliances, not nations. What is even more complicating, these alliances—while both include a superpower as the dominating element—are not symmetrically structured at all. Third, land, air and sea components of the respective military postures have evolved to their present quantities and qualities in markedly different ways. Political functions and military missions of NATO forces in peacetime, crisis or war can only rarely be considered as mirroring the roles of WTO forces and vice versa. Fourth, geography is an eminent political factor, influencing the perception of all actors on the European scene. It has become commonplace to refer to that factor using 'geostrategical' or 'geopolitical' language (e.g., interior lines versus exterior lines, maritime versus continental alliances, containment versus encirclement, etc.). Although this language is not completely inadequate in expressing the global understanding of competitive great powers, it appears to be necessary to take a closer look at the usually incongruent mental maps of European peoples. These maps represent the multitude of political and military realities which, taken together, constitute the complexity of what above has been called the confrontational military system in Europe.

There is a prismatic concentration of this complexity in Central Europe. It may therefore be valuable to begin the discussion and evaluation of possible consequences of a United States withdrawal right in that area.

II. The structure of the analysis

Assessing the share that US forces hold in contributing to NATO's deterrent in Central Europe requires weighing their impact on the overall posture of the alliance. Moreover, it is equally important to understand how the Soviet Union perceives that contribution. As it is the primary objective of NATO to preserve peace through a balanced combination of inter-alliance co-operation and the maintenance of sufficiently credible deterrent capabilities, priority has to be given to reliable and unequivocal communication to the other side of exactly this political strategy in peacetime and, if circumstances require, in times of crisis. In fact, while it is true that in actual war the military effectiveness of the alliance forces assumes much greater significance for defence and (intra-war) deterrence, political messages conveyed to the enemy would still be critical factors in the process of avoiding cataclysmic developments. That is why in the assessment of missions or mission gaps one should not restrict the analysis to dealing with the military performance of a given force structure in a certain theatre of operations or—even more narrowly defined—battlefield. Instead, one has to address political, strategic and operational issues alike.

Three separate, in substance interconnected questions are therefore asked:

1. What is the general political message that the withdrawal of US forces from Central Europe would send to the Soviet Union?

2. What would be the overall strategic message relevant to peacetime and crisis?

3. How would the Soviet Union perceive the military risk of an attack by its forces in the Central European theatre?

The three questions may at first glance seem straightforward and well defined. Yet, they are not. Any realistic thinking about the suggested US move—the 'withdrawal'—as well as the Soviet move—the 'attack'—requires clarification.

First, it appears reasonable to consider the two withdrawal options posited in this study. The radical approach would be to pull back all US troops currently deployed in Europe, together with the equipment operated by them. This is the 'gone for good' option (option A), here called W^1. In this case there would be no arrangement whatsoever for later reinforcements by US military power (air, sea or land forces). The less radical approach includes keeping the US commitment to reinforce Europe in times of crisis and in war while still withdrawing the largest part of European-based forces to the continental United States (option B). This option is here called W^2.[4]

Of no lesser importance for the results of the analysis is the set of assumptions on the nature and the scale of a supposed Soviet or WTO attack against Central Europe. Strategic analysts in the West are becoming more and more aware of the fact that the assessment of force balances and relative capabilities depends heavily upon the feasible scenarios to be considered.

This implies a close relationship with the process of force planning: just as air force armament planning cannot do without taking into account distinctly different future air defence environments of the enemy, so does any attempt to evaluate current force capabilities require looking at a range of options that might be consistent with the enemy's political goals.[5] Hence, at the operational level the assessment of mission gaps in Central Europe has to follow a basic game-theory approach in that the two withdrawal options W^1 and W^2 are confronted with two attack options of WTO forces, thereby providing a payoff matrix in which the resulting two times two campaigns are evaluated.

One attack option, here called A^1, would see a deliberate surprise assault of combat-ready forward-deployed Soviet divisions, minimizing warning time for NATO forces and capitalizing on the national coherence in doctrine, training, leadership and command and control, albeit at the expense of sustainability and overall force levels.[6] An opposite option, here called A^2, would include all WTO divisions deployed in Eastern Europe for a massive, well-prepared attack on a broad front backed by immediate mobilization of available reserve divisions in the Soviet Union, granting NATO considerable warning time and decisively increasing the role of NATO's crisis management.

It is the purpose of this chapter to relate systematically the three above-mentioned questions to the hypothetical situations described by two withdrawal options and the two attack options. Section III deals with the political respectively strategic messages which the two forms of US withdrawal (to be specified further) would be likely to convey in peacetime or crisis. Section IV is—on the basis of a more detailed concept of attack options—devoted to assessing the military-operational risks that the Soviet Union might perceive it would face in attacking the central front. Finally, section V sums up the results of the analysis and contributes to setting the stage for the overall European scenario as well as suggesting a theoretical framework for a feasible European defence alternative following a US withdrawal.

III. Political choices: gaps in deterrence

US forces in Europe assure allied countries of the permanent and institutionalized support given by the people of the United States to secure not just the abstract ideological scheme of the 'free world', but rather the steady striving to realize the common moral goals of freedom and justice. US troops are the highly visible symbols of the resolve to join the United States in any European conflict and war in which those goals would be at stake. Indeed, an additional qualification to that engagement is the fact that the US military presence in Europe requires US forces to fight in Europe—not in some distant waters or remote areas of secondary conflict, nor in a gigantic space battle over the heads of Europeans ducking below. Hence, the clear political message of the United States has two addressees: the European NATO partners are told that the major post-World War II stabilizing power intends to maintain its basic geopolitical priorities by remaining physically present in Europe. The Soviet Union, on the other hand, is told that it is not allowed to become the dominant *political* power in Europe in spite of the fact that it has been the predominant *military* power— in conventional terms—since World War II.

It is hardly conceivable that a US military withdrawal from Europe would not radically change the European political balance. However, balance is a metaphorical term; there is a need to look at the realities that lie beneath it. If the USA gave up its weight in or influence on the military posture of NATO, whose member it is still assumed to be, West European economic and technological co-operation as well as inter-alliance *détente* could, in fact, make up for the loss of military 'weights', which are but one element of the clearly multidimensional system of power relations. Such hypothetical political compensation of a military disengagement rests on the somewhat fatalistic assumption that the USA—in relation to other powers—is already a declining world power,[7] which has lost its original capability to impose its order and principles on its allies, thereby allowing them to take over its role in shaping Europe's future.[8] It follows logically that the political message a US withdrawal would send to the USSR need not be devastating to the texture of European power relations if—and only if—that message is being sent by a West European community of nations that has grown more and more confident of its own status as a world power.

A word of caution is warranted, however. While it is true that the decline of US power has taken place in *relative* terms and is likely to continue within the global pentarchy system often referred to by historians and politicians,[9] the United States still has the strongest economy in *absolute* terms and is extensively linked to the economies and societies of Western Europe. In trade and international finance as well as in technology and culture, transatlantic communications and exchange are powerful means that tend to protect both parties from isolationism and parochialism, respectively. A US military withdrawal could, in fact, be conducive to, if not already indicative of, the weakening of those essential transatlantic ties. This danger will certainly be greatest in case of the 'gone for good' option (W[1]). Under circumstances of irrevocable withdrawal, the Atlantic Ocean would no longer, in a crisis or war in Europe, be a vital lifeline to preserve US and European interests. In short, the geographical distance would turn into political distance, the conservation of a formal alliance notwithstanding.

The situation appears different for the 'commitment retained' option (W^2). In order to make such an arrangement work, vital parts of specifically US infrastructure (headquarters, training areas, air base logistics, storage sites, communications and intelligence networks) would have to remain in Europe. Moreover, some troops might stay, as it would be necessary to have a permanent avant-garde stationed in areas where reinforcements would be received. A reasonable assumption in that regard would see one brigade-size unit per corps remaining. In other words, on the basis of current deployments and today's reinforcement plans, US V and VII Corps would each leave one brigade or Armoured Cavalry Regiment in the Federal Republic of Germany. These formations—apart from their operational significance, discussed below—would serve the purpose of signaling to the Soviet Union that the United States intends to stick to its obligations to defend peace and freedom in NATO Europe.

Such a commitment radically changes crisis management in Europe. Whereas today the peacetime forward deployment of large land and air forces allows for sensitive, finely tuned and utterly discriminate measures to manage crises, flexibility would decrease considerably in the withdrawal case W^2. Early reactions and decisions would be required if political significance were to be given to military measures. Major visible steps such as calling up reserves in the USA, preparing for large transoceanic shipments or getting Wartime Host Nation Support (WHNS) arrangements under way could unleash mutual reactions early in the game. Decisions toward de-escalation would, under these conditions, become more difficult since the entire machinery would be less flexible. The risks associated with assessment errors or misinterpretation of enemy intentions are bound to be relatively high compared to the situation today, when NATO enjoys the robustness of diversified, sizeable in-place forces.

Today's defence arrangements in Central Europe preclude the possibility of a war being fought in Europe without US involvement. This not only means that US forces currently deployed in Europe would immediately be facing any attacking Eastern forces. US involvement rather confronts the Soviet Union with an extended threat; looking at the world map from a Soviet perspective, the Moscow leadership has to take into account the fact that US military power can almost immediately be brought to bear on the Soviet homeland and on areas of strategic interest from all directions, using a broad spectrum of military means—from naval and naval air power in northern waters or in the Mediterranean to Navy and Air Force assets in East Asia.

It is not even necessary to consider these options in the context of NATO strategy, as if an attack in Europe triggered some pre-planned sequence of intentional escalatory actions by US or Allied forces; the mere existence and availability of US military assets around the northern half of the globe forces the Soviet Union to be prepared for the possibility or even feasibility of US global reaction to any Soviet move into Central Europe.

As much as the Soviet Union needs a military counter-potential to become convinced of the futility of any aggressive actions, it is obvious for its leaders that they cannot rule out the risk of a very large-scale confrontation with the fellow superpower. It is this awareness that is at the heart of the meaning of deterrence; being but a part of the entire military arsenal of the United States, US troops deployed in Europe do not suffer but rather take advantage from that very fact: *deterrence by distribution.*

There is, however, a second complexity with which a potential attacker in Central Europe would have to cope. Under present circumstances it is hardly conceivable that war in Central Europe could remain non-nuclear for long. NATO's force structure today combines conventional and nuclear capabilities, by maintaining stockpiles of US nuclear warheads (under US custody) as well as dual-capable and forward-deployed delivery systems, by keeping nuclear force elements integrated in the overall command structure, and, most importantly, by further developing and refining the strategic doctrine of flexible response.

Historical experience suggests that in times of crisis the availability of nuclear weapons can contribute to caution in conflict management by putting a very high premium indeed on prudent behaviour—active as well as reactive. Recent analysis of the proceedings of the Executive Committee assembled by John F. Kennedy during the Cuban missile crisis emphasizes that general assumption.[10] However, there is also another side to that coin. It is commonly accepted that crisis stability decreases with the vulnerability of high-value assets such as nuclear weapons and nuclear-related installations. Moreover, command and control of nuclear forces appear to require decision-making procedures that do not always contribute to avoiding equivocal messages to the opponent. Preparing nuclear weapons for ready use may well be perceived on the other side as the first step towards an irreversible chain of events leading to nuclear war, thus provoking time pressures to strike first.[11] However, taking into account the unforeseeable consequences—in scale and scope—of using nuclear weapons as well as the chances of accidental and inadvertent use, both sides would most likely be very much inclined to contain any political crisis and to avoid war breaking out at all: *deterrence by uncontrollability.*[12]

Given the present role of US forces in Europe, a complete withdrawal could clearly upset the existing balance to the effect that there would no longer be an immediate, in-place counter-threat distributed over a very large geographical area—from Turkey to the northern waters. A speculation might then arise as to the likelihood of the USA not entering a European war at the outset, a situation vaguely resembling initial World War I and World War II scenarios, when it took the USA two or more years to decide on its involvement. European NATO members would probably act more visibly in unison than they do today; but a redefinition of French and British nuclear policy would be inevitable, thus considerably changing the risks and responsibilities within the alliance.

As a political perspective, it is not so much the final state of military affairs following a US withdrawal—e.g., the lack of an important part of the conventional and nuclear potential—that counts in the deterrence equation in Europe; it would rather be the actual *process* of withdrawal, with all the unavoidable explanations and declarations that go with it—a process that would destroy the familiar web of existing security structures in all of Europe, encompassing not only co-operative intra-alliance ties but also confrontational inter-alliance relations. After all, the remnants of a stabilizing symmetry within the generally asymmetrical alliance structures in East and West would disappear; the fellow superpower to the Soviet Union would no longer be in a parallel military role in Europe, and the Soviet Union would conceive of having to deal with a polycentric and politically less predictable conglomerate of West European states incomparable to its own European alliance structure. This Soviet perception may well be among the reasons why the Soviet Union, in the course of 1989, expressed more clearly than ever before its interest to keep the United States involved in European security affairs. Logically, not even the prospect of German unity and its yet unknown effect on the dual alliance structure have prevented the Soviet Union from suggesting US-Soviet troop parity in the present Federal Republic of Germany and German Democratic Republic.

If the United States retained its commitment to return troops in crisis or war, the provisions and preparations for adapting US and European troops to that commitment would be highly visible, costly and clearly indicative of NATO's common resolve to maintain its forces and dispositions for countering any attempt to seize Western Europe by military force.

While crisis management might become more difficult, as noted above, the deterrence value of a massive 'Reforger' (Return of Forces to Germany) annual exercise regime is not to be underestimated. It goes without saying that the operational consequences could turn out to be very substantial indeed—obviously to NATO's detriment. This is dealt with in the next section. At the same time, the peacetime strategic message would largely consist of the forceful reassertion that the USA will reinforce Europe if necessary and will not shy away from heavy costs and massive management efforts to fulfil its obligation.

IV. Military choices: gaps in forward defence

The US presence in NATO Europe, after culminating in 1955 at about 430 000 men, reached a low in 1973 at 200 000 as a consequence of the Viet Nam War. Subsequently, a steady increase has occurred, until in 1987 the strength of US forces in Europe, including naval personnel, was about 354 000, of which 327 000 are permanently stationed in territories of European NATO nations.[13]

The US presence in the FRG and West Berlin, which is the corner-stone of the defence contribution to the Central Front, has been subject to similar fluctuations: during the Berlin crisis of 1961–62 it rose to 280 000, decreased to 244 000 in 1973, and rose again to a high of 256 000 in 1982. Since then there has been a slight decrease to about 249 000—roughly 208 000 Army and 41 000 Air Force personnel.

Of the total number of 1264 US military bases world-wide, 374 are located overseas. Of these, 80 per cent (224) are in the FRG.[14] This share clearly emphasizes the significance attributed to the defence of the central region by US strategic planners.

Organizationally, US forces in Europe—Army, Air Force and Navy—are commanded by the Commander-in-Chief (CINC), US European Command (USEUCOM), with headquarters in Stuttgart. CINC, USEUCOM is at the same time NATO's Supreme Allied Commander Europe (SACEUR) at the Supreme Headquarters Allied Powers Europe (SHAPE) in Mons, Belgium. Similarly, two other subordinate commanders wear more than one hat: the Commander, US Air Force Europe (USAFE), with some 93 000 airmen in the FRG, Great Britain, Belgium, the Netherlands, Iceland, Spain, Portugal (Azores), Italy, Greece and Turkey, is also Commander, Allied Air Forces Central Europe (AAFCE); the Commander, US Army Europe (USAREUR) is also Chief of the 7th US Army, with headquarters in Heidelberg, and of NATO's Central Army Group (CENTAG), with headquarters at Seckenheim.

The current deployment of US forces designated to the defence of the central region is characterized by a mixture of (*a*) forward-deployed land and air forces, (*b*) a contingent of strike air power stationed in the UK, and (*c*) land and air forces earmarked for the reinforcement of Central Europe (possibly augmented by a US Marine Corps contingent for help in the defence of Jutland). Before assessing the likely gaps a US withdrawal would bring about, it is appropriate to take a closer look at the military potential that the above-mentioned three components would represent quantitatively and qualitatively under current circumstances.

Forward-deployed land and air forces[15]

The US 7th Army with V and VII Corps—flanked by FRG forces, in the north by III Corps and in the south by II Corps—covers roughly half of the Central Army Group (CENTAG) area of responsibility for forward defence of the Central Front. Each US Corps has one Armoured Division, one Mechanized Infantry Division and one brigade-sized Armoured Cavalry Regiment. The VII Corps includes an additional mechanized brigade which serves as forward detachment of a US-based Mechanized Infantry Division.

There is one more USAREUR brigade outside CENTAG: the 3d Armored Brigade, with headquarters in Garlstedt near Bremen—the vanguard of the III Armored Corps from Fort Hood, Texas, which is earmarked to reinforce the Northern Army Group (NORTHAG) and to serve as a major Army Group operational reserve in the defence of northern FRG.

A significant part of the 7th US Army combat power is contributed by additional combat support and support troops. These include: (*a*) the 32d Army

Air Defense Command (AADC, Darmstadt), with Improved Hawk and Patriot surface-to-air missiles (30 and 18 batteries, respectively) that represent roughly 30 per cent of AAFCE's integrated land-based air defence in the central region; (b) the 56th Field Artillery Brigade (Schwäbisch-Gmünd), engaged in 1989 in the process of dismantling the Pershing 2 system to be withdrawn and destroyed under the provisions of the Treaty between the USA and the USSR on the elimination of their intermediate-range and shorter-range missiles (the INF Treaty); (c) the 18th Engineer Brigade (Karlsruhe); (d) the 21st Support Command (Kaiserslautern); and (e) numerous other signal, intelligence, logistic and special purpose commands and installations.

In addition, the strong organic artillery and engineer support of the two US corps should be mentioned. The V Corps commands two artillery brigades with four battalions each, while the VII Corps has three artillery brigades. New technology is playing an ever increasing part here with the deployment of more than 200 Multiple Launch Rocket System (MLRS) vehicles, a system of some growth potential and a major complement to the traditional US fire support scheme based on tube artillery. Another important element of the force structure is a contingent of some 240 attack helicopters—more than the total number of anti-tank helicopters used by the FRG Army.

While numbers still count, if one is to evaluate the likely performance of US troops in war, more attention has to be given to qualitative aspects of the force posture. While a number of technology-related factors—command, control, communications and intelligence (C^3I) improvements, from computer-assisted fire support on the battalion level to the complex fusion of intelligence and surveillance data to imminent breakthroughs in terminal guidance of submunitions—represent typical advantages of the military potential of a high-tech nation, the continuous upgrade of the level of motivation and training of US forces over the last 10–15 years may still be the most significant feature of the US contingent's defensive power in Central Europe.[16]

This assessment is likely to be valid for the US Air Force as well.[17] While the main forward-deployed component of USAFE to serve the needs of air power on the Central Front is the 17th Air Force, headquartered at Sembach, the 3d Air Force, with headquarters at RAF Mildenhall, UK, will very likely support combat in Central Europe using its unique long-range F-111 fighter-bombers and close air support A-10 aircraft, which are rotated regularly to the FRG in order to acquaint the pilots to their likely mission areas as well as to the cooperative needs of the combat troops to be supported in war. The 17th and 3d Air Force also comprise electronic combat (EF-111 and F-4G 'Wild Weasel') and tactical reconnaissance (RF-4, SR-71 and TR-1) aircraft and are also supported by tactical and strategic airlift capabilities (including tanker aircraft) provided by other major commands. The bulk of the tactical fighter arsenal is characterized by recent state-of-the-art aircraft such as the F-15, F-16 and F-111, with less modern F-4s decreasing in numbers. Several modernization programmes for all types of aircraft provide for continued adaptation to the enemy's capabilities.

Moreover, as is the case for the US Army in Europe, USAFE effectiveness has been upgraded systematically by force multipliers such as improved C^3I systems, ammunition modernization, enhanced electronic warfare training, US- and NATO-funded hardening and sheltering programmes for air bases and, not least, agreements on extended air defence of bases—that is, the Roland/Patriot agreements between the USA and the FRG.[18] Doctrinal developments also contribute to a more efficient use of available resources: although the AirLand Battle Doctrine is meant to represent a US—as distinct from NATO—view of the operational art of war,[19] inter-service co-operation in the European theatre will undoubtedly benefit from US Air Force officers having ben educated and trained to take a more complex and integrated view of joint warfare.

It is probably not mere coincidence that in recent years a doctrinal debate has gained ground in the FRG Army. This debate has focused on the limits and benefits of rediscovering operational thinking (*operatives Denken*), which, located conceptually between strategy and tactics, emphasizes the necessary flexibility in corps and, even more, army group operations as well as the synergistic effects of joint (army and air force) operations.[20] The particular deployment pattern of USAFE in the central region can in fact help to put these US–FRG conceptual similarities into effect. With the exception of one fighter squadron with F-15 aircraft stationed in the Netherlands supporting the 2 ATAF, almost all of the assets of the 17th Air Force are located in the CENTAG/4 ATAF region, thereby concentrating US ground and air forces in the southern half of the FRG and giving joint operations of the respective commands an idiosyncratic bilateral momentum.

Not the least important forward-deployed assets of the US force posture in Europe are its nuclear elements. With the exception of the independent British and French arsenals, the USA controls all nuclear warheads in Western Europe. NATO formulated its basic policy regarding nuclear weapons in the European theatre in the 1983 Montebello decision of the Nuclear Planning Group. While predating any formulation of the premises of what later became the INF Treaty, the conceptual principles of this decision are still valid, entailing the modernization of remaining nuclear systems (including dual-capable systems) and, possibly, the future reduction of warheads as determined either by nuclear weapon requirement studies carried out by SACEUR or through successful negotiations on shorter-range missile systems in conjunction with a success of the Conventional Armed Forces in Europe (CFE) Negotiation in Vienna.

US air power based in the United Kingdom

As mentioned above, the 3d Air Force (USAFE) is based in the UK. While it would certainly support not only the Central Front and the southern part of CINCNORTH's area of responsibility—the Baltic Approaches (BALTAP) area—its three tactical fighter wings (two equipped with F-111 and EF-111, one with A-10 close support aircraft) would make a difference in the overall air balance in Europe. Range, payload and all-weather capacity, survivability (owing to the rear deployment) and penetration characteristics give the F-111 force a particular value for SACEUR's flexible tasking of nuclear and conventional missions. TR-1 and SR-71 reconnaissance aircraft add considerably to NATO's capabilities to carry out combined air operations in Central Europe.

US forces earmarked for the reinforcement of Central Europe

Six US divisions currently stationed in the continental USA are assigned to reinforce AFCENT within 10 days. Under this Rapid Reinforcement Program some 90 000 additional personnel would roughly double the combat potential of the US Army in the FRG. The reinforcements would strengthen both V and VII Corps in CENTAG and form the additional III Corps to become NORTHAG's operational reserve.[21]

Apart from the existence of the six divisions (and II Corps support) in the United States, this reinforcement scheme rests on three vital prerequisites:

1. Annual 'Reforger' exercises, which since the late 1960s have brought part of the reinforcements overseas to their likely combat areas for training;
2. The building and maintenance of large depots and warehouses, in which the heavy equipment of those forces is stored in order to reduce airlift/sealift requirements and travel times to a minimum. These POMCUS (prepositioned materiel configured to unit sets) sites are dispersed (four in the Netherlands, two in Belgium and two in the FRG) and operated by the Army Combat Equipment Group Europe (CEGE) as part of the above-mentioned 21st Support Command in Kaiserslautern.[22]
3. The Wartime Host Nation Support (WHNS) Agreement between the United States and FRG signed on 15 April 1982. Under this agreement the FRG has agreed to set up a reserve force of nearly 100 000 men, with an active component of 1000 men, trained and equipped to relieve the US reinforcements of a wide range of logistical, medical, and other service support functions.[23]

AFCENT would probably not be the only receiver of US reinforcements. If circumstances require and world-wide commitments of the US Marine Corps allow, this particularly versatile and flexible service might well come to support the defence of Jutland and the Danish Straits, thus complementing British forces earmarked for that task. Force ratios in the BALTAP region are currently considered so critical—although heavily scenario-dependent, of course—as to give British and US reinforcements a significance greater than their size in numbers of soldiers, armoured personnel carriers or aircraft suggests.

USAFE and NATO as a whole are preparing to receive and support US-based air forces and to integrate them rapidly and effectively into the European force posture. With some 300 combat aircraft permanently stationed in the FRG and roughly 300 more in other West European countries, the USAF Rapid Reinforcement Plan would increase the share USAFE holds in peacetime, by 600–1000 within 10 days. There are even sources that mention over '2000 combat aircraft' as the size of the rapid reaction force—if only after 30 days of mobilization time.[24]

However, this enormous influx of additional combat aircraft in a short period of time requires that great attention be given the task of sheltering, maintaining and operating that force far away from home bases. Owing to the limited receiving capacities of US air bases in Europe, current NATO plans provide for 70 US–European Collocated Operating Bases (COBs) to serve as host installations equipped with the necessary infrastructure and supported by additional reserve personnel. The aim is to complete the programme by the early 1990s.

Taking together the deployment of all US forces that are to contribute to the defence of the Central Front, it is not unfairly imprecise to state that in peacetime, before a crisis might provoke large-scale mobilization and reinforcement, the USA provides around 20 per cent of NATO's conventional defensive deterrent in Central Europe. The Army share is likely to be a little higher than 20 per cent, the Air Force a little lower, taking into account the hypothetical commitment of the 3d Air Force outside the AFCENT/BALTAP region. Given this baseline assessment of the US contribution, a comment of political importance is in order here. The familiar estimate of NATO's force buildup in crisis and war suggests that mobilization and reinforcement would lead to a considerably higher share of US forces committed to the central region.[25] This means that the allied *defence* in Central Europe is as dependent upon the USA as is nuclear *deterrence* and, for that matter, that the European NATO countries seem to pay less attention to supporting an enduring conventional defence than does the USA Here, a basic East–West asymmetry ought not to be overlooked. As Soviet forces in other WTO countries account for well over 50 per cent of the combat-ready forces even in peacetime,[26] the transition from crisis to war does not affect the principle of Soviet preponderance. Thus, both ends of the spectrum of aggression—the standing start A^1 option and the prepared attack A^2 option—would undoubtedly represent campaigns in which Soviet interest and influence were equally dominant. The basic fact of post-World War II Europe has long been and is very likely to remain the military overweight of the USSR within the WTO, whereas the USA has never enjoyed the privilege nor carried the burden of dominating NATO in its European military posture.

This is not true for nuclear weapons, obviously. Since they would certainly go as well in the case of a US troop withdrawal from Europe—and as their political and strategic function is discussed in section III above—it is now appropriate to ask the topical question of the required analysis: To what extent do the military risks of Soviet aggression change, in fact decrease, if the United States withdrew the above described forces? Due regard will have to be paid to the two withdrawal options W^1 and W^2.

Summary of US missions

Four principal missions of allied forces that include US contributions in defense of the central region are to be distinguished: (*a*) holding and regaining ground in order to achieve the essential political goal of securing territorial integrity of the FRG and Denmark; (*b*) executing effective offensive air operations to support forward defence; (*c*) executing superior defensive air operations to deny the enemy success of his (offensive) air operations; and (*d*) maintaining effective combat support, especially reconnaissance, electronic warfare and C^3.

The mission of securing the territorial integrity of the two countries that would be subject to initial Soviet or WTO offensive operations—the FRG and Denmark—places a heavy burden on allied land forces. In the case of a very short-warning surprise attack, given no commitment for withdrawn US forces (the A^1/W^1 case), a significant difference would arise between the LANDJUT (Allied Forces Schleswig-Holstein and Jutland)/NORTHAG) sector and the CENTAG sector.

With no US land forces in CENTAG left and 4th ATAF radically weakened in offensive as well as defensive capabilities to roughly a third of the current potential, the southern half of the FRG would have to be defended by two (instead of four) German corps and a Canadian brigade only. This results in force-to-space ratios far below any realistic threshold that might give some hope for defensive success.[27] Even if French forces count as possible operational reserves, as they do today, their combat potential and the resulting general force ratios in the CENTAG sector, including the air balance, would in all likelihood preclude successful forward defence there.

On the other hand, the situation in the northern sector could possibly be stabilized since US withdrawal (W^1) would only remove one brigade of US in-place land forces. However, major deficiencies would result from the changes of the air balance:

1. The offensive potential of the 3d Air Force and the F-15 squadron stationed in the Netherlands as part of the defensive air assets of 2d ATAF would no longer be available.

2. The weak air and air defence arm in the CENTAG/4th ATAF region gives the WTO forces a good opportunity to deploy comparatively substantial amounts of offensive air against NORTHAG and LANDJUT. Even BALTAP's naval activities could be suppressed more effectively and their contribution to the integrity of Danish and Schleswig-Holstein territory degraded.

3. Unique USAFE capabilities in reconnaissance and electronic warfare would be lost.

Whether the A^1 option would in fact be a rational option in the calculus of the WTO leadership must remain an open question. For this option to become a strategic and, eventually, a political success it would certainly require that France, after the likely rapid defeat of CENTAG forces and the imminent threat to its territory, be prepared to give in and not be threatened to escalate to nuclear war. For it would be French nuclear weapons alone that in the W^1 scenario would remain as a theatre nuclear deterrent. After all, the unquestionable dominance that the Soviet Union would enjoy in Europe does not prevent the opposite nuclear power from breaking the rules and using its arsenal in unexpected ways.[28]

As devastating as the A^1/W^1 option would be to the FRG, still greater concern would be raised by any larger-scale attack of WTO forces in case the United States withdrew for good (the A^2/W^1 option). All the above-mentioned weaknesses and deficiencies would multiply, with an inevitable rapid subjugation of essential parts of Western Europe inevitable. NORTHAG would find itself massively outnumbered as well since CENTAG's weakness allows for directing massive forces to attack in the northern region. A successful defence of Denmark would be very unlikely.

Summing up, it appears plausible that in the W^1 case, NATO would not be able to take advantage of the longer warning time if the WTO chose to embark on a larger-scale attack: the success of a sustained defence becomes *less* likely with more warning.

If the possibility is considered that the USA remains committed to return to Europe in a crisis or war, however, a strategically and politically different perspective is offered. In this case, the success of a sustained defence would become *more* likely with increased warning. Leaving aside the fundamental question of the rationality of any war fought in Europe, it would probably be the WTO's most rational option militarily in the W^2 scenario to try to pre-empt in Central Europe. Soviet forces achieving strategically significant goals such as seizing Bremerhaven (or even Rotterdam), Frankfurt and Stuttgart in a blitzkrieg-like operation would give US-reinforced NATO counter-offensives very little chance of success. Thus, it is a fair assumption that the A^1/W^2 scenario would be the least stable situation from a NATO perspective, since the attacker would be forced to press on as rapidly as possible and to take more risks compared to the more predictable A^1/W^1 situation, where no US commitment was imminent. Much would depend, however, on whether the USA decided to keep at least some residual, combat ready and survivable nuclear potential with the forward-deployed contingents typical to the W^2 option.

If, on the other hand, the WTO opted in favour of a larger-scale mobilization and attack (A^2/W^2), which would give NATO several days or weeks of warning and allow for the preparation of receiving installations (first and foremost airfields, POMCUS sites and WHNS), then the attackers' calculations would become less straightforward and NATO's expectations less gloomy. The success of a sustained forward defence could no longer be excluded. Of course, this hope clearly rests on the assumption that crisis management in NATO is perceived by the opposite leadership as being timely, resolute and utterly effective—insensitive, moreover, to efforts of intimidation or, for that matter, deceptive signals of last-minute appeasement.

Operational risks for NATO remain extremely high, nevertheless. Reinforcements would need days of adaptation and re-integration into the NATO command structure. Early WTO breakthroughs would be hard to avoid. But the visible and massive commitment of US-based forces is likely to change what initially would be a very unfavourable balance of forces to the better.

This commitment can indeed be assumed to reflect and symbolize a US foreign policy that relies on world-wide flexibility and discriminate interventionism. It might be this basic foreign policy approach more than anything else that would convince WTO leaders of the absolute priority, in time and effort, of achieving a quick and complete military victory over NATO forces if war seemed unavoidable. Following that logic, the WTO would have the greatest advantage over NATO's defensive options if they were to execute a surprising, deep-reaching attack operation into Western Europe, designed to deny NATO the orderly, pre-planned integration of overseas reinforcements into its defensive posture. Therefore, early transition from an intended A^2/W^2 into a modified A^1/W^2 situation is likely to occur, with a surprise attack being backed from the outset by a larger mobilization force and the addition of Polish, Czechoslovakian and Hungarian forces, which would not necessarily be committed in the theatre battle.

The somewhat paradoxical conclusion of the above analysis suggests, therefore, that in the calculus of military rationality, a WTO attack against the weaker NATO posture (W^1) would require the larger, better prepared and more sustainable potential on the part of the aggressor, whereas in an attack in the W^2 case the WTO would be forced to pre-empt by surprise and momentum, employing a smaller force of combat-ready, forward-deployed, mostly Soviet divisions—on balance: more weight against the weaker West European force, fewer attackers against the potentially stronger defending force.

US force multipliers

There is one more question that must be addressed with regard to the military consequences of a US withdrawal. Do US forces possess critical capabilities that work as force multipliers or elements making the whole of NATO's potential more than the sum of its parts?

Command, Control and Communications

At corps level and below, C^3 is a national responsibility. Army Group headquarters and higher level C^3 are planned, funded and operated by NATO as a whole, not by individual member nations. A US withdrawal would thus have almost no effect on the functioning of the remaining forces as far as communications are concerned. To be sure, minor rearrangements would be necessary, for example, compensation for the withdrawal of a US signal battalion that supports CENTAG headquarters, and additional European efforts to replace the contribution of the US European Defense Network to NATO's Terrestrial

Transmission System, which is to complement the satellite communication facilities. However, the significance of US C^3 assets can be considered uncritical.

Reconnaissance

The above judgement on C^3 is not altogether justified with regard to tactical or strategic reconnaissance. While the contribution of US satellite reconnaissance in peacetime, crisis and war remains and would remain in the relative darkness of intelligence channels, the current lack of any such European capabilities suggests that the United States—particularly in the W^1 case—would leave the European forces relatively blind and at least partially deaf, even if some substrategic electronic intelligence (ground- and air based) were to remain available to European decision-makers. In tactical reconnaissance, the very high-flying US TR-1 and SR-71 aircraft would be painfully missed, not only in the W^1 case but also in the W^2 case, as reconnaissance would hold special importance in a crisis and early phases of a war, before reinforcement plans had even been set in motion. Moreover, it is necessary to add that the synergistic function of reconnaissance can hardly be overestimated: knowledge of where and how the enemy chooses to place his centre(s) of gravity and to allocate follow-on-forces is a prerequisite for effectively planning the reactive defensive battle; this is particularly important for a generally outnumbered defender, who is forced to avoid strategic and operational surprise to the greatest possible extent. Even a partial loss of US reconnaissance capabilities or a late arrival of these forces would certainly critically reduce the remaining defensive power.

Electronic warfare

Although European forces such as those of the FRG, UK, France and Italy dispose of airborne electronic such as support and counter-measure capabilities, US forces in Europe still have the most sophisticated arsenal for electronic warfare. Of special significance are the EF-111 Raven (stationed in the UK) and the F-4G Wild Weasel (stationed in the FRG) squadrons, which add to the penetration capability and survivability of attacking aircraft. Their withdrawal would hit the integrated air forces in AAFCE hard.

High technology

The technology of electronic warfare is but one example out of a wide array of front-end high-tech areas in which the United States is the leading Western or world power. Radar, terminal guidance, thermal imaging technologies and, not the least, a clear lead in computer hardware and software—both as to their technical standard, affordability and military applicability—are significant elements of US military power in Europe. This technological advantage has been a driving factor in the establishment of a European armaments industry, which while slowly catching up in recent years, is not yet able to match, across the board, the capabilities of the transatlantic partner. Consequently, without the shining shield of US forces, the deterrent effect of the European armies, navies and air forces would probably be perceived by the Soviet Union as being less dynamic and less challenging.

In some not unimportant areas, therefore, the Soviet Union would correctly feel relieved of some of the concerns it is facing as long as the United States remains what it is today: a highly competitive military power with an explicit priority given to maintaining a strong presence in Europe.

V. Europe without US troops: the zero-based assessment

Europe is not yet a world power. It is divided politically, rich in rivalries, proud of traditions and longing for an understanding of unity. The economic superstructure of 1992 and beyond is not, in the short term, decisively going to change the distribution of political power.

The USSR is a European power. Its military potential will continue to weigh on the continent to a degree that is mainly defined by its capability to maintain a military presence in the Czechoslovakia, GDR, Hungary and Poland. This very presence objectively provides the USSR with military options that require balancing. Since World War II, that balance has been guaranteed by a US force presence in Western Europe. Underneath this power structure the European post-war order has been established and has developed from confrontation to elements of mutual acceptance and, finally, of limited co-operation, to the unevenly distributed benefit of both sides.

A US withdrawal would radically change that power structure. However, whether it would also change the present political order in Europe is a question that must remain unanswered within the framework of this study. What can be predicted, however, is the change of stability of the still existing European confrontational system: if a major political crisis were to occur and the employment of military instruments could not be excluded, the absence of US forces would give the USSR an unmatched advantage owing to the simple fact that it would be the dominant military power on the old continent. The USSR will clearly remain dominant militarily even after the reductions of Soviet troops announced by General Secretary Michail Gorbachev in his UN speech on 7 December 1988. To be sure, the announcement itself has 'altered the *political* landscape upon which the United States and its Allies must formulate NATO military posture and arms control proposals in the months and years ahead'.[29] If and when executed as declared, the unilateral steps hold 'the potential of a meaningful reduction in the Warsaw Pact's short warning threat'.[30] However, as long as the analysis is based on present military realities, the basic assessment of mission gaps in Central Europe does not have to be modified.

There is one more reason for the USA not to disturb the existing balance in Europe. The USA wants to remain a world power and to exert influence on many regions in order to secure or promote its interests. Keeping Europe free of turbulence and unrest will always be a prerequisite for USA to advance global political ambitions.

On strategic accounts a US withdrawal from Europe would probably have an effect that would raise the most serious concerns—for Americans and Europeans alike—in the nuclear area. A regionalization of nuclear deterrence would in fact almost be unavoidable. It goes without saying that this effect would not be at all consistent with the familiar NATO strategy. Moreover, existing conventional imbalances would no longer be checked by non-inferior, that is, not fundamentally inefficient, nuclear deterrents.

In the narrow sense of the mission analysis of US forces in Europe, it has been argued that the principal advantage the Soviet Union could draw of a US withdrawal would be the opportunity to choose optimal attack options—thus maximizing calculability or minimizing unforeseeable risks, including nuclear.

In sum, under *present*—and for the sake of hypothetical zero-based assessment *frozen*—circumstances, a US withdrawal from Europe would on balance have detrimental consequences for NATO, involving strategic, political and military risks.

VI. Postscript

The currency of power in Europe has long been the military—or so it seemed. Political relations were expressed accordingly: the cold war had dominated most of the last four decades. Now, after the East Europeans in the autumn of 1989 took to the streets, won freedom and began to build up democratic institutions, the concept of a possible WTO aggression is becoming absurd. The military and political mission of US—and all NATO—troops is changing. The single most important strategic decision of the Soviet Union has been to build down its troop presence in the Eastern part of Europe to a level that is unexpectedly low and thus cutting deep into traditional offensive capabilities. Military attack options that require balancing by the other superpower may disappear altogether. What is not likely to disappear, however, is the delicate balance of influence that both the Soviet Union and the United States want to keep over the future development of Central Europe, particularly the present two Germanies. In this context, a US withdrawal of the kind considered in this chapter would probably be contingent on negotiations in the framework of the Conference on Security and Co-operation in Europe (CSCE). These negotiations would have to be conducive to establishing in all of Europe a new security order that was designed to devalue the military currency in European relations. To assess the remaining strategic, political and military risks involved in such a future requires a radically different analysis. While transcending the scope of this chapter such analysis will soon be necessary.

Notes and references

[1] *Conventional forces in Europe: the facts* (NATO: Brussels, 25 Nov. 1988); 'Force comparison 1987—NATO and the Warsaw Pact' (Press and Information Office, West German Ministry of Defence: Bonn, May 1988); IISS, *The Military Balance* (International Institute for Strategic Studies: London, successive years); the comparison of WTO forces published in *Pravda* on 30 Jan. 1989. On the other hand, even the 1988 issue of the DOD annual *Soviet Military Power* (US Government Printing Office: Washington, DC, 1988) contains comparisons of army and air force potential that refer to complex calculations of a mix of quantitative and qualitative factors determining the balance, pp. 110–12. Another example is the report prepared by Senator Carl Levin (D–Mich.) entitled *Beyond the Bean Count: Realistically Assessing the Conventional Military Balance in Europe*, Senate Armed Services Subcommittee on Conventional Forces and Alliances, US Senate, 99th Congress (US Government Printing Office: Washington, DC, 20 Jan. 1988). The most thought-provoking debate on the European balance has recently taken place in a number of issues of *International Security* (see in particular 'The European conventional balance: policy focus', vol. 12, no. 4 (spring 1988), pp. 152–202).

[2] The opportunity of a late (Feb. 1990) review of the chapter, which had originally been finished in early 1989, requires some remarks on the notion of the 'confrontational system' used as a general political assumption in the text. While the revolutionary changes in the Eastern part of Europe have drastically reduced the probability of an armed conflict between NATO and the WTO, they have not yet made the alliances obsolete. The envisaged Conference on Security and Co-operation in Europe (CSCE) summit meeting in late 1990 may well produce a new scheme for a collective security system in Europe; at the same time, however, the Conventional Armed Forces in Europe (CFE) Negotiation in Vienna is not likely to go further than to rearrange the *military* balance in Central Europe massively in favour of NATO while aiming at parity in the Atlantic-to-the-Urals (ATTU) region with respect to tanks, infantry fighting vehicles, artillery aircraft and helicopter. In fact parity can still be considered a solution of the numbers game in the remaining confrontational context.

[3] An interesting approach to this problem can be found in Biddle, S. D., 'The European conventional balance: a reinterpretation of the debate', *Survival*, vol. 30, no. 2 (Mar./Apr. 1988), pp. 99–121.

[4] The Feb. 1990 US–Soviet agreement on Central European troop ceilings of 195 000 indicates the strength of the continuing commitment to Europe.

[5] See Cohen, E. A., 'Toward better net assessment: rethinking the European conventional balance', *International Security*, summer 1988, pp. 50–89, esp. p. 84 ff.

[6] SACEUR does not exclude this option: 'I can't predict whether it could be a short warning surprise attack using simply the Soviet forces that are in place'. SHAPE Public Information Office, *ACE Output*, vol. 6, no. 7 (July 1988), p. 2. See also Dean, J., 'Military security in Europe', *Foreign Affairs*, vol. 66, no. 1 (fall 1987), p. 28: 'The contingency which continues to cause the greatest concern to NATO leaders is that the Soviets, in the hope of securing their objectives before the United States could become fully involved in conventional battle or decide to use nuclear weapons, would launch a minimum preparation attack against NATO's Central Front in Germany'. However, Secretary General Gorbachev's withdrawal announcement before the United Nations on 7 Dec. 1988 may reduce the feasibility and probability of that contingency. Meanwhile, the Soviet Union is going much further. As of 1990, Soviet withdrawals from Czechoslovakia, Hungary and probably Poland—parallel to multilaterally negotiated reductions of Soviet troops in the DGR—will reduce attack options significantly.

[7] This assumption has recently been given much attention; otherwise the spectacular success of Paul Kennedy's *The Rise and Fall of the Great Powers* (Random House: New York, 1987) could not be explained. Actually, a brief and concise analysis of the world power status of the United States had been given a few years earlier by Dieter Senghaas, in *Die Zukunft Europas* (Suhrkamp: Frankfurt, 1986), pp. 23 ff. For a different view of US power, see Nye, J., *Bound to Lead: The Changing Nature of American Power* (Basic Books: New York, 1990).

[8] For a sceptical view on the political turbulences facing Europe if the USA withdrew, see Joffe, J., 'Europe's American pacifier', *Foreign Policy*, no. 54 (spring 1984), pp. 64–82. See also Freedman, L., 'Managing alliances', *Foreign Policy*, no. 71 (summer 1988), pp. 65–85.

[9] It is interesting to note that authors as different as Carl Friedrich von Weizsäcker, in *Wege in der Gefahr* (Hanser: Munich/Vienna 1976), and Fred C. Iklé and Albert Wohlstetter, editors of *Discriminate Deterrence: the Report of the Commission on Integrated Long Term Strategy* (US Government Printing Office: Washington, DC, Jan. 1988), refer to similar concepts of the global power distribution.

[10] See Blight, J. G., Nye, J. S. Jr and Welch, D. A., 'The Cuban missile crisis revisited', *Foreign Affairs*, vol. 66, no. 1 (fall 1987), pp. 170–88.

[11] See Bracken, P., *The Command and Control of Nuclear Forces* (Yale University Press: New Haven/London, 1983).

[12] Recently, Paul Bracken has elaborated on the 'more bizarre' role of nuclear weapons today, as compared to the earlier decades of the nuclear age. See Bracken, P., 'Do we really want to eliminate the chance of accidental war?', *Defense Analysis*, vol. 4, no. 1, pp. 61–63.

[13] IISS, *The Military Balance 1987–1988* (International Institute for Strategic Studies: London, 1987) gives recent strengths of US forces in the FRG and elsewhere in Europe. Historical figures are from DOD sources referred to in Baldauf, J., 'Zur amerikanischen Kritik an der US-Truppenpräsenz in Europa', working paper no. SWP-AP 2529 (Stiftung Wissenschaft und Politik: Ebenhausen, 1987), pp. 16 ff.

[14] See the *Report of the Defense Burdensharing Panel of the Committee on Armed Services*, US House of Representatives, 100th Congress (US Government Printing Office: Washington, DC, 1988), p. 52.

[15] A major source for the following information is IISS (note 13). See also 'United States Air Forces in Europe', *Air Force Magazine*, May 1987, pp. 133–36; 'Die 7. amerikanische Armee und die 17. Air Force: Gliederung, Dislozierung, Ausrüstung', *Österreichische Militärische Zeitschrift*, no. 6 (1985), pp. 567–80.

[16] See the interview with General Otis in *Armed Forces Journal*, Jan. 1987.

[17] *Air Force Magazine* (note 15), p. 133.

[18] See FRG Ministry of Defence, *Mitteilungen an die Presse*, no. I/47, 12 July 1984.

[19] Otis (note 16).

[20] An early source of the discussion is Luttwak, E. N., 'The operational level of war', *International Security*, vol. 5, no. 3 (winter 1980/81), pp. 61–79. In the FRG, General von Sandrart, former Chief of the Army Staff and current CINCENT, has been among the proponents of *operatives Denken*.

[21] It was in the fall of 1987 that the US-based forces of III Corps for the first time took part in a large field exercise in northern FRG.

[22] See Manes, S., 'POMCUS stationing in the Netherlands', *Army Logistician*, Jan./Feb. 1987, pp. 18–21.

[23] For a good assessment of WHNS see Kaldrack, G., 'Wartime Host Nation Support—Zusammenarbeit auf der Basis des Vertrauens', *Europäische Wehrkunde/WWR*, no.1, 1986, pp. 21–25.

[24] Welch, L. D., 'NATO Rapid Reinforcement—a key to conventional deterrence', *NATO's Sixteen Nations*, Feb./Mar. 1987, pp. 114–16.

[25] This and other important data on the Central Front balance is well documented in Mako, W. P., *US Ground Forces and the Defense of Central Europe* (Brookings Institution: Washington, DC, 1983). Although the figures are dated, the analysis presented in this study is still valid.

[26] This is a fact that is not going to change if and when Soviet forces are reduced as promised by General Secretary Gorbachev on 7 Dec. 1988. It will no longer be true, however, as soon as Soviet troops Czechoslovakia, Hungary and possibly Poland (see note 6).

[27] This assessment will probably remain valid even if the Soviet Union were to withdraw (and dismantle) four tank divisions from the Group of Soviet Forces in Germany (GSFG), as announced in the Gorbachev speech.

[28] Bracken (note 12).

[29] Nunn S., 'Gorbachev's reductions present NATO with challenges' (US Information Service: Bonn, Dec. 19, 1988), p. 11; emphasis added.

[30] Nunn (note 29), p. 13.

17. The military implications for NATO of a US withdrawal from the southern region

George E. Thibault
Booz Allen & Hamilton, Washington, DC

I. Purpose and approach

The purpose of this chapter is to examine the military mission gaps, excluding the loss of C^3I assets and the US 6th Fleet that would be created by a US withdrawal from NATO's Southern Flank. The chapter assumes that all US personnel and facilities are withdrawn, that the NATO Treaty remains intact and that the Supreme Allied Commander Europe (SACEUR) is a European.

The military mission areas identified here are the official missions of the Commander-in-Chief, Allied Forces Southern Europe (CINCSOUTH). These are: (*a*) to defend NATO territory; (*b*) to maintain the Mediterranean sea lines of communications (SLOC); and (*c*) to attain strategic leverage—defined as the ability of Allied Forces, Southern Europe (AFSOUTH) to influence the battle in the central region.[1]

The military mission gaps which are discussed are divided into two categories. First, there are the traditional mission areas and the first-order military support functions that affect the missions directly: (*a*) operational environment; (*b*) credible military presence in the region and peacetime tasks; (*c*) operational and logistical support; (*d*) reception of external reinforcements; and (*e*) transit to out-of-area contingencies. Second, there is a set of no less important contributions of the US military which cut across mission areas. A US pull-out would leave significant gaps that would affect NATO military performance or capability in the region. These contributions include: (*a*) regional leadership; (*b*) influence on national focus and will; (*c*) the nuclear option; and (*d*) US weapon modernization.

Rather than addressing the military gaps which would be left by departing US forces as if each would have to be covered by remaining NATO forces, the existence and location of US facilities or troops in specific parts of the southern region need to be placed in perspective. Strategic good sense and tactical necessity are, of course, the primary reasons for the placement of military forces. However, they are not the only reasons US forces are located where they are today throughout NATO's southern region. To assume that would be to assume that political, economic and social factors did not also influence military force decisions over the more than four decades of NATO's existence. Many times transitory political needs, the anticipated effects of an economic shot-in-the-arm, the degree of xenophobia or the ease by which the introduction of a base or facility could be accomplished, rather than the best military judgement, have dictated where a base or weapon stockpile or communications facility would be located.

Furthermore, the relative weight of those considerations reflected the situation at the time the decision was made. The same decision made today might reflect other imperatives and be resolved quite differently. Consequently, the effect of the departure of US forces on military mission capability in the southern region cannot be generalized. It must be assessed theatre by theatre. In one theatre it might make little difference; in another, the effects could be profound. It is important then to look at the southern region through the eyes of the military commander.

II. Commander's strategic estimate[2]

Unlike the central and northern regions, which are unified strategically and will be defended by integrated NATO forces, the Southern Flank comprises five geographically separated theatres, whose NATO and national leaders all see the threat differently and who plan to defend them individually with national forces alone or in combination with US forces.

Writers divide the region differently. As military entities, however, the region has four land theatres: northern Italy, northern Greece/Greek Thrace, western Turkey/Turkish Thrace and eastern Turkey; and a fifth, entirely maritime theatre—the full sweep of the Mediterranean, Adriatic, Aegean and Black seas. In addition there is an air component which some consider a separate entity because of the considerable regional challenge of attaining air superiority. The five theatres vary considerably in history, view of the region, available resources, size, political orientation, strength of military organization and in terms of what would be required to stop an attack from the WTO. Excluding the Commander, US 6th Fleet in the Mediterranean Sea theatre, the senior military commanders in each theatre see theirs as geographically and strategically separate from neighbouring theatres, and expect each to be fought as a separate, largely autonomous war zone.

The assumption that it is unlikely that the Soviet Union alone or in conjunction with all or parts of the WTO would launch a simultaneous attack against the entire northern rim of the region is probably sound. The length of the front line and the difficulty of maintaining a cohesive front would make an offensive very difficult to manage. The geographic diversity of the five theatres makes it difficult for an aggressor to focus its forces meaningfully or to coordinate effectively the timing of a co-ordinated attack. The ability of the Soviet Union to gain the co-operation or acquiescence of Albania and Bulgaria, to say nothing of Yugoslavia, in a unified offensive against the whole Southern Flank short of repelling a NATO-initiated offensive from the south is probably very low. Finally, the lack of a single strategic objective which would require taking on the entire region simultaneously makes that task unnecessary.

Instead, one or possibly two theatres (in the case of Greek and Turkish Thrace or Turkish Thrace and eastern Turkey) would probably be selected for specific objectives, such as drawing or distracting allied forces from the Central Front or testing NATO's unity. WTO commanders would apply to those theatres only the forces necessary to accomplish their limited objectives. The attempt to separate this region from the rest of NATO before or early in a war is often cited as a reasonable Soviet objective. It does seem to offer an easy target, given the perception in the region of being separated, if not isolated, from the rest of NATO and of being considered by others in the alliance as less relevant to NATO's or the United States' immediate concerns.

The two exceptions to this intra-regional separation of theatres within the Southern Flank are the two Turkish theatres and the Mediterranean Sea theatre. An attack against either or both Turkish theatres would probably be countered by the full weight of Turkish capabilities, as an incursion on either front would represent a serious threat to Turkey as a whole. An attack on one or both Turkish theatres could be met with a transfer of forces between Turkish Thrace and eastern Turkey.[3]

The results of a threat to NATO's free access to and use of the Mediterranean would have to be taken seriously by all regional members of NATO. The Mediterranean represents one of the two 'glues' keeping the region together. Every nation in the region depends on it for resupply, reinforcement and the transfer of its power projection capability within as well as from outside the region. The other regional 'glue' is the US strategic capability discussed in section IV below.

Looking across the southern region from west to east, the major installations which would be most affected by a withdrawal of US forces are listed below under their respective theatres.

Portugal and Spain: supporting theatres

Because of their remoteness from both Central Europe and the WTO, it is not likely that either Portugal or Spain would be in danger of early or direct invasion in any current scenario. None the less, the United States uses installations in both countries that would be vital to the Southern Flank—if not to all of NATO—in wartime. Their loss would severely handicap any military efforts in the region.

While Portugal hosts no US or NATO installations on the mainland (aside from some minor port and stand-by command facilities in Lisbon), the United States operates a number of important facilities on the Azores, including the Lajes Air Force Base (see figure 17.1).

In Spain the United States operates four major bases: the naval base at Rota, and the air bases at Torrejón, Morón and Zaragoza (see figure 17.2). Spain is not a member of the military wing of NATO, and the bases, which are used

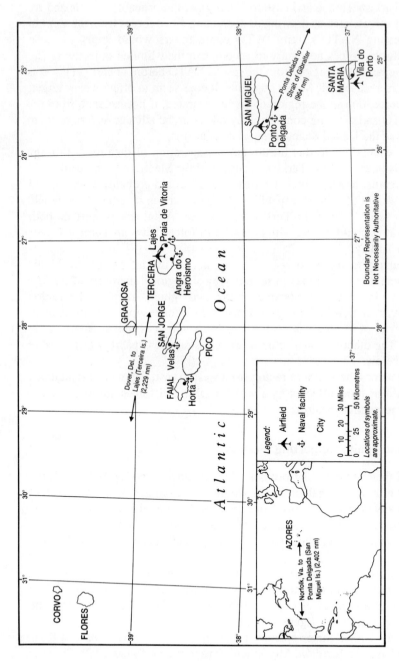

Figure 17.1. Major US military installations on the Azores

Source: Congressional Research Service, *US Military Installations in NATO's Southern Region*, Report prepared for the Subcommittee on Europe and the Middle East of the Committee on Foreign Affairs, House of Representatives, 99th Congress, 2nd Session (US Government Printing Office: Washington, DC, 7 Oct. 1986), p. 9.

jointly by the US and Spanish military, are only available to the USA as a result of bilateral agreements dating from 1953. A new agreement was signed in September 1988. Although the USA has continued access to all four bases, it is complying with the Spanish desire to withdraw its wing of 72 F-16 aircraft from Torrejón.[4]

The Azores

Anti-submarine warfare (ASW) and logistics staging and support are the two main military missions accomplished from installations in the Azores. Secondarily, the US Defense Communications System (DCS) is supported by installations in the Azores.[5]

The ASW mission is part of a broader effort to monitor Soviet submarine activities in the eastern Atlantic and western Mediterranean: 'the Azores are strategically located for the support of such a mission. From facilities in the island chain it is possible to track Soviet submarines located within a 1,000-mile radius . . . [and] to keep watch over the midpoint of the 4,000-mile sea lane that links the U.S. Sixth Fleet in the Mediterranean with its major supply depots in the American east coast'.[6] Lajes Air Base supports this effort and is a principal origination point for specific anti-submarine searches in the vicinity of the Straits of Gibraltar.[7]

Lajes Air Base is also a major high-capacity refuelling base for US military transport aircraft (especially C-5A and C-141) flying between the United States and Mediterranean and other European ports. The island of Terceira is the location of major ammunition and fuel storage as well as communications facilities used by Lajes.

Rota

The naval base at Rota is the most important US base in Spain. Rota provides the 6th Fleet with a major air base supporting ASW operations and ocean surveillance over the western approaches to the straits, the western end of the Mediterranean and significant parts of the eastern Atlantic. The naval base at Rota can provide anchorage, dockage and repair for large deep-draft warships as well as submarines close to the entrance to the Straits of Gibraltar, and is an important storage site for petrol, oil, lubricants (POL) and ammunition for the fleet surface and air assets. Rota's DCS terminal is a major link with the 6th Fleet, other parts of Spain, Greece and Morocco.

Torrejón

Torrejón Air Base remains the headquarters of the 16th Tactical Air force of US Air Forces, Europe (USAFE). It is a major staging, reinforcement and logistics airlift base and communications centre for US forces throughout the southern region. Aircraft from Torrejón rotate between Torrejón and Aviano Air Base in

374 EUROPE AFTER AN AMERICAN WITHDRAWAL

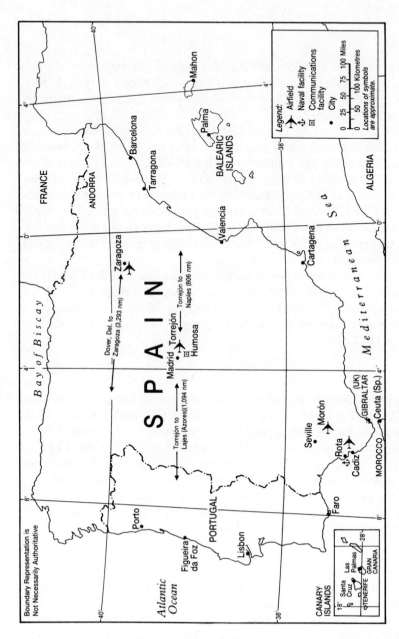

Figure 17.2. Major US military installations in Spain

Source: Congressional Research Service, *US Military Installations in NATO's Southern Region*, Report prepared for the Subcommittee on Europe and the Middle East of the Committee on Foreign Affairs, House of Representatives, 99th Congress, 2nd Session (US Government Printing Office: Washington, DC, 7 Oct. 1986), p. 20.

Italy and Incirlik Air Base in Turkey. The 76 F-16 fighter aircraft now based at Torrejón are to be transferred to Crotone, Italy.[8]

Zaragoza and Morón

Zaragoza Air Base is used for tactical fighter training and as a home base for five aerial refuelling aircraft. Morón Air Base provides support for USAFE, is home base for 15 aerial refuelling aircraft and hosts a communications facility that provides a link with Rota, Greece and Morocco.

Northern Italy

Italy is not contiguous to any Warsaw Pact country, but is vulnerable to ground assault by the Warsaw Pact in the northeast, through Austria or Yugoslavia. An attack through Austria would have to be mounted from the Soviet's Western TVD [Theatre of Military Operations] with forces already in or staged through Czechoslovakia. The mountainous terrain of Austria and northern Italy would make this a difficult operation, though Austrian forces could not offer substantial resistance. The Italians would have good defensive positions in and south of the Brenner pass.

The question is whether, assuming simultaneous war on the western front, the Soviets would be willing to free enough forces from the Western TVD to open an Italian front. There would not seem to be any pressing need for them to do that. That is, northern Italy does not pose a substantial threat to their Western TVD operations; and the prize for the Soviets of conquering Italy does not seem so great as to warrant early action at the risk of weakening the Western TVD's thrust towards West Germany.[9]

In other words, it is assumed that Italy would not stand in immediate threat of general assault in the early stages of a conflict. None the less, the nature of US facilities in Italy and the level of dependence of the rest of the region on those facilities could make them in themselves the targets of concentrated attack.

Italy has been one of the United States' staunchest allies for many years—trusting, accepting and supportive. Over those years, Italy has agreed to permit the United States to use significant numbers of Italian facilities, to base forces on its soil, and to assemble large amounts of war stores. Figure 17.3 shows the location of the main US installations in Italy.

Naples

Naples is headquarters for the Commander-in-Chief, Allied Forces Southern Europe (CINCSOUTH), for the US 6th Fleet (actually less than an hour's drive up the coast at Gaeta), for the Commander, Task Force 67 (CTF67) or Fleet Air Mediterranean and for the Commander, Task Force 69 (CTF69), who directs nuclear-powered attack submarines (SSNs) in the Mediterranean. Naples is also the home port of a US destroyer tender and the site of a major DCS terminal.

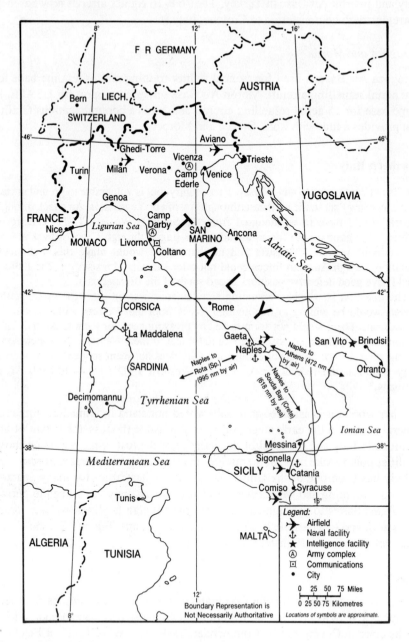

Figure 17.3. Major US military installations in Italy

Source: Congressional Research Service, *US Military Installations in NATO's Southern Region*, Report prepared for the Subcommittee on Europe and the Middle East of the Committee on Foreign Affairs, House of Representatives, 99th Congress, 2nd Session (US Government Printing Office: Washington, DC, 7 Oct. 1986), p. 28.

US WITHDRAWAL FROM THE SOUTHERN REGION   377

Sicily

Aerial ASW operations are staged out of the naval air facility in Sigonella. The air station also serves as home station for VR-24, the carrier on-board delivery (COD) squadron that provides for the air delivery of spare parts, special maintenance personnel and logistic support for the US Navy at sea in the Mediterranean. A POL and ammunition storage facility is located at Augusta Bay, Syracuse.

Sardinia

La Maddalena, on the island of La Maddalena off Sardinia, is home port for a tender servicing US attack submarines operating in the Mediterranean. The Italian air base at Decimomannu in the south provides an instrumentation and training range for air-to-air combat for US and other NATO air forces.

Aviano

The Aviano Air Base on the Veneto–Friuli plain is used as a home base by the US rotational tactical fighter group from Zaragoza. In wartime, strike missions would originate here.

Camp Ederle and Camp Darby

Camp Ederle, near Vicenza, is the home base for the 5th Allied Tactical Air Force and the 4th Airborne Battalion of the 325th Infantry, part of the Allied Command Europe (ACE) Mobile Force. Camp Darby, near Leghorn, is the location of the 8th Logistics Command, a major command headquarters and storage facility for the US Army Europe (USAREUR). The Southern European Task Force (SETAF), parts of which are located at both Darby and Ederle, is a skeletal structure which would provide support to Italian ground forces in wartime and serve as a logistics base for all operations in the region. An intricate network of agreements has been set up that would govern the operations of SETAF in wartime.

Greek Thrace

Like Turkish Thrace, Greek Thrace is vulnerable along the length of its northern border with Bulgaria. The contiguous nature of Greek and Turkish Thrace makes it likely that an attack might not differentiate between the two, especially if the Soviet Union saw advantages in including sea-borne forces from the Black Sea. There is no strategic depth to the theatre for defence, and reinforcements that do not come via the Aegean would have to funnel into the area from the East and West Flanks, making them highly vulnerable to flank attacks before reaching positions where they might be useful. In addition, Greek forces, positioned as they are to counter an attack from Turkey, are poorly placed to meet an attack from the north.

Major Greek–US air bases are distant enough from Thrace to make air support of the land battle less than a simple undertaking (Iraklion on Crete is over 600 km away, and Athens is 400 km away; see figure 17.4). A substantial number of aircraft is therefore required to ensure adequate density over targets.

Politically, the USSR might feel it wise to isolate its efforts against either Greek or Turkish Thrace so as not to provide the catalyst which could bring the two countries together to face a common threat. On the other hand, in a unified attack against all of Thrace, the USSR might bet on Greece and Turkey waiting each other out to see which was getting the worst of the battle, and then holding back to let the USSR do to the other what neither would dare. However, whatever the Soviet strategy, Thrace and its access to deep water, ports and some of NATO's most important operational facilities are a prize worth fighting for. (See below under Turkey for further discussion on the defence of Thrace.)

Souda Bay

Located on the north-western coast of Crete, Souda Bay is the location of a major support centre for all operations in the eastern Mediterranean, especially those of the 6th Fleet. POL stores, wartime ammunition and fleet spares are stockpiled here. The bay is nearly large enough to anchor the entire 6th Fleet, with pier space to accommodate small combatants and supply ships. The airfield at Souda Bay can accommodate up to and including C-5A transport aircraft; however, the taxiways cannot accommodate either C-141s or C-5As. The airfield is used as a base for maritime patrol operations and general support.

Hellenikon

Hellenikon Air Base near Athens is the headquarters for US Air Force communications and support in Greece. It serves as an staging base for USAFE transport operations and supports the US Military Airlift Command (MAC). It has limited but important facilities for supporting transit aircraft, and is among the nine US bases in Europe due for closure.

Other installations in Greece primarily serve to support communications and intelligence collection in the region.

Turkish Thrace

Turkish Thrace is vulnerable by land on its western border with Bulgaria and by sea from the Black Sea to the north. It is cut off from the rest of Turkey by the Bosporus and contains one major city, Istanbul. The terrain lacks any meaningful defensive positions. The odds are very high that the Soviets would want very much to wrest control of the Bosporus and the Dardanelles from Turkey in any full scale war with NATO . . . The following assumptions apply.

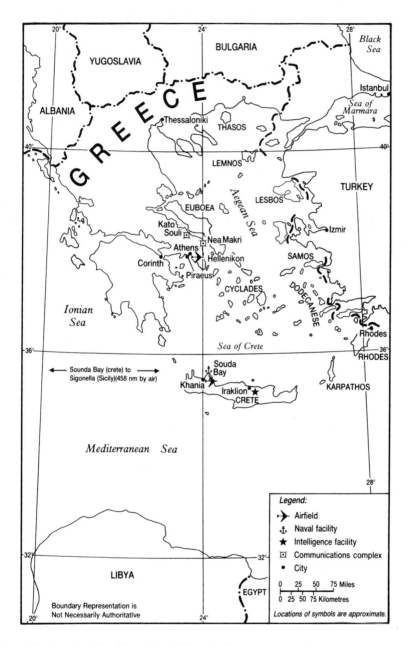

Figure 17.4. Major US military installations in Greece

Source: Congressional Research Service, *US Military Installations in NATO's Southern Region*, Report prepared for the Subcommittee on Europe and the Middle East of the Committee on Foreign Affairs, House of Representatives, 99th Congress, 2nd Session (US Government Printing Office: Washington, DC, 7 Oct. 1986), p. 41.

- A war against Turkish Thrace could occur separately from a general war between NATO and the Warsaw Pact.
- The Soviets might assign most of the Southwestern TVD's [Theatre of Military Operations'] forces to this theater, but would have to retain several divisions in each Hungary and Rumania for internal security and against an Italian–Yugoslav counter-attack.
- The Bulgarians would fight, but would leave 3 divisions at home for internal security and defense against a Greek attack.
- NATO would have considerable warning since the Soviets and Bulgarians would have to mobilize.
- The mobilized Warsaw Pact forces would be of poor quality—a 90% effectiveness factor applies. Soviet forces operating more than 500 miles from normal bases would have 90% effectiveness.
- Turkish divisions are enough larger than Warsaw Pact to warrant a 1.5 : 1 ratio.
- Only Turkish forces assigned to the Thrace theater would be able to be brought to bear in time. The Turks must retain the equivalent of 2 divisions in reserve for the threat of an amphibious assault behind their lines. However, they will be able to defeat such an assault by a combination of mines, artillery and submarine attacks.
- Turkish divisional effectiveness is only 90% due to antiquated equipment.
- The Turkish Air Force is well trained, but only 90% effective in support of the Army due to lack of good coordination.[10]

Incirlik

Incirlik Air Base near the south-central coast (see figure 17.5) is the most important tactical fighter air base in the eastern Mediterranean. It is a key deployment base for USAFE aircraft. The fighters that rotate to Incirlik from Aviano (and, until last year, Torrejón) are the most forward-deployed land-based tactical fighter aircraft in the eastern Mediterranean.

Iskenderun and Yumurtalik

These are the sites of the most important POL and storage facilities for US forces in the eastern Mediterranean.

Ankara, Izmir and Cigli

The air station and US Logistics Group (TUSLOG) in Ankara in central Turkey is the central logistical and support command for all US military supply services in Turkey. Izmir on the west-central coast is an air support base and headquarters of Allied Land Forces, Southeastern Europe (LANDSOUTH-EAST) and of NATO's 6th Allied Tactical Air Force (SIXATAF). Cigli is a Turkish tactical air base, but is used by the US Air Force for exercises.

Eastern Turkey

Because of the severe nature of the terrain in the Caucasus, the course of the campaigns that have unrolled there . . . have followed a highly repetitive pattern.

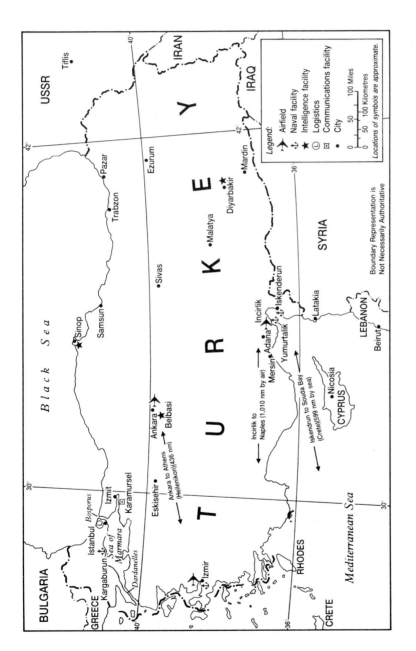

Figure 17.5. Major US military installations in Turkey

Source: Congressional Research Service, *US Military Installations in NATO's Southern Region*, Report prepared for the Subcommittee on Europe and the Middle East of the Committee on Foreign Affairs, House of Representatives, 99th Congress, 2nd Session (US Government Printing Office: Washington, DC, 7 Oct. 1986), p. 55.

'Traditional invasion route' is . . . a cliché . . . [but] in the context of the Caucasus it has inarguable force. There are . . . only three routes by which the Caucasus can be 'traversed' by large bodies of men in the east–west axis. The first is the coastal route along the Black Sea, a narrow corridor dominated on the landward side by ranges rising to over 10,000 feet. The second, [with] two points of entry at Kars and Kagizman, leads along the valley of the River Aras (Araxes) to Erzurum. The third, which leads to Malazgirt and Lake Van, is not open throughout its length, but requires the crossing of a saddle at Diyadin.[11]

While the Turkish Army does not have the most modern equipment, its men are famed for being well-trained, strong, determined and deeply patriotic. One need go no further than accounts of Gallipoli in World War I for evidence. Because any invasion of eastern Turkey would have to come in relatively narrow columns through the defiles described above, by simply positioning forces where the defiles open into broader plains, digging in and fighting the enemy in detail as they debouch, Turkey could make any attempted aggression time-consuming and very costly.

The road system also favours the defender. It is very poor near the border and improves as one moves west where the defenders would be. Consequently, the aggressor would be impeded by a poor road network as he mounted his campaign—with the implication that delays and complications would provide ample warning to the Turkish forces and better opportunity for targeting—and the defender would have much better roads and greater openness to supply and reinforce his lines.

None the less, the Soviet Union has considerable force massed on its side of the border. Despite no more than a seven-month operational window for army operations on the ground, when freezing weather and deep snow would not be blocking the passes, the Soviet Union has over a dozen air bases strung from the Black Sea coast to the River Aras at the south-eastern end of the Caucasus range from which to launch air strikes at eastern Turkey.

The geography is formidable, and the prize—aside from the possibility of fragmenting NATO by demonstrating that it might not come to the aid of an ally struck at the alliance's easternmost extremity—is questionable. Eastern Turkey might very well be overflown and staged from by the Soviets to support Middle East operations, but the likelihood of the Soviet Union deciding to attack and occupy Turkey is small.

Aside from a number of key installations devoted to electronic intelligence gathering and navigation, there are no major US operating bases in eastern Turkey.

The Mediterranean

NATO has over 2,000 miles of sea lanes to defend between Gibraltar and its most remote ally, Turkey.[12] The narrowness of the Straits of Gibraltar is a major advantage for NATO and should enable NATO to prevent reinforcement of Soviet naval forces in the Mediterranean once the war starts, or the escape of forces inside the Mediterranean. NATO also has the advantage that land-based air power can be brought to bear on any point in the Mediterranean for attack or search, and against Soviet land-based air forces which must overfly NATO territory to gain access to the Mediterranean.

Soviet forces pose three distinct threats: (1) a surprise attack by Soviet surface ships from the Mediterranean Squadron which are trailing NATO forces at the commencement of hostilities; (2) submarine attacks from predeployed Soviet submarines; and (3) attacks by Backfire aircraft from the Soviet Union.

The following assumptions apply:

– Combat in the Mediterranean Theater will be part of a general war between NATO and the Warsaw Pact.

– U.S. Striking Forces South [NATO wartime designation of US 6th Fleet] cannot count on having 2 aircraft carriers available. With only one it would be extremely risky to operate in the eastern Mediterranean. Any damage that stopped flight operations would almost certainly be fatal.

– The Soviets will attempt surreptitiously to augment their submarine force prior to hostilities as long as there is even one carrier in the Mediterranean.

– The Soviets will trail carriers in the Mediterranean and sacrifice them if necessary in the expectation of getting in a first shot.

– Sea control of the Mediterranean for resupply and reinforcement of Italy, Greece and Turkey is not vital in the early weeks of a general war. There are no military forces the United States would move to this theater in competition with the Central Front in the early stages of war. Nor would the United States have war or civilian supplies to spare for this front, or sealift in which to send them in the early weeks.

– The importance of sea control in the Mediterranean is for any possible projection of power missions. Those would be only aircraft or missile strikes, largely from the carriers.

– There is no mission for amphibious projection in this theater in a general war.[13]

On balance then, while interrupting or closing the Mediterranean SLOCs would be a high Soviet priority, it is unlikely that Soviet/WTO naval forces would be able to do so. Early in a war NATO naval forces would be fully engaged in clearing the Mediterranean of any Soviet submarines in the area and attaining air superiority over the sea. How quickly this could be accomplished would depend on the focus and intensity of the land battle and the number of air assets the Soviet Union felt it could spare against naval assets. However, even after air superiority has been achieved some naval offensive power would have to be applied in order to maintain that superiority. It is therefore unlikely that NATO naval forces would ever be able to devote themselves totally to the support of the land war in the region.

The geographic separation of the land theatres in Italy, France, and Turkey, coupled with the lack of funds for defense, particularly in Greece and Turkey, make the mobility and air power of the Carrier Battle Groups absolutely crucial, both to achieve an integrated air defense and also for offensive missions against land targets. . . .

Much of the logistic support to sustain any campaign must be carried by sea through the Mediterranean. The most potent threat to these comes from submarines. ASW in the Mediterranean is complicated by an exceptionally difficult environment. Only the

US possesses the quantity and quality of hardware and the operational expertise to be successful. Other nations can contribute, but they cannot substitute SSN's, towed arrays, P3C's [ASW patrol aircraft], etc. . . .

Soviet/Warsaw Pact land and air capability could be brought to bear in any one or all of the land theatres. Uniquely, however, NATO, with a formidable US contribution, dominates the Mediterranean and compensates for the weak and fragmented situation on land. If one were considering an order of importance of US assets in the region, I consider the most important (and last to cut) should be the 6th Fleet.

After the Americans, the French have the most powerful forces in the region, particularly maritime, with SSN's, two carriers, and, of course, an independent nuclear option. If a US withdrawal were coupled with some resurgence of French integration into some formal military structure, they would provide a useful compensation. However, I would judge that their willingness to be committed to contingency plans, in support of Turkish Thrace, for example, is a long way off. Even given such willingness, their total capability is but a fraction of that of the 6th Fleet.[14]

III. The traditional military mission gaps

Operational environment

The withdrawal of US forces from the region could cause a deterioration in the quality of the operational environment and, ultimately, the physical ability of the environment to support certain levels of military operations. The infrastructure required to support US operations, the skilled personnel and the depth of regional capabilities to support and sustain a major war with the Soviet Union would all be weakened, unless member nations in the region raised their individual and combined levels of activity and support. The necessity to maintain regional exercises, regular relations with NATO headquarters and forces outside the region, frequent port visits and regional and NATO-wide planning conferences, simulations and symposia, all contribute to the level and quality of the operational environment. Each benefits substantially from the active involvement of US forces and would suffer from their absence.

As with a loss of bases in either Japan or the Federal Republic of Germany, a loss of bases in the Southern Flank could, for the United States, 'bring with it the elimination of the strategic requirement to project power'.[15] The US inclination to defend the bases it would need should additional power be required in the region will decrease as the US need to project such power decreases. This does not mean, however, that the United States could totally lose interest in the region even after a withdrawal.

Certain . . . bases serve American strategic requirements . . . not directly related to local defense. Bases that support the US Navy as it tries to put Soviet ballistic missile submarines at risk, facilities that allow the US convenient communications with strategic reconnaissance satellites, or that allow the US to recover strategic bombers may still be useful even if the United States is less interested in defending countries near the Soviet Union.[16]

Two other changes may occur as a result of losing bases. First, the USA may use the money it has been spending on those overseas (naval) bases on capabilities which will permit the Navy to perform without them in the future.

More communications relay satellites and more systems with ultra-long ranges may be necessary, which may force a restructuring of [the] Navy and Air Force . . . Second, in the days before the United States possessed overseas bases, its military foresaw the need to seize and hold forward bases from the enemy or other powers in the event of war. This was the genesis of the modern amphibious mission of the marine corps. Loss of bases could lead to a resurgence of analogous forces . . . Once again, this could lead to the need to restructure [US] armed forces.[17]

Credible military presence and peacetime tasks

A credible military presence is the foundation of deterrence in Europe. Indigenous national forces form a large part of that deterrent presence in each military theatre. However, US forces, especially sea and air forces, are able to range the full length of the region as well as quickly deploying across national boundaries to display a heightened NATO presence or interest in times of crisis. Political differences among allies in the region and the general lack of surplus reserves for regional tasks today prevent any other NATO country from being able to provide the same regional coverage in peacetime or in crisis.

The United States—and other NATO allies—are concerned that the reduction of US military forces in Spain—including the F-16 wing at Torrejón—will reduce the ability of the United States to maintain a credible military force presence in Europe in the period following the implementation of the INF treaty with the Soviet Union. The INF treaty will eliminate medium and short-range nuclear missiles from Europe. The Spanish bases facilitate the implementation of a variety of missions of US and NATO military forces located elsewhere in Europe, especially those in NATO's Southern Region. In particular, the F-16's at Torrejón are reportedly to deploy to Italy and Turkey in wartime. From these countries they could conduct combat strikes against aggressor forces with nuclear or conventional weapons . . . for the defense of the Southern Region.[18]

As they move further away, the ability to use these F-16s flexibly throughout the southern region could be lost, as could their potential for early and rapid reinforcement to central as well as southern Europe.

Operational and logistical support

The crux of the problem of support, especially external support, is the amount of power which can be delivered over time. If needed, the argument could be made that, even with the withdrawal of the 6th Fleet from the Mediterranean, the Fleet could return to the theatre relatively quickly in time of warning. Because it would have on-board stores, it would not be dependent immediately on shore facilities, and would be able to provide a certain amount of power over time nearly anywhere in the region.

This applies to the short term. However, only so much resupply is organic to the Fleet. Without forward stores of munitions, the infrastructure to repair damaged ships and the ability to refuel and rearm them, its ability to deliver power would drop off quickly after the initial crisis. Improvised maintenance and repair of ships in the region would be impossible for lack of parts or would take too long for the Fleet to be able to maintain the necessary levels of power. All supplies would have to come from the farthest point, the United States, with the added time that would take and the added risk of bringing them in from outside the region. It is also unlikely that stores which might be available on the central front or the Northern Flank could be tapped to support the Mediterranean. Even if they could be spared, the distances involved hardly constitute an improvement. Consequently, the ability to reinforce by sea would be reduced very quickly.

The situation is the same for aircraft reinforcements. Without bases with reasonably complete supplies of vital spare parts and trained and experienced support personnel already in place to receive and service the aircraft, these could not simply be flown into the theatre. They would have to be preceded by ground crews who were familiar with the aircraft and associated munitions and who could communicate easily with US air crews. Spare parts would also have to be flown in in advance as there would be aircraft that would need replacement and repairs on arrival in theatre. Consequently, while what will be needed most at the time will be combat personnel and munitions, the first aircraft to arrive in the region would need to carry support personnel and repair and servicing equipment. In addition, even if this occurs during a crisis, before conflict begins, during the several weeks it will take for personnel and spares to arrive and prepare to receive the aircraft, the enemy will be doing all it can to hinder the accomplishment of those tasks.

If conflict has already begun, the absence of maximum available aircraft in the area from the first day will make the task of gaining air superiority more difficult and more drawn out. The level of hostile air coverage may delay further the possibility of bringing in support personnel and equipment to receive needed reinforcements. Consequently, the region would be weaker initially and weaker in the long run. Local commanders might actually never get capabilities in place because of the full-time effort which will be required just to catch up. If stores are immediately used up every time they are brought into an area, it will be impossible to get above the threshold where commanders can begin to establish a surge capability. This would be a critical problem in particular in the beginning of a conflict.

Another essential element of the allied effort is control over the Mediterranean chokepoints—Gibraltar, Messina, the Dardanelles, the Bosporus and the Suez Canal. Without timely reinforcements in theatre these could quickly be lost. The military axiom that it is easier to hold on to territory than to get it back is especially true here, where the correlation of forces would be in NATO's favour if it were in a position of defending these vital passages rather than of having to retake them.

Tied closely to the ability to reinforce are timing and the will to reinforce. A possible consequence of a US withdrawal could be a reduced commitment on the part of the United States to support the theatre under all conditions. If the Soviet Union could attack the Southern Flank without attacking US forces or facilities and if no US citizens would be at risk, the threshold for the United States to enter the conflict early could be significantly raised, despite all pledges to the contrary. As the mingling of allies in the southern region is already at a minimum, the United States—and the rest of NATO for that matter—could play a selective game. To restate a long-held concern of some NATO planners that is no less valid today, if the Soviet Union were to make it clear that it were not attacking any country but Turkey, would NATO respond? NATO says yes, and that is probably what would happen. However, the United States is Turkey's major military backer. Without US forces in place, Turkey is left considerably more vulnerable to a singling-out strategy by the Soviet Union or the WTO—just as Norway is on the Northern Flank.[19]

Reception of external reinforcements

More than other regions, Southern Region deterrence requires a demonstrable ability to reinforce. The vast majority of these [reinforcements] can only come from the US — 401st TFW [Tactical Fighter Wing], 1 Mech[anized] Brigade, 82nd Airborne Div[ision], Amph[ibious] Forces. In place, US manned reception facilities are essential for credibility.[20]

In many ways, this problem is tied to the one discussed above. A principal difficulty of US re-entry into the southern region for reinforcement relates to timing and political will, not just of the USA, but of the receiving or host country. It is always more difficult to commit out-of-region forces to a potential or actual conflict or to agree to accept foreign forces on one's soil than to utilize forces already at the site of the conflict. Will the sitting US president recognize the threat to US interests, come to the conclusion that those interests must be protected with the commitment of military force and be able to make a case to that effect to Congress and the American people quickly enough to get forces to the region in adequate time for them to make a difference? 'The decision to commit . . . forces would be a function of political will as much as the availability of resources and logistical management . . . it is questionable whether they can be depended upon to arrive early enough in the conflict to affect the military outcome. The more ambiguous the warning, the greater potential for delay in coming to the decision that could place great political strains on European allies.'[21] A Soviet/WTO thrust into NATO's Southern Flank would also be seen as a threat to the Central Front. In such a situation, NATO would have little reason to reinforce the Southern Flank with the limited

US forces allocated for NATO reinforcement. This is all the more likely taking into consideration that any forces that in such a withdrawal scenario are returned to the USA from NATO and that are not ordered into stand-down or reserve status would likely be used to fill already pressing NATO reinforcement commitments now being met with double- or triple-committed forces.

The other half of the problem, that of reaching an agreement with the proposed host nation for the reception of these forces, could be just as vexing. Today there are detailed agreements with host nations for the support and use of facilities. Existing facilities would presumably be vacated if the US forces left, which raises a number of questions about how forces would be reintroduced on short notice, should that be in NATO's interest. It would not be surprising if host nations were unwilling to receive the forces of another nation on their soil without clear limits being spelled out beforehand. Unfortunately, the large number of standing WTO forces on the Greek and Turkish borders makes the possibility of a lightning thrust south, with limited political and military objectives, quite real. In such an instance, out-of-region forces would have little chance of affecting the outcome of the engagement before it is over.

While the economic implications are beyond the scope of this chapter, 'it is estimated that the cost of relocating the 76 F-16's with their associated personnel from Torrejón to elsewhere in the region [or presumably out of the region entirely] as a result of the US–Spanish base negotiations, will cost from $300 to $500 million'.[22] Naturally these figures would not apply to a reintroduction of the aircraft and their supporting infrastructure to the region in time of conflict. However, they do reflect the substantial support required for them in peacetime. Actually the dollar costs of reintroducing them in wartime could be substantially higher as haste, inexperience and an inevitable no-holds-barred wartime attitude took their toll on men and equipment in the process.

Can the region do without these forces, or could they be replaced by regional forces? Either option seems unlikely. The defence spending of the United Kingdom, the FRG and France—the three European NATO countries with the strongest economies and the ones most willing to support substantial military expenditures—are projected to fall relative to that of the United States between now and the turn of the century, from 43 per cent to 32 per cent of US defence spending.[23] It does not seem very likely, therefore, that the European NATO countries would have the financial ability, nor perhaps even the will, given the changes in the WTO in the course of the autumn/winter 1989/90, to support major replacements of US forces and facilities in the Southern Flank.

However, assuming that the United States is expected to contribute to the defence of the region even if the substitution of regional air forces would do the job in peacetime, this would not mitigate the economic and military problems of reintroducing US aircraft, with the peculiarities of support and servicing requirements they would bring with them, should that be needed as a crisis or conflict escalates.

Further complicating the problem of re-entry are three factors which may join to permit the Soviets to allocate more of some kinds of forces to a regional war than they have been able to in the past. For example, it is expected that the level of Soviet submarine activity targeted against Mediterranean SLOCs could increase substantially in the next decade. First, as a result of technological developments, Soviet submarine noise may soon approach that of ambient noise. Less noise will make them harder to detect, requiring fewer attack submarines (SSNs) to protect their ballistic missile submarines (SSBNs), thereby freeing more SSNs to attack NATO SLOCs. Second, as the Soviets move more to mobile land-based missiles, the survivability of SSBNs will become less critical, permitting them to use them earlier in the war, reducing their need for long-term protection. Finally, if reliance on strategic ballistic missiles is reduced as a result of the US Strategic Defense Initiative (SDI) or through cuts in strategic arsenals as a result of treaties, this too will reduce the requirement to protect the SSBNs.[24]

Out-of-area contingency transit

Proximity to the Middle East has made the out-of-area issue more relevant to the southern region than to the central or northern regions. There is concern that Turkey could be dragged into a Middle East conflict because of contiguous borders with the Soviet Union, Iran, Iraq and Syria. Military deployments to South-West Asia by some of the major NATO powers could have a negative impact on NATO readiness. The use of any NATO base for out-of-area operations by the United States or any other ally could involve NATO in a Middle East conflict. 'An airlift/tanker "bridge" from Europe to the Persian Gulf, for example, would impose excessive strains on NATO's logistic infrastructure'.[25]

The Portuguese Government reserves the right to insist that the facilities in the Azores, such as Lajes Air Base, be used only for NATO-related contingencies. However, Portugal has been noticeably co-operative in letting the United States use these facilities for other purposes. Portugal was the only NATO ally to permit the United States to stage resupply to Israel during the 1973 war.[26]

Mediterranean members of NATO, while anxious for the protection of the Alliance as a whole and especially of the United States, are reluctant to be dragged automatically into quarrels arising from American policy in the Middle East. Turkey faces a particularly delicate problem because it could be the base from which American forces might threaten the flank of Soviet interventions in such areas as Iran. Equally, Turkish airspace could be valuable to the Soviet Union in such ventures.[27]

Installations in Greece have historically been denied for out-of-area contingencies. It is unlikely that this policy would change in the foreseeable future.

In 1973, Greece remained neutral in the Arab–Israeli war and made it clear that its bases were not available to facilitate the US effort to resupply Israel. More recently, the Greek Government denied American requests to use Greek installations to resupply both US forces serving in the Multinational Force in Lebanon as well as those of the

Lebanese Government ... Greece, in recent years, has placed a priority on maintaining good relations with the Arab world in part because of its dependence on Arab oil.[28]

Turkey has always supported the use of its bases for any NATO-related operation. However, in general it has interpreted this rather strictly and, while not precluded by treaty, has not been in favour of its bases being used for out-of-area contingencies. As Turkey is the only Muslim country in NATO, it has been sensitive to political perceptions of its bases being used for US excursions in the Middle East. None the less, in 1983 a Transit Terminal Agreement was signed by Turkey and the United States which permits Incirlik to be used to support the US contingent of the Multinational Force in Lebanon with transportation of personnel, non-military supplies and medicines.

IV. The functional military gaps

US leadership

The Southern members of NATO, unlike their Nordic partners, have recent histories of domestic political instability. Perhaps more importantly, they are deeply divided from each other politically as well as geographically. In particular, Greece and Turkey are at daggers drawn over Cyprus, over rights in the Aegean to territorial waters and to oil, and are estranged by the memory of centuries of mutual antagonism. As a result, the command structure of NATO on the Southern Flank is fragmented, and it is not much of an exaggeration to regard the area as a series of bilateral alliances between each local ally and the United States.[29]

Coupled with this is the absence in Europe over the past 10 or more years of any significant movement towards co-ordinated European defence and foreign policies. The current French–FRG co-operation on some operational military issues is encouraging but does not in itself constitute evidence that it will lead to a unified European policy.[30]

All of this suggests that there seems to be no ready or able candidate to take up the leadership role now played in NATO by the United States. While this could present a problem in the northern and central regions, it is especially serious for the southern region where national animosity and rivalry sometimes threaten to fracture the alliance altogether.

The senior US military commanders in the south is the Commander-in-Chief, Allied Forces Southern Europe (CINCSOUTH), who is also the senior NATO commander in the southern region and one of the three major commanders in NATO reporting to SACEUR. He is a US Navy (USN) four-star admiral, with headquarters in Naples, Italy. Of the five principal commanders reporting to CINCSOUTH, two are US officers: the Commander, Naval Striking and Support Forces Southern Europe (COMSTRIKEFORSOUTH), a USN three-star admiral who is also Commander, US 6th Fleet (COMSIXTHFLEET), and the Commander Allied Air Forces Southern Europe (COMAIRSOUTH), a US Air Force three-star general, also with headquarters in Naples.

US commanders in the southern region have provided cohesive intra- and inter-regional leadership; have guaranteed regional priorities in NATO councils as a result of US alliance-wide involvement; have provided, often through national channels, the needed strategic, air and sea forces; have ensured a high level of allied and US national interest in the region which has provided tangible benefits; and have largely been responsible for addressing out-of-region problems not provided for in the North Atlantic Treaty.

Beyond the leadership responsibilities now assumed by the United States, the importance of US support staff should also be considered. 'When considering a withdrawal, the impact of cuts in headquarters staff and communications personnel should not be underestimated. Some aspects such as reduced intelligence and C^3 are obvious: competent, English speaking, staff officers with experience of operating as a first class power are irreplaceable'.[31]

Influence on national focus and will

The presence of the US 6th Fleet in the Mediterranean in the 1950's was tacitly accepted by Yugoslavia as a factor which indirectly enhanced its security. Belgrade, despite the fact that the threat of Soviet invasion receded after Stalin's death in 1953, still needed an effective NATO nuclear umbrella of which the 6th Fleet was one component. Not until 1964 did Yugoslavia become openly critical of the US and NATO presence in the Mediterranean and supportive of Soviet aims in the area. This policy shift was influenced by Yugoslavia's relations, first with Moscow, and second with the Arab world.[32]

While it is not suggested here that NATO or the United States actively cooperates with the WTO or other non-NATO countries in their relations with the Soviet Union, it is reasonable to attribute to NATO and to the United States—and especially to the 6th Fleet and US air assets in the region—a balancing role. To the extent that US forces influence this balance, measures will need to be taken following a US withdrawal to redress the resulting imbalance. Given the traditionally separate national approach to the threat taken by southern region member states, this may not be easy or even possible in the short term.

The US presence in the southern region permits Greece and Turkey to have the best of all worlds, somewhat at the expense of the rest of NATO. Each views the other as a greater potential threat than the USSR. Their forces are deployed accordingly, rendering them poorly positioned to meet a strike from the WTO. The presence of US forces in the region permits Greece and Turkey to assume this east–west defence posture with impunity, and without appreciably increasing the risk to the region. With a US withdrawal they would have to choose the greater threat. Chances are that their tactical deployment would not change. To what extent would the Soviet Union see this as an invitation to attack? It could be very seductive if limited objectives were in mind. To the extent that this were the case, the regional risk of a US withdrawal increases.

Control of the Turkish Straits in wartime is especially important if the maritime threat is to be controlled, reduced and, in time, eliminated. The Soviet Union has not found a way around the 1936 Montreux Convention for the transit of submarines,[33] so it brings all its submarines to the Mediterranean from the Northern and Baltic Fleets.[34] Up until now, the Turkish Government has not had any difficulty in enforcing the provisions of the Convention. This could change, however, in times of crisis or war. The Convention's enforcement provisions are ambiguous and neither the Soviet Union nor—for that matter—any other nation is above bullying its neighbours if and when it feels that its national interests are at stake. To the extent that the presence of US forces in the region helps Turkey to stand up to the Soviet Union, these would have to be replaced. Again, this would be difficult for any power not willing or able to devote considerable resources to intra-regional defence.

The Soviet Union's potential use of bases—especially air bases—in North Africa has long been of concern to NATO planners. At present, the threat is expected to come from the north or north-east. NATO's defences are mainly pointed in those directions, and some conservation of resources is possible with this limited threat axis. If the Soviet Union were able to establish operating bases on the coast of Africa, in Tunisia or Libya, for example, NATO land as well as maritime forces would be threatened from all points of the compass. This has obvious implications for the orientation and adequacy of air defences in the region, the vulnerability of the SLOCs, the ease of inter-regional transfers of forces and supplies, the control of vital straits and passages and the safety of reinforcements. The in-place strength that would be needed to defend the region from a Soviet standing-start attack from 360 degrees and with little warning would be overwhelming.

The loss of the nuclear option

The removal of nuclear forces from the region will mean forfeiting any deterrent value those forces have had, and might preclude their reintroduction when they are needed. When US nuclear custodial units withdraw from the southern region, all nuclear weapons (primarily Air Force bombs and Army Lance missiles) would be removed as well. This would leave in the region only French nuclear weapons, which are not at the direct disposal of NATO. Although the southern region is usually thought of as a contiguous reach of NATO nations from the Portuguese Azores on the Atlantic approaches to eastern Turkey, the southern region is in fact a group of ethnically, culturally and politically dissimilar nations, with little holding them together militarily except for the region-wide reach of the US 6th Fleet and air forces in the region. As even these forces are strained to cover the breadth of such a large geographic expanse, the real strategic 'glue' of the region, as measured by regional deterrence of the Soviet Union/WTO, has been provided by US strategic, not conventional, forces.

It is unlikely that nuclear weapons would be used in the region before their general release on the Central Front. Both flanks would be assessed in light of the military situation on the Central Front. If the Central Front was not under direct attack or was holding with conventional forces, any Soviet/WTO successes in the southern region would be seen as reversible over time, and would not motivate crossing the nuclear threshold. It is precisely because of this that the immediate availability of nuclear weapons in the southern region is so important as a deterrent. As long as they are in the region, the Soviet Union/WTO cannot plan on them not being used in a situation where NATO seems to be losing conventionally. It is also reasonable to assume that Soviet/WTO planners calculate that NATO would find it easier and less escalatory to use nuclear weapons on the flanks first, thereby making nuclear weapons a stronger deterrent on the flanks than in the centre.

Even more than with conventional forces, however, once taken out of the region by the United States, it is less likely than nuclear weapons will be reintroduced quickly. The political decision to reintroduce them in time of crisis will be seen in the United States as possibly escalatory. The reintroduction of nuclear weapons will be seen as the first step towards their probable use, a step which could, in effect, introduce a certain logic and momentum to the release process, making it easier to actually release them. As such, they will not be permitted to leave the United States, except in an unambiguous and broadly supported situation where US interests are seen to be *in extremis*.

Furthermore, even if the political decision were made to reintroduce nuclear weapons in a crisis—the word 'political' is used advisedly, as the military will not prevail in the councils where and when that decision is being made—it should be kept in mind that transport and off-loading, movement to storage or ready-for-use sites and the re-establishment of units to receive, manage and protect them will all take a great deal of time. One month would be an optimistic estimate. This delay will destroy their deterrence value as, for the reasons given above, it is unlikely that a decision would be made in the United States to reintroduce them before they are actually needed in battle. If on the other hand the decision to redeploy is delayed until it is clear that no other solution is possible, the conflict will certainly be fully joined by all sides, with NATO losing conventionally. In such circumstances, the WTO can be expected to fight openly and aggressively to prevent the reintroduction into the region of forces that could turn the tide against it.

Decoupling NATO from US weapon modernization

Possibly the greatest danger facing the European NATO countries as a result of a withdrawal of US forces and equipment, and potentially the one which will produce the most significant mission area gap, is the possibility that they fall behind both the Soviet Union and the United States technologically as new weapons are developed and introduced on the battlefield. If this were permitted to happen, Europe could find itself facing the Soviet Union with new but irrelevant weapons, tactics and battlefield doctrine and unable to assimilate US forces into NATO's forces.

Naturally, this would not happen overnight. The USA and Europe would continue to share technology as it is introduced into their respective military structures. It is inevitable, however, that as systems and doctrines evolve, production and doctrinal thinking will always lag somewhat behind, even in the country where the technology is being developed. For both economic and practical reasons the producing country will supply its own forces before sharing new technology with its allies. In the process, weapon support, expertise and doctrinal mismatches are to be expected, some with potentially serious consequences. The following discussion suggests the breadth of changes which the future may bring, along with some first-order battlefield mission implications.

A recent US military forecast identified 70 'breakthrough' technologies and system concepts with the potential to revolutionize the battlefield of the early 21st century. Most would relate to conventional operations within the earth's atmosphere. They include low-cost tactical drones, multi-mission remotely piloted vehicles (RPVs), aircraft or missiles with advanced 'stealth' and long-endurance capacity and high-altitude unmanned aircraft. Advances that would make existing systems much more capable include optical processing, 'smart skin', ultra-light airframes, ultra-structured materials, the 'super cockpit' (involving vast improvements in man–machine interaction), smart built-in test capability, advanced manufacturing technology, geometric improvements in guidance capability, high-performance turbine engines, unified life cycle engineering, robotic telepresence and full-spectrum ultra-resolution sensors.[35]

As military equipment becomes increasingly dependent on state-of-the-art technology, costs will increase and the number of manufacturers with the requisite production capability will decrease. For that reason alone, some equipment will be unavailable outside the producing country. As a natural consequence, allies will be unfamiliar with the use of the equipment in question and will be unable to integrate it into their force structures and tactical doctrine or to service and repair it close to the battlefield. Costs alone may preclude European manufacturers from moving to the next generation of weapon systems unless they can compete for overseas sales. The United States and the Soviet Union may be the only countries able to afford to make systems such as the advanced medium-range air-to-air missile (AMRAAM), the Patriot anti-aircraft guided missile and the Advanced Technology Bomber.[36]

Implications of technology for conventional forces

There is evidence that the Soviet Union believes that conventional weapon systems will benefit most from emerging technologies.

Development of long range, highly accurate and remotely guided combat systems, remotely piloted vehicles, and qualitatively new electronic control systems will permit 'reconnaissance–destruction complexes' that link sensor, communications, and fire systems in real or near real time in the execution of fire missions at depths up to 200 km and deeper throughout the enemy tactical, operational, and even strategic rear. Consequently, the technologies of greatest interest to the Soviets over the next 10–15 years appear to be those associated with microelectronics, automated decision support systems, telecommunications, and enhanced munitions lethality.[37]

Soviet writings express the belief that considerable 'emerging technology' is beginning to appear on today's battlefield. Because of normal budgetary restraints and the transition time needed to develop and field new tactics and doctrine, however, their full military impact will not be appreciated by either side until well into the 1990s.[38] They see future weapons based on new physical principles, permitting conventional weapons to have greatly increased range, accuracy, destructiveness and speed.

Nuclear energy might be used less for destruction than to power this new generation of conventional weapons. This new class of weapon would change the whole dynamism and tempo of war, bringing conventional war closer to the present model of nuclear war. For example, a cruise missile with self-guided cassette munitions would be as destructive as a 1-kiloton neutron bomb.[39] Troop control facilities could become major targets in this new model. Command and control centres, radars and weapon guidance networks already appear prominently on Soviet target lists for these kinds of weapon.[40]

These new systems have rather reinforced than changed the Soviet belief in fast offensive operations whose objective is to break through allied defences in one or two days and achieve strategic objectives within the first month of battle.[41] Timelines will be shortened if anything, providing less time for reinforcements to be brought into the area and moved up to the front lines. There will be less time to accommodate for a different mix of systems on the battlefield, many potentially new to European forces, to say nothing of their supporting tactics. Soviet planners estimate that with new technologies, the 'detection–destruction cycle' of fire missions could be reduced by a factor of 10 or more.[42]

Conventional weapons with the effect of today's nuclear weapons could keep the war at the conventional level and increase its duration at that level. Soviet military doctrine since 1979 has stated that 'the sharp increase in the combat range of conventional weapons makes it possible to immediately envelop not only the border regions but also the territory of an entire country with active combat operations, which was impossible in wars of the past'.[43]

The Advanced Technology Fighter (using 'stealth' technology) will enable NATO to attack Soviet assets at their point of origin. Increased survivability of the weapon or aircraft in the face of Soviet integrated air defences will increase the probability of the weapon actually reaching the target and will reduce the risk to the pilot.[44]

Robotics will proliferate on the battlefield to augment forces, to do mundane, time-consuming or high-risk tasks, and to detect, classify and attack a whole range of targets.[45]

Biotechnological weapons—weapons which attack other weapons rather than people—have long been discussed and are now very attractive potential force-multipliers. This involves technology that would destroy some element of a given weapon system without which it cannot function. One example would be a gas which is harmless to humans but which will change the chemical composition or state of engine fuel. Another would be a gas which will harden and crumble the plastic or rubber mountings, shock absorbers, connectors or treads on vehicles.[46] Without continuing close working ties with US forces and employment planning processes, it is unlikely that these kinds of system—the effectiveness of which largely depends on the surprise element—would be shared with other NATO forces.

Unless NATO can keep pace developing and fielding counters to Soviet advances in this rapidly changing area of technological development, NATO will not have meaningful forces in a NATO–Soviet/WTO conflict. The conventional imbalances that worry NATO today will be wished for in comparison. It is here, perhaps more than anywhere else, that NATO is in danger of becoming 'defenceless' without US forces in place in Western Europe.

As the USA and the USSR as well as other countries move into space over the next 5–10 years, these new systems will fundamentally change the nature of the battlefield. Even modest defences have the potential for revolutionizing the way wars are fought. Space weapons and space defences could render many existing military systems ineffective. Space technology is expensive, but the leverage it provides the first nation to develop and field a new technology is priceless. It is unlikely, however, that the European NATO countries will be able to afford or politically be able to spend the amounts which will be required to keep up with the USA and the USSR. Furthermore, as the USA and the USSR attempt to keep costs under control, decisions will be made to limit the numbers and kinds of system produced and the coverage envisioned. With US troops out of Europe, the political, economic and social pressures to do only what is necessary for the United States may be powerful. As a result, Western Europe may be left on the outside of the tent looking in.

Implications of manpower for conventional forces

In population, Europe is one of the slowest growing major regions in the world. Although several of the European NATO allies counted among the top 10 most populous countries in 1950, none has remained on the list. By 2010 the national youth pool in Western Europe will drop below current levels and remain there, making it more difficult for NATO to maintain current force structure levels and producing strong demographic pressures against either denuclearization or conventional force buildup.[47] Available manpower in the United States and the

Soviet Union will also be constrained in the short term, but will rebound to exceed current levels by 2005. 'The number of 18-year old males in the Soviet Union will decline until about 1995, then increase; but by 2000 the total size of the pool will still be less than it was in 1979'.[48]

V. Conclusion

The USA may at times be seen as a difficult and even unheeding partner by its southern region allies. Its out-of-region interests and involvement may sometimes seem dangerous to those of them close to the action. At the same time, some of the military contributions the USA makes in the region are unique.

Any NATO partner or group of partners could contribute the forces and infrastructure which the USA now provides, given the same commitment of funds and effort. So most of the gaps that would be created in the traditional military mission areas by a US departure from the region, except perhaps the nuclear contribution, could be filled from regional resources. The gaps which would be difficult to fill, at least without major political realignments and the regional agreement in specific nations assuming quite new regional roles, would be those functional gaps discussed in the second part of this chapter: leadership, influence on national will and the tie to US weapon system modernization.

It would be difficult for any nation in the region to assume cross-regional leadership. To do so would require the tacit acknowledgement that one partner enjoys a special region-wide relationship not shared by others in the region. To most that would be acknowledging something that does not exist and to some that would require giving up a degree of regional independence to a neighbour. The United States, with bilateral relationships across the region, with equally strong ties to NATO's north and central regions and with world-wide interests which in most instances coincide with regional interests, offers a brand of leadership which is at the same time workable, yet not threatening.

The United States, both as an external power and as a recognized world power, also provides strong support for national power in the region. The deterrent and warfighting capability of each US ally in the region is strengthened because its forces must be counted, not alone, but with the forces of the United States in any correlation of force computations done by the Soviet Union or its allies. No other single ally in the region can provide the degree of capability enhancement that flows automatically from an alliance with the United States. Consequently, the withdrawal of the United States from the region would reduce the effect which the presence of US forces now has in strengthening individual national deterrence and warfighting capability.

Further, the presence of US forces helps some nations to put aside differences with NATO neighbours in preparing to meet a common threat. In some cases the United States represents the only functioning channel of co-operative communication between rancorous neighbours. This role, vital to the integrity of the entire alliance, could not be served by any other NATO country.

Finally, one of the most serious difficulties faced by perhaps the majority of NATO partners since the inception of the alliance has been the problem of obtaining and fielding the most modern equipment. Modern here simply means equipment comparable in range, fire-power and response to the enemy equipment against which it is expected to compete. The absence of US forces in the region will uncouple European NATO from that automatic modernization which has until now ensured the presence of US troops with advanced weaponry and the inevitable sharing of those advances within the alliance. While US technology would undoubtedly still be available to NATO, even without the actual presence in Europe of US forces, its availability and adoption by NATO will inevitably be slower and less uniform. Over time, NATO forces could fall seriously behind and, despite possible increases in actual force structure, find itself less and less capable against more modern, conventional Soviet/WTO forces.

Notes and references

[1] Krause, M. D. (Lieut. Col.), 'Joint and combined operations at the operational level of war, AFSOUTH', paper presented in completion of course, Naval War College, Newport, R.I., 1988.

[2] Most of the conclusions reflected in this section derive from the author's personal experience as Executive Assistant and Senior Aide to the Commander-in-Chief, Allied Forces Southern Europe, in Naples, Italy, from 1975 to 1977, and from extensive discussions on the southern region in Feb. and Mar. 1988 with Admiral Stansfield Turner, US Navy (Ret.), former Commander-in-Chief, Allied Forces Southern Europe.

[3] Turner, S. (Adm. USN, Ret.), 'Net Assessment of NATO's Southern Flank', informal note prepared for discussion with representatives of the Office of the Secretary of Defense, US Department of Defense, Washington, DC, 1988.

[4] Duke, S., SIPRI, *United States Military Forces and Installations in Europe* (Oxford University Press: Oxford, 1989), chapter 11; Grimmett, R. R., Congressional Research Service, *U.S.–Spanish Bases Agreement*, Major Issues System Issue Brief, order code IB88010 (US Library of Congress: Washington, DC, 22 Feb. 1988).

[5] Congressional Research Service, *US Military Installations in NATO's Southern Region*, Report prepared for the Subcommittee on Europe and the Middle East of the Committee on Foreign Affairs, US House of Representatives, 99th Congress, 2nd Session (US Government Printing Office: Washington, DC, 7 Oct. 1986), p. 4. This and subsequent information on major military missions served by installations in the region was taken largely from this excellent compilation prepared mainly by Dr Richard Grimmett of the Congressional Research Service of the Library of Congress.

[6] Congressional Research Service (note 5), p. 5.

[7] The maps used in this section on installations were taken from Congressional Research Service (note 5).

[8] Duke (note 4); 'United States Air Forces in Europe' (Reports from the Major Commands), *Air Force Magazine*, May 1989, p. 98.

[9] Turner (note 3), pp. 1–2

[10] Turner (note 3), pp . 1–2.

[11] Keegan, J. and Wheatcroft, A., *Zones of Conflict: An Atlas of Future Wars* (Simon & Schuster: New York, 1986), pp. 20–21.

[12] Vice Admiral Sir Patrick Symons, KBE, Senior British Officer at AFSOUTH, Naples, in his most helpful 15 Apr. 1988 review of this paper, 'Comments on Captain Thibault's Paper',

makes the important point that sea control is relevant not just to power projection missions but to early reinforcement as well.

[13] Turner (note 3), pp. 1–3.
[14] Symons (note 12), p. 2.
[15] FSEWG, 'Sources of change in the future security environment', paper prepared by the Future Security Environment Working Group for the Committe on Long Term Strategy, US Department of Defense, Washington, DC, Apr. 1988.
[16] FSEWG (note 15).
[17] FSEWG (note 15).
[18] Grimmett (note 4), p. 8.
[19] Interview with Mr Mark Herman, Senior Associate, Booz Allen & Hamilton, Inc., Washington, DC, 7 Mar. 1988.
[20] Symons (note 12), p. 2.
[21] Snyder, J. C., *Defending the Fringe: NATO, the Mediterranean, and the Persian Gulf* (Westview: Boulder, Colo., 1987), pp. 3–4.
[22] Grimmett (note 4), p. 9.
[23] FSEWG (note 15).
[24] FSEWG (note 15).
[25] Remarks by Diego Ruiz Palmer, Manager, European Security Studies, BDM Corp., NATO Southern Region Working Group, Rapporteurs' Report of Session I: 'The military balance and strategy in the southern region' (Center for Strategic and International Studies: Washington, DC, 29 Feb. 1988), p. 4.
[26] Congressional Research Service (note 5), pp. 7–8.
[27] Martin, L., *NATO and the Defense of the West* (Holt, Rinehart & Winston: New York, 1985), p. 35.
[28] Congressional Research Service (note 5), pp. 39–40.
[29] Martin (note 27), p. 33.
[30] FSEWG (note 15).
[31] Symons (note 12), p. 2.
[32] Vego, M. N., *Yugoslavia and the Soviet Policy of Force in the Mediterranean Since 1961*, Professional Paper 318 (Center for Naval Analysis: Washington, DC, 1981), p. 1.
[33] 'The Montreux Convention, signed 20 July 1936, lays down the régime applicable to the Turkish Straits: the Dardanelles and the Bosphorus. While merchant vessels enjoy free transit, transit of vessels of war is subject to more restritictive rules . . . [A]s far as submarines are concerned, the provisions of the Convention prohibit the USSR from building up a submarine force in the Black Sea; in fact she can only let them pass through to the Mediterranean in dribs and drabs'. Vignes, D., commentary to chapter 15: '1936 Montreux Convention regarding the régime of the Straits', ed. N. Ronzitti, *The Law of Naval Warfare. A Collection of Agreements and Documents with Commentaries* (Martinus Nihoff: Dordrecht, 1988), p. 438.
[34] Roberts, S. S., *The Turkish Straits and the Soviet Navy in the Mediterranean*, Professional Paper 331 (Center for Naval Analysis: Washington, DC, 1982), p. 14.
[35] FSEWG (note 15).
[36] FSEWG (note 15).
[37] FSEWG (note 15).
[38] FSEWG (note 15).
[39] FSEWG (note 15).
[40] FSEWG (note 15).
[41] FSEWG (note 15).
[42] FSEWG (note 15).
[43] FSEWG (note 15).
[44] FSEWG (note 15).
[45] FSEWG (note 15).
[46] FSEWG (note 15).
[47] FSEWG (note 15).
[48] FSEWG (note 15).

18. The contribution of US C³I to the defence of NATO

Paul Stares
Brookings Institution, Washington, DC

John Pike
Federation of American Scientists, Washington, DC

I. Introduction[1]

The provision of intelligence and communications facilities is not the most obvious contribution of the United States to the defence of NATO. Unlike the men and *matériel* that make up the more visible manifestations of the US military presence in Europe, command, control, communications and intelligence (C³I) assets are rarely if ever included in the frequent stock-takings of the US commitment to NATO. A major reason for this is that command and control in general is rarely accorded the status and attention it deserves as a vital component of military power. As a result, it has largely been ignored in assessments of the conventional balance in Europe. In fairness, however, there are major methodological barriers to assessing the military value of command and control. While the technical performance of C³I systems can be assessed in the abstract, their overall contribution to combat effectiveness does not lend itself to meaningful measurement. The role of a particular intelligence sensor, computer system or communications link and, for that matter, the information that is respectively collected, processed and transmitted by them cannot be quantified in a precise way. Nevertheless the role of command and control in wartime is too important to be ignored; some judgement has to be made about the role and adequacy of the systems in place. The purpose of this chapter is threefold: to identify the US C³I systems that contribute to the defence of NATO, to evaluate the consequences of their withdrawal from the European theatre and to consider how the European NATO countries might cope thereafter.

Major problems confront the analyst in trying to discuss the level of US intelligence and communications support to NATO. For obvious reasons, immense secrecy surrounds the US intelligence-gathering effort in Europe. How intelligence derived from US-controlled sensors contributes to NATO's defence is particularly difficult to determine since there are multiple paths by which it can be disseminated to the allies. The most obvious are the official intelligence channels, whereby information is exchanged on a multilateral basis among all alliance members. In addition, the United States also has special

THE CONTRIBUTION OF US C³I TO THE DEFENCE OF NATO

bilateral arrangements with certain NATO allies, notably the United Kingdom, with the result that some receive more US intelligence than others. Furthermore, there are indirect ways by which NATO benefits from US intelligence. Frequently intelligence is collected by or passed on to local US commanders who in turn make it available on an *ad hoc* basis to nearby NATO forces that they believe will benefit from it. In wartime, this kind of sharing is expected to be quite common.

Gauging the level of European dependence on US communications systems also presents problems. While there are dedicated US military communications networks in Europe, it is not clear how much use is made of these systems by NATO in peacetime or how much would be likely in wartime. Assessments depend in part on detailed calculations of the relative vulnerabilities of US and allied networks under a range of different scenarios. Similarly, the extent to which Europe relies on US personnel to maintain and operate commonly funded NATO communications networks is also unclear.

This raises a more fundamental problem concerning the nature of the postulated US withdrawal from Europe. If the USA removes all personnel and transportable equipment from Europe but maintains its commitment to return and defend Europe when called upon, then the overall impact will clearly be much less than if US security guarantees are also withdrawn. This would most likely entail the complete dismantlement of all US-owned facilities in the region. As is demonstrated in this chapter, the consequences of this more extreme form of US disengagement could be quite serious in certain cases.

Sections II and III are descriptive summaries of US intelligence and communications assets either located in Europe or contributing in some way to its defence. Separating intelligence from communications assets is somewhat arbitrary since many command and control systems perform both functions. The AWACS and forthcoming JSTARS aircraft are examples of this blurring of missions. Section IV assesses the consequences of a US withdrawal from Europe under various scenarios and examines how the rest of NATO might adapt. Before continuing, however, it is important not to overlook the US contribution to NATO's command organization. Should the USA withdraw completely from Europe, major changes to NATO's current chain of command would be necessary.

The NATO command structure

As befitting the United States' role within the alliance, US personnel fill—on a more or less permanent basis—many of the senior positions in NATO's command hierarchy. Two of the three major NATO commands—Allied Command Europe (ACE) and Allied Command Atlantic (ACLANT)—are headed by US officers. Besides providing the Supreme Allied Commanders for both these commands—SACEUR and SACLANT respectively—many of the most important subordinate command slots are also filled by US officers. These include: the Commander, Central Army Group (CENTAG); the Commander, Allied Air Forces Central Europe (AAFCE); the Commander-in-Chief, Allied Forces Southern Europe (AFSOUTH); and the Commander, Allied Air Forces Southern Europe (COMAIRSOUTH). These officers are all 'dual-hatted' in the sense that they also hold positions in the US European Command (USEUCOM) hierarchy. Given the US involvement in NATO's command structure, US officers also make up a large proportion of the administrative personnel at the main NATO headquarters.[2]

If NATO retains the same command structure following a US withdrawal from Europe, then these positions would obviously have to be transferred to European officers—at least during peacetime. It is more likely that NATO would use the opportunity provided by a US withdrawal to restructure its command organization. However, provision would have to be made for the re-integration of US forces should they return to defend Europe. Failure to make such provision is likely to cause immense problems in the period leading up to and during a conflict.

II. Intelligence assets

This section discusses the various intelligence collection and analysis systems operated by the USA that would be used in a NATO theatre conflict. These may be divided into two groups: those of primary relevance to AirLand Battle scenarios on the European mainland, and those principally supporting the Maritime Strategy on the flanks. Further, they are organized according to the levels at which the intelligence assets are managed. How intelligence is disseminated to both US and NATO forces is also discussed.

With regard to intelligence assets in support of AirLand Battle, national systems are those managed by organizations such as the Central Intelligence Agency (CIA) or the National Security Agency (NSA). Echelon-Above-Corps, here referring to the US European theatre command infrastructure, include US Air Force Europe (USAFE), US Army Europe (USAEUR), the 2nd Fleet in the Atlantic and the 6th Fleet in the Mediterranean. Corps (and numbered Air Force) and Division levels include those collection assets organic to these levels.

Primary emphasis is given to national and Echelon-Above-Corps systems. These levels operate intelligence assets that in many instances are unique to the USA, such as satellite systems, and are a major source of intelligence for agencies of other NATO allies at similar levels as well as for US forces at lower levels. In addition, many of the assets available at US Corps and Division levels are organic to these echelons and are replicated in the force structure of allied units. Finally, the national and Echelon-Above-Corps assets are among the more highly classified and less widely understood elements of the US intelligence system.

AirLand Battle: national systems

Space systems constitute a unique US contribution to NATO. Satellites are increasingly used to directly support military forces on earth, serving as 'force multipliers', supplying the information needed to increase the military effectiveness of terrestrial weapon systems. In general, US satellite systems provide intelligence products that are not otherwise available to European forces. In particular, their coverage not only includes the whole of Eastern Europe but also extends to the full depth of the Soviet Union, giving an unparalleled appreciation of Soviet intentions and capabilities. Missions include photographic and electronic intelligence collection, warning of missile attack, meteorology and nuclear explosion detection.

Space-based imaging intelligence

The National Reconnaissance Office (NRO) has primary responsibility for operating US photographic and electronic intelligence satellites. These satellites provide the primary—if not the sole—source of intelligence concerning Warsaw Treaty Organization (WTO) military activity in areas removed from the immediate borders of WTO member states.

In recent years, the photographic intelligence satellites have included the film-return KH-8 Hexagon Low Altitude Surveillance Platform, the KH-9 Gambit Big Bird as well as the digital-return KH-11 Kennan, which can return images in near real time.

A more advanced system, the KH-12, which will combine the high resolution of the film-return satellites with the real-time transmission capabilities of the KH-11, has entered service. The KH-12 system involves a constellation of four satellites orbiting simultaneously and operating continuously. This will provide more comprehensive coverage than the two-satellite KH-11 system, particularly as the KH-12 has a far longer operational life span than its predecessor. The KH-12 uses the Milstar communications satellites to relay data to central processing facilities in the USA, as well as to tactical users in the field.

The KH-12 is most probably an extremely agile satellite, with frequent manœuvres a routine aspect of its operations, able to respond rapidly to emerging situations, as well as pass unpredictably over a target, thereby frustrating evasion/deception efforts on the ground. Frequent manœuvring also permits shorter intervals between coverage of individual targets, as several satellites can manœuvre for repeated passes over the target area.

Despite these advances, the KH-12 suffers the shortcoming common to all photographic intelligence satellites, the inability to see through clouds. With much of the Soviet Union and other areas of interest frequently covered by clouds, this has always posed a problem for intelligence collection. However, in the past, this problem was primarily one of directing the satellite's coverage towards cloud-free areas and awaiting improved visibility in cloudy regions.

While this procedure may have been adequate for peacetime operations, it is clearly inadequate for wartime target acquisition.

As space-based imaging radar can see through clouds, utilization of synthetic aperture radar (SAR) techniques can potentially provide images with a resolution approaching that of photographic reconnaissance satellites. A project to develop such a satellite was initiated in late 1986 by Director of Central Intelligence George Bush.[3] This effort led to the successful test of the Indigo prototype[4] imaging radar satellite in January 1982.[5] Although the decision to proceed with an operational system was very controversial, development of the Lacrosse system was approved in 1983.[6]

The distinguishing features of the design of the Lacrosse satellite include a very large radar antenna, and solar panels to provide electrical power for the radar transmitter. Reportedly, the solar arrays have a wing-span of almost 50 metres,[7] which suggests that the power available to the radar could be in the range of 10–20 kilowatts, as much as 10 times greater than that of any previously flown space-based radar. It is difficult to assess the resolution that could be achieved by this radar in the absence of more detailed design information, but in principle the resolution could be better than 1 metre. While this is far short of the 10-centimetre resolution achievable with photographic means, it would certainly be adequate for the identification and tracking of major military units such as tanks or missile transporter vehicles.

Like the KH-12, the operational concept for the Lacrosse systems involve a constellation of four satellites of each type orbiting simultaneously and operating continuously, providing much more comprehensive coverage than afforded by previous systems.

Following the first Lacrosse, launched on 2 December 1988, two KH-12s were launched on the space shuttle from the Eastern Test Range in 1989 an 1990. Although the 57-degree and 62-degree inclination orbits of these satellites will preclude coverage of the northernmost reaches of the Soviet Union, these orbits provide enhanced coverage of the European theatre, as well as Soviet ICBM deployment areas along the Trans-Siberian railway. In 1990–91 additional pairs of KH-12 and Lacrosse satellites will be launched on Titan 4s from Vandenberg Air Force Base in California into the polar orbits traditionally used by US reconnaissance satellites. These will supplement the coverage of the lower-inclination satellites, as well as provide coverage of northern regions such as the Kola peninsula and the polar ice pack.

These satellites are the centre-piece of a new effort aimed at providing satellite imagery to tactical users. The Tactical Exploitation of National Capabilities (TENCAP) programme is designed to 'facilitate tactical use of national intelligence systems within an operational framework'.[8] TENCAP's purpose is to provide satellite imagery to battlefield commanders and weapon systems.

THE CONTRIBUTION OF US C³I TO THE DEFENCE OF NATO   405

Space-based signals intelligence

Satellites are also used to intercept radar and radio signals for analysis and interpretation. This intelligence is used to assess the identity, location, readiness and disposition of WTO combat elements throughout the entire depth of WTO territory. Interception of radar signals provides important information concerning the location and capabilities of WTO air-defence systems. Although radio transmission can be encoded, it is not possible to hide the transmitters from detection, and these satellites are a major source of information concerning the electronic order of battle of Soviet forces.

All of these satellites are built and operated by the NRO, with technical design support from the CIA. The USA currently operates three general classes of electronic intelligence satellites, distinguished by the orbits they use.

The geostationary satellites are the best known of these systems. The first-generation Rhyolite, and the much larger Chalet, are approaching the end of their operational lives. The first of the current generation of satellites, Magnum, was launched on the space shuttle in January 1985, and additional launches on the shuttle and Titan 4 are anticipated.

The Jumpseat[9] satellites operate in the so-called Molniya orbit, a highly elliptical orbit with a period of 12 hours and an inclination of 63 degrees. The first Jumpseat was launched in 1975, with the fifth launch in February 1987.

Down-links for signals intelligence (Sigint) data are located in the United States, Australia and Europe. Menwith Hill in the UK is reported to be the principal NATO theatre ground-segment node for the higher-altitude electronic intelligence satellites.[10] Although the facility is jointly operated with the UK's Government Communications Headquarters (GCHQ), the latter is not privy to the intelligence down-linked to Menwith Hill, since tapes containing the data are returned to the USA for analysis. Thus it is questionable whether this facility would be turned over to GCHQ in the event of US withdrawal from Europe.

Space-based meteorology

The United States Air Force currently orbits two of its own weather satellites as part of the Defence Meteorological Satellite Program (DMSP). The two most recent launches occurred in June 1987 and December 1988. These are supplemented by two similar satellites of the civilian National Oceanic and Atmospheric Administration (NOAA), which are an integral part of the military programme.[11] During the Falkland/Malvinas conflict, both of the military satellites were inoperative, and the British Navy was forced to rely entirely on the civilian satellites. The information from meteorological satellites has proved to be invaluable to the conduct of military operations, particularly the planning of air strikes and maritime manœuvres.

Weather data from DMSP are processed at the Air Force Global Weather Central at Offut Air Force Base, Nebraska, and at the Fleet Numerical Oceanographic Central in Monterey, California, for subsequent dissemination to tactical users. No permanent ground segment associated with DMSP is located in the European theatre, although aircraft-carriers do have the AN/SMQ-10 terminal, and a number of Mark III and Mark IV mobile receivers are operated by the Air Force.[12] The Air Weather Service 2d Weather Wing at Kapaun Air Station, Federal Republic of Germany, provides meteorological support to a variety of NATO functions.[13]

Space-based early warning

Early-warning satellites of the Defence Support Program (DSP) carry sensors that detect the heat from a rocket's engines, monitoring missile launches to ensure treaty compliance, as well as providing early warning of a nuclear attack. These satellites, along with radars of the Ballistic Missile Early Warning System (discussed below) would provide the primary (if not the only) source of warning for NATO of attack by long-range Soviet nuclear missiles. The degree of integration of these sensors with NATO defence preparations remains unclear. However, these systems could be very important for alerting air forces to scramble prior to missile impact.

The United States maintains a constellation of three active satellites in geostationary orbit, with an additional two older satellites in reserve. The first launch of the Titan 4 in June 1989 carried the first of the more sophisticated DSP-I (Improved).

The ground segment for the DSP includes a Simplified Processing Station (SPS), located at Kapaun Air Station, FRG.[14] Operated by the 6th Detachment of the 1st Space Wing, the SPS became operational in late 1982. Although located in Europe, data from the station are transmitted to Space Command headquarters in the USA for analysis and fusion with data from other sensors.

Space-based nuclear burst detection

The USA uses a number of different satellite systems to detect and locate nuclear explosions in space and in the earth's atmosphere. These systems would constitute a major source of intelligence concerning the location and yield of Soviet nuclear attacks and could assist in the planning and assessment of NATO nuclear strikes. The Vela satellites, which were dedicated to this function,[15] were deactivated in early 1985. This mission is now accomplished by nuclear detection sensors that are hosted on the DMSP[16] and DSP satellites. The Navstar navigation satellites currently entering service carry an improved nuclear burst detection system. This system, formerly known as Integrated Operational Nuclear Detection System (IONDS) and now referred to simply as the Nuclear Detection System (NDS), will relay this information to widely dispersed mobile ground terminals, enabling battle managers to identify the targets missed by defective missiles or warheads and to assign further strikes.

Airborne systems

Almost three decades after the U-2's intelligence mission was compromised with the downing of Francis Gary Powers over the Soviet Union, the United States continues to rely on this aircraft for strategic intelligence collection.[17] Despite the advent of the high-performance SR-71, the U-2 continues to offer advantages over the SR-71, in terms of dwell-time and the ability to maintain continuous coverage of an area of particular interest.[18] Production of the more capable U-2R began in 1967, with a number of these aircraft currently based at the RAF Akrotiri facility in Cyprus, as part of the 3d Detachment of the 9th Strategic Reconnaissance Wing. These aircraft (totalling between 14 and 17 in number) provide high-resolution imagery as well as electronic intelligence on WTO activities unmatched by other NATO collection assets.[19]

Twenty-five years after its first flight, the high-flying SR-71 remains the world's fastest air-breathing aircraft,[20] with a maximum speed of over Mach 3.2 (3900 km/h) at an altitude in excess of 26 kilometres. Of the 32 SR-71s originally produced, 20 are still in existence, with 9 operational at any one time, including 2 based at RAF Mildenhall (UK). Although these aircraft are subordinate to the US Strategic Air Command (SAC), they are responsive to theatre tasking requirements. The SR-71 carries a variety of cameras (some capable of a resolution of 75 cm at a range of 100 km), as well as passive electronic intelligence sensors and high-resolution radars that can transmit data to ground terminals in real time. The SR-71 is capable of covering an area of over 250 000 square km in one hour's flying time.[21]

In addition to the U-2 and SR-71 aircraft the SAC also operates a fleet of dedicated airborne electronic intelligence systems, such as the RC-135, from European bases. These aircraft, carrying a variety of passive electronic sensors, monitor Soviet and WTO communications, air defence radars and other active emitters. Bases that are host to RC-135 elements of the 55th Strategic Reconnaissance Wing include the 306th Strategic Wing at RAF Mildenhall and the 992d Support Squadron at Hellenikon Airport (Greece).[22]

Ground-based missile warning radars

The United States operates a world-wide network of large radars to provide warning of long-range missile attack. Along with data from the DSP satellites, this radar network is the principal source of warning of Soviet missile attack for both the United States and NATO.

The USA is now upgrading the existing Ballistic Missile Early Warning System (BMEWS), replacing the present fixed-array and mechanically-steered radars with phased-array radars. This replacement process was completed in 1987 at the Thule, Greenland, site and is under consideration for the Clear, Alaska, site. The Fylingdales Moor site in the UK will also be upgraded in the next few years, with a phased-array radar of the Pave Paws type, having three faces instead of two. All three sites are receiving new data-processing equipment and missile attack assessment computers. Although the Thule and Fylingdales sites are under the nominal control of British and Danish forces, the radars themselves are operated by US Space Command personnel.

Ground-based space-track systems

In addition to these early-warning radars, the USA maintains a global network of satellite tracking sensors. Many of these radar sensors are currently being upgraded to increase their range and sensitivity, greatly enhancing US ability to monitor events in the geostationary orbit. The Ground-based Electro-Optical Deep Space Surveillance (GEODSS) network of sophisticated telescopes will track very small objects in high, geostationary orbits.

Elements of this network in the European theatre include: (*a*) the 19th Surveillance Squadron, with two AN/FPS-17 detection and one AN/FPS-79 tracking radars at Pirinclik, Turkey, with support facilities at Diyarbakir, Turkey; (*b*) a Baker-Nunn tracking camera at San Vito dei Normanni, Italy; and (*c*) a solar observatory (part of the Solar Electro-Optical Network), also at San Vito dei Normanni. Planned upgrades include:(*a*) a GEODSS site in Portugal (probably the Azores);[23] and (*b*) a new deep-space imaging radar at Pirinclik.[24]

These space-track capabilities provide vital information to enable NATO tactical forces to evade the prying eyes and ears of Soviet photographic and electronic intelligence satellites, which can be used to locate and identify terrestrial combat units in peacetime as well as wartime.

Ground-based signals intelligence

The NSA operates a variety of ground-based electronic intelligence facilities in the European theatre. These systems are a major source of information concerning the activities of WTO air forces. The largest of these systems are the AN/FLR-9 Wullenweber Circularly Disposed Antenna Arrays (CDAAs).[25] These large circular arrays, with a diameter of about 265 metres, are used to locate and intercept low-frequency (such as submarine) and high-frequency (including radio-telephone) communications. CDAAs are located at Menwith Hill and Chicksands in the UK, San Vito dei Normanni in Italy, and in the FRG. The NSA also operates detachments in Berlin and Augsburg.

AirLand Battle: Echelon-Above-Corps systems

As mentioned above, the term Echelon-Above-Corps refers in this study to the US European theatre command infrastructure. This comprises USAREUR, USAFE and US Naval Forces Europe (USNAVEUR), which all fall under the overall authority of USEUCOM, with headquarters at Stuttgart. US intelligence assets controlled at this level of command apparently consist entirely of airborne imagery and Sigint collection systems.

THE CONTRIBUTION OF US C³I TO THE DEFENCE OF NATO   409

Reconnaissance aircraft

The most important of these assets is the TR-1, an updated derivative of the U-2 intelligence aircraft, operated by the SAC under the control of USAFE. The USA, however, has declared that these aircraft are NATO intelligence assets in that they will be available to reconnoitre areas besides those of immediate US interest. With a maximum ceiling of 21 km, these aircraft are 'the primary high-altitude reconnaissance force in the NATO area'.[26] A total of 12 TR-1s are flown by the 95th Reconnaissance Squadron of the 17th Tactical Reconnaissance Wing, based at the RAF Alconbury facility in the UK.

The TR-1 platform is host to the Advanced Synthetic Aperture Radar (ASARS-II),[27] which will locate fixed WTO command posts, transportation choke points and surface-to-surface missile sites, using both low-resolution search patterns and high-resolution spot beams, and has an imaging range of about 50 km beyond the forward line of own troops (FLOT).[28] Imagery from the radar will be down-linked directly in digital format to a Tactical Radar Correlator (TRAC) for subsequent dissemination.[29]

Besides the TR-1 aircraft, other theatre-level collection assets include the Comfy Levi and Coronet Solo airborne Sigint platforms.[30] These specially converted EC-130 aircraft are primarily intended for intercepting WTO communications.

By the mid-1990s the United States also plans to begin deployment of the Joint Surveillance Target Attack Radar System (JSTARS), a joint Army/Air Force large airborne imaging radar for identifying ground targets such as tanks and trucks at long ranges. The initial E-8A (formerly C-18) JSTARS platform will consist of two specially modified Boeing 707-320 aircraft, while a variety of other platforms are under consideration for the operational system.[31] JSTARS, which builds on earlier experimental systems, will be operated by the Air Force, although the Army is the principal user of the system's product.

The Grumman-built radar, which is mounted in a 7.5-metre pod on the side of the forward fuselage of the aircraft, 'would in some instances be able to distinguish tanks, which would be high-priority targets, from less-threatening trucks'.[32] The radar can operate in a synthetic-aperture mode to detect fixed targets, while mobile targets would be detected by operating the radar in a doppler mode.

Data from the aircraft will be down-linked to a mobile Ground Support Module, built by Motorola, for further analysis. Additional JSTARS connectivity is provided through the Joint Tactical Information Distribution System (JTIDS) and the Single Channel Ground and Airborne Radio System (SINGARS).[33] The data can also be relayed to Joint Tactical Fusion All-Source Analysis System and to NATO Battlefield Information and Collection System centres (all of which are discussed in section III). Planning is under way to make JSTARS compatible with the ground segments of the French ORCHIDEE (Observatoire Radar Cohérent Heliporté d'Investigation des Eléments Ennemis) and the British ASTOR (Airborne Standoff Radar), although both of these systems use much less capable radars carried, on small, low-altitude platforms. However, doubts remain over whether the data from JSTARS will be transmitted directly to allied forces.[34] Each aircraft will be able to monitor an area roughly 480 km by 320 km.[35] A total of 22 aircraft and 107 Ground Support Modules are planned, with the first aircraft operational in Europe by 1994.[36]

The area of coverage of JSTARS is limited by concerns over the aircraft's survivability. According to JSTARS Deputy Programme Manager, Colonel G. Sidney Smith:

Over the battlefield, everything is vulnerable. People confuse survivability with immortality. AWACS faces the same problem, but no one worries about it because it is so protected. Our layered defenses are there to protect AWACS, and will be there to protect the Joint STARS aircraft. And this aircraft won't be sitting up at the FLOT [forward line of own troops], it stands back. If it is in danger, it can move still further back and still provide usable pictures ... [JSTARS] will be able to stand back almost 190 [nautical miles] from the forward edge of the battle and be capable of looking 20-30 [nautical miles] beyond the FLOT.[37]

Remotely piloted vehicles (RPVs) can also be expected to make a significant contribution to NATO's intelligence assets in the future. Although many of the NATO allies have developed a number of RPV systems, for the most part these are short-range craft of limited endurance dedicated to direct support of lower-echelon forces.[38] In contrast, the USA has long focused on long-range high-altitude RPVs for strategic intelligence collection, beginning with the supersonic D-21 in the mid-1960s.[39] The YQM-94A and YQM-98A Compass Cope prototypes, which flew in the mid-1970s, were very large (wing-spans of 25–27.5 metres), operating at altitudes above 17 km.[40] With a flight duration in excess of 24 hours, these platforms would have supported imaging radar collection at ranges of up to 200 km, and signals intelligence at ranges up to 500 km.[41] Although these RPVs never entered operational service, the 1986 US Department of Defense Remotely Piloted Vehicle Roadmap called for the development of three classes of RPVs, including a long-range vehicle to support theatre intelligence collection.[42] The US Defense Advanced Research Projects Agency (DARPA) Director Robert Duncan noted that the Teal Cameo project was developing a 'high altitude theatre UAV [Unmanned Air Vehicle] aimed at providing alternatives for an unmanned successor to the TR-1 and JSTARS platform'.[43]

In addition to the nationally controlled theatre intelligence assets, the USA also provides personnel to man and maintain the NATO-operated E-3 Sentry AWACS, one of the pricipal components of the alliance's air defence system. The purchase and maintenance of the 18 NATO AWACS is funded under the NATO Infrastructure Programme, and thus would remain in place following a US withdrawal from Europe. However, the USA would remove its own AWACS aircraft operated by Tactical Air Command from the theatre.

AirLand Battle: Corps and lower-echelon systems

For the Army at the Corps level, intelligence support is managed by a Combat Electronic Warfare and Intelligence (CEWI) Group, which operates a Tactical Operations Center to co-ordinate 'intelligence systems from other services or from strategic assets normally associated with echelons above corps'.[44] The CEWI Group includes an Aerial Exploitation Battalion, which controls the Guardrail and Quicklook airborne collection systems.[45] Guardrail is an airborne communications intelligence (Comint) system mounted on a Beechcraft RC-12D. It intercepts ground-based communications, calculates the location of emitters, and transmits these data to a transportable ground station.[46] Quicklook is a tactical electronic intelligence system mounted on the Grumman RV-1D Mohawk aircraft, used for detection and localization of opposing forces' radars.[47] The CEWI's Tactical Exploitation Battalion controls ground-based assets such as Quick Fix (AN/ALQ-151), Teampack (AN/MSQ-103A), Trailblazer (AN/TSQ-114), Tacjam (AN/MLQ-34), Tacelis (AN/TSQ-112) and Rembass.[48]

The US Air Force operates a number of organic reconnaissance aircraft, such as the RF-4. However, these aircraft are similar in capabilities to those of other European forces, and their withdrawal would not, in and of itself, have a significant impact on the military situation.

Intelligence assets for naval warfare

All NATO naval forces have organic intelligence collection systems that monitor their immediate environment. However, the US Navy operates a variety of assets that monitor the entire NATO maritime theatre of operations. In addition to the space-based systems described above, the White Cloud electronic intelligence (Elint) satellite network is primarily dedicated to the maritime mission. The land-based P-3 Orion anti-submarine warfare (ASW) aircraft also perform Elint functions. Sea-based sensors include the Sound Surveillance System (SOSUS) and the Surveillance Towed Array Sensor System (SURTASS). All of these sensors are important for locating Soviet fleet elements that are not currently engaged with NATO forces. These systems are particularly important for the support and protection of convoys moving forces from the USA to Europe, and they also will assist in operations in the Baltic and Mediterranean seas.

White Cloud

The Navy, in conjunction with the NRO, operates the White Cloud Naval Ocean Surveillance Satellites, which fly in low-altitude orbits. These satellites, used to locate naval forces and air defence radars through triangulation, are launched in clusters of one primary satellite and three sub-satellites. Five of these clusters were thought to be operational in 1989. The first launch of White

Cloud on the new Titan 2 booster took place in September 1988. The Navy Security Group Facility at Edzell, Scotland, is one of five world-wide ground stations for the White Cloud system.[49] This facility has been expanded in recent years with several additional domes for tracking antenna.

SOSUS

The principal US and NATO system for detecting submerged submarines in the open ocean is the Sound Surveillance System. SOSUS is a global network of submerged hydrophone arrays that detects the noise generated by submarines. The data generated by these arrays are transmitted via cables to coastal stations, which relay the information to centralized processing facilities for analysis and dissemination. SOSUS hydrophone arrays in the NATO theatre form a barrier across the Greenland–Iceland–UK (GIUK) Gap, as well as between Norway and Bear Island in the Barents Sea, along the Atlantic approaches to the Straits of Gibraltar, and at critical locations in the Mediterranean Sea and other coastal areas.[50] SOSUS is used both for detection and localization of Soviet submarines, for attack by P-3 Orion patrol aircraft and other forces. SOSUS sensors can reportedly determine the location of a submarine with an accuracy of a few tens of km[51] at ranges of several thousand km.[52] The system has been constantly improved since the initial introduction of the Caesar system in the early 1950s.[53] The Integrated Undersea Surveillance System is the focus of these efforts.[54] Planned upgrades to SOSUS include the introduction of fibre-optic cables to link acoustic sensors to shore stations under the Fixed Distributed System, also known as Ariadne.[55] SOSUS is an integral part of NATO's strategy of denying Soviet submarines access to the North Atlantic during a conflict. As such it makes an important contribution to the maritime resupply of Europe from the United States.

SURTASS

In addition to the fixed SOSUS system, the US Navy is also deploying a fleet of ships carrying the Surveillance Towed Array Sensor System, which is 'designed to be a mobile replacement for SOSUS'.[56] Ten of the initial 18 Stalwart class ships (designated T-AGOS), displacing 2285 tons, are currently operational, with delivery of the last ship planned for 1990. An additional 9 ships, with a small waterplane twin hull (SWATH) configuration and displacing 4200 tons, are to follow.[57] The ships tow long hydrophone arrays which can detect and localize targets over extended ranges.

According to Admiral James Watkins, the long sensor array of the T-AGOS ships 'permits us to work them in synergism with the SOSUS arrays' with submarines being detected at ranges in excess of 1000 miles (1600 km). These ships also 'provide a very survivable platform as well. We know that the Soviets have targeted our SOSUS arrays. And they will be gone probably before conflict starts by either sabotage or some other manner'.[58]

P-3 Orion patrol aircraft

Along with a number of other NATO allies, the United States Navy operates the P-3 Orion patrol aircraft, used for localization and destruction of hostile submarines. These aircraft carry acoustic sensors as well as torpedoes and nuclear depth charges.[59] In addition, a small number of these aircraft have been specially modified (the EP-3E) for electronic intelligence collection.[60] With an operational radius of 1000–1500 nautical miles, these aircraft patrol the entire Mediterranean Sea, the North Atlantic Ocean and the coastal waters of the Soviet Union. Bases supporting P-3 operations in the NATO theatre include Lajes (Azores), Sigonella (Sicily), Rota (Spain), and Keflavík (Iceland). Additional out-of-theatre P-3 bases supporting NATO include Roosevelt Roads in Puerto Rico, Bermuda Island and Brunswick (Maine).[61] The Navy has long sought a more capable replacement for the P-3.[62] Under current plans, the first Long Range Air Antisubmarine-warfare Capable Aircraft (LRAACA)[63] would fly in 1990, and 125 would be produced by the end of 1998.

Data fusion systems

Although collection assets are the most visible element of the US intelligence presence in Europe, the most vital component of the intelligence process is analysis. The United States currently operates several data fusion systems to correlate the intelligence collected by theatre and other sensors, and several major upgrades are under development. Unfortunately, the central importance of these intelligence correlation systems is matched by the paucity of descriptive material in the open literature. The precise functioning of these systems is among the more closely held aspects of the US military establishment. An assessment of the importance of these systems is further complicated by the present flux in their development. The adage 'if it works, its obsolete' aptly describes the current state of affairs. Major uncertainties surround the timing and capabilities of most of the systems that are slated for deployment in the early 1990s.

AirLand Battle systems

The United States has initiated a number of programmes to improve the dissemination of intelligence derived from national systems to combat forces. These include the Intelligence Communications Architecture, the Imagery Acquisition and Management Plan, and the Defence-Wide Intelligence Plan. The operational expression of these plans is the Tactical Exploitation of National Capabilities (TENCAP). This programme is designed to 'facilitate tactical use of national intelligence systems within an operational framework'.[64] TENCAP will provide imagery and electronic intelligence from national systems, particularly satellites, to battlefield commanders and weapon systems.

Initiated by the Army in 1973, TENCAP met with sufficient success that Congress in 1977 'directed each service to establish a [. . .] TENCAP office to improve military use of national systems'.[65]

DOD formed the Defence Reconnaissance Support Programme (DRSP) to specifically address, across all Services, the use of space-based systems from stand-alone acquisition to sharing of national space programs, and research and development toward new capabilities. Also, each of the Services has organized offices to address the operational concepts and the employment of space systems in direct support of tactical operations.[66]

TENCAP systems will rely on the combination of the KH-12 and Lacrosse satellite intelligence and the Milstar satellite communications system. Field commanders will be able to request imagery via Milstar and then receive it within minutes either directly from KH-12 and Lacrosse, or via Milstar. A number of shortcomings remain, however:

The problems of national systems support to contingency forces have generally fallen into three major categories. First, the systems themselves must respond to collection priorities set at the national level. These priorities may or may not be responsive to collection requirements for contingency operations of in Third World areas. The second category of problems compounds the first, in that the general capabilities of tactical units to request and receive information from national intelligence systems are frequently lacking. Third, are the limitations of the Armed Services for using information from national sources. In other words, the inability to properly consume and exploit information from national technical means.[67]

In an effort to remedy these problems, the Joint Service Imagery Processing System (JSIPS) was initiated to co-ordinate Service development TENCAP hardware. 'JSIPS provides tactical commanders with a deployable, modular intelligence support system that processes and exploits imagery data from national, tactical and strategic sensors and platforms'.[68] The Air Force and Marine Corps initiated JSIPS in November 1985, with the Army joining in January 1987. JSIPS elements include the Air Force ADDISS (Advanced Deployable Digital Imagery Support System), the Marine Corps ASIP (All Source Imagery Processor) and the Army IPDS (Imagery Processing and Dissemination System). FIST (Fleet Imagery Support Terminal), the Navy equivalent of the JSIPS effort, is expected to achieve an initial operational capability by December 1988 and a full operational capability by October 1990.[69]

Current Air Force systems for TENCAP applications include the Operational Application of Special Intelligence System (OASIS) at the Air Force's Tactical Fusion Center collocated with the Allied Forces Central Europe (AFCENT) Static War Headquarters (SWHQ) at Börfink, FRG.[70] OASIS will provide a 'complete integrated air and ground situation assessment, based on both NATO's operations data and data from United States special intelligence sources'.[71] This includes two-way links between OASIS and the USAFE Combat Operations Intelligence Center (COIC) at Ramstein AB, FRG.

THE CONTRIBUTION OF US C³I TO THE DEFENCE OF NATO

US Corps-level headquarters use a variety of analysis assets. The Electronic Processing and Dissemination System (EPDS) is a ground-based, computer-assisted electronic intelligence correlation facility for theatre and national sensors. The Enhanced Tactical User Terminal (ETUS) is a processing and visual display for ELINT and imagery intelligence support to the Corps.[72]

The principal Corps-level data fusion improvement programme of the US Army and Air Force, however, is the Joint Tactical Fusion Programme (JTFP).[73] The All-Source Analysis System (ASAS) is the Army element of this programme, and the Enemy Situation Correlation Element (ENSCE) is the Air Force equivalent. ASAS is intended to be the standard data fusion system used by US Echelon-Above-Corps, Corps and Division level headquarters. The Air Force will deploy the ENSCE as part of the Tactical Air Control Center of the numbered Air Forces.[74]

ASAS/ENSCE will correlate data from organic sensors, lower-echelon intelligence assets, as well as higher-echelon sensors, including national systems. 'The ASAS/ENSCE is the analytical hub for intelligence fusion and dissemination in the Army Corps and division and the numbered air forces . . . The Army contributes about 90 per cent of the programme funding and the Air Force about 10 per cent'.[75] This programme will 'develop a single automated system that would correlate, analyze, and disseminate high volumes of time-sensitive, multi-sensor intelligence data. ASAS/ENSCE is to provide tactical commanders with precise location and structure of the opposing forces and near real-time battle situation displays'.[76] 'The system will also facilitate tasking of organic assets and requests for intelligence data from national assets'.[77]

'These functions, which can take days to perform with current systems, can be done in minutes, if ASAS performs as expected'.[78] However, the programme has experienced serious cost growth and schedule slippage, primarily due to problems with mission software, with an initial operational capability expected in April 1993. Procurement of over 100 modules is anticipated.[79]

Design of ASAS/ENSCE began in fiscal [year] 1984. Development of system modules began in fiscal [year] 1985. The system was designed to incorporate two other intelligence fusion systems: the Technical Control and Analysis Center-Division (TCAC-D, also often referred to by its military designation as the AN/TSQ-130V) and the Battlefield Exploitation and Target Acquisition (BETA) system.[80]

The TCAC system has been deployed with 5th and 7th Corps, and BETA has become the core of the Limited Operational Capability Europe (LOCE), discussed below.

Maritime systems

The Navy Ocean Surveillance Information System (OSIS) presently provides 'near real-time, all-source indications and warning, positional and movement information, and over-the-horizon targeting support . . .'. It includes two primary subsystems, the Sea Watch programme and the OSIS Baseline

Subsystem (OBS). A Naval Ocean Surveillance Information Center (NOSIC) in Suitland, Maryland and five other world-wide centres, including two in Europe, operate the OSIS and disseminate its automatic contact and tracking reports.[81]

Data from SOSUS sensors are relayed to a number of stations located on the coasts of NATO countries. The data are combined with data from other intelligence sources at NOSIC in the United States and at Fleet Ocean Surveillance Information Centers (FOSIC) in Rota, Spain, and London, UK.

The Navy's increasing reliance on space intelligence is notable. 'Today we use over three-quarters of the tactical data gathered by National space systems', according to Dr E. Ann Berman, Deputy Assistant Secretary of the Navy. She notes the need to expand the Navy's tactical use of space for surveillance, 'first, by more effective leveraging of national assets; and second, by designing, developing, acquiring, and operating tactical space systems dedicated to our warfare commanders'.[82] Naval Space Command Commodore Richard H. Truly concurs, for he considers direct fleet support the pre-eminent mission of his command, and this includes 'exploring methods for improving national systems support through such programs as Tactical Exploitations of National Capabilities'.[83]

Dissemination of US intelligence to NATO allies

As briefly discussed in the introduction, intelligence derived from US-controlled sources is made available to the NATO allies in a variety of ways. These are best described in terms of the peacetime and wartime arrangements.[84]

Peacetime arrangements

While intelligence collection in NATO remains a national responsibility, the allies are obliged to contribute to the regular alliance-approved threat assessments of the WTO and, moreover, provide timely warning of hostile indications. Nations distribute their own finished assessments through their intelligence liaison officers or 'national intelligence cells' at Supreme Headquarters Allied Powers Europe (SHAPE) and other major headquarters. Due to the widespread dissemination of the intelligence input at this level, the USA reserves its most sensitive intelligence—particularly communications intercepts—for distribution through the relatively recent Special Handling Detachments (SHD) at SHAPE and NATO headquarters. While access to and use of this high-grade intelligence is more restricted, it does provide an invaluable contribution to NATO decision-making.

US intelligence also feeds into NATO's Indications and Warning (I&W) system that monitors on a 24-hour basis the operational readiness of WTO forces. As part of the US Worldwide Warning Indicator Monitoring System (WWIMS), the major US commands in Europe operate a sub-network of I&W centres known as the Warning Indications Systems, Europe (WISE). At the apex of this network is the I&W board, run by USEUCOM in Stuttgart, that feeds and receives information from the I&W centres at the subordinate command headquarters, such as the one operated by USAREUR headquarters in Heidelberg. Any significant change in the alert status or location of WTO forces is reported to the I&W staffs at SHAPE, AFCENT and NATO headquarters for comparison with other intelligence.

While these organizational arrangements are adequate for exchanging finished intelligence products in peacetime, and may suffice during times of crisis, the large volume of intelligence data that must be processed and disseminated to the relevant users in wartime has led to several efforts to reorganize and automate US–allied intelligence exchange arrangements.

Wartime arrangements

The only current semi-automated system for distributing US intelligence to NATO allies is the Limited Operational Capability Europe (LOCE). As one description states, LOCE 'currently provides intelligence collection, analysis, and distribution to NATO users over secure communications systems, for early warning, situation assessment and targeting activities. Information provided will assist early target nomination for strike or reconnaissance, and support threat assessment'.[85] The LOCE deployment in Europe consists of a Central Automated Data Processing Facility (Correlation Center) located within the Combat Operations Intelligence Center (COIC) at Ramstein AB, FRG. This provides the hub that links a network of Sensor Interface Modules at key intelligence sensor collection centres to various user display terminals at major NATO headquarters.[86]

Sensor Interface Module locations included (as of 1981) 26th Tactical Reconnaissance Wing, 6911st Electronic Security Squadron, 330th Company, V Corps, VII Corps and TAC Fusion Center. Remote display locations include AAFCE, Northern Army Group/2d Allied Tactical Air Force (NORTHAG/2d ATAF), CENTAG/4th ATAF, EUCOM, USAREUR, V and VII Corps.

The USA hopes that LOCE will provide the basis for a current NATO programme to automate the exchange of data between national intelligence systems. Known as the Battlefield Information Collection and Exploitation System (BICES), it has been described as:

[A] concept for using a combination of NATO and individual nations' [automatic data processing] systems to provide timely intelligence support to commanders . . . By the very nature of certain sources, intelligence on events in the area of one nation's forces will inevitably be collected by others, and a rapid, automatic means of transmitting such material across national boundaries is required. Internetting national systems within the BICES framework will also allow this.[87]

The architecture of BICES has yet to be established, as 'the alternatives being considered range from a "big intelligence factory processing intelligence for all commanders" to a decentralized network with localized data bases permitting direct transmission of intelligence among intelligence systems'.[88]

The initial operational capability for BICES is planned for 1992, with a full operational capability by 2000.

While the primary function of BICES will be to exchange data relevant to the prosecution of ground operations, parallel programmes are also under way for exchanging tactical air and naval intelligence among NATO users.[89] However, these systems are being designed to exchange intelligence to only the SECRET level of classification. Higher grade intelligence will most likely be fed into NATO channels via the principal points of contact: SHAPE's Static Wartime Headquarters located next to the peacetime facility near Mons, Belgium; AFCENT's SWHQ at Börfink through the collocated US-operated Tactical Fusion Center; and the CENTAG SWHQ at Ruppertsweiler, FRG, via links to the newly created Echelon-Above-Corps Intelligence Center (EACIC) run by USAREUR. Planning is also underway to link the latter to the main intelligence centres supporting NORTHAG and the FRG armed forces.

III. Communications assets

The USA operates an elaborate network of communications systems for the support of the European Command as well as various intelligence facilities located within the theatre. In terms of direct communications support to NATO, the US contributes one of two signals battalions that make up the CENTAG Signal Support Group. This is responsible for linking CENTAG HQ with the national corps units under its command. In addition, the USA contributes approximately 28 per cent to NATO's commonly funded command and control infrastructure budget.[90] Last but not least, US personnel also operate jointly managed NATO C^3I facilities such as ground-based air defence radars and AWACS (Airborne Warning and Control System) aircraft.

An essential prerequisite for any assessment of the impact of a US withdrawal from Europe for NATO command and control is obviously a clearer understanding of the present US communications facilities in Europe and NATO's dependence on them. These can be divided up according to whether their *primary* mission is peacetime support of forces garrisoned in Europe or wartime operational control, although some serve both functions.

Peacetime communications support

The principal US communications network in Europe is the Defence Communications System (DCS) that serves US forces and government agencies worldwide. This consists of terrestrial over-the-horizon and line-of-sight microwave radio links extending from Iceland to eastern Turkey as well as a network of satellite communications terminals that utilize the Defense Satellite Communications System (DSCS).[91] Since the European DCS is primarily an analog transmission system, it is currently being converted to carry digital communica-

THE CONTRIBUTION OF US C³I TO THE DEFENCE OF NATO

tions under the Digital European Backbone System (DEB) modernization programme.[92] Fibre optics are also being added as part of this modernization.[93]

The European DCS provides the transmission links for the three main US military communications switching networks: AUTOVON (Automatic Voice Network) is the world-wide US military telephone system for non-secure communications while AUTOSEVOCOM (Automatic Secure Voice Communications) provides the same service for secure communications.[94] AUTODIN (Automatic Digital Network) is used solely for passing data and record traffic. These communications sub-systems of the DCS are all being upgraded (essentially digitalized), AUTOVON by the Defence Switched Network (DSN), AUTODIN by the Integrated AUTODIN System (IAS), and AUTOSEVOCOM by the Secure Voice System (SVS).[95] Approximately 7000 of DCS Europe's circuits are owned directly by the DOD while the remaining 3000 are commercially leased.[96]

In addition to AUTOVON, the US military also operates the European Telephone System (ETS), which is designed to replace several separate non-secure voice networks in the NATO region (USAREUR Dial Service Assistance and DOD Network, the USAFE Dial Network and the USAFE Ringdown Network).[97] Some of these local networks may still be operational. Using equipment purchased from a consortium led by Siemens, the ETS currently serves US forces in the FRG, the Netherlands, Belgium and Italy. Plans exist to extend this network to other NATO countries.[98]

The DCS and ETS networks are used primarily for day-to-day administration and support. While they would also be used for alerting and managing US forces during a crisis, their operational role once war starts is limited. In addition to the US forces base networks, the US intelligence agencies operating in Europe also have separate communication channels, although they appear to use the DCS for transmission purposes. The NSA, for example, uses the Defense Special Security Communications System (DSSCS) which ties each of the listening posts in Europe to NSA Headquarters at Ft Meade outside of Washington, DC.[99] The CIA presumably uses US Embassy communications.

Peacetime/wartime operational control

A separate category of military communications systems provide operational control of US armed forces in Europe, both in peacetime and during war. Some of these, however, have only an indirect relevance to the defence of NATO. They are also under control of the individual services.

Air Force communications

The US Air Force operates the following communications facilities in Europe:

The North Atlantic Relay System (NARS). This consists of a chain of troposcatter transmitters between North America and Great Britain via Greenland,

Iceland, and the Faeroes. NARS essentially links together the air defence and ballistic missile early warning radars at those locations. The US Navy apparently makes use of this system as well for transmitting information acquired from its underwater SOSUS ASW sensors.[100]

Scope Signal III. Established originally under the Giant Talk programme, Scope Signal III provides high-frequency (HF) communication links to SAC forces—principally bombers—operating from or transiting through the European theatre. Several HF transmitters are located in the UK, Turkey, Spain and the Azores.[101] SAC has also established its own communications system to handle information acquired from its airborne intelligence operations. Known as Rivet Switch, there are a variety of ground facilities in Europe that receive data down-linked from RC-135 Rivet Joint and Cobra Ball reconnaissance aircraft.[102]

The USAFE Primary Alert System (PAS). Formally a non-secure voice communications system linking Air Force tactical command posts via military and leased circuits, PAS has been upgraded in recent years to permit secure teletype service.[103]

The USAFE Inform Net. A more extensive HF radio network for command and control has also been established throughout Europe to support USAF operations. Transmitters are located at all the principal USAFE operating bases.[104]

Navy communications

The US Navy operates an extensive command and control system in Europe consisting of the following facilities: (*a*) the Scottish Microwave Network, linking US naval facilities in northern Scotland; (*b*) HF transmitters at various locations in Europe providing communications back-up to satellite links; (*c*) Very Low Frequency (VLF) stations supportint submarine operations; (*d*) FLTSATCOM (Fleet Satellite Communications System) ground terminals at main Navy facilities; and (*e*) Loran-C radio navigation stations.[105]

Army communications

The US Army operates the USAREUR Tactical Alert Net (TAN) which, like the USAFE PAS, provides non-secure voice communications over leased circuits. The US Army also has its own mobile communications facilities, organic to Division and Corps units. These are currently undergoing major modernization with the TRI-TAC/Mobile Subscriber Equipment (MSE) programmes.

Wartime operational control

The final category of US communications systems in Europe are those solely for wartime use—principally the command and control of nuclear forces. This includes four EC-135 H airborne command posts—code-named Silk Purse—for use by CINCEUR/SACEUR. These are based at RAF Mildenhall, with associated communications ground entry points at other locations in Europe.[106] To maintain connectivity with nuclear storage facilities, custodial units, and deployed nuclear capable forces, CINCEUR can use either a HF radio net known as Regency Net or a UHF satellite link known as Flaming Arrow.[107]

IV. Impact of a US withdrawal

The consequence of a US withdrawal from Europe would hinge, as noted above, on whether the USA maintain its commitment to defend Europe in the event of an attack on NATO. This study posits two different US withdrawal options. In the less sweeping scenario, option B, all US military personnel and transportable systems are withdrawn, with fixed facilities either 'mothballed' or handed over to European personnel. Option A envisages comprehensive withdrawal of all US personnel; equipment with fixed facilities is closed or dismantled, and the USA is no longer committed to defend Europe. The USA remains committed to return to Europe if required. The impact of a withdrawal of US C^3I assets from Europe is discussed below, starting with option B.

Option B

The basic objective of this form of disengagement would be for the USA to retain the necessary infrastructure to defend NATO without there being an overt US military presence in Europe. Periodic exercises could also be carried out to ensure that if US forces are required to return there will be a minimum of disruption and friction. The absence of such exercises would greatly complicate the task of re-integrating US C^3I support to European forces.

This scenario appears the least damaging to European security interests. While NATO would lose the contribution of US airborne intelligence sensors operating from European bases in peacetime, the United States can still make available to its allies intelligence gathered from its imaging and Sigint satellites. This would include valuable status reports on the WTO forces in Eastern Europe and the western districts of the USSR. The major problem concerns the fixed intelligence facilities such as those operated by the NSA or the US Navy. The USA could conceivably transfer these facilities to European control. Given the sensitivity of the equipment this would be very difficult, but it is not inconceivable. The same applies to the intelligence fusion and distribution centres, such as the Navy's FOSIC.

In contrast, the status of US communications assets in Europe under this arrangement presents fewer problems. The various transmission systems such as the DCS troposcatter and microwave towers, SATCOM terminals and base telephone networks could be maintained on a caretaker basis by European personnel or US civilian contractors. Since much of this equipment supports US peacetime operation in Europe, some of it could be dismantled with minimum consequences, should US forces return at a later date. Some of the equipment could also be put to peacetime use by the Europeans, without any of the problems associated with the transfer of more sensitive intelligence-gathering facilities.

Under this scenario, the USA would presumably continue to contribute to the NATO Infrastructure Programme, which funds NATO's communications projects. While the US signals unit supporting CENTAG would leave, its equipment could be transferred to European control. Similarly, the loss of US personnel from jointly-run NATO headquarters and other C^3I facilities would have to be redressed by increasing the numbers of European personnel. US-supplied equipment at these facilities would presumably also require periodic maintenance by US civilian contractors. Examples include the Central Region Command and Control Information System (CCIS) that uses the same data processing equipment as the US Worldwide Command and Control System (WWMCCS). Similarly, the E-3 AWACS airframe is manufactured by Boeing, the main computers are built by IBM, engines by Pratt and Whitney and the radar by Westinghouse. Currently these companies are under contract to provide support for the NATO AWACS aircraft.[108]

Option A

Under this scenario the European members of NATO cannot count on the support of US intelligence and communications assets that under option B could either return or be reactivated in a serious crisis. Of the two types of disengagement, this would clearly be more damaging to NATO.

The most serious gaps would appear in the area of intelligence gathering. As noted above, US intelligence assets provide a unique service to NATO in being able to observe WTO activities well beyond the immediate border areas. The loss of satellite-derived imagery, providing indications and warning, would be particularly hard felt since this is by far the most vital function of US satellites. To a great extent, however, the on-site monitoring arrangements that are expected to be instituted as part of a Conventional Armed Forces in Europe (CFE) agreement will help in this regard. European photoreconnaissance satellites may also feel some of the gap. Pictures taken by the French SPOT (Système Probatoire d'Observation de la Terre) satellites would provide some information, but not of the detail currently available. The French Helios photographic reconnaissance satellite, under development with Italian and Spanish participation,[109] is based on the SPOT remote-sensing satellite. The first of two planned satellites is slated for launch in 1993 or 1994. However, Helios will not provide the high-resolution imagery available from US systems, nor would it be survivable in the face of Soviet anti-satellite operations. In wartime, as Brigadier-General Howard D. Graves has observed, satellites 'offer unique real-time capabilities to see to the full depth of the enemy's forces and their supporting bases and to help us attack as deep as necessary to disrupt his ability to execute his plans—the keystones of the Airland battle Doctrine'.[110]

Signals intelligence would be another area of major loss. While the European allies—principally the UK and the FRG—have some ground-based Sigint collection facilities whose coverage presumably extends into Eastern Europe, the loss of the impressive range of US national and theatre-controlled Sigint assets would cause serious gaps in coverage. This might be offset by the possible availability of the first British Zircon electronic-intelligence satellite, although official British Government statements have asserted that this programme has been cancelled.

Withdrawal of monitoring activities close to the Forward Edge of the Battle Area (FEBA), where the Europeans are less dependent on US intelligence assets, would still be hard felt by NATO. The exception would be the seven British E-3s[111] along with France's four AWACS aircraft.[112] The Europeans have no equivalent to the TR-1 aircraft that patrol the border areas, nor do they possess the range of US airborne Sigint assets. While the planned procurement of such systems as the electronic intelligence variant of the Tornado aircraft, the Castor system, and long-range RPVs will improve the situation, the Europeans will mourn the loss of their US counterparts. The biggest gaps will obviously be created in the areas occupied and immediately adjacent to US forces. However, the loss of the intelligence assets under the control of US III Corps, which is nominally committed to the defence of the NORTHAG region, would also be considerable. These assets are more than the combined total of all of NORTHAG.

Similarly, the US naval sensor systems provide a broad area-surveillance capability that is not matched by other NATO assets. SOSUS and SURTASS have no allied equivalent, and the P-3s are significantly more capable than their allied counterparts. In the absence of US area-surveillance capabilities, NATO ASW forces would be largely reduced to prosecuting Soviet submarine targets as they present themselves to NATO naval forces. This would seem to concede the initiative to the Soviet Union, and call into question the effectiveness of NATO's ASW effort.

In contrast to the loss of US intelligence assets, the complete removal of US communications facilities from Europe would not be so damaging. NATO's existing commonly funded communications infrastructure is quite extensive and, in many respects, duplicates the functions of the US systems that would be dismantled.

NATO's counterpart to the DCS network in Europe is the NATO Integrated Communications System (NICS). Like the DCS it comprises a variety of transmission and switching systems. The backbone of the network is the Ace HIGH system of troposcatter and microwave links that stretches across Europe from Norway to Turkey.[113] This is being modernized as well to take digital communications. A separate microwave network, known as CIP-67, has also been constructed to serve the central region of Allied Command Europe. Furthermore, NATO has its own communications satellite programme and associated ground terminals that will shortly be upgraded following the purchase of British Skynet IV satellites. These transmission networks support the various dedicated NATO switching systems: the Telegraph Automatic Relay Equipment (TARE) for data traffic, the Initial Voice Switched Network (IVSN), for telephone communications, and the NATO Secure Voice Network (NSVN).[114]

To a varying extent NICS is interoperable with the US DCS network although this is often achieved through a cumbersome and time-consuming system of 'gateway' links.[115] This ensures some degree of redundancy in the event that parts of either network are damaged or destroyed in wartime. The dismantlement of the US-controlled network, therefore, would deny NATO the use of those systems should the NICS be degraded in some manner. One exception, however, would be the use of US communications satellites which presumably would not be affected by a general US withdrawal from Europe.[116]

The loss of US nuclear-related communications facilities would have even less of an impact since a US withdrawal would presumably leave only British nuclear forces under the nominal control of a European SACEUR. In any event, NATO operates two HF communications nets: Bright Dawn and the HF Broadcast Net for the use of SACEUR.[117]

V. Conclusion

In conclusion, US C^3I assets make a substantial and growing contribution to the defence of NATO. This is most evident in the provision of intelligence that cannot be gathered by European systems. While the United States could still provide such intelligence to its European allies without a peacetime presence in Europe, this would not be the case under the more extreme form of disengagement outlined above. Under this scenario, the European members of NATO would need to redress their deficiency in long-range reconnaissance and surveillance systems, or alternatively, adopt a defence strategy that places less emphasis on tactics that require the support of such systems. Given the already extensive network of NATO communications system, the loss of US communications assets from the region would, in contrast, not pose a serious problem to the defence of Europe.

Notes and references

[1] This chapter was relied on heavily by Simon Duke for his chapter on C^3I in the first volume to emerge from the SIPRI 'Europe After an American Withdrawal' project; *United States Military Forces and Installations in Europe* (Oxford University Press: Oxford, 1989).

[2] For further details on the US contribution to NATO's command structure, see US General Accounting Office, *Relationships between US and NATO Military Command Structures—Need for Closer Integration*, report no. GAO/LCD-77-447 (US Government Printing Office: Washington, DC, 26 Oct. 1977).

[3] Woodward, B., 'At CIA, a rebuilder goes with the flow', *Washington Post*, 10 Aug. 1988, p. A8.

THE CONTRIBUTION OF US C^{3}I TO THE DEFENCE OF NATO   425

[4] Kenden, A., 'A new US military mission', *Journal of the British Interplanetary Society*, vol. 35 (1982), pp. 441–44.

[5] Kenden, A., 'Unusual military payload', *Spaceflight*, Nov. 1982, p. 410.

[6] Casserly, J., 'Reading between the pages, or why we can't keep a secret', *Armed Forces Journal International*, Aug. 1988, p. 60.

[7] 'Satellite with 150-ft. span set for launch on mission 27', *Aviation Week & Space Technology*, 7 Nov. 1988, p. 25.

[8] US Air Force, *FY 1984 RDT&E Program Element Descriptive Summaries* (US Department of the Air Force: Washington, DC, 1984), p 690.

[9] Richelson, J., *The U.S. Intelligence Community* (Ballinger: Cambridge, Mass., 1985), p. 122.

[10] Ball, D., *Pine Gap* (Allen & Unwin: Sydney, 1988), p. 61.

[11] US General Accounting Office, *Weather Satellites, User Views on the Consequences of Eliminating a Civilian Polar Orbiter*, report no. GAO/RCED-86-111 (US Government Printing Office: Washington, DC, Mar. 1986).

[12] Ball, D., 'The Defense Meteorological Satellite Program', *Journal of the British Interplanetary Society*, vol. 39 (1986), pp. 43–45.

[13] 'A high-tech eye on Europe's weather', *Jane's Defence Weekly*, 19 Dec. 1987, p. 1439.

[14] Ball, D., *A Base for Debate* (Allen & Unwin: Sydney, 1987), pp. 58–62.

[15] Ball, D.,'The U.S. Vela Nuclear Detection Satellite (NDS) system: the Australian connection', *Pacific Defence Reporter*, Mar. 1982, pp. 15–19.

[16] 'Metsat RFP due with short response time', *Military Space*, 6 July 1987, pp. 3–4.

[17] van der Aart, D., *Aerial Espionage* (Airlife: Shrewsbury, 1985), pp. 40–44.

[18] *Department of Defense Appropriations for Fiscal Year 1986*, Hearings before a Subcommittee of the Committee on Appropriations, House of Representatives, 99th Congress (US Government Printing Office: Washington, DC, 1985), Part 2, p. 377.

[19] Streetly, M., 'US airborne ELINT systems, part 2: the US Air Force', *Jane's Defence Weekly*, 16 Feb. 1985, pp. 273–76.

[20] Crickmore, P., *Lockheed SR-71 Blackbird* (Osprey Publishing: London, 1986).

[21] Streetly, M., 'US airborne ELINT systems, part 4: The Lockhead SR-71', *Jane's Defence Weekly*, 13 Apr. 1985, pp. 634–35.

[22] Streetly, M., 'US airborne ELINT systems, part 3: the Boeing RC-135 family', *Jane's Defence Weekly*, 16 Mar. 1985, pp. 460–65.

[23] 'AF plans installation of final GEODSS site', *Aerospace Daily*, 23 Mar. 1988, p. 451.

[24] US Air Force (note 8), p. 457.

[25] Richelson (note 9), pp. 126–27.

[26] 'TR-1s provide high altitude reconnaissance and surveillance', *Aviation Week & Space Technology*, 5 Aug. 1985, pp. 59, 62.

[27] *Department of Defense Authorizations for Appropriations for Fiscal Year 1986* (note 18), part 4, pp. 2079–82.

[28] 'TR-1 reconnaissance aircraft', *C^{3}I Handbook 1986* (EW Communications: Palo Alto, Calif., 1986), p. 111.

[29] Carlucci, F. C., *Support for NATO Strategy in the 1990's, A Report to the United States Congress in Compliance with Public Law 100–180* (US Department of Defense: Washington, DC, 1988), p. V-5.

[30] 'Reconnaissance and special duty aircraft', *Air Force Magazine*, May 1985, p. 154; Joint Chiefs of Staff, *Military Posture for FY 1989* (US Department of Defense: Washington, DC, 1988), p. 80.

[31] Leopold, G., 'Service asks for plans to expand JSTARS role', *Defense News*, 28 Mar. 1988, pp. 1, 12.

[32] Grier, P., 'High-tech flying sentries may soon keep watch over battles', *Christian Science Monitor*, 17 June 1986, p. 7.

[33] Leopold, G., 'STARS poses challenge of producing software for complex system', *Defense News*, 6 Apr. 1987, p. 3.

[34] Leopold (note 31), p. 12.

[35] Broadbent, S, 'Joint-STARS: force multiplier for Europe', *Jane's Defence Weekly*, 18 Apr. 1987, pp. 729–32.

[36] Scarborough, R., 'Air Force may scrap JSTARS 707', *Defense Week*, 19 Jan. 1988, pp. 1, 12.

[37] Walker, K., 'Joint STARS: a soldier's spy', *Flight International*, 22 Nov. 1986, pp. 37–39.

[38] Munson, K., 'RPVs—who are the real remote pilots?' *Jane's Defence Weekly*, 24 Aug. 1985, pp. 360–64; *Jane's Defence Weekly*, 31 Aug. 1985, pp. 411–13.

[39] 'D-21 drone features similar to SR-71', *Aviation Week & Space Technology*, 31 Oct. 1977.

[40] *Department of Defense Appropriations for Fiscal Year 1976*, Hearings before a Subcommittee of the Committee on Appropriations, House of Representatives, 99th Congress (US Government Printing Office: Washington, DC, 1975) part 4, pp. 4001–4.

[41] O'Lone, R., 'Compass Cope seen in anti-tank role', *Aviation Week & Space Technology*, 13 June 1977, pp. 60–62.

[42] 'Pentagon encouraging use of UAVs, family of vehicles planned', *Defense Daily*, 4 Mar. 1986, p. 11.

[43] 'Work under way on "more stealthy" Joint STARS platform', *Aerospace Daily*, 8 Aug. 1986, p. 217.

[44] Gourley, S., 'Tactical intelligence is key to the AirLand Battle scenario', *Defense Electronics*, Feb. 1988, p. 44.

[45] Rawles, J., 'U.S. military upgrades its battlefield eyes and ears', *Defense Electronics*, Feb. 1988, pp. 56–70.

[46] Rawles J., 'The mighty five', *Defense Electronics*, Nov. 1987, p. 60.

[47] Rawles (note 46).

[48] Rawles (note 46), p. 74

[49] Richelson (note 9), p. 141.

[50] Wit, J., 'Advances in antisubmarine warfare', *Scientific American*, Feb. 1981, pp. 31–41.

[51] Biddle, S., CDI, 'U.S. undersea area surveillance: an examination of SOSUS, SURTASS and RDSS', mimeo (Center for Defense Information: Washington, DC, 31 Aug. 1979), p. 14.

[52] Booda, L., 'Antisubmarine warfare reacts to strategic indicators', *Sea Technology*, Nov. 1981, p. 12.

[53] Bussert, J., 'Computers add new effectiveness to SOSUS/CAESAR', *Defense Electronics*, Oct. 1979, pp. 59–64.

[54] Dugdale, D., 'Navy plays a listening game in search for Soviet subs', *Defense Electronics*, Mar. 1986, pp. 68–74.

[55] Cushman, J., 'Navy weaves perilous web for Soviet submarines', *Defense Week*, 30 July 1984, pp. 1, 10–12.

[56] *Department of Defense Appropriations for Fiscal Year 1985*, Hearings before a Subcommittee of the Committee on Appropriations, House of Representatives, 99th Congress (US Government Printing Office: Washington, DC, 1984), Part 6, p. 439.

[57] Thomas, V. (ed.), *The Almanac of Seapower 1988* (Navy League of the United States: Arlington, Va., Jan. 1988), p. 181.

[58] *Department of Defense Appropriations for Fiscal Year 1986* (note 18), Part 2, pp. 912–13.

[59] Seabrook Hull, E. W., 'Search and destroy', *National Defense*, Jan./Feb. 1987, pp. 280–83.

[60] Streetly, M., 'US airborne ELINT systems, part 1: The US Navy', *Jane's Defence Weekly*, 12 Jan. 1985, pp. 69–70.

[61] Breemer, J., *U.S. Naval Developments* (National and Aviation Publishing Company: Annapolis, Md. 1983), p. 58.

[62] Kovit, B., 'New anti-sub aircraft', *Space/Aeronautics*, Feb. 1966, pp. 58–71.

[63] Rawles, J., 'Can the Navy avert an air ASW airframe shortfall?' *Defense Electronics*, Mar. 1988. p. 59–74.

[64] US Air Force (note 8).

[65] US Air Force, *FY 1987 RDT&E Program Element Descriptive Summaries* (US Department of the Air Force: Washington, DC, 1987), p. 746.

[66] Latham, D., 'Space-based support of military operations', *Armed Forces Journal International*, Nov. 1987, pp. 38–46.

[67] Allen, J. (Capt.), 'National intelligence support a vital necessity', *Amphibious Warfare Review*, Aug. 1985, pp. 130–34.
[68] *Department of Defense Appropriations for Fiscal Year 1988*, Hearings before a Subcommittee of the Committee on Appropriations, House of Representatives, 99th Congress (US Government Printing Office: Washington, DC, 1987), Part 6, p. 77.
[69] US Navy, *FY 1988/89 RDT&E Program Element Descriptive Summaries* (US Department of the Navy: Washington, DC, 1988), p. 910.
[70] 'Martin Marietta to lead tactical intelligence effort', *Defense Electronics*, Jan. 1985, p. 28.
[71] US Air Force, *FY 1985 RDT&E Program Element Descriptive Summaries* (US Department of the Air Force: Washington, DC, 1985), pp. 714–17.
[72] Carlucci (note 29)
[73] 'Martin Marietta to lead tactical intelligence effort' (note 70).
[74] Salisbury, A. (Brig. Gen.), 'Beyond BETA: tactical intelligence fusion systems in transition', *Signal*, Oct. 1983, p. 20.
[75] US General Accounting Office, *DOD Acquisition Programs—Status of Selected Systems*, report no. GAO/NSIAD-87-128 (US Government Printing Office: Washington, DC, Apr. 1987), pp. 39–44.
[76] US General Accounting Office (note 75).
[77] Rawles (note 45), p. 72.
[78] US General Accounting Office, *Battlefield Automation—Army Command and Control Systems Acquisition Cost and Schedule Changes*, report no. GAO/NSIAD-88-42FS (US Government Printing Office: Washington, DC, Dec. 1987), p. 8.
[79] US General Accounting Office (note 75).
[80] Rawles (note 46), p. 72.
[81] 'Army, intel officials note TENCAP uses', *Military Space*, 29 Oct. 1984, p. 6.
[82] Berman, A., 'Space and maritime strategy', *IEEE An Anatomy of Space—86*, Sep. 1986, p. 23.
[83] 'Above the battle: the naval space command', *Sea Power*, Oct. 1985, p. 59.
[84] Information for this section was obtained in numerous interviews that were conducted on the understanding that they would not be for attribution.
[85] Carlucci (note 29).
[86] Myers, E., 'Joint tactical fusion—limited operational capability Europe', *IEEE EASCON 83*, pp. 327–29.
[87] Meier, A. L. (Brig.), 'BICES: a central region perspective', *International Defense Review*, no. 10, 1986, pp. 1445–50.
[88] Marcus, D., 'NATO intelligence system faces first major review', *Defense News*, 15 Feb. 1988, pp. 1, 28.
[89] Respectively the Air Command and Control System (ACCS) and the NATO Maritime Operational Intelligence System (NMOS).
[90] Leopold, G., 'Former NATO C3 expert remains warily optimistic about progress', *Defense News*, 1987.
[91] For a comprehensive assessment of the European portion of the DCS, see DCS, *Defense Communications System / European Communications System Interoperability Baseline*, internal report (US Defense Communication Agency: Washington, DC, 1 Feb. 1981).
[92] DCS (note 91), pp. 7.1–7.2; DCS, *FY 1987 Annual Report* (US Department of Defense: Washington, DC, 1986), p. 23.
[93] US Department of Defense, *Report of the Secretary of Defense Caspar W. Weinberger to the Congress for Fiscal Year 1988* (annual report), (US Government Printing Office: Washington, DC, 1987), p. 243.
[94] SIPRI, *SIPRI Yearbook 1986*: World Armaments and Disarmament (Oxford University Press: Oxford, 1986), p. 511.
[95] US Department of Defense, *Report of the Secretary of Defense Caspar W. Weinberger to the Congress for Fiscal Year 1987* (annual report) (US Government Printing Office: Washington, DC, 1986), pp. 250–51.
[96] DCI (note 91), p. 7.1.

[97] DCI (note 91), pp. 1.2–1.3; Hanson, R. T. and Strassman, K., 'The US armed forces European telephone systems', *Signal*, Oct. 1986, pp. 61-63.

[98] McKnight, C. E. (Maj. Gen.), 'The Army's national role in NATO communications', *Signal*, Dec. 1980, p. 46.

[99] Bamford, J. V., *The Puzzle Palace* (Houghton Mifflin: Boston, Mass. 1982), pp. 103–4.

[100] Cambell, D., *The Unsinkable Aircraft Carrier: American Military Power in Britain* (Michael Joseph: London, 1984), pp. 86–87.

[101] Cambell (note 100), p. 184; Arkin, W. and Fieldhouse, R., *Nuclear Battlefields: Global Links in the Arms Race* (Ballinger: Cambridge, Mass., 1985).

[102] Arkin and Fieldhouse (note 101), p. 220. Known locations include Hellenikon AB (Greece), Döbraberg AB and Hof AB (FRG).

[103] US Department of Defense, *Report on the Nuclear Posture of NATO* (SECRET/partially declassified) (US Department of Defense: Washington, DC, 1 May 1984), p. IV–9.

[104] Arkin and Fieldhouse (note 101).

[105] Cambell (note 100), pp. 225–29.

[106] Cambell (note 100), pp. 185–86.

[107] Blair, B., 'Alerting in crises and war', eds A. Carter, J. Steibrunner and C. Zraket, *Managing Nuclear Operations* (Brookings: Washington, DC, 1987), p. 102. For details of Regency Net and Flaming Arrow see 'Regency Net contract', *Signal*, Apr. 1984, p. 72

[108] Walsh, B., 'An eagle in the sky', *Countermeasures*, July 1976, pp. 30–32, 60–63.

[109] 'Helios to deliver imagery to 3 nations', *Military Space*, 21 Nov. 1988, pp. 1–3.

[110] Graves, H. (Brig. Gen.), 'Army directions in space', *EASCON 85*, Oct. 1985, p. 341.

[111] 'British opt for seventh AWACS', *Aerospace Daily*, 17 Nov. 1987, p. 251.

[112] 'France orders fourth AWACS from Boeing', *Defense Daily*, 21 Aug. 1987, p. 284.

[113] Wentz, L. and Hingorani, G. D., 'NATO communications in transition', *IEEE Transactions on Communications*, vol. 28, no. 9 (Sep. 1980).

[114] Hingorani, G. and Brand R., 'Architectural framework for the evolution of NATO integrated communication system', *Signal*, Oct. 1985, pp. 55–65. Bowman, R C. (Maj. Gen., Ret.), 'NATO command and control', *Armed Forces Journal International*, Feb. 1982, pp. 59–65.

[115] Wentz, L., 'NATO national strategic defense communications systems: interoperability and rationalization: rhetoric or reality', *Signal*, Oct. 1985, pp. 83–89.

[116] NATO satellite terminals are interoperable with US satellites.

[117] US Department of Defense (note 103), p. IV–9.

Part IV
Coping with a US withdrawal

19. Could NATO cope without US forces?

Martin Farndale

Former Commander, Northern Army Group (NATO Central Europe).

I. Introduction

NATO force structure and strategy have evolved over the past 30 years to meet the perceived threat from the Warsaw Treaty Organization (WTO). National contributions to that structure have varied according to the financial and manpower capabilities of the NATO member states. The result is a complex matrix, resulting from years of political, economic and military negotiations, agreements and events. Each NATO member nation is an independent and sovereign voice; each has produced its own armed forces, which remain under national command until the transition to war process begins, to meet its own estimation of the alliance's needs against very general force goals set out by NATO. In calculating its force structures, each nation has nevertheless been able to allow for the capabilities of its neighbours. This means that a member nation can take risks in a certain capability, as long as it knows that an ally possesses a similar capability that can fill it. The result today is an interwoven set of armed forces specifically matched to each nation's role in the alliance.

Some member nations have additional forces not committed to NATO, to meet commitments outside the alliance, but most of these would become available to NATO in war. All these various categories of force dovetail into the whole in a way that makes the removal of any one nation's contribution far more complicated than the process of extracting it. This is particularly the case for the larger members of the alliance and above all for the United States.

If any nation unilaterally either alters significantly or removes any major part—let alone all—of its forces from the alliance, therefore, the effect is out of all proportion to the size and shape of the forces themselves. The effect is to leave a complicated series of gaps affecting the strategy of the alliance as a whole and the concepts of operations in the various theatres. A withdrawal of US forces from Iceland, for example, has implications quite different from those of a withdrawal from Greece, Italy, the Federal Republic of Germany or the United Kingdom. All would be affected in a myriad of ways, some major and critical, others relatively minor, but collectively these become very important to the credibility of the whole. Multilateral reductions as a result of negotiations between NATO and WTO do not have the same effect since reduced defence against a reduced threat can be maintained as in the Conventional Armed Forces in Europe (CFE) Negotiation, although reductions beyond a certain point can threatened the credibility of defence.

Although the military implications for NATO of a withdrawal of US forces are quite different under the two scenarios posited in this study—these are examined below—it must be stated from the start that despite the dramatic events of late 1989 and early 1990, unless WTO (and especially Soviet) capabilities are significantly reduced, a complete withdrawal under either category would be catastrophic for NATO. This would be the case for the solidarity of the alliance, for the defence of Europe and the USA and above all for the credibility of deterrence and the prevention of war. The reasons for this are also explained below.

The aim of this chapter is to assess the effect of a US withdrawal under both scenarios and to see if the European NATO countries alone could create a defensive system strong enough to provide a credible deterrence, at least as effective as they have now, and which can cope with the foreseeable future.

Details of US bases and defence establishments in Allied Command Europe are discussed in the previous chapters, and a full account is presented in a companion volume[1]. This chapter will look at the main war-fighting and mission gaps that would be left in both scenarios, as of February 1990. At the same time, the reader should keep in mind that many lesser, but nevertheless important gaps, also listed, can never be filled. When added together, these could also have a significant and serious effect on the cohesion of NATO defence.

Assumptions

Although the reader will be familiar with the assumptions by this stage, it is important to restate three of them here: (*a*) the Soviet threat to NATO remains, despite the events of 1989–90, that is, a threat remains after conclusion of a CFE agreement and an agreement on shorter-range nuclear forces (SNF); (*b*) NATO nations retain in the foreseeable future a level of funding for defence sufficient to sustain their contributions to the alliance; (*c*) Flexible Response and the threat of first use of nuclear weapons remain an essential component of NATO strategy.

Scope

This chapter covers the subject in five broad areas, defined by the following questions:

1. What are the gaps created by US withdrawal in both scenarios, and what is the effect on NATO strategy?

2. Can these gaps be filled by European NATO forces in such a way as to provide a defence at least as credible as it is today?

3. Assuming that a one-for-one replacement is both impossible and undesirable, what action could the European NATO countries take in terms of

restructuring their forces and changing conceptual thinking, methods of recruitment and purchase of arms to compensate for a US withdrawal?
   4. Would current NATO strategy survive, or would it have to change?
   5. What command structure would be necessary in each scenario?

Before tackling these issues it is first necessary to describe briefly the broad principles for how NATO currently plans to meet the military threat before the results of the current arms control measures take effect. This includes the deployment of US forces, should, for whatever reason, war break out. This then becomes the baseline against which the European allies would have to provide a credible defence on their own, adjusted to meet the threat as it evolves.

II. The threat

Any military threat consists of the hardware with which to go to war and the intention to use it. The intention to use the massive hardware still remaining in the WTO area is clearly not present. Even when the currently planned defence cuts are complete, however, the WTO will still have a massive military arsenal of conventional and nuclear weapons, albeit not all in the area from the Atlantic to the Urals. With the sympathies of the East European countries shifting to the West, the evacuation of Soviet forces from these WTO allies and the likely prospect of a united Germany within NATO, the practicalities of a Soviet attack westwards have diminished considerably. This is not to say, however, that the threat has disappeared.

Few could have predicted in mid-1989 the events of the subsequent autumn and winter; likewise, no matter what the future looks like in light of these changes, no one can predict how it will turn out. A series of events, incomprehensible today, could once again combine to threaten Western Europe, especially as the necessary hardware remains—and will continue to remain—available after the planned reductions have been completed. As the situation remains extremely volatile and unpredictable, this is not the time to tinker with a system that has been so successful for so long.

It is first necessary to look at the military threat which could develop, in spite of all the rhetoric on the world stage today. No matter how unthinkable a NATO–WTO war might be in 1990, it is this that NATO must be able to deal with if US forces leave Europe under any scenario. Modern war is now so complex that should NATO drop its guard and should US forces depart, there will be no time to prepare for a threat once it has developed. Here is not the place to say whether such a threat will or will not develop, but rather to say what must be done to ensure that it does not.

Official data from both NATO (November 1988) and the WTO (January 1989) show how at that time conventional and nuclear (quantitative) superiority was in favour of the WTO.[2] What the raw numbers did not show, however, was that NATO's one-time qualitative advantage was also eroding. This is probably still the case. The problem is further compounded by the fact that the WTO enjoys the advantage of operating from interior lines while NATO must operate from exterior lines. This means that the USSR could, in war, swing its forces at will between theatres. NATO cannot do this to any significant degree. Each NATO theatre has to fight its campaign with its own in-place forces reinforced, after a pause for mobilization, according to plans which are different for all theatres and therefore not inter-changeable. These forces would later be reinforced from the USA. Apart from US reinforcements, it would be virtually impossible, in political as well as practical terms, to swing other NATO forces on any scale between regions. Even if the main Soviet thrust came in the central region and no attack developed in Turkey, for instance, it would be hard to imagine a move of Turkish troops to that region. In other words, simply adding together the total WTO forces and the total NATO forces will always give a misleading picture, since NATO forces would have a much greater area to cover and could not be moved around to face the main threat.

Interior lines permit the USSR to accept large force reductions and still retain the ability to concentrate overwhelming force at a chosen point of main effort and to change this point at will. Thus a major combined land, sea and air thrust into the Northern or Southern flanks, to draw NATO reserves, can be changed to a major thrust in the central region. This is an enormous strategic advantage and one that greatly enhances the value of any residual WTO superiority. Because NATO would fight on exterior lines, and as the European NATO nations with few execeptions only provide forces for themselves, NATO is denied this advantage in war. In addition, Soviet forces withdrawn to the east of the Urals only have a 1200 kilometre march across Russia and Eastern Europe to be able to reinforce any forces remaining there. US forces must first cross the Atlantic Ocean, a distance of almost 6000 km.

In June 1987 the WTO declared that its military doctrine was strictly defensive, a claim repeated many times since by Mikhail Gorbachev and by Soviet and East European senior military officers and supported to some extent by force restructuring and changes in the conduct of military exercises.[3] It is thus difficult to imagine what might prompt WTO planners to select the war option. Soviet military planners accept, however, that some offensive capability must remain even in a defensive concept.[4] The task here, however, is not to argue why the WTO would go to war in the first place, but rather to assess what strategy and concept of operations would give it the best chance of success in the event of war.

Having, for whatever reason, made the momentous decision to go to war, what would be the USSR's best option? Soviet planners must recognize that if the war goes nuclear, they stand to lose everything. Somehow the USSR must achieve annihilation of NATO's defences before NATO can use its nuclear weapons. Thus, before going to war Soviet planners would have to be absolutely convinced that victory was certain, perhaps because they believed that NATO had allowed its forces to decline sufficiently, or that the anti-war lobby in the West had become so strong that NATO would not fight if attacked. In either case they must be convinced that by a combination of surprise, deception, concentration of force and offensive action they could achieve a rapid victory before NATO could organize its defences. For example, they must believe that in the central region they could get massed armour onto the Rhine within 72 hours, before NATO could respond with effective conventional or nuclear weapons. Although Soviet force reductions, whether unilateral or as a result of the CFE process, will make a WTO surprise attack extremely difficult to mount and a standing start attack almost impossible to achieve, surprise is such a vital principle of war that every effort will be made to achieve it. Such an attack, however unlikely it may seem, can never be entirely ruled out.[5]

Before Mikhail Gorbachev came to power in the USSR the WTO had built up, and is still continuing to modernize, massive conventional forces. While the kind of cuts being proposed at CFE (a reduction to 195 000 Soviet and US forces in the central region) would certainly undermine a short warning attack capability, any Soviet commander would try to achieve a sudden surprise start to the fighting. Given its advantage of interior lines, the USSR, even with the proposed reductions, would be difficult to defeat in a conventional war in the 1990s. This is because NATO in the past 30 years has made little effort to equate the size of its conventional forces to those that the WTO could bring to bear in any one region, relying instead primarily on nuclear deterrence. This policy, combined with the fact that NATO cannot move conventional forces in any significant number from region to region to meet an attack, has allowed the WTO to build up sufficient superiority to avoid conventional defeat. Therefore, if some or all of NATO's nuclear forces are removed by negotiation, the Soviet chance of victory in some future conflict is improved considerably. Soviet planners also know from their study of NATO force structures, equipment, organization and training that NATO has no major offensive capability other than the use of nuclear weapons. NATO's maximum conventional capability is only sufficient to conduct limited offensive operations as part of an overall defensive battle.

In preparing to attack, the USSR would adopt deception strategies involving a combination of bogus negotiations, exercises and red-herring situations over a period of time. Thereafter, there are many possible start scenarios, and the USSR would be free to select the moment to strike. Assuming that the world situation evolved in such a way that East European members of the WTO agreed to go to war against Western Europe—if only to give passive support such as allowing the passage of armed forces—how might such a war develop?

Without a declaration of war, large-scale pre-emptive conventional air attacks could initiate hostilities right across the NATO front to conceal for as long as possible the point of main effort. These would be followed up with armoured, artillery, airborne and sea attacks. The size of WTO conventional forces (in 1989) enabled them to mount such attacks and retain sufficient strength to follow up at their selected point of main effort. This will still be possible after all current arms control measures have been completed, but there will no longer be a capability to attack more than one NATO region at once. In the north the aim would be to seize the whole of Norway and as much of Sweden and Finland as is necessary to free the northern naval and air fleets for operations in the North Sea, the Atlantic and against the UK. In the south there are many options, but probably the most effective would be to seize Yugoslavia and Albania to pose a major threat to northern Italy and Austria. Why antagonize Greece and Turkey? They would not counter-attack for fear of each other.

If the USSR chose to attack in the central region, violent attacks would probably be made against Denmark—to secure the Baltic exits—and against all NATO forces in the FRG under massive aerial and artillery bombardment. The very nature of such attacks against the conscripts and half-trained reservists that compose much of NATO's forces would have a devastating effect. History has shown that only highly trained and motivated professionals can withstand the initial shock and horror of battle on a large scale, and even they can have problems. This is less so for an attacker who knows he has the advantage of surprise and the initiative. It is almost impossible to recognize the sheer terror and shock of the first attacks and this is particularly so with young men brought up in a modern society. In the Falklands/Malvinas War, British professional soldiers had the whole voyage south to prepare themselves for combat. In war, the USSR would see to it that very little time was available and that men would have to cope with going from total peace to total war in a matter of hours. The initial bombardments would be designed to exploit this factor. It takes years of iron discipline and hard training to withstand such a barrage, and most NATO forces do not receive such training.

Once committed it would be imperative for Soviet forces to maintain their speed, momentum, fire-power and weight of attack by exploiting any breakthrough or collapse of any part of the NATO front. If there was any difficulty here, the concentrated use of chemical weapons should ensure that the breakthrough occured. This would in turn be exploited by the continual commitment of fresh reserves to battle, either in further-echelon attacks or as operational manœuvre groups, if necessary coming from east of the Urals. These must in turn be supported by continual aerial and artillery attack to ensure that the momentum carries them on through the still unprepared or unreinforced NATO positions, exploiting the shock and horror of war to the full. It is not necessary here to pursue the details of such an attack except to say that even with all the planned reductions, Soviet superiority at the selected point of main effort could still be up to 2 : 1 or, in places, 3 : 1, and simultaneous airborne attacks would be occurring in depth throughout.

III. Meeting the threat

As long as the hardware enabling the USSR to conduct such an attack remains, NATO is under threat, since the intention to use it can change overnight.

Nevertheless, even before the cuts announced in December 1988 and the latest CFE proposals, the USSR would have had enormous difficulties in conducting operations of this kind even without NATO opposition. A number of factors come into play here. First, there is the question of the reliability of the non-Soviet WTO countries, and therefore of the security of Soviet lines of communication with the front. Second, there is the sheer magnitude of the task of controlling troop movements and of co-ordinating all arms on such a scale and at such a pace. Third, there is the problem of maintaining logistics and mobility on unknown, possibly hostile, terrain. When NATO forces, both conventional and nuclear, are superimposed on this situation, the problems for the USSR are compounded even further.

Current NATO strategy and concepts of operations aim not only to exploit these problems but also to react quickly to the operational and tactical errors which the Soviet commanders would inevitably make. To be seen to be able to do this is one of the greatest facets of deterrence.

Assuming some deterioration in East–West relations and at least a short period of tension before the outbreak of hostilities, NATO commanders would look to their political leaders to authorize deployment from peacetime locations, to start the outloading of combat supplies, the preparation of defences, mobilization and reinforcement before the initial attacks occur. Enemy deception efforts would be aimed at preventing these decisions from being made on time. Hopefully before hostilities start at least a proportion of these measures will have been taken, but very few reinforcements will have arrived. Many of the in-theatre forces have long distances to travel from Belgium, the United Kingdom and the Netherlands, and within Italy and Norway. In addition, these forces have already been much reduced and few are trained for instant combat. Efforts have been made to create in-theatre operational reserves, but this has only been possible by double earmarking, at the expense of forces holding the forward positions. These reserves have very limited capability and they include US in-place forces as well as reserves from the United States which may or may not be there in time. On the other hand, they are currently backed by a full spectrum of nuclear weapons, and this makes the whole concept credible in the eyes of Soviet planner. The removal of any of these nuclear weapons could undermine the credibility of NATO's ability to deter a WTO attack. If in crisis everything goes exactly according to plan, if all intelligence indicators are correctly interpreted, if all political decisions are made by all 16 nations in time (that is, in less than 48 hours), if all deployment, outloading, preparation of defences and the start of reinforcement occur without a hitch, then NATO's conventional forces can probably halt and destroy the initial surprise attacks by the lead echelons and hold out for a few days until their supplies are depleted. However, it is hardly likely that such momentous decisions and plans will be made and executed in so short a time and without a hitch. After major force reductions, it is probably true that the danger of a surprise attack will be reduced. On the other hand, this means that fewer NATO forces will have to cover the same ground. As they are already overstretched, the likelihood is therefore that the nuclear threshold would be lower.

The strategy of Forward Defence and Flexible Response has been imposed politically on military commanders. They have therefore had to produce concepts of operations within that strategy and within the financial, manpower and logistic constraints that go with them. It is pointless to speculate here on alternative strategies which might have grown up over the years. Whenever operational commanders have wanted to adjust their concepts to make more use of modern technology or match changes in the threat, however, they have been constrained by political issues. Peacetime operational concepts are always a compromise in a democratic system. The NATO forces must therefore always commence fighting with a disadvantage that does not apply to an aggressor who gives highest priority to military requirements. What is more, NATO has maintained roughly the same force levels as existed for the trip-wire strategy that preceded Flexible Response. Under the former strategy, conventional forces operated in support of nuclear weapons which were to be used from the start of hostilities. The withdrawal of French forces from the NATO command structure in 1966 significantly reduced NATO's conventional force levels. The growth of weapon capability since then, although significant, has not compensated for such reductions or for the improved Soviet forces.

Thus, despite repeated US efforts to persuade its allies to improve their conventional forces since NATO adopted the doctrine of Flexible Response in 1967, there has been little increase in immediately available combat power since the days of Massive Retaliation. NATO commanders would therefore be asked to fight the battle against numerically superior WTO forces with their own forces reduced and without using nuclear weapons. If in addition the forces of the major force contributor—the USA—are removed, the capability of the whole is reduced to the point where a cohesive conventional defence is impossible unless European forces are increased to compensate.

NATO's conventional forces, no matter how strong, can deter the the USSR only as long as they are backed by the full spectrum of nuclear weapons. Since both sides know how to make them, it must be assumed that nuclear weapons will always be available in war. Nevertheless, in the strategy of Flexible Response conventional forces have a critical role to play. First, they must halt and contain the enemy's attempts to achieve a quick breakthrough; second, they must buy time and postpone the the crossing of the nuclear threshold for as long as possible so that attempts to stop the fighting have a chance of success. This would be very important if the USSR began to realize that its gamble to achieve victory before NATO had organized its defences was failing and the war was about to go nuclear.

NATO commanders have developed various concepts of operations. The details do not concern us here; suffice to say that they are all based on common factors. They all take their political direction from the NATO Council. They all rely on early mobilization and reinforcement and they all rely on fighting the initial battles with in-place forces which are manifestly too small to succeed alone. Finally they all rely on nuclear weapons to take over after very few days if the fighting cannot be stopped. Thus NATO envisages fighting in four phases, each different in detail in the three regions but co-ordinated by the Supreme Allied Commander as follows:

1. First comes the containment, holding and surveillance phase. The aim here is to contain the initial attacks and locate the enemy's point of main effort. During this phase the long-range or Follow-on Forces Attack (FOFA) begins. The aim is to prevent a breakthrough using in-place forces.

2. Next comes the commitment of in-place reserves to complete the defeat of the initial attacks. NATO forces cannot afford to get involved in a battle of attrition which they could not hope to win.

3. The third phase will again be one of containment as the enemy commits its next echelon. Mobilized reinforcements should be arriving to assist the hard-pressed in-place forces who will by now be getting short of supplies. The WTO forces will realize that time is short if they are still to win before nuclear weapons are used. This could cause them to use chemical weapons to force a breakthrough.

4. Finally comes the commitment of external reserves (from the USA) to convince the aggressors that the gamble has failed and that a nuclear fire-plan is now ready unless they agree to a cessation of hostilities.

This concept would work today provided that NATO's range of conventional and nuclear weapons is maintained and modernized. It is essential at this stage to understand how NATO planners see the fundamental and critical role of nuclear weapons in the prevention of war. We have to accept that, having been invented, nuclear weapons are forever available to an aggressor who possesses them or, if they have been removed, knows how to make them. Therefore military commanders must always assume that they might be used. However, NATO military commanders accept that they will not be allowed the use of such weapons—unless the enemy uses his first—until they have used up all their conventional forces and are in the process of losing the war.

The most difficult decision would be how to make the first nuclear strike in the war. In the conventional fighting does not stop, even if NATO forces are successful in holding the initial conventional attacks, then, once all reserves and combat supplies are used up, NATO will have to resort to nuclear weapons to avoid defeat. It is not likely that the NATO Council will authorize the initial use of strategic weapons because of a battlefield crisis. Tactical weapons are not suitable for initial use because of their short range and small size. The ground-launched cruise and Pershing missiles were ideal, and their deployment made the war option unthinkable, but they have now been withdrawn under the provisions of the Treaty between the USA and the USSR on the elimination of their intermediate-range and shorter-range missiles (the INF Treaty). In 1989 NATO planners felt an urgent need to fill the INF gap by introducing an air-delivered stand-off system and a longer-range replacement for the Lance missile, such as the US Army Tactical Missile System (ATACMS).[6] Unless this is done the WTO might just consider that the risk of war is worth taking.

However, military men must think further. What do they do if the use of these intermediate weapons still does not stop the fighting? By this time enemy conventional forces will be breaking through to victory, and all—or most—of NATO's conventional capability will be exhausted. The only weapons which can stop them at this stage are tactical weapons (shells and Lance replacement ATACMS). Without these, the only options would be to use strategic weapons or to surrender. This argument does not invalidate the political achievement of agreeing on a reduction in nuclear weapons—there are probably more nuclear weapons in the world than are necessary to prevent war. However, as NATO stated in May 1989, it is essential to retain some in each category (the exact figure is open to negotiation) to avoid nuclear blackmail.[7]

IV. Filling the gaps

In option A, all US forces, bases, headquarters and other defence-associated establishments are withdrawn from Allied Command Europe, with no option to return. In this scenario it would still be possible for US forces to visit Europe in order to pursue US foreign policy objectives in the Middle East, for example, by sending the 6th Fleet back into the Mediterranean. In such an event, however, there would be no US bases from which they could operate. In this scenario the USA remains in the Atlantic Alliance on much the same terms as France, leaving the European allies uncertain of the US response in war and forcing NATO's military commanders to plan for a defence that does not include US participation or assistance.

The departure of US forces on such a scale would inevitably and fundamentally change the shape, size, nature and cohesion of what was left of NATO. It would have an enormous and profound effect on the United States' ability to face the Soviet Union as it would mean accepting the possibility of Soviet domination of Europe either by conquest or by 'Finlandization'. In addition, if war did occur, the United States would be giving up the opportunity to fight it well away from its own shores. It would in effect be committing itself to nuclear operations only, since it would not easily be able to bring its conventional forces to bear where it mattered most.

In option B, the US forces depart in the same way but specific plans are made for their return in the event of a crisis. In this case some US-based forces would have to remain to look after stockpiles and to act as advance parties for returning forces. In addition, US forces would have to return frequently for training. The net effect of this would probably be that some US forces would always be on some form of training exercise in Europe. This would be an exceptionally expensive option since all forces would have to be retained active in the United States and, in addition, there would be the cost of movement to, and training in, Europe. The scale of training necessary for modern armed forces is considerable. It involves more than the much-publicized large-scale annual exercises of the Reforger (Return of Forces to Germany) type—hundreds of movements would be necessary for conferences, studies, reconnaissances, briefings (held every time a commander at any level is changed), command and signal exercises and many other events. All this would be very expensive and it would also be necessary to keep the stockpiled equipment modernized and battle-worthy.

Of the two options, option B would be the worst. The operational effect would be much the same as for option A. It would signal to the Soviet Union that a rapid, no-warning WTO attack, before US reinforcements could arrive and before agreement to use nuclear weapons could be achieved, was possible. It would cause great uncertainty in NATO planning because the European allies would never know when option B was going to change into option A. It would enable the USSR to concentrate specifically designed military forces to prevent US forces from ever reaching Europe intact or in time. Above all the United States would, in this case as well, be committing itself to nuclear operations only, as the war would almost certainly have gone nuclear before its forces—air power excluded—could be landed in Europe.

Finally, the effect of withdrawal in either case on other NATO nations must not be underestimated. Many of them have just as acute political, financial and demographic problems as the United States and many would deduce that if the USA can pull out, then they too could reduce their commitments to NATO, particularly in the new political climate developing in Europe.

It is neither necessary nor possible to list all the US defence installations that would disappear from Europe under these options.[8] The task here is to clarify, region by region, what broad defence missions filled by the USA would be lost and what could be done to replace them.

The northern region

Losses resulting from US withdrawal in the northern region would mainly occur under option A and would essentially entail land, sea and air capability to reinforce the region in time of crisis. There would also be the loss of US early-warning and anti-submarine warfare (ASW) functions and defence infrastructure throughout the area. The special role of Iceland and Greenland would probably remain because of their importance to the defence of the continental USA, but it is assumed here that as they form part of Europe (Greenland as an autonomous region of Denmark), US forces and installations are withdrawn from them as well. Officially, the USA has no military personnel in Norway and Denmark in peacetime, but it has more than 3000 in Iceland and almost 300 in Greenland. Apart from the loss of these forces, the loss, in option A, of war-time arrangements to reinforce the region especially Norway, with US forces would be extremely serious and would require some form of compensation.

In war, the USSR would try to seize Norway, Denmark and Iceland as quickly as possible in order to dominate the Norwegian and Baltic Seas and the Atlantic sea lines of communication. Replacement of US forces would, therefore, be particularly important in option A. The loss of US forces in the UK would also have considerable effect on naval and air operations in this region, particularly in option B when, in order to bring US forces back, it would be necessary to re-open the Atlantic sea lines of communication without US bases in Europe. Norway would continue to need considerable reinforcement to resist attack. Today it receives land forces from four sources:

1. The Allied Mobile Force (AMF). This force is also earmarked for missions on the Southern Flank and might well would not be available.
2. The NATO Composite Force. This is a new unit set up to replace the Canadian Air-Sea Transportable (CAST) Brigade, until 1988 earmarked as a reinforcement force for Norway. Made up of 2500 troops from the United States, Canada, the FRG and Norway itself, the Composite Force has only half the personnel strength of the CAST Brigade.
3. The UK/Netherlands Landing Force, a brigade-sized unit which could also have other missions in the area, such as defending the Baltic approaches.
4. The US Marine Expeditionary Brigade (MEB). This is a force of some 5000 men. It could be followed by a second sea-borne MEB within some 35 days. These brigades have great fire-power and are, aside from the Composite Force, today the only land forces totally dedicated to the region. These would be lost under option A.

In addition, Norway is earmarked to receive the following air reinforcements: *(a)* four squadrons with the AMF; *(b)* four squadrons with the MEB (64 aircraft, almost the size of the Norwegian Air Force); *(c)* two squadrons of US Hawk air defence missiles; *(d)* seven additional US aircraft squadrons; and *(e)* up to at least 100 carrier-based aircraft. These come from the Strike Fleet and may or may not be available.

NATO Europe replacements for US forces and capabilities: option A

Even with a reduced threat as a result of arms control measures it will be necessary to have plans to reinforce the Northern Flank because current plans are less than what is needed and we must assume some cuts in Norwegian and Danish in-place forces.

With regard to ground forces, it might be possible for NATO Europe to find 5000 marines. In practical terms, these would probably have to come from an increase in British and the Dutch force levels. They do not have to be marines but they would need much special training and equipment. This would involve extra costs, preventing NATO Europe from maintaining level defence budgets.

Replacement of air forces would be impossible without an increase in European air power and therefore also in defence budgets. Eleven squadrons

COULD NATO COPE WITHOUT US FORCES? 443

would have to be created as there would be none to spare from the other regions, where air power is already short, and this numer could not be reduced.

US naval forces could never be replaced on a one-for-one basis. However, under this option it would not be necessary, as the Atlantic lines of communication would be of less significance. Nevertheless, it would not be easy to recreate all the requirements of a successful anti-submarine warfare campaign without US intelligence and hardware. Some expansion of European naval strength would therefore be essential. This would probably fall to the UK, with its existing blue-water fleet. If the United States remains a member of the alliance under this scenario, it should be possible for the European allies to receive US naval intelligence.

NATO Europe replacements for US forces and capabilities: option B

In terms of ground forces, option B involves few changes in the north. Without bases in Europe, however, it would almost certainly take longer for reinforcements to become operational in war.

For the air forces the situation is again much as now, but it will take longer to transport reinforcements to Norway without US air bases in the UK.

As for naval forces, provided that naval and intelligence losses elsewhere in Europe can be replaced, naval operations should be much as now.

If the USA left Greenland and Iceland, and if the early warning of attack currently provided by the US facilities there is to be maintained, extra manpower would be required for an anti-submarine warfare command and a maritime patrol base in Iceland as well as an early-warning station in Greenland.

There are no specific US reinforcements earmarked for Denmark but the absence of US forces in the region would have an adverse impact on Danish morale and will to fight.

Strategic and/or conceptual changes

The terrain in the northern region is of such a special nature that there is only one way to defend it. Indeed, the area is so vast and the effect of poor communications, high valleys with few lateral routes and poor weather for low flying, is so great that even today's force levels could not hope to last for long. Only a man-for-man replacement would compensate for the loss of US reinforcements and this would still leave a precarious situation in the region. Even counting earmarked US forces, the present force levels are dangerously low. The exception might be in terms of naval operations under option A, which, because of the lack of need to keep the Atlantic communications open, could be done with fewer forces. This would mean taking a risk, however, as Soviet forces would then have easier access to the Atlantic and could, unless checked, mount a considerable maritime threat to the western shores of Europe.

Summary

In option A, Europe would have to find 2 new brigade-size formations, 11 squadrons of aircraft and some naval forces. There would also be the need for extra manpower and aircraft for a maritime patrol base in Iceland. In addition there would be the need for countless minor units and establishments to replace the command, control and intelligence resources so vital to the conduct of modern war. All these would have to be additional to current orders of battle as there would be none to spare from other regions. In option B there would be little change to the current situation, except for early-warning and ASW functions in Iceland which would have to be performed in peacetime by European NATO units. However, both situations would be less credible than the current situation because of the loss of other US bases in Europe, notably in the UK. It should be possible to maintain the flow of intelligence from US forces in both cases, but not to the same extent.

The central region

The situation in the central region is very different. Here the United States provides a major part of the vital in-place forces which will have to meet and destroy the initial attacks. It provides a major part of the air forces, the command and control system, early-warning and intelligence-gathering machinery and virtually all the nuclear forces. The US European Command (USEUCOM), the highest US headquarters in Europe, located in the region, commands all US sea, land and air forces in Europe. Its Commander-in-Chief is also Supreme Allied Commander Europe (SACEUR).

The Central Army Group (CENTAG)

Of the 217 000 US Army personnel in Europe, over 210 000 are stationed in the central region, practically all in the FRG. Before transfer to NATO command in the transition to war, these forces form the 7th US Army. In war their commander becomes Commander Central Army Group (COMCENTAG), when in addition to his own two corps he assumes command of two West German corps and a Canadian brigade to form CENTAG, which defends over half of the region's frontier with the WTO.

The US contribution to CENTAG is extensive. It includes:

1. The V US Corps, consisting of one armoured and one mechanized division and an armoured cavalry brigade.
2. The VII US Corps, consisting of one armoured and one mechanized division, one armoured cavalry brigade and one mechanized brigade—the advance party for a mechanized division located in the USA in peacetime.
3. Corps troops, consisting of all the combat and logistics support troops necessary in war. They include: (a) 7 field artillery brigades; (b) 4 artillery

groups; (*c*) 9 missile battalions; (*d*) 30 Hawk air defence batteries; and (*e*) 18 Patriot air defence batteries.
  4. The 21st Support Command, providing all logistics above corps level.
  5. The 4th Transport Command.
  6. The 18th Engineer Brigade.
  7. Numerous communications and other units.

Plans exist, and are practised, to reinforce CENTAG during the transition to war by some six divisions, in peacetime located in the USA. The equipment for these formations is stockpiled in the FRG. The troops available for rapid reinforcement amount to some 90 000 men, making a total army strength in CENTAG of some 295 000 men. It should be noted that both US corps, with little more than two divisions each, are already the minimum size to be operationally ready.

The Northern Army Group (NORTHAG)

US forces provide nuclear warheads and custodial troops for both CENTAG and NORTHAG. In addition, US forces reinforce NORTHAG in war. To do this, one armoured brigade is located in northern FRG and acts as the advance party for the III US Corps, whose equipment is also stockpiled in Europe. This Corps consists of another two armoured divisions and one mechanized division and a full set of corps combat and logistics support troops. It is also located in the United States in peacetime.

Together US CENTAG and NORTHAG forces amount to about 25 per cent of all ground forces in the central region. They possess massive fire-power and they exercise control over all land-based nuclear weapons.

Air forces

The USA provides some 320 air defence and attack aircraft in peacetime in the central region (about 15 per cent of the peacetime total). There is a similar number located in the UK which could be used in this theatre. In war some 30 more squadrons arrive from the USA (about 600 aircraft). To support this force there is a large air force infrastructure of air bases, command, control and intelligence-gathering facilities and logistics depots. The bulk of these air forces make up the 4th Allied Tactical Air Force (4 ATAF), which is designed to support CENTAG.

Can these forces be replaced by NATO Europe?

In broad terms, the extensive US contribution to the central region encompasses: 50 per cent of ground and air forces in CENTAG; 20 per cent of ground forces in NORTHAG; all land-based nuclear weapons and their custodial arrangements; the Commander and some 50 per cent of the staff of CENTAG; the Commander and most of the staff of Allied Air Forces Central Europe

(AAFCE); the Supreme Commander and a high proportion of his staff; a major source of intelligence; and much of the ground-to-air defences in the region.

Without fundamental changes in the WTO force posture, the effect of removing such a high proportion of NATO forces in the central region would be close to disaster. As a large part of the main battle in this region would most probably have to be fought by in-place forces before the bulk of the reinforcement occurs, even after all currently planned arms control measures are effective, the effect of such a move would be similar in both scenarios. Either there will have to be a fundamental change of operational concepts or the West Europeans will have to do their best to replace as many US forces in CENTAG as possible with in-place forces, since any military concept demands at least enough manpower and equipment to cover the ground. In 1989, as the CFE Negotiation began, NATO had the absolute minimum forces needed to cover the ground and to support a concept which already relies on the early use of nuclear weapons.[9] Perceived gaps were covered by modern technology, but not sufficiently to hold for long. Thus to make the conventional phase of the battle last longer, and to make the defence more credible, more forces are needed, not less.

The main threat is probably against NORTHAG, where few US troops are currently deployed and which is essentially unaffected by either withdrawal option, except for the all-important loss of the Army Group reserve corps. If NATO Europe could replace this corps at least NORTHAG would be little worse off then it is today. NORTHAG is already taking grave risks, with almost 50 per cent of its forces poorly located. In either case some significant changes to operational concepts are likely to be necessary.

Option A in NORTHAG. The problem here is to replace the III US Corps. Assuming that the United States would leave behind the stockpile of equipment located in Europe, the Europeans would have to provide the manpower, that is some additional 60 000 men. The problem would be their status. They would have to train with the equipment if they were to be any good at all in war, and the equipment would therefore have to be activated in peacetime. This could not be done effectively by reservists, although some might be involved in manning. The equipment is currently in modern storage but more barrack accommodation would be necessary close to the stores. At a cost, the United Kingdom ought to be able to man one of these armoured divisions, particularly after its withdrawal from Hong Kong, and the FRG a second, each with a share of corps troops. If France could man a third, a truly European Armoured Corps would have been created. Its value as a reserve force would be questionable for many years, however, and there would be many problems of command and control to solve. Of course, if the US forces did not leave the equipment behind there would be a major outlay to replace it.

Option B in NORTHAG. In this scenario there would be little change from the status quo. NORTHAG would get its reinforcements from the USA in the same way as it does now, although different reception arrangements would have to be made and some US troops would probably have to remain in the area to act as a training, liaison and reception team.

Option A in CENTAG and 4 ATAF. The situation in CENTAG and 4 ATAF is much more serious. The problem here is the replacement of a US Army of two corps and corps troops, some 300 aircraft and the ability to reinforce in wartime. Not even a major change in strategy and concept, coupled with major changes in recruitment and allocation of troops to task, nor the return of France back into what would be a very different NATO alliance would compensate for such a loss without massive increases in expenditure. France would first have to agree to deploy troops much further east than is contemplated under current plans. Even so the 1st French Army would only go a little way in replacing the 7th US Army, as a French Army is not much larger than a NATO corps, and it is in terms of equipment and capability is much less powerful than an equivalent US force. In option A the stockpiles of US equipment might be used to strengthen French forces, but there should be no illusion that this would be anything but a second-best solution, resulting in a less effective and less credible defence.

Likewise the French Air Force could take over from the US Air Force in 4 ATAF, but again it would need to be improved considerably. The next problem would be the creation of reserves, so vital to making the whole strategy viable. The only untapped European source is Spain. Could it be persuaded in these circumstances to allocate a corps to Allied Forces Central Europe? Aside from France and Spain, there are no realistic sources of men and equipment. Large-scale mobilization is no good in modern war unless large quantities of modern equipment exist in peacetime and unless the men are trained to use it. NATO cannot plan on being given time for this to occur once war is imminent. Even so, NATO planners must recognize that any solution along these lines would be much weaker than what already exists.

Option B in CENTAG and 4 ATAF. Although the US forces are expected to return in a crisis, a credible array of *in-place* forces would still have to be created as in option A. The French and Spanish forces would thus be needed in this scenario as well. However, the arrival of US reinforcements would again only be decisive if they can return in time and if they have carried out sufficient training in Europe to go into action instantly in a crisis. They would thus have to have stockpiled equipment. The early arrival of air forces, which should be possible, would be vital. The stockpiled US equipment would not be available to the Europeans forces, however, adding outlays for replacement stocks to the overall cost increase.

The sequence of battle

Before summing up the situation in the central region it is necessary to look at the effect of each case on the possible sequence of battle.

Option A. With all US forces gone the remaining in-place European forces would be insufficient to cover the ground along any front which developed in war, especially in the CENTAG sector. Even if French and Spanish land and air forces were made available, there would be no operational reserves of any consequence, unless France and Spain spent exhorbitantly to create, train and equip them. These forces would be of little use unless they were located in-theatre. The initial conventional phase of battle, already precarious, could not last more than a few days before nuclear weapons would become necessary—there may not even be time to use them before an enemy breakthrough had occurred. Much worse, such a defence would not appear credible either to the WTO or to the NATO defenders themselves. The position could be improved marginally and over a long time if NATO achieved more interoperability and integration of its European defence effort. The situation would be much more serious if the European NATO countries have themselves reduced their forces as a result of arms control negotiations.before the US forces depart.

Option B. Initially the position would be the same as in option A. It is marginally better because the defence plans would be backed by the arrival of US forces in a crisis. Indeed, US air forces could arrive very quickly and would be a great asset. Nevertheless, the USSR would be bound to conclude that its best chance of victory would be a quick attack before these reinforcements could arrive. There would be a very great danger that the war would turn nuclear just as US forces were arriving. Further integration of European forces would help, but would not compensate for the lack of in-place US forces.

Summary

Although it is conceivable that in the very long term, under some form of political union, Europe could stand alone in defence, any major withdrawal of US forces from the central region would so adversely effect the capability and credibility of that defence that the risk of war would become much greater. If war broke out before some form of political union had been established in Western Europe, it would either turn nuclear or NATO Europe would very quickly be overrun, even if efforts were made to compensate for the US withdrawal. This would certainly be the situation in option A. In option B the battle might be prolonged a little because of the early arrival of US air power, but this alone could not halt the attacker long enough for US ground forces to arrive, build up and get into action. It has been said that the European allies should concentrate on manpower and land forces and let the USA provide the bulk of the air power. It would be unwise for the European allies to assume, however, that such a commitment on the part of the USA could not be subject to change. It is therefore vital for NATO Europe to retain balanced land, sea and air forces of its own. Measures should be taken to bring France into some kind of European NATO defence grouping, to persuade Spain to play a more active part in forward defence and for all nations to integrate their defence efforts further.

The real danger is that Europe will so reduce its own forces as a result of arms control negotiations that, should the USA pull out completely under any scenario, what is left will simply not be viable on its own. Massive increases would be needed to create a credible minimum and this could only be done on an all-NATO European basis.

The southern region

The effect of the withdrawal of US forces from the southern region would take a different form. First there would be the physical loss of in-place forces and bases, notably naval and air assets, but in this region in particular there would be significant gaps in regional leadership, co-ordination and influence on national forces and nuclear force capabilities and in weapon production and modernization. The southern region is not a single unified entity like the other two; it comprises five separate theatres of operation, each of which faces its own threat and plans to meet it individually, in conjunction with US forces. These theatres are linked by sea and air power, most of which comes from the United States.

There are other, more remote US installations the loss of which would seriously affect the defence of the region. Loss of bases in the Azores, Spain and Portugal would deprive the region of vital intelligence, naval and air power early in the campaign. In the Mediterranean region, the loss of sea and air bases in Italy would be serious. On the other hand, there are no major land forces involved. Similarly it would be the loss of naval and air power as a result of closing down the US air bases in Greece and Turkey that would be so serious.

In such a disparate and dispersed region the loss of US infrastructure and co-ordination would be missed as much as the massive power of the 6th Fleet, the US Air Force and the nuclear forces themselves. Furthermore, the loss of the bases would make US reinforcement in war extremely difficult. Lastly there is the loss of US command and control, signals intelligence and satellite tracking capabilities which would have to be replaced somehow.

Option A

With the total loss of all US forces from the region and no plans to return, the situation would indeed be serious. It may be that some extra air power could be made available from Spain, but this would need considerable updating if it had to replace the US capability. Air power is so critical to the region that it would have to be of very high quality to be credible. To replace the 6th Fleet is a much more difficult task. Even though it would not need to be replaced on a one-for-one basis, a sizable fleet would be required to maintain a credible counter-weight against the powerful naval forces that the Soviet Union could bring to bear in the Mediterranean Sea. This would have to become the main role for the French, Italian and Spanish navies. However, it would be the replacement of the co-ordinating effect of US leadership in the region that would be so difficult to replace. No single European nation could easily wear that mantle. It may be that the role could be best assumed by France as the European ally nearest to the region but far enough away not to be too involved with intimate problems of the area. France, being a nuclear power, could also go a little way to replace the US nuclear deterrent in the region. Although some British naval units might be available for operations in the Mediterranean, they are more likely to be fully committed in the Atlantic Ocean and North Sea.

Option B

The situation here would be better, provided that NATO Europe could maintain the bases needed for US forces as they returned in an emergency. The vital air force units could return very quickly and it would not take long for the 6th Fleet to be back on station. These could return with their nuclear weapons so that it would only be necessary for NATO Europe to provide for the ground forces. However, to be credible to prevent war in the first place it would still be necessary to cover the area with West European (French and Spanish) naval and air forces in option A.

Summary

It is probable that the physical presence of US forces in the southern region could be replaced by the Europeans, albeit with a number of gaps. However, the loss of the US stabilizing and cohesive influence would be almost impossible to replace and this alone could spell disaster in maintaining deterrence and in meeting aggression in war. Inevitably there would be considerable cost increases for European NATO.

V. Changes in NATO strategy and operational concepts

Replacement of nuclear forces

A West European spectrum of nuclear weapons would be essential in the absence of US forces. The British and French navies could provide a small strategic system based on their current submarine fleets, but an intermediate-range system would have to be created, with all warheads under direct Euro-NATO control, along with existing tactical nuclear weapons.

Strategic changes

Much has been written about possible changes in NATO strategy, and whether such changes could in some way compensate for a US withdrawal. The FRG is unlikely to accept any change that weakened the Forward Defence doctrine even if it must be interpreted differently in the future. If this is so, the Forward Defence doctrine must remain a corner-stone of any new strategy. Without a deterrent US force presence, nuclear weapons will be even more important, the only alternative to Flexible Response being Massive Retaliation, or the re-adoption of a 'trip-wire' philosophy. This is not an option, given public opinion on nuclear matters. Furthermore, whatever NATO does it must remain credible in the eyes of the USSR. This means that the latter must believe that without US forces it runs a high risk of losing any war with the new alliance. Thus NATO numbers, weapons, training, morale, will to fight, strategy and operational concept must all be seen to be credible.

To achieve this without the strategic depth provided by the US force presence would be virtually impossible, since NATO Europe must somehow sustain operations at sea, on land and in the air in any theatre, under any conditions, either conventional or nuclear, wherever or however Soviet forces attacked. Unless balanced by massive Soviet reductions, this will be even more difficult if Euro-NATO has already reduced forces as a result of negotiations prior to a US withdrawal.

Changes to operational concepts

Operational concepts define how commanders in the field use their resources to defeat the enemy within a given strategy. Any concept based on a perception of what might happen in a future war, and on a specific system of defence or set of assumptions, will almost always fail in battle. This is because a defender is making it quite clear in peacetime how, where and when he intends to fight in war. The aggressor studies this in his own time, plans accordingly and strikes in a way which the defender cannot match because he has deprived himself of capabilities and flexibility.

Much is written about 'defensive defence', a concept which in its extreme version sees lightly armed infantry with anti-armour weapons spread in a frontier zone designed to take a heavy toll of enemy armour as they cross.[10] In some versions these men are reservists. An attacker who applies the principles of war would find little difficulty in defeating such a defence, even if it initially involves accepting high casualties. That is of course not to say that there is no place on the battlefield for light infantry equipped with anti-armour weapons—there is, and they are already a part of modern defence concepts. Senior Soviet commanders agree with this philosophy.[11]

Thus when the time, size and nature of an attack cannot be predicted, a defender must retain balanced forces which enable him to respond to take action to seize the initiative. This is vital if he is outnumbered in the first place. Unless he does this he cannot win. More important, it can be seen that he cannot win, and thus he cannot deter war. Successful military concepts must be built up around the principles of war. These have nothing to do with specific situations, ground or threats. The only way a commander can meet such a situation is by the availability of well-trained reserves ready to conduct any type of operation. It is, after all, his use of reserves that gives him the only chance of seizing the initiative and winning. If they are to stand any chance in the totally unpredictable operations that they will be asked to conduct, reserve troops cannot be drawn from mobilisable out-of-theatre forces, with no knowledge of the ground, nor can they be part-time soldiers or reservists.

West European governments in NATO must nevertheless improve the way they field their forces and provide them with equipment, roles and tasks. Much thinking and work is already going into these problems and it is difficult to see any remaining area where this might be achieved on the scale needed to compensate for the loss of US forces. The increased use of helicopters and airmobile forces, the creation of an effective long-range attack capability, the greater use of electronics and computers and improved surveillance equipment are all being introduced in the next few years. These force-multiplier measures will compensate for current limits on manpower, but they will not cope with the loss of US forces as well. Developments in electronic warfare, directed energy weapons and 'smart' munitions are already affecting operational concepts. All will help, all are under way and all are needed anyway.

There is one political area which would ensure a better and more coordinated defence of the European pillar of the alliance, however. It would not alone compensate for US withdrawal but it could help, certainly in option B. This would be the creation of a more integrated European defence system. No sovereign nation can afford to give up a major defence function if there is any chance that it might, at some future date, have to defend itself or engage in combat somewhere without the Alliance. Thus, such action is not possible until a West European political union is established and until there is an agreed foreign policy for West European armed forces to support. Nevertheless, there are some ways in which Euro-NATO could get better value for its defence money, by taking the first steps towards the establishment of a unified European force structure. Such a force would still be composed of independent national contingents, but these would come much closer together in certain areas such as conceptual thinking, equipment research, development and production, training, logistics and command and control. Already the Army Groups work to integrated concepts agreed by the nations that compose them. This itself is a great step forward but further interoperability and possibly some integration between nations is required to make them work properly and remain credible. Such measures would produce a more cost-effective European defence and also be more convincing to the USA. Still, it would not compensate for the loss of US forces.

Provision of manpower

It is sometimes argued that European armies produce very small armed forces compared with the available number of men between 18 and 30 years of age. There is some truth in this. There is no doubt that Europe could produce the numbers to replace the US forces but this could only be done at the expense of the manpower needed to maintain their individual economies and defence efforts. Even if the manpower were made available, involving the mobilization in peacetime of men up to 30 years of age, further large sums of money would have to be spent on equipping them. European countries could not then sustain their defence efforts and remain independent nations. The task of maintaining 14 separate economies is very different from the USA's task of maintaining one. In any case, NATO must now accept manpower ceilings in the Atlantic-to-the-Urals area, and must ensure that if the US forces withdraw, the ceilings remain even if they cannot be completely filled by European forces.

VI. Command and control

Not only does the USA exercise a co-ordinating influence over the southern region but it does so over the alliance as a whole. It alone in NATO can match the USSR, and this gives the USA a special position of leadership which is more or less acceptable to all allies, and which none of them could replace. This makes it difficult for the remaining alliance partners to agree on an effective military command structure where one is the undisputed leader. The only possible alternative system is likely to be some kind of power-sharing and rotational arrangement, never a good system for the conduct of war.

Replacement of the US commanders in the NATO command structure would pose major problems. The posts in question are, in order of seniority, as follows and one possible way of replacing them with Europeans is given:

1. SACEUR would have to be a European in each of the scenarios posed in this book. There are three possible systems for selection: (*a*) allot the post to one country (the question is which: France, the FRG, Italy and the UK would be contenders); (*b*) allot the post to these countries on a rotational basis; (*c*) select the best individual for the job every three years as is currently done for the Chairman of the military committee.

2. The deputies to SACEUR could be appointed in a similar way, with candidates from any of the countries above that do not provide the SACEUR.

3. The Allied Command Europe (ACE) Chief of Staff could be an officer from the country not represented on the posts above or it could be open to officers from all member countries.

4. The Commander-in-Chief Allied Forces Northern Europe (AFNORTH) would remain a British officer.

5. The Commander-in-Chief Allied Forces Central Europe (AFCENT) would remain an officer from the FRG.

6. The Commander-in-Chief Allied Forces Southern Europe (AFSOUTH) would be a French officer.

7. The Commander NORTHAG would remain a British officer.

8. The Commander CENTAG could be an officer from the FRG or the post could alternate between the FRG and France.

In addition, there would probably be a need for three NATO naval commanders:

1. The Commander-in-Chief Atlantic Approaches would be a British officer.
2. The Commander-in-Chief Baltic and North Seas would be an officer from the FRG.
3. The Commander-in-Chief Mediterranean could be a French officer or the post could alternate between France and Italy.

Air Force commands could be reorganized as follows:

1. The Commander-in-Chief UK Air would remain British, with command extending to cover Norway and the Baltic Approaches in war.
2. The Deputy Commander-in-Chief AFCENT could remain as an air force post rotated between France and the UK,
3. The post as Commander AAFCE should be deleted.
4. The Commander 2 ATAF could remain British.
5. The Commander 4 ATAF should be French if COMCENTAG goes to the FRG. Alternatively the post could rotate between the two countries.

There are many ways of allocating these posts and the principle should be flexibility, so that the best man for the job is selected wherever possible. However, there will always be a need to maintain a balance of interests. It should also be a principle that commands be given to officers from nations within the regions concerned. There would therefore be little change in the northern and southern regions, where the national sectors are commanded by officers from the nations that provide the forces in them.

There would be little problem with this kind of plan in option A but in option B problems might begin to arise when large-scale US forces returned in war, and would then come under European command. This would be the price that the USA would have to pay. In option B its forces would not fight under overall US command (SACEUR, Region and Army Group Commanders would be Europeans). There could be no question of having US commanders in Europe in peacetime after all US forces had gone, even if they were planning to return. Equally, it would be impossible to change the high-level command structure just as the fighting started and US forces began to arrive.

VII. Conclusion

NATO has from the start been built up around the forces of its members, including the USA. To remove the forces of any one country would court disaster. If that country were to be the USA, NATO would quite simply collapse. Europe can probably build a new and different defence structure in the long term, but this would demand much expenditure, the mobilisation of more manpower and a high degree of integration of national armed forces. To attempt this with the WTO countries in a state of turmoil and instability would be folly of the highest order.

Notes and references

[1] Duke, S., SIPRI, *United States Military Forces and Installations in Europe* (Oxford University Press: Oxford, 1989).

[2] For NATO figures, see 'Conventional forces in Europe: the facts' (NATO Press Service: Brussels, Nov. 1988); for WTO figures, see Committee of the Ministers of Defence of the Warsaw Treaty Member States, 'On the relative strength of the armed forces and armaments of the Warsaw Treaty Organization and the North Atlantic Treaty Organization in Europe and adjacent water areas', *Pravda*, 30 Jan. 1989.

[3] Communiqué of the Political Consultative Committee of the Warsaw Treaty member states, 1 June 1987; 'On military doctrine of Warsaw Treaty Member States', statement by the Political Consultative Committee of the Warsaw Treaty member states, 1 June 1987.

[4] Visit to the Soviet Union by the author and a team from the Royal United Services Institute for Defence Studies (RUSI), and discussions at the Soviet Ministry of Foreign Affairs and Ministry of Defence.

[5] Smith, R. J., 'CIA Chief sees reduced risk of surprise attack', *International Herald Tribune*, 14 Dec. 1988; Forsberg, R. et al., *Cutting Conventional Forces: An Analysis of the Official Mandate, Statistics and Proposals in the NATO/WTO Talks on Conventional Forces in Europe*, East–West Conventional Force Study (Institute for Defense and Disarmament Studies: Brookline, Mass., July 1989); Karber, P., testimony to the US House Armed Services Committee, House of Representatives, US Congress, 3 Mar. 1989.

[6] On 3 May 1990 President Bush announced cancellation of work on a new ground-launched follow-on to Lance (FOTL) and on the modernization of nuclear artillery. See Fitchett, J., 'Missile initiative by Bush signals acceleration of European agenda', *International Herald Tribune*, 4 May 1990.

[7] *A Comprehesive Concept of Arms Control and Disarmamement, adopted by Heads of State and Government at the Meeting of the North Atlantic Council in Brussels on 29th and 30th May 1989* (NATO Information Service: Brussels, May 1989).

[8] See Duke (note 1) for complete listings.

[9] At the 'Open Skies' meeting in Ottawa in February 1990, the 23 CFE Foreign Ministers agreed that US and Soviet troop levels in the central region of the Atlantic-to-the-Urals (ATTU) zone should be reduced to 195 000, with an additional 30 000 US troops permitted in the ATTU periphery. See US Arms Control and Disarmament Agency, *CFE Negotiations on Conventional Armed Forces in Europe* (ACDA Office of Public Affairs: Washington, DC, March 1990). See also Riddell, P., 'Ottawa agreements show decline in Soviet power', *Financial Times*, 15 Feb. 1990; Pringle, P., 'US plays down its triumph at Ottawa arms meeting', *The Independent*, 15 Feb. 1990.

[10] For a summary and contrasting views of various 'defensive defence' concepts, see Dean, J., 'Alternative defense: answer to NATO's Central Front problems?', *International Affairs*, vol. 64, no. 1 (winter 1987/88), pp. 61–82; Flanagan, S. J., 'Non-offensive defense is overrated', *Bulletin of the Atomic Scientists*, vol. 44, no. 7 (Sep. 1988), pp. 46–48.

[11] Note 4.

20. The United States coping without bases in Europe

Robert E. Harkavy
Pennsylvania State University, University Park, Pennsylvania

I. Introduction

For a number of years, approaching a crescendo in the late 1980s, newspapers and journals in the United States have been filled with dire warnings about what appeared a gradual, relentless unravelling of the hitherto global US basing structure.[1] Numerous points of access—among them Jamaica, Libya, Madagascar, the Maldives, Malta, Sierra Leone, South Yemen, Trinidad and Tunisia—had of course already been lost because of the collapse of the British, French and other European empires.[2] Defections from the US political orbit via coups, revolutions, wars, political radicalization, changed alignments, and so on, had caused some further loss of access in such disparate locales as Ethiopia, Iran, Madagascar, Morocco (now partially regained), Pakistan, Thailand (now partially regained), the Seychelles and Viet Nam, among others.

More recently, nuclear/environmental issues have led to the curtailment of US access to New Zealand[3] and have become points of controversy in Australia, Belau, Denmark, Japan and elsewhere.[4] By the late 1980s, however, US basing access diplomacy was mainly focused on a number of key, hitherto dependable allies—Greece, the Philippines, Portugal, Spain and Turkey—that no longer fully appeared to accept the mutuality of security concerns with the United States, and which were—to varying degrees—requesting higher rents and tighter restrictions on US access.[5]

Still, the United States managed to cope. Some newer points of access were acquired or others expanded: China, Diego Garcia, Egypt, Kenya, Norway, Oman, Somalia and others.[6] And the ongoing march of technological development—longer-range ships and aircraft, space-based C^3I in lieu of land-based facilities—allowed for easy replacement of some lost basing assets.[7] The loss of intelligence listening posts in Iran, for instance, was compensated for by a combination of newer, advanced satellites and substitute access in China, Turkey and Norway and perhaps also in Pakistan.[8]

Today, however, the growing talk about a 'Europe without America' portends a qualitatively different sort of problem.[9] Western Europe has, of course, long hosted a massive array of US installations and bases, long thought of as secure and semi-permanent within the NATO alliance structure. Indeed, almost forgotten during the past generation by those lacking in sufficient historical perspective was the fact that the long-term granting of access between sovereign, albeit allied, states such as had existed since the late 1940s within NATO is historically unique. There is no apparent precedent, ever, not even during periods of global or regional ideological polarization and concomitantly stable alliances.[10] Hence, the beginnings of a move towards a 'Europe without America' might be perceived as inevitable in a historical context, or rather as a return to what historically has been the norm.

The timing of the debate is interesting of course, coming as it does in the wake of a number of other tidings of change: the end of the Reagan era, *glasnost* and *perestroika*, the signing of the Treaty between the USA and the USSR on the elimination of their intermediate-range and shorter-range missiles (the INF Treaty) and the prospect of follow-ons, the seeming imminence of a truly more integrated Europe, the apparent death and subsequent rebirth of orthodox Chinese communism, the supersession of US global economic leadership by Japan and, of course, the stunning events in Eastern Europe and the USSR in 1989–90. Indeed, as this chapter was being written, Henry Kissinger, appearing on the widely viewed US *Nightline* television news programme, spoke solemnly of the advent of a wholly new era of international relations, of an evolving new international system.[11] In that regard, he spoke of an ineluctable and inevitable drawing apart of the USA and Western Europe and of a similar inevitable loosening of the Soviet grip on Eastern Europe.

These broader contextual issues are noted just because it may be important to avoid a 'static' analysis (the USA withdraws from Europe, but everything else stays the same save for the impact on the central European military balance). Indeed, the hypothesized US withdrawal from Europe might later be viewed both as cause and effect in relation to numerous, interlocking secular shifts in the structure of the international system. In the most general sense, this may involve fundamental shifts towards multipolarity, power diffusion, the decline of ideology as a basis for alignments and rivalries, a trend towards more rapidly shifting alliances, and so on; that is, somewhat reminiscent of the so-called 'classical system of diplomacy' of the 18th and 19th centuries.[12]

On a more specific level, one might speculate upon the implications for US interest in defending the Persian Gulf area, a possible enlargement of US attention to Latin America, or the impact on a more threatened Israel (following a US withdrawal from Europe). These are but examples. There may be others not now in view.

The guide-lines for the studies in the present volume call for the consideration of two options: a US withdrawal without a commitment to return (option A), and a withdrawal of the peacetime presence but with a commitment to return in a crisis situation, implying the retention of pre-positioned equipment (or POMCUS: Pre-positioned Organizational Material Configured to Unit Sets) and skeleton staffs to facilitate the return of US troops (option B).

The implications of these options are clear and unambiguous enough as they concern the US contribution to the defence of Western Europe itself. However, as regards the other functions of US forces and bases in Europe, that is, the nuclear defence of the United States itself and the projection of power into the Third World, there are some areas of ambiguity. Some of these problems are noted elsewhere in this volume, but they are worth underscoring here just because of their importance from the perspective of US security interests, some of which do not dovetail with those of Western Europe.

One problem is, of course, whether some US C^3I assets might be retained in Europe along with POMCUS stocks, in some cases involving assets which would be necessary if US forces were to return to defend Western Europe. There is also the question of the future of the 6th Fleet, with or without varying degrees of access for refuelling, rest and recuperation (R&R), and so on. The whole problem of residual US aircraft overflights remains, particularly as it applies to 'out-of-area' operations.[13]

Geographically speaking, Europe is here defined as including all NATO countries and their constituent territories. That is, it includes Iceland, Greenland, the Faeroes and the Azores, but not such colonial remnants as Bermuda, Ascension Island, Diego Garcia or French possessions in the Pacific. It is worth noting in this context, however, that a full US withdrawal from Europe might well put greater pressures on US access to some of those remnants. Some—Diego Garcia, Ascension and Bermuda—involve crucial US assets in the areas of anti-submarine warfare (ASW), aircraft staging, satellite communications, space-tracking, and so on.

What the United States would lose

The total—or almost total—withdrawal of US forces from Europe and the concomitant loss of basing access (leaving open the residual questions of limited naval access, i.e., port visits, refuelling, etc., and of aircraft overflights) would have an impact on the US defence posture in several general areas. They are: (*a*) the US nuclear deterrent posture *vis-à-vis* the USSR, including capabilities for monitoring existing or prospective arms control agreements; (*b*) the US capability for maintaining the option of assisting the defence of Western Europe, should it again become necessary; (*c*) US capabilities for 'out-of-area' conventional operations, particularly in the Middle East, potentially involving either direct interventions with US forces, coercive diplomacy or arms resupply to clients or friends engaged in conflict; and (*d*) a variety of horizontal escalation scenarios, involving wars or crises initiated in the Far East or Middle East and involving the possibility of spread to Europe, or the threat of same.

The preceding chapters cover category (*b*), involving a residual commitment, retention of POMCUS, and so on. Of course, the United States may be said to be 'impacted' or 'affected' by the loss of capability in this area to the extent that keeping Western Europe free of Soviet control may be said to be in the interests of the United States. If this is a concern, however, why withdraw or be forced to withdraw? Here one enters the metaphysical realm: Should the United

States feel a great sense of loss for no longer bearing the responsibility and cost of the defence of Western Europe?

II. Bases for strategic nuclear forces

If the US military were to withdraw from Europe, assured nuclear deterrence would presumably become, more than ever, the very centre-piece of US security policy; indeed, one might reasonably expect that the nuclear component of the US defence budget would inherit some of the 'dividend' provided by (presumably) reduced conventional budgetary expenses. Nevertheless, the US nuclear posture would obviously be affected by the loss of a number of bases in Europe. How severe and irreparable might that loss be? To what extent could it be compensated for by technological fixes, the absorption of some loss of redundancy, acquisition of alternative facilities elsewhere outside of Europe, if not in the United States and Canada? Furthermore, what might be the impact on the Strategic Arms Reduction Talks (START), and on other negotiations involving nuclear test bans, nuclear weapon-free zones, anti-satellite (ASAT) weapons, strategic missile defence, and so on?

The current US dependence on nuclear-related European bases can be analysed in either or both of two ways. It can be examined by function or system, utilizing the familiar breakdown used in SIPRI studies of foreign military presence: air, naval, ground, missile, C^3I, research and environmental monitoring facilities and various sub-categories within these.[14] This dependence can also be gauged on a country-by-country basis. Allowing for some exceptions or mitigations, this means looking at what happens if, for instance, the United States can retain access only to Greenland, Bermuda and the Azores; or if it can retain port visit rights in Italy and Turkey; or if it can retain either regular or contingent access to air bases in the United Kingdom, and so on.[15]

As it turns out, the current US dependence on European bases for strategic nuclear launching platforms is not very substantial—nevertheless, it does matter. The land-based intercontinental ballistic missiles (ICBMs) are all based in the continental USA (and the INF Treaty has caused the removal of the Pershing and ground-launched cruise missiles (GLCMs) which, although designated as theatre weapons, are strategic in the sense that they could reach targets within the Soviet homeland). What remains is the residual dependence on Holy Loch, Scotland, for basing of some ballistic missile submarines (SSBNs);[16] the land facilities afforded 6th Fleet carriers in the Mediterranean which can launch nuclear-armed attack aircraft with reaches into the USSR as well as Eastern Europe; *en route* tanker refuelling bases for home-based B-52 and B-1 bombers;[17] and the forward air bases for US F-111 as well as nuclear-capable F-16 and F-15 aircraft, all of which can reach targets within the USSR proper. In other words, the United States relies considerably on forward basing for the efficacious deployment of two of the three legs of its strategic triad.

Naval assets

The US submarine base at Holy Loch was inaugurated in 1960. According to Duncan Campbell: 'Its initial strength grew from one to five as submarines of the *George Washington* class were completed. Then it built up to ten'.[18] Later in the 1960s, as the Polaris force grew to 41 submarines, other forward bases at Rota, Spain, and Guam were made operational. Meanwhile, in the course of the past 20 years the range of the Polaris missiles was increased from the original 2800 kilometres to the 4800 km of the follow-on Poseidon missiles, armed with multiple independently targetable re-entry vehicles (MIRVs).[19] Still later came the Trident I missile with a range of 6400 km, thereby removing the necessity for forward bases.[20] The SSBN bases at Rota and Guam were closed down, and the United States was reported to be planning to withdraw the remaining Poseidon force from Holy Loch.[21] That has not yet happened. Meanwhile, the ongoing START talks, if consummated, could result in significant reductions of SSBNs and submarine-launched ballistic missiles (SLBMs) on both sides, which would further reduce the importance of the remaining US presence at Holy Loch.

The future of the 6th Fleet in the wake of a US withdrawal is indeterminate. The Fleet now normally deploys two carrier battle groups in or around the Mediterranean which utilize NATO host-nation ports such as Naples, Souda Bay, Rota, Izmir, and so on.[22] Each carrier will carry three nuclear-armed attack squadrons, with some 24 A-7E or F/A-18 and 10 A-6E aircraft. These carry three, two and four nuclear bombs apiece, respectively, with respective aircraft ranges of 2800, 3200, and 1000 km.[23] This amounts to a capability for targeting some 200 nuclear weapons on the USSR (not counting reloads or repeat missions). Granted, many scenarios for all-out war in Europe see the allegedly vulnerable 6th Fleet withdrawn from the Mediterranean at the outset, and its aircraft would no doubt have other missions.[24] Nevertheless the forward-based systems are there for a purpose.

Air Force assets

The US strategic bomber force now comprises some 260 long-range bombers (some 20 B-1B and 240 B-52G/H aircraft) and some 55 medium-range FB-111A bombers.[25] This force, which carries over 5000 nuclear warheads, is based primarily in the continental United States, but can also utilize numerous pre-attack or intra-crisis 'dispersal' bases, as well as post-attack recovery bases in other locations. Many of these are outside the United States, including some in Canada, such as Cold Lake (Alberta), Goose Bay (Labrador), Namao (Alberta) and Whitehurst (Yukon).[26] There are others in Europe: in the UK at Brize Norton, Marham and Fairford; in Greenland at Søndre Strømfjord; and in Spain at Morón and Zaragoza.[27] Campbell has noted the frequent peacetime rotation of B-52s and FB-111s to British bases, in preparation for possible wartime forward deployment, and suggests that the entire 55-aircraft FB-111 force might be rotated forward in a major emergency.[28] Hence, if the USA were to lose this access, its strategic bomber force would lose a significant amount of its forward dispersal capability. That in turn would presumably create the need for planning new bomber routes, or perhaps a higher level of airborne alert at various levels of crisis for US home-based bombers.

Similarly, the USA is availed of numerous tanker bases for its Strategic Air Command (SAC) bomber force. Many of the tanker aircraft have dual use and would be expected to play a major role in any reinforcement of Europe. If, in a major crisis with Moscow, the latter option was denied as a result of a withdrawal, this might ironically allow for more tanker assets to be used by SAC. There are tanker bases in Canada (Namao, Goose Bay), in the UK (Mildenhall, Fairford), Spain (Zaragoza), Iceland (Keflavík) and Greenland (Thule). Others, such as those on the Azores and Bermuda, could presumably still be utilized as tanker dispersal bases.[29] To the latter can be added Diego Garcia, also slated as a prospective tanker base for B-52 operations, which, as noted above, might be jeopardized by a US withdrawal from Europe.[30] In recent years, as a result of the development of long-range stand-off weapons—the air-launched cruise missile in particular—the USA has begun to plan for more omni-directional bomber approaches to the USSR with the resultant requirement of more dispersed tanker bases. Loss of the European bases would nevertheless hurt, particularly the loss of those in Iceland and Greenland. It might force the USA into greater reliance on a narrower swath of transarctic approach routes (it is assumed that the USA will retain overflight rights over Canada and Greenland or would assert those rights, complaints notwithstanding, in a serious crisis).

Loss of European bases would also entail removal of other US forward-based nuclear systems, that is, fighter-attack aircraft. This involves the F-111 E/Fs based in the UK and the F-16s based in the FRG. Some 72 F-16s and some 150 nuclear weapons are based at Hahn and Ramstein air bases in the FRG; a similar number of F-4Es (to be replaced by F-16s) are based at Spangdahlem with another 150 weapons stored.[31] Aviano in Italy also has some 200 weapons said to be stored in connection with 72 F-16s now rotated forward from Torrejón in Spain to Italy.[32] The base at Incirlik in Turkey can also support forward-rotated nuclear-capable F-16s.[33] Loss of these bases would not jeopardize overall US second-strike deterrence (many are really slated for tactical use in Eastern Europe), but it would reduce the number of nuclear weapons at least theoretically targetable on the USSR.

It is further worth noting, as a parenthesis, that there are a large number of nuclear weapons in Europe with the forces of European NATO countries, involving a 'two-key' system and US custodial control. This involves nuclear-armed F-104 G/S, F-4E/F and F-16 aircraft in Belgium, the FRG, Greece, Italy, the Netherlands and Turkey.[34] Similarly, there are 'two-key' Lance missile warheads, artillery shells, Nike-Hercules air defence missiles, anti-submarine warfare (ASW) aircraft with nuclear weapons, and so on.[35] These would all be withdrawn as part of an overall US withdrawal from Europe—in either of the two scenarios—leaving only the residual British and French assets.

The US ASW effort (discussed further below with regard to SOSUS) would clearly be damaged by removal of US nuclear-armed ASW aircraft from Europe, even in the case of a US–Soviet conflict not directly involving Europe (although the task of securing sea lines of communication, SLOCs, during a reinforcement of Europe may be disregarded here). US P-3 Orion ASW aircraft are today based at or staged through British bases at St Mawgan and Machrihanish, with some 63 nuclear depth bombs stored at each.[36] Other US P-3s with forward-stored nuclear depth charges (or contingency plans for same) are based in Europe at Keflavík (48 bombs), Sigonella on Sicily (63 bombs) and Rota (32 bombs).[37] US P-3s are also rotated through Andøya and Bodø in Norway, Souda Bay (Crete), Montigo (Portugal), and 32 nuclear depth bombs are stored at Lajes for wartime use.[38] In a withdrawal scenario in which the United States were to retain access for the 6th Fleet in the Mediterranean, the question of access for ASW aircraft as well becomes of immediate interest, lest the Fleet be left highly vulnerable to Soviet attack submarines. As a possible replacement force, one could envisage expanded European P-3 operations in the Mediterranean, building on current Italian flights of Breguet Atlantique aircraft out of Sigonella and Catania, from which also British Nimrods operate.[39] This would still involve US custodianship of nuclear weapons (and hence custodial units), however, as would the option of operating Canadian ASW aircraft out of bases in Nova Scotia (Greenwood) and British Columbia (Comox). Neither option could therefore be maintained in either of the two withdrawal scenarios.[40]

The United States would also lose some air defence capability, related mostly to SLOC control but with some implications for its nuclear deterrent posture. Iceland now hosts one squadron of F-15 aircraft which could be utilized against Soviet long-range bombers.[41]

Early-warning and space-surveillance assets

Loss of access to Europe would, of course, seriously affect a variety of functions falling under the general heading of early warning and space surveillance, both of which are important components of the US nuclear deterrent posture. Many are, of course, dual-purpose in relation to nuclear and non-nuclear roles. Some relate primarily or solely to the defence of Western Europe. Some are important to the nuclear defence of the United States itself.

For early warning of nuclear attack, and assessment of such attacks, there are the ground down-links for early-warning satellites; the Ballistic Missile Early-Warning System (BMEWS); and several strings of warning radars across the Arctic region—linking up with NATO systems in Iceland and Greenland—to detect *en route* bombers.

While the main ground link for the Defence Support Program (DSP) East infra-red satellite early-warning system is located at Nurrungar in Australia (from there it is linked via several alternate channels to Buckley Air Force Base in Colorado and on to the North American Air Defense Command (NORAD) Cheyenne Mountain Complex), a back-up so-called Simplified Processing Station (SPS) deployed at Kapaun in the FRG.[42] The USA is currently moving towards deployment of mobile ground terminals (MGTs) as well, so that while the closure of the Kapaun facility would presumably represent a significant loss, it would by no means be a fatal blow to US early-warning capabilities.[43] There are also DSP satellites located over the Indian and Atlantic oceans, called DSP West, but these do not depend on down-links on foreign soil.

The closure of two of the three US BMEWS radars—those at Thule in Greenland and Fylingdales Moor in the UK—would presumably be a major loss, one which would come at a time when the USA is upgrading some of these radars in the face of Soviet complaints about arms control violations. The BMEWS radars, with a range of 4800 km, detect and track satellites as well as ICBMs and intermediate-range ballistic missiles (IRBMs).[44] It is not clear just what residual capability would be maintained by the remnant station at Clear, Alaska, or the extent to which coverage lost could be made up by adding to the DSP satellite system or installing radars elsewhere. Some analysts cast doubt on the efficacy of BMEWS, because of the vulnerability of its three facilities to pre-emptive interdiction by low-flying bombers, cruise missiles or SLBMs, as well as to jamming.[45] At best, the redundancy of the overall US early-warning networks would be reduced. US dependence on facilities in Australia would perhaps also be increased as it would be by the loss of Kapaun.

For warning of the approach of bombers, the USA has long relied on a series of radar picket lines across the Arctic known as the Distant Early Warning (DEW) and Pinetree lines. These have been located across Alaska, Canada and Greenland, with a few additional outposts in Iceland and the Faeroes.[46] These are now being superseded or upgraded by the North Warning System (NWS), utilizing many of the old DEW Line stations. There will be 13 Seek Igloo AN/FPS-117 radars in Alaska, 13 in Canada, 4 in Greenland, 1 in Scotland and 2 in Iceland.[47] A US withdrawal from Europe, as defined here, would result in the loss of the seven easternmost stations, presumably creating a big gap in coverage of bomber approaches. This might perhaps be compensated for by the use of submarines or surface ships as radar picket vessels. The degree to which these could provide complete compensation is not known to the author.

Space tracking is another field in which the USA relies heavily on foreign-based facilities—the BMEWS posts are components of this network. Indeed, although the USA is now planning a space-based satellite surveillance system, for the time being the US mechanisms for tracking and detecting Soviet satellites are almost entirely ground-based. The varied ground-based systems fall under the overall rubric of SPADATS (Space Detection and Tracking System). According to the typology provided by Jeffrey Richelson, this incorporates three types of sensor—dedicated, collateral and contributing—involving sensors whose primary missions are, respectively, space surveillance,

SAC sensors with a secondary space-surveillance role and non-SAC sensors with a secondary space-surveillance role.[48]

For a long time the heart of the dedicated sensor system was a group of Baker–Nunn optical cameras with the ability to see an object the size of a basketball more than 50 000 km in space.[49] Until recently the USA had six of these spread around the globe, one of them at San Vito dei Normanni, Italy.[50] These are now being replaced by the Ground-based Electro-Optical Deep Space Surveillance (GEODSS) system. This will have five stations spread about the globe, with one in Portugal.[51] A US withdrawal would presumably leave a gap in coverage. This could perhaps be filled by acquiring a facility in Morocco or, in imitation of Soviet practice, by the use of shipboard systems.[52]

Some other sets of dedicated sensors—the Naval Space Surveillance system (NAVSPASUR), optical radars in the United States, the Pacific Radar Barrier (PACBAR), the Cobra Dane phased-array radar in Alaska, the Pave Paws phased-array warning system and the Enhanced Perimeter Acquisition Radar Attack Characterization System (EPARCS)—do not involve European access.[53] An AN/FPS-79 radar at Pincirlik/Diyarbakir, Turkey, is utilized as a contributing sensor, however,[54] as is BMEWS.[55] The same is true for some radars primarily used as part of the US satellite tracking and data network (STADAN) of installations used to track and monitor US space activities, including the down-range trajectory of launches. Included among these facilities are those in the UK (Winkfield), the Canary Islands (Tenerife)[56] and Ascension and Bermuda (possibly retained following a US withdrawal).

Generally speaking, US loss of access to Europe would also subtract some capabilities in satellite control. Concerning DSP early-warning satellites, for instance, it has been reported that there are seven substations linked to the master station at Sunnyvale, California, three of which are in the USA.[57] Outside the USA there are satellite control facilities at Oakhanger, UK, and at Thule, Greenland. In addition, Paul Stares has noted a number of other US satellite control and receiver sites at European locations, all of which play roles in the US nuclear deterrent scheme, as shown in table 20.1.

Clearly an analysis of the impact of the loss of each of these on the US nuclear defence posture would be beyond the scope of this chapter—a full analysis of gaps created would presumably require access to classified data. Assuming a full US withdrawal, the question again arises whether assets would not be replaced by ship-borne systems.

With regard to ocean surveillance, loss of the US base at Edzell, Scotland would presumably be important, but perhaps more in the context of SLOC warfare involving reinforcement of Europe than of a central nuclear war *not* involving Europe.[58] Aside from Classic Wizard satellite ground stations, there are additional land-based high-frequency direction-finder (HF/DF) facilities in Europe at Rota (Spain), Keflavík (Iceland) and Brawdy, Wales.[59] If SLOC warfare were to involve interdiction of Soviet surface cruise missile carriers in the Atlantic, a loss of these facilities would hurt. So too would the loss of bases

Table 20.1. US satellite control and receiver sites in Europe, 1987

Category/programme	Control or receiver site
Signals intelligence (Chalet, Magnum, Jumpseat)	Menwith Hill, UK
Ocean reconnaissance (White Cloud or Classic Wizard)	Edzell, UK
Early warning (Defense Support Program)	Kapaun, FRG
Communications (DSCS III)	Landstuhl, FRG
Fleet Satellite Communications (FLTSATCOM)	Bagnoli, Italy
Navigation (Navstar/GPS)	Ascension Island (owned by the UK)
Meteorology	Thule, Greenland

Source: Stares, P., *Space and National Security* (The Brookings Institution: Washington, DC, 1987), p. 189.

for aircraft (EP-3E, EA-3B) utilized for ocean surveillance, some of which have operated from bases such as Rota as well as from aircraft-carriers.

Anti-submarine warfare assets

Concerning ASW, the United States would suffer a major degradation of its Sound Surveillance System (SOSUS) network with the loss of access to Europe. Again, this might pertain primarily to the contingency of monitoring the Greenland–Iceland–UK (GIUK) or Greenland–Iceland–Norway (GIN) SLOCs in case of a war in Europe requiring US reinforcement. However, while it is true that Soviet SSBNs can easily utilize firing stations in the so-called 'bastions' of the Norwegian and Barents seas, Moscow has long moved SSBNs forward into the western Atlantic, and the loss of US SOSUS hydrophone networks and shore-based computer processing terminals in Europe would presumably allow the Soviet Union to do this in a way less easily monitored by the United States. Of the some 22 major US SOSUS facilities, several are located in Europe: Scatsa (Shetland) and Brawdy in the UK; two in Turkey; one each in Bermuda, Norway, Iceland (Keflavík), the Azores (Santa Maria), Ascension, Italy, Denmark, Gibraltar, as well as on the European-owned territories of Diego Garcia and Ascension.[60]

In addition to the impact (presumably devastating) on US coverage of the Atlantic, there would be a big impact on US ASW capabilities in the Mediterranean, with loss of facilities in Turkey, Italy and Gibraltar. This, along with the loss of P-3 bases, would make 6th Fleet operations in the Mediterranean much less tenable.

Nuclear-detection assets

Withdrawal from Europe would also entail some loss of capability in the field of nuclear detection. This is relevant to three general areas of security policy: monitoring nuclear detonations during wars, which provides guides to further targeting; monitoring or verifying arms control agreements such as a limited or total test ban; and monitoring the activities of potential aspirants to nuclear status. Regarding the first function, Paul Stares reports that:

[T]he space-based US Nuclear Detection System (NDS) would complement the data supplied by the DSP early warning satellites in important ways. Though the decision time would be slim, the NDS sensors could, for example, 'record the detonation of Soviet SLBMs on US territory some 10 to 20 minutes before the expected arrival of the more accurate 'silo killing' Soviet ICBMs. The US leadership would have additional information to use in making the dangerous decision of whether to save the threatened ICBMs by launching them promptly.'[61]

The United States now has an elaborate global network of ground-based seismological stations, some of which could be used to monitor detonations within the USSR. Some of these are in Europe; at Torrejón (Spain), Iraklion (Crete), RAF Lakenheath (UK), Wiesbaden (FRG), Belbasi (Turkey) and, most notably, the Norwegian Seismic Array (NORSAR), consisting of seven sub-arrays, centred on Hamar and Karasjok.[62] The latter have reportedly become more vital (along with other arrays acquired in China) following the United States' loss of facilities in Iran in 1979. There may be some additional unmanned sites, and there are others in Canada, for instance, such as the Regional Seismic Test Network at Yellowknife, Northwest Territories.[63]

Ground stations are not the only vehicles for nuclear test detection. The Air Force Technical Applications Center (AFTAC) can presumably avail itself of numerous airfields open to the United States around the world. In addition, the newer GPS (Global Positioning System) satellites are reported to have a nuclear-detection mission in addition to those of navigation and targeting. These satellites rely on several external control or receiver sites, one of which is on Britain's Ascension Island.[64]

It is difficult to say just how irreplaceable the loss of the ground seismological stations would be for the United States. Presumably the GPS satellites, once the full system of 18 satellites is in place, will be able to handle the role of nuclear detection during wars, although their reliance on ground stations remains.[65] It would appear the United States would above all lose capability in the area of arms control verification, which might provide one rationale for Europeans to retain these US installations even if much else were to go.

Communications assets

Some global communications systems closely involved with the US strategic nuclear posture would also be affected by a US withdrawal from Europe. In this context it is important to note that in some cases, such communications systems are dual-usable for strategic nuclear and conventional purposes. This is a large subject, and only a quick sketch and some examples can be given here.

Communications with submerged submarines (SSBNs, SSNs), for instance, involves a number of systems, some now utilizing ground facilities in Europe. There was earlier heavy reliance on Loran (long-range navigation) C/D transmitters and monitoring systems for positioning of submarines. In Europe this involved facilities in Bermuda (Witney's Bay), the Faeroes (Ejde), Greenland (Angissoq), Iceland (Keflavík, Sandur), Italy (Crotone, Lampedusa, Sellia Marina), Norway (Jan Mayen Island), Spain (Estartit), Turkey (Kargabarun), the UK (Shetland Islands) and the FRG (Sylt).[66] Earlier, the United States also utilized its global network of Omega very-low-frequency (VLF) transmitter stations, one of which was in Norway at Brattland.[67] Many of the Loran stations were operated by the US Coast Guard, and the Omega stations tended to be operated jointly by host nation personnel.

For communicating with submarines the United States has also used a number of other VLF and LF transmitters. In Europe, these have been located at Grindavík (Iceland), Guardamar del Segura (Spain), Helgeland (Norway), Kato Souli and Nea Makri (Greece), St George's (Bermuda), Thurso (UK) and Tavolara (Sardinia).[68]

Finally, as a supplement to land-based VLF and LF transmitters, the United States also deploys some 18 TACAMO ('take charge and move out') radio-relay aircraft. These are retrofitted C-130 transport aircraft which are aloft at all times over the Atlantic and Pacific oceans.[69] They utilize forward bases outside the United States—on the Atlantic side this involves Bermuda (St George's) and the UK (Mildenhall).[70]

In the present period, however, the United States is moving towards reliance on the GPS system for its accurate positioning of submarines, and on US-based extremely-low-frequency (ELF) grids in Michigan and Wisconsin for long-range communications with submerged submarines.[71] This should at least serve to mitigate what otherwise would appear to be the serious consequences of a withdrawal from Europe in this domain of strategic communications.

Surface naval communications also involve HF receivers, some of which are in Europe (primarily for communications in European waters). They include facilities in Azores (Cinco Pinco, Vila Nova), Bermuda (South Hampton), Greece (Nea Makri, Kato Souli), Iceland (Grindavík, Sandgerdhi), Italy (Naples, Licola), Spain (Guardamar del Segura, Rota) and the UK (Edzell, Thurso).[72] Loss of these would, again, primarily be felt in connection with SLOC control and 6th Fleet operations in the Mediterranean; less for US strategic nuclear deterrence *vis-à-vis* Moscow. Again, a US withdrawal and a reduction of the US role in the conventional defence of Europe (or deterrence of a conventional attack) would also mean reduced C^3I requirements, although it is difficult to measure to what extent.

The dependence of the SAC B-52 fleet on overseas tanker and dispersal bases has already been noted. These forces also utilize 'fail-safe' communications (otherwise described by the term 'positive control', meant to imply that the bombers go ahead with attack missions only if given 'execution instructions' once aloft and nearing the USSR), largely dependent on transmitters in foreign locations. As they enter the Arctic region the aircraft are supposed to receive orders over the SAC 'Green Pine' network, with transmission sites mostly in Canada, with one at Grindavík, Iceland.[73] In addition, SAC's bombers and tankers also utilize a global Giant Talk/Scope Signal III network of some additional 14 stations below the Arctic. Among those facilities which might be affected by a US withdrawal from Europe are those at Ascension, Thule, RAF Croughton and Mildenhall, Lajes, Torrejón and Incirlik.[74] Again, this might have the effect of channelling SAC's operations north over Canada, making the bombers more vulnerable to Soviet air defence.

The US strategic forces also rely heavily on the Air Force Satellite Communications (AFSATCOM) and Fleet Satellite Communications (FLTSATCOM) systems. The former is a UHF network primarily devoted to strategic nuclear-related purposes. Rather than being carried on dedicated satellites, it is carried in transponder packages on the Defense Satellite Communication System (DSCS) III, Satellite Data System (SDS), FLTSATCOM and other satellites.[75] In Europe there is an AFSATCOM-related facility at Landstuhl, FRG, also described as a control site for DSCS III.[76] The FLTSATCOM system, with four satellites parked in geosynchronous orbit, operates at UHF.[77] One of its six control or receiver stations is at Bagnoli, Italy; another is at Diego Garcia.[78]

The DSCS III satellites are also involved in strategic nuclear-related communications—including emergency-action messages from the President to US nuclear forces. It has European control or receiver facilities in Belgium (Kester), Diego Garcia, Turkey (Diyarbakir), the UK (Thurso, Oakhanger), Iceland (Keflavík), Cyprus (Episcopi), the Azores (Cinco Pincos, Villa Nova), and the FRG (Landstuhl, Stuttgart).[79] Kester is the master control station for NATO.[80]

Signals intelligence assets

The United States would suffer the loss of a large number of facilities utilized for signals intelligence should it withdraw from Western Europe, although to what extent such intelligence relates to operations outside Europe is not easy to ascertain. Several types of facility are included: ground stations, satellite downlinks, bases for specialized electronic-intelligence aircraft and ports used by spy ships. This is a complex and elaborate subject. As outlined by Richelson, the term Sigint (signals intelligence) subsumes a number of activities, with a

THE USA COPING WITHOUT BASES IN EUROPE   469

Table 20.2. US signals intelligence collection sites in Europe, by country, 1985

Country	Site
Azores (Portugal)	Lajes (Terceira)
Cyprus	Akrotiri, Nicosia, Karavas, Mia Milia, Yerolakkos, Pamavia, Aphendrika
Denmark	Bornholm Island
FRG	Augsburg, Hof; others too numerous to list here
UK	Cheltenham, Wincombe, Morwenstow, Menwith Hill, Chicksands, Alconbury, Kirknewton, Brawdy
Greece	Iraklion, Nea Makri, Hellenikon, Levkas Island
Iceland	Keflavík, Stokksnes
Italy	San Vito dei Normanni, Vicenza, Treviso, Naples
Spain	Rota, Torrejón, El Casar del Talamanca
Turkey	Pirinclik, Adana, Istanbul, Kunia, Sinop, Diyarbakir, Samsun, Karamursel, Antalya, Agri, Kars, Edirne, Ankara
Norway	Vardø, Vadsø, Viksjøfjellet

Source: Richelson, J. T. and Ball, D., *The Ties That Bind* (Allen & Unwin: Boston, Mass., 1985), appendix.

commonly utilized breakdown between communications intelligence (Comint) and electronic intelligence (Elint). The latter is further subdivided into primarily radar intelligence (Radint), telemetry intelligence (Telint), and foreign instrumentation signals intelligence (Fisint). Signals are intercepted all along the electromagnetic spectrum.[81]

A number of US satellite programs are devoted to Sigint—Rhyolite, Argus, Ferret and Magnum Aquacade. The Chalet and Magnum systems, according to Stares, have one of their control/receiver sites at Menwith Hill, UK (others at Ft Meade, Maryland, and Pine Gap, Australia). The impact of the loss of this site cannot be forecasted.[82]

There are of course a large number of ground-based Sigint collectors in Europe, as well as elsewhere around the globe. They are of diverse types: AN/FLR-9 HF and VHF intercept and direction-finding systems with the CDAA (Circularly Disposed Antenna Array); combined VHF–UHF–SHF telemetry interceptors; FPS-17 detection radars; FPS-79 tracking radars utilized in connection with missile launches, and so on. Identified sites in Europe for such facilities are listed in table 20.2.

Some of these have been deemed particularly important, but again, with European scenarios involving US forces in mind. Karamursel is used to cover Soviet fleet movements through the Turkish Straits.[83] Iraklion in Greece monitors the telemetry of Soviet missile launches at Kapustin Yar and Tyuratam.[84] Those at Keflavík and Rota are used for ocean surveillance;[85] and that on Denmark's Bornholm Island to monitor missile launches from Plesetsk.[86] Turkey hosts a whole array of important US Sigint facilities, some vital for verification of arms control agreements. Norway, too, hosts a number of important Sigint facilities—including some that monitor Soviet satellite communications with military installations around the Kola Peninsula—that clearly bespeak important functions related to nuclear deterrence.[87]

US spy ships utilize ports around the world for refuelling, provisioning and R&R—these presumably include European ports.[88] The high-flying U-2 and SR-71 spy planes, known mostly for photographic intelligence, also perform Sigint operations, now under the so-called Senior Ruby Elint system. The U-2s operate out of RAF Mildenhall, UK, and Akrotiri, Cyprus.[89] In addition, European bases are used to stage flights by other aircraft used for Elint operations, such as the EC-135, RC-135 and EC-121 (the latter has recently been retired).[90] The RC-135, a modified Boeing 707 with Comint and Elint capability, is the most important today. The 18 aircraft now in use have utilized bases at RAF Mildenhall, UK, and Hellenikon, Greece, in addition to others outside of Europe,[91] but the Hellenikon base is being closed down at any event.

Outside the realm of classified analysis it is not easy to say just how irreplaceable the loss of all of these facilities for the United States would be. Presumably the loss would be considerable, perhaps even massive and crucial. Satellites and ship-borne systems (themselves somewhat dependent on access, i.e., to ground stations and ports) would need to be expanded to compensate. How much could be done—and at what cost—is hard to say.

Research and environmental monitoring assets

Less visible but just as important are the numerous types of research and environmental monitoring facility that the United States operates around the world, including some in Europe and in territories controlled by European nations. Some of these are related to the US nuclear posture. A few examples will serve to illustrate the implication of the loss of such facilities.

The US Eastern Test Range for strategic missiles launched from Cape Canaveral towards the South Atlantic utilizes tracking facilities on Bermuda and Ascension.[92] Bermuda also hosts a test facility (at Tudor Hill) for long-range sonar propagation.[93] The United States is also involved in a NATO ASW research centre at La Spezia, Italy.[94]

The United States utilizes a number of facilities in Europe for practising bombing or the firing of air-to-surface and surface-to-surface missiles (in Spain, in the UK, in the Netherlands and in Crete), but these relate solely to the US mission in Europe itself.[95]

Greenland hosts a seismic research site testing the suitability of an ice cap location for underground nuclear test detection seismic arrays.[96]

Particularly relevant to the US strategic nuclear posture are weather stations (forecasting cloud cover for the operations of reconnaissance satellites, predicting low-altitude weather for low-flying bombers, etc.), a number of which are located in the FRG, in Iceland and in Spain.[97] There is a tracking and receiving station for the Defense Metereological Satellite System (DMSP) meteorological satellites at Banns, FRG, and a related control facility at Thule, Greenland.[98] In addition, there are weather intercept facilities (for monitoring other nations' weather broadcasts) in, among other places, Greece (Nea Makri), Spain (Torrejón) and Turkey (Incirlik).[99] Finally, the United States has a global network of facilities for monitoring solar flare emissions—the Solar Optical Observing Network (SOON). One of these facilities is at San Vito dei Normanni, Italy, another on Ascension Island.[100] Again, a global network might be left with gaps in coverage in the case of a withdrawal from Europe.

Country-by-country breakdown of US force assets in Europe

Cutting across the above analysis, one may also conduct a country-by-country analysis to indicate where, in relative terms, the USA would lose basing access applicable to its strategic nuclear posture. (Because the scenarios in this study involve across-the-board withdrawal from Europe, the relative importance of the individual countries matters less here.) Almost all of the various European allies provide some important points of access. Some, particularly the UK (the 'unsinkable aircraft carrier'), are more important than others. But Turkey, Italy, Iceland, Greenland, Norway, and the Azores are also vital.

The UK hosts a profuse variety of nuclear-related facilities, with a not unexpected close integration of US and British forces and bases.[101] There are F-111E bomber bases, SAC standby bases, and those for GLCMs as well as the Poseidon SSBN base at Holy Loch. There are air bases and nuclear storage sites for P-3 ASW aircraft, other bases for EF-111 'Raven' electronic warfare and command and control aircraft, and for tankers and TACAMO aircraft. There is a profusion of C³I and space-related facilities: SOSUS, Giant Talk/Scope Signal transmitters for SAC, nuclear test detection stations, Loran-C, BMEWS, stations in the Automatic Digital Network (AUTODIN), a satellite control facility and weather intercept control units. A number of important headquarters are also located in the UK, such as that of the US Naval Forces Europe/US Commander Eastern Atlantic. For the US nuclear posture access to the UK is crucial, in some ways perhaps irreplaceable.

Norway, with its critical location on NATO's Northern Flank and in near proximity to the Soviet cluster of naval and air bases in the Kola Peninsula area, hosts a variety of important nuclear-related facilities: bases for P-3 ASW aircraft, Sigint and nuclear detection facilities, VLF communications and a forward base for Airborne Warning and Control System (AWACS) aircraft.[102]

The FRG, while hosting the large US 7th Army and its associated facilities, is of course also vital to the US nuclear posture. The FRG hosts numerous nuclear-armed strike aircraft as well as numerous tactical and battlefield nuclear systems such as Lance missiles, nuclear-capable howitzers and atomic demolition mines. The Pershings and GLCMs have been removed. There are also vital Sigint and communications facilities, the early-warning satellite down-link at Kapaun, bases for AWACS and U-2s, Loran-C and a host of important headquarters and command centres. Generally speaking, however, the FRG's importance to the United States in this respect is concentrated in the areas of theatre and tactical nuclear capability, that is, with regard to the defence of the FRG and of Europe itself.

Spain hosts major air bases of importance to KC-135 tanker aircraft and for SAC bomber recovery as well as for P-3 ASW operations. Spain also provides facilities for naval HF and SAC 'fail-safe' communications, weather intercepts, nuclear bombing practice, nuclear detection and satellite communications. The F-16 aircraft now removed from Torrejón were previously considered crucial to compensate for possible loss of theatre and tactical missiles through agreements such as the INF Treaty.

Portugal's mainland provides only a part-time P-3 rotational base, but on the Azores Islands it hosts a number of crucial nuclear facilities: serving SOSUS, missile tracking, satellite control, P-3 basing and HF transmission functions. Crucial are the functions related to ASW and ocean surveillance, for obvious geographical reasons paralleled by facilities on other islands such as Diego Garcia, Ascension and Bermuda.

Italy is a crucial hub of US strategic nuclear activities. This involves F-16 rotational bases and nuclear storage, a vital attack submarine base on Sardinia, another for P-3 and other ASW aircraft, a variety of communications and surveillance functions, Loran-C, a SOON observatory, and a home port and headquarters for the 6th Fleet's carrier battle group. It is the hub of NATO's operations in southern Europe and the Mediterranean and ranks third in nuclear warheads deployed in overseas countries, after the FRG and the UK. It also deploys Lance missiles and nuclear artillery under US custodianship.

Greece, under Papandreou politically somewhat estranged from the United States, has still provided access to the United States for vital nuclear facilities, although this presence was being reduced in 1990. There are tactical fighters armed with two-key nuclear weapons, although the Nike-Hercules and Honest John missiles have been phased out. Vital are a P-3 aircraft rotational base, a NATO missile training range, HF and LF naval transmitters (some used for submarines), facilities for KC-135 tankers and strategic reconnaissance aircraft, weather intercepts, an AWACS forward base and some headquarters functions. Rival Turkey is equally if not more important, hosting five bases for nuclear-armed strike aircraft, a DSCS satellite communications down-link, Giant Talk/Scope Signal III stations, an AWACS forward base, critical Sigint stations and radars vital for missile telemetry and satellite tracking, early-warning radar and NATO Air Defense and Ground Environment (NADGE) tactical radars, nuclear detection and Loran-C facilities. Incirlik is a crucial air staging base for US Middle Eastern operations as well as a communications hub. Pirinclik hosts facilities critical for monitoring Soviet missile tests.

Numerous recent analyses in US newspapers and journals have, in the light of recent and continuing problems with Greece, Turkey, Spain and Portugal, underscored the losses that would be incurred by the United States if this position should unravel piecemeal, particularly in the Mediterranean. In these analyses, fall-back positions within Europe are usually pinpointed, that is, loss of facilities in Greece is made up for in Turkey; loss of those in Spain by augmented access to Italy, and so on. Here too, of course, the political problems addressed in this volume tend to be overshadowed by the economics of security assistance, as the United States attempts to overcome the tendency on the part of some of its allies to regard the hosting of US forces and installations more as an issue of rents and payments than of shared security perspectives. It is shown, however, that wholesale loss of access cannot so easily be compensated for by increased access to neighbours' territories. Other avenues must also be sought.

III. Conventional out-of-area problems after a US withdrawal

Rivalling in importance the impact on the strategic nuclear competition is the effect of a US withdrawal from Europe on a variety of conventional military issues. This pertains to the use of US bases in Europe for 'out-of-area' operations.[103] In geopolitical terms, this applies in particular to the so-called 'arc of crisis' stretching across northern Africa, the Horn of Africa, the South-West Asia/Persian Gulf area and South Asia.

Generally speaking, several related kinds of problem are involved here: (*a*) possible actual US military interventions (or joint NATO interventions) in this broad region, involving numerous possible scenarios; hence involving stockpiled or pre-positioned *matériel*; (*b*) contingent US access to bases in order to deter Soviet conventional military operations outside Europe, that is, in South-West Asia; (*c*) air staging and overflight rights in connection with arms resupply of US clients involved in conflict; (*d*) the movement of 'surrogate' or 'proxy' forces to Third World conflicts, and related logistics; (*e*) coercive diplomacy—the use of bases in Europe for various manifestations of gunboat diplomacy, or for subtler but similar forms of deterrence; and (*f*) 'presence'—naval and other visits, to 'show the flag' as an aspect of ongoing diplomatic competition.

The post-war era has up to now seen numerous examples of the USA's use of its European bases for one or another of these purposes. A comprehensive history is beyond the scope of this chapter; some illustrations are given, particularly as they relate to the prospect of loss of access and of future capabilities.

The USA's use in April 1986 of bases in the UK for mounting an air raid on Libya was one salient example of the use of European access for out-of-area military invention, one which saw US access to *en route* use of airspace denied by its allies France and Spain.[104] Much earlier, the 1958 US intervention in Lebanon was mounted from bases in the FRG, utilizing *en route* staging access to Turkey.[105] The US hostage raid in Iran was known to have involved access to facilities in Egypt and Oman; presumably use was made of some European facilities as well in marshalling men and equipment for the operation.[106]

Regarding arms resupply operations, the airlift to Israel in the 1973 war is a classic case, with most European nations denying the USA access under pressure from OPEC (Organization of Oil Exporting Countries) member Arab states. In this instance the USA made use of its staging access to Lajes Air Base in the Azores, is rumoured secretly to have operated tankers out of air bases in Spain during the crisis, and is reported to have moved *matériel* to Israel from US 7th Army stocks in the FRG, host government denials notwithstanding.[107] At the time it was also alleged by Egypt that US SR-71 reconnaissance aircraft based in Cyprus had provided intelligence information to Israel which the latter used to plan its crossing of the Suez Canal in the later stages of that war.[108] The USA has had some earlier diplomatic imbroglios with Spain and Portugal over the ferrying of fighter aircraft to nations in the Middle East other than Israel.[109] US aircraft based in Europe helped ferry Belgian, French and Moroccan troops (and their *matériel*) to Zaire in various crises during the late 1970s, utilizing *en route* access to airfields in Senegal, Gabon and Liberia.[110]

With regard to coercive diplomacy, one may note the use of the 6th Fleet for deterring a Syrian invasion of Jordan in 1970, another instance in which US and Israeli interests and actions dovetailed. And more recently, US AWACS aircraft, presumably staged through Europe, have been moved to Sudan and to Egypt as warnings to Libya, aimed at deterring Libyan aggression against Chad and Sudan.[111]

US fleet units based in and home-ported in Naples regularly make courtesy calls to nations such as Morocco, Tunisia, Egypt and Israel, as well as transiting the Suez Canal for port visits around the Indian Ocean. This is a traditional aspect of ongoing diplomacy.[112]

Regarding conventional deterrence *vis-à-vis* the USSR, to the extent this is separable from nuclear deterrence, the most salient example is that of US access to some half a dozen air bases in eastern Turkey, intended to help deter a Soviet invasion of Iran and a subsequent expansion into the Persian Gulf area.[113] This access to Turkey was expanded at the time the USA was developing its Rapid Deployment Force (now designated the Central Command), which involved expanded access to Oman, Egypt, Somalia, Kenya, Diego Garcia and Morocco.

It has been noted that the United States in recent years has moved towards forward shipboard pre-positioning of *matériel* (off Norway, in the Mediterranean and in the Indian Ocean) and has also formed several Marine Amphibious Brigades (MABs) to marry up with this *matériel* in the case of conflict. These deployments in Norway pertain solely to the eventuality of a central war with the USSR; those in Diego Garcia do not directly involve Europe. But the pre-positioning and forward shipboard stationing of Marines in the Mediterranean are, obviously, related to possible 'out-of-area' conflicts in North Africa or South-West Asia.[114]

Looking to the future, then, and in the context of full or partial withdrawal of US forces from Europe, it is clear that the United States (and some of its allies and/or clients) would be subjected to a considerable loss of capability in some areas, again crucially involving the Middle East. The degree of that loss would depend on the degree of access retained by the 6th Fleet—during normal times or amid crises; on the degree of access retained by the USAF particularly in Turkey (but also arguably involving tankers in Spain and fighter aircraft in Italy); and on such disparate factors as air staging rights in the Azores. It would further depend on just how a loss of access in Europe would in turn affect access to air and naval bases in Morocco, Egypt, Israel, Sudan, Somalia and Oman, among others. In some circumstances, the United States would become more dependent on the latter states.

Although the USA has continued to bargain for continued access to Lajes, it is widely assumed that in the case of another Arab–Israeli war requiring an arms resupply of Israel, such access would not be granted the next time around,[115] nor would the use of tankers operating out of Zaragoza in Spain, as is alleged to have taken place in 1973.[116] Since 1973, however, the USA has greatly increased its capacity for staging *matériel* from the USA non-stop to Ben-Gurion airport in Tel Aviv. Of course, without the use of Lajes there would be a penalty regarding trade-offs between fuel and cargo, hence increasing the cost of such an airlift. There is of course some question about whether Spain and Morocco jointly could attempt to prevent US aircraft from overflying the Straits of Gibraltar during such a crisis. As noted above, a Europe from which the USA has withdrawn its forces is less likely to support US policies, particularly as these apply to the Arab–Israeli conflict. Over it all looms the spectre of nuclear war or the use of both atomic and chemical weapons.

The capacity of the Central Command to conduct long-range logistics operations to the Persian Gulf might be affected somewhat, but only if access for tankers to Lajes and Zaragoza were denied (perhaps not likely if the operation were on behalf of Saudi Arabia and/or other Gulf OPEC states). Otherwise, *en route* aerial access to Kenitra, Morocco, and to Cairo would presumably suffice for a Persian Gulf contingency. Alternatively, Italian bases such as Sigonella, Sicily, might well be made available on an *ad hoc* or contingency basis, following a US withdrawal.

For unilateral operations in southern Africa, the United States apparently does not require access to European facilities, with the possible exception of British-owned Ascension. In operations on behalf of 'moderate' African regimes, the United States could probably bank on access to Liberia's Roberts Field and perhaps to Dakar (Senegal), Kinshasa (Zaire) or Libreville (Gabon) in certain circumstances.

Regarding coercive diplomacy or quick retaliatory operations well short of major military interventions, the United States could well suffer a considerable loss of capability if loss of access for the 6th Fleet were to cause its full withdrawal from the Mediterranean. No longer would it be so easy to conduct bombings or battleship bombardments against Lebanese factions or to conduct a repeat of the earlier raid on Libya or the coercive movements of AWACS to Egypt and Sudan. To what extent would such a withdrawal represent a loss of security for Western-aligned nations in North Africa and the Middle East? How would Libya and terrorist groups respond to such a development? Would Western Europe pick up any of the slack (or could it do so if so inclined)? If not, would there be an overall shift towards greater Soviet influence or towards more power to radical states in the region? Or would the removal of the threat of US forces on the contrary lessen tensions in some areas? The possible permutations of this particular withdrawal scenario are too complex to offer a basis for any meaningful prediction in this context. One factor that should not be overlooked in this regard, however, is the impetus a US withdrawal could give the threshold states in the region—in particular Israel—to come forward with their own means of deterrence.

IV. Horizontal escalation involving crises arising outside of Europe or the Persian Gulf

In recent years the recognition has dawned on US defence planners that in the future, dangerous crises—perhaps even those with potential to escalate to the nuclear level—might originate according to 'scripts' outside the 'standard scenarios' that have hitherto formed the basis for the endless war games in the Pentagon and elsewhere. The two 'standard scenarios' have been: (*a*) a war originating in Central Europe, whether out of an escalating crisis relating to the German question or to one of the East European states (or, alternatively, a premeditated onslaught by Moscow on Western Europe), and (*b*) a crisis over the Persian Gulf involving a Soviet thrust through Iran which would then escalate, horizontally, to Central Europe. The latter escalation could be either at Soviet or Western initiative, or might conceivably result from an all-out naval conflict which then spread to land.

There is growing recognition, however, that there are many other possible scenarios that might ultimately involve Europe or raise the issue of US use of bases in Europe.[117]

For instance, another round of the Korean War might envisage Chinese support of South Korea spreading to an all-out Sino-Soviet conflict, which may or may not be restricted to conventional weapons. The United States might end up coming to China's aid, perhaps with nuclear assistance. Such a scenario sometimes envisages a horizontal escalation to Europe, also involving the United States and the Soviet Union. Would the course of the scenario be different if the United States had previously withdrawn from Europe and lost access to its bases? Would Moscow retaliate against US support of China by attacking Europe in such circumstances? The US presence itself would be absent as a target; on the other hand, Europe might be perceived as a more vulnerable sphere of US influence. Recent events in Europe may substantially have altered these calculations, however.

Another scenario starts with a war between Pakistan and India, perhaps involving nuclear weapons, which then leads to Soviet action against Pakistan, to US support of Pakistan and onward to horizontal escalation towards Europe. Even scenarios far afield—a Grenada-type situation in Vanuatu, Fiji, Kiribati or Belau leading to a global conflict—may be considered along these lines. Again, the key question is whether Europe is less or more likely to end up as a battlefield with or without a US presence. What might be the alternative consequences for the United States?

Some global conflict scenarios have until recently involved tit-for-tat exchanges of attacks on rival superpowers' overseas bases, as in a situation in which the USSR strikes at Subic Bay and the United States retaliates against Cam Ranh Bay (or Aden, or Santiago de Cuba). Such exchanges may represent efforts on either side to shift the strategic nuclear balance by diminishing the assets of the other side, but there is always the risk that such a strike may also engender a pre-emptive response.[118] US SOSUS facilities in places such as Iceland or Norway often figure in such scenarios, along with other key C^3I facilities. Would such exchanges become less likely were the United States to withdraw from Europe, or would a withdrawal have a converse effect by making it more likely that bases in the continental United States would more rapidly become involved up the ladder of escalation? This is an obscure and highly hypothetical subject. Another conundrum: How might escalatory situations involving anti-satellite exchanges between the United States and the Soviet Union be affected by a reduced or eliminated US presence in Europe?

V. Ways in which the United States can compensate for loss of access to Europe: general approaches

A 1987 report by the Hudson Institute lists several means by which the United States might compensate for what is perceived as a gradual unravelling of its global basing system.[119] Although the relative applicability may differ, the arguments are worth re-stating also in a less-than-global context, in particular as they apply to Europe.

The several alternative strategies listed by the Hudson Institute were: (*a*) to pay higher rents, that is, offering increased security and economic assistance; (*b*) to develop technological fixes; (*c*) to change overall strategies to fit altered political circumstances, that is, reduced basing access; and (*d*) to find newer, alternative points of access, that is, new client states.

The first of these alternatives, paying higher rents, is one the United States has been using in the recent past to deal with the poorer NATO nations—Spain, Turkey, Greece and Portugal—all of which have complained bitterly in recent years about what they perceive as insufficient aid (even as the United States has been requesting increased 'reverse aid' in the form of offset payments from the FRG). Increasingly, however, economic criteria have come to be outweighed by political ones, particularly in the case of Spain. At any rate, the withdrawal scenarios given for this project exclude the option of retaining bases in Europe, whether by higher rents or any other means.

Technological fixes (including alternative military systems not necessarily involving new technologies) provide perhaps the most promising route for compensating for loss of bases. Use of satellites can be expanded in numerous areas of C^3I—nuclear detection, communications relays, Sigint, and so on.. In imitation of Soviet practice, increased use could also be made of ships for communications, satellite down-links, detection of submarines, satellite tracking, early-warning relays, and so on.

Indeed, the Hudson Institute (and also an earlier RAND study) has also looked into the possibility of building artificial islands for these purposes.[120] As yet, however, this remains pure hypothesis.

Increased use of ships in lieu of land-based facilities applies not only to C^3I functions. Press speculations about a possible US carrier strike against a Libyan chemical plant—in the face of obvious European reluctance to provide springboards for land-based bombers or permission for overflights—provides one example. Indeed, one possible result of US loss of access to European airfields and ports might be increased pressure on the United States to increase the number of its carrier groups.

Regarding changed strategies to match decreased access, it would appear this might be most applicable to power projection contingencies. Loss of access to European bases might, for instance, lead to virtual abandonment of the strategy earlier embodied by the Carter Doctrine, which committed the United States to the defence of the Persian Gulf area. It might also lead to a reduced propensity to retaliate against Libya in connection with terrorism. However, US nuclear strategy is not likely to be altered substantially by reduced access to Europe, unless the US posture were so significantly jeopardized that it would lead to a declaratory policy of LUA/LOW (launch-under-attack/launch-on-warning). That does not appear to be the case.

Finally, the United States could seek new or enhanced access elsewhere to compensate for loss of access to Europe, even though the options may be limited. Use of C^3I facilities in China, for instance, could be expanded, as they were earlier in response to US loss of access to Iran. Morocco, Tunisia and Egypt could, if so inclined or if the price were right, provide naval, air and C^3I access in lieu of some of what might be lost further north. The United States could also ask Israel for expanded access for its 6th Fleet, obviously at the price of worsening US relations with the Arab world. Some assets, of course, such as the SOSUS facilities in Iceland, the UK and Norway, could not be replaced by alternative land-based systems in other countries.

VI. Summary

It is not easy to bring this analysis down to the bottom line, to answer yes or no to the question of whether the United States can cope with losing its bases in Europe. There are too many ifs, and a wide range of uncertainty about compensatory technological and political fixes and about how all of this would work its way through the budgetary process in Washington. Furthermore, any answer must be broken down along the divide of strategic nuclear defence of the US homeland and of power projection to the Third World. Whereas the former obviously involves a crucial, irreducible national security interest, the latter—in the context of 'coping'—involves differing and often ephemeral perspectives about what is and what is not vital and worth defending. To complicate matters further, some of the answers about technological fixes to deal with loss of access can only be found in classified material, particularly as this applies to nuclear-related capabilities. In general terms, however, one can give a halting answer. The USA could cope, easily so, but not without some loss of capabilities and not without costs and the need for building new capabilities.

Regarding nuclear deterrence—perceived as maintenance of assured second-strike retaliatory capability and as maintenance of rough equivalence in targetable warheads—it is clear that all three legs of the strategic triad will easily survive loss of access to Europe. The ICBMs are all in the USA, the SSBN force—shifting towards a smaller number of Tridents—will no longer require Holy Loch, and the bombers are already all based in the USA. The newer, evolving composition of the US SSBN force will also reduce the importance of Europe-based navigation systems such as Loran-C and Omega. Future expected or possible changes in the composition of the US strategic nuclear force—more MX, Midgetman and Trident/D-5 missiles and the B-2 'stealth' bomber—will not change this picture as it applies to basing problems.[121] Whatever degradation of launcher capabilities will occur as a result of loss of access to Europe would appear to involve tanker bases and fail-safe communications facilities which, as noted above, might cause some alteration in bomber routes and some reduction in the possible avenues of attack on the USSR. That might at least raise the question of whether the USA would in turn decide to shift some capabilities away from the bomber force towards the sea- and land-based missile components of the triad. The outcome of the START negotiations may further complicate this picture. Indeed, if Washington knew it would lose access to Europe (also involving its forward-based systems), that knowledge might indeed affect its START negotiating position *vis-à-vis* Moscow.

If the impact of loss of access to Europe is minimal as concerns the launchers themselves, that of loss of C^3I assets would appear more serious. Here the discussion enters into a highly classified realm, and one in which the main questions appear to be about the implications of loss of redundancy in some areas, and how that could be compensated for by technological fixes or by putting more eggs in other baskets. Fail-safe communications have been noted. Ground-based Sigint, space-tracking and seismological functions already have redundant capabilities on satellites. Presumably these could be expanded, although it is an open question—the resolution of which would require access to classified data—whether given specific ground-based capabilities could be performed by satellite. One area which apparently would involve a major loss of capability not easily compensated for is in the ASW capabilities of SOSUS in the northern seas, in the eastern Atlantic and in the Mediterranean (loss of such capability across the GIUK gap might increase the risk that US home-based bombers not on aerial alert could be destroyed by close-in Soviet SSBNs). Compensating for that would presumably require a huge increase in ship-borne Surveillance Towed-Array Sensor Systems (SURTASS), perhaps also involving carrier-based ASW aircraft. A major degradation of ASW capabilities (also involving loss of P-3 aircraft access in several European countries) might be the single most damaging and least easily compensated blow to the US nuclear deterrent posture. Another area of concern would be early warning, involving BMEWS, the DSP down-link at Kapaun and the easternmost outposts of the North Warning System radar picket line. Here as well, however, satellites and perhaps new shipborne systems could presumably take up most if not all of the slack.

Yes, the United States could cope. To the extent that some loss of capability might result, however—even if pertaining only to redundant capabilities—several implications then bear examination. One involves the cost of compensating systems, not the least with regard to how this might relate to maintenance of capability to return to the defence of Europe (note the US insistence that NATO pay for moving the F-16s from Torrejón to Italy). Second, if the United States must move towards ever-greater reliance on satellites in lieu of land-based systems there is the question of whether it then becomes more vulnerable to Soviet ASAT capabilities at the outset of a conflict (and how that would affect US planning for such a conflict). Finally, if it is the case that US early-warning and ASW capabilities are degraded, does it nudge the United States at least towards a more unstable pre-emptive posture involving launch-on-warning or launch-under-attack? Probably not, not in the face of the current Soviet capabilities.

Regarding power projection, it is clear that the United States' ability to project power into the Middle East/South Asia area would be degraded, as would its ability to support countries such as Egypt, Israel, Jordan, Kenya, Oman, Pakistan, Somalia, and so on, by arms resupply, intervention, coercive diplomacy and other means. On the other hand, this will presumably depend upon *ad hoc*, situational European decisions about access, a situation that indeed differs little from the present, where US access has by no means been automatic, as witnessed by its difficulties at the time of the Libya raid and during the earlier 1973 war.

Quite probably loss of access, particularly in the Mediterranean, will make it less likely that the United States will intervene in the Middle East, simply because it will become more difficult. After 1973 the USA did expand its capability to airlift *matériel* to Israel without use of Lajes. Still, a complete loss of access to Europe (which also means the loss of the option of moving earmarked stocks out of Europe) will no doubt have an impact on US and Israeli postures. The possible implications include: a more likely Israeli conventional pre-emptive posture, Israeli efforts at enhanced stockpiling of *matériel*, and perhaps a greater likelihood of Israeli resort to weapons of mass destruction.[122] There may be parallel if less dramatic impacts on some of the other US clients mentioned above, although in cases in which the USA is called upon to support Jordan, Oman, Pakistan, Saudi Arabia or any other country in the region, it may well be able to count on access (ports, air staging, overflights) in 'moderate' Arab states astride the logistics lines between the USA and the Middle East, such as Morocco, Tunisia and Egypt. Generally speaking, even in the absence of immediate conflict or crisis in the Middle East, a primary question is the extent to which access for the 6th Fleet, if denied by European Mediterranean powers, would or could be compensated for by access to Morocco, Tunisia, Egypt and Israel, and whether the USA would perceive maintaining the Fleet in the Mediterranean on such a new basis worthwhile.

Another important, albeit vague, question is whether US loss of capability to intervene, or coerce or deter, in the Middle East/South Asia area will embolden some (unfriendly) nations to attack US allies, either directly or through terrorism or other forms of low-intensity warfare. That is both an important and unanswerable question as it pertains to US access. No core US interests are involved, however, save Persian Gulf oil, the latter perhaps more a vital European or Japanese interest. But yes, the United States can cope in this regard. Some of its friends in the Middle East may find it more difficult.

Finally, in closing, one might merely raise the question of whether US arms control verification capabilities might be degraded by loss of access to Europe. That would above all involve Sigint stations in relation to telemetry intercepts from missile tests and seismological verification of nuclear test bans. Similarly, it might also involve loss of access for photoreconnaissance aircraft used to monitor the Camp David agreement and, presumably, any possible follow-on. Would the Europeans themselves take on added responsibilities in these areas? If not, might it lessen US inclinations to engage in arms control, or decrease the chances for treaty ratification in a perhaps more reluctant US Senate?

Notes and references

[1] 'Shrinking power: network of U.S. bases overseas is unraveling as need for it grows', *Wall Street Journal*, 29 Dec. 1987, p. 1; Snyder, J.,'Security slips at NATO's South Flank', *Wall Street Journal*, 18 Nov. 1987, p. 33; '5 Allies want U.S. soldiers out', *Centre Daily Times*, 30 Nov. 1987, p. A7.

[2] Harkavy, R. E., *Great Power Competition for Overseas Bases* (Pergamon: New York, 1982), ch. 4.

[3] Alves, D., *Anti-Nuclear Attitudes in New Zealand and Australia*, A National Security Affairs Monograph (National Defense University Press: Washington, DC, 1985); 'New Zealand rebuff: a baffling furor', *New York Times*, 7 Feb. 1985, p. A10.

[4] See, regarding Denmark, 'Danish voting Tuesday centers on military issues', *New York Times*, 9 May 1988, p. A9; 'Slicing NATO too thin in Denmark', *New York Times*, 9 May 1988, p. A18; 'NATO: alliance a la carte?' *Newsweek*, 23 May 1988, p. 27. Regarding Belau, see 'Vote to end Pacific islands' atom-arms ban is challenged', *New York Times*, 13 Oct. 1987, p. A19; 'Pacific isle blocks nuclear accord with U.S.', *New York Times*, 18 Sep. 1986, p. A21.

[5] 'Shrinking power: network of U.S. bases overseas is unraveling as need for it grows' (note 1).

[6] Regarding US acquisition of access to China, see Richelson, J. T. and Ball, D., *The Ties That Bind* (Allen & Unwin: Boston, Mass., 1985), p. 323; Bamford, J., *The Puzzle Palace* (Houghton-Mifflin: Boston, Mass., 1982), p. 201; 'U.S., China discuss tracking Soviet tests', *International Herald Tribune*, 20 Mar. 1986, p. 1.

[7] See Harkavy (note 2), ch. 1, for a discussion of the interplay between technological development and the politics of basing access.

[8] Regarding China, see Harkavy, R. E., SIPRI, *Bases Abroad: The Global Foreign Military Presence* (Oxford University Press: Oxford, 1989), pp. 183, 198; Bamford (note 6), p. 201; Richelson and Ball (note 6), p. 323. On Turkey, see Bamford (note 6), pp. 159–60; Harkavy, pp. 182–83; 'How important are those U.S. bases in Turkey?' *Christian Science Monitor*, 11 Aug. 1975. Regarding Norway, see Richelson and Ball (note 6), p. 188; Hersh, S., *The Target is Destroyed* (Random House: New York, 1986), p. 4. Regarding Pakistan, see Khalid, Z. A., 'A new round of American installations in Pakistan', *Asian Defence Journal*, May 1982, pp. 29–34.

[9] This topic was earlier bruited in Palmer, J., *Europe Without America?: The Crisis in Atlantic Relations* (Oxford University Press: Oxford, 1987). From the perspectives of individual European nations this is also addressed in a variety of ways in Rudney, R. and Reychler, L. (eds), *European Security Beyond the Year 2000* (Praeger: New York, 1988).

[10] Harkavy (note 2), chs 1, 4 and 5.

[11] ABC 'Nightline' television programme, 11 Aug. 1988.

[12] These themes, familiar to students of international relations theory, are dealt with in, among other works, Gilpin, R., *War and Change in World Politics* (Cambridge University Press: Cambridge, 1981); Waltz, K., *Theory of International Politics* (Addison-Wesley: Reading, Mass., 1979).

[13] For a general analysis of overflights, see Dadant, P. M., 'Shrinking international airspace as a problem for future air movements—A briefing', Report R-2178-AF (RAND Corp.: Santa Monica, Calif., 1978). See also *Discriminate Deterrence*, Report of The Commission On Integrated Long-Term Strategy (US Government Printing Office, Washington, DC, Jan. 1988), esp. pp. 24–25.

[14] See Harkavy (note 8), Introduction.

[15] See Arkin, W. and Fieldhouse, R., *Nuclear Battlefields: Global Links in the Arms Race* (Ballinger: Cambridge, Mass., 1985), appendix A; Duke, S., SIPRI, *United States Military Forces and Installations* (Oxford University Press: Oxford, 1989).

[16] Campbell, D., *The Unsinkable Aircraft Carrier* (Paladin: London, 1984), esp. pp. 205–23.

[17] See Harkavy (note 8), chs. 3, 8; Arkin and Fieldhouse (note 15).

[18] Campbell (note 16), pp. 211–12.

[19] Campbell (note 16), p. 213.

[20] Campbell (note 16), p. 230.

[21] Campbell (note 16), p. 230.

[22] IISS, *The Military Balance: 1986–1987* (International Institute for Strategic Studies: London, 1986), p. 28.

[23] IISS (note 22), p. 23; Cochran, T., Arkin, W. and Hoenig, M., *U.S. Nuclear Forces and Capabilities* (Ballinger: Cambridge, Mass., 1984), pp. 207–10, 223–25.

[24] This was discussed at the conference out of which this volume grew. See also Tritten, J. T., *Soviet Naval Forces and Nuclear Warfare* (Westview: Boulder, Colo., 1986), pp. 187–88; Wegener, E., *The Soviet Naval Offensive* (Naval Institute Press: Annapolis, Md., 1975), pp. 45–48.

[25] IISS (note 22), p. 20

[26] SIPRI data; Arkin and Fieldhouse (note 15).

[27] SIPRI data; Arkin and Fieldhouse (note 15).

[28] Campbell (note 16), p. 268.

[29] SIPRI data; Arkin and Fieldhouse (note 15).

[30] Arkin and Fieldhouse (note 15).

[31] Arkin and Fieldhouse (note 15).

[32] Arkin and Fieldhouse (note 15).
[33] Arkin and Fieldhouse (note 15).
[34] Arkin and Fieldhouse (note 15).
[35] Arkin and Fieldhouse (note 15).
[36] Arkin and Fieldhouse (note 15).
[37] Arkin and Fieldhouse (note 15).
[38] Arkin and Fieldhouse (note 15).
[39] Arkin and Fieldhouse (note 15).
[40] Arkin and Fieldhouse (note 15).
[41] Arkin and Fieldhouse (note 15).
[42] Ball, D., *A Base For Debate: The U.S. Satellite Station at Nurrungar* (Allen & Unwin: Sydney, 1987), pp. 59–62.
[43] Ball (note 42), pp. 62–65.
[44] Richelson, J., 'Technical collection and arms control', ed. W. Potter, *Verification and Arms Control* (D. C. Heath: Lexington, Mass., 1985), p. 193.
[45] Ford, D., *The Button*, (Simon & Schuster: New York, 1985), p. 73.
[46] Arkin and Fieldhouse (note 15), pp. 76, 79.
[47] Arkin and Fieldhouse (note 15), pp. 76, 79. See also, 'In remote outposts, U.S. is upgrading Vigil', *New York Times*, 12 Feb. 1986, p. A27; 'Update defenses, Canada is urged', *New York Times*, 28 Jan. 1985, p. A5; 'Echo in Arctic tundra of new military goals', *New York Times*, 6 July 1987, p. 2.
[48] Richelson (note 44), pp. 190–94.
[49] Richelson (note 44), p. 192.
[50] Richelson (note 44); Harkavy (note 8), p. 186.
[51] Richelson (note 44); Harkavy (note 8), p. 186.
[52] Regarding the Soviet shipboard systems, see *Soviet Space Programs: 1976–1980*, Hearings before the Committee on Commerce, Science, and Transportation, US Senate, 97th Congress (US Government Printing Office: Washington, DC, 1983); Stuart, G. and Taylor, L., 'The Soviet naval auxiliary force', eds B. Watson and S. Watson, *The Soviet Navy* (Westview Press: Boulder, Colo.,1986), p. 109.
[53] Richelson (note 44).
[54] Richelson (note 44), p. 194.
[55] Richelson (note 44).
[56] Richelson (note 44); SIPRI data; IISS (note 22), p. 20.
[57] Stares, P., *Space and National Security* (The Brookings Institution: Washington, DC, 1987), p. 189.
[58] Classic Wizard, including the facility at Edzell, is discussed in Richelson and Ball (note 6), pp. 214–16; and in Karas, T. H., Occasional Paper No. 25, 'Implications of Space Technology for Strategic Nuclear Competition' (The Stanley Foundation: Muscatine, Iowa, July 1981).
[59] Richelson and Ball (note 6), appendix, under 'The UKUSA SIGINT Network'; Richelson (note 44), p. 188.
[60] Richelson (note 44), p. 189.
[61] Stares (note 57), pp. 68–69. Quote from Carter, A. B., 'The command and control of nuclear war', *Scientific American*, vol. 252 (Jan. 1985), p. 35.
[62] Richelson (note 44), p. 198; SIPRI data; *International Herald Tribune*, 13 Aug. 1985, p. 6.
[63] Arkin and Fieldhouse (note 15), p. 79; Richelson (note 44), p. 198.
[64] Stares (note 57), p. 188.
[65] Richelson (note 44), p. 196.
[66] Harkavy (note 8), pp. 161–62. For discussion of Loran, see also Arkin and Fieldhouse (note 15), pp. 128, 148; Blair, B., *Strategic Command and Control* (The Brookings Institution: Washington, DC, 1985), p. 170; Hersh (note 8), pp. 7–8.
[67] Harkavy (note 8), p. 161; Arkin and Fieldhouse (note 15).
[68] Harkavy (note 8), p. 161; Arkin and Fieldhouse (note 15). The role of Nea Makri is noted in 'U.S. gets ready to lose its bases in Greece', *Daily Telegraph*, 16 Feb. 1985, p. 6.
[69] Harkavy (note 8), p. 87; Arkin and Fieldhouse (note 15), p. 80 and appendix A; Blair (note 66), pp. 156–57, 169–75, 198–201.

[70] Arkin and Fieldhouse (note 15), p. 127 and appendix A.
[71] Arkin and Fieldhouse (note 15), pp. 24, 80.
[72] Harkavy (note 8), p. 163; Arkin and Fieldhouse, (note 15).
[73] Arkin and Fieldhouse (note 15), p. 80; Harkavy (note 8), p. 164; Blair (note 66), pp. 103–4.
[74] Arkin and Fieldhouse (note 15), pp. 80, 128.
[75] Arkin and Fieldhouse (note 15), pp. 77–78.
[76] Arkin and Fieldhouse (note 15); Stares (note 57), p. 188.
[77] Blair (note 66), p. 203.
[78] Stares (note 57), p. 188.
[79] 'Satellite delays may erode U.S. warning systems', *Washington Post*, 12 May 1986, p. A16.
[80] Arkin and Fieldhouse (note 15), p. 216 and appendix A.
[81] Richelson (note 44), pp. 178–79.
[82] Stares (note 57), p. 188; Richelson (note 44), p. 181; IISS (note 22), p. 21; Burrows, W., *Deep Black: Space Espionage and National Security* (Random House: New York, 1986), pp. 192, 198.
[83] Richelson (note 44), p. 185; Harkavy (note 8), p. 182; Richelson, J., *The U.S. Intelligence Community* (Ballinger: Cambridge, Mass., 1985), p. 127.
[84] Harkavy (note 8), p. 182; Richelson (note 83), p. 127. Iraklion's role of monitoring Soviet fleet movements in the Mediterranean is also noted in 'Anti-Americanism menaces Crete bases', *International Herald Tribune*, 26 Aug. 1975, and in 'U.S. forces keep low profile in Greek areas', *International Herald Tribune*, 20 Feb. 1976.
[85] Harkavy (note 8), p. 182; Richelson (note 44), p. 188.
[86] Harkavy (note 8), p. 182.
[87] Richelson and Ball (note 6), p. 188.
[88] Bamford (note 6), pp. 212–35.
[89] Richelson (note 44), pp. 182–83. A follow-on to the SR-71, with 'stealth' capabilities, is discussed in 'New uncertainty on Lockheed,' *New York Times*, 20 Jan. 1988, p. D8.
[90] Richelson (note 44), p. 183.
[91] Burrows (note 82), p. 171. See also Hersh (note 8), p. 9.
[92] Harkavy (note 8), p. 234; Arkin and Fieldhouse (note 15).
[93] Harkavy (note 8), p. 237; Arkin and Fieldhouse (note 15), p. 216.
[94] Harkavy (note 8), p. 238.
[95] Harkavy (note 8), p. 238; Arkin and Fieldhouse (note 15).
[96] Harkavy (note 8), p. 239.
[97] Harkavy (note 8), p. 243.
[98] Stares (note 57), p. 188; Arkin and Fieldhouse (note 15), p. 237.
[99] Arkin and Fieldhouse (note 15), appendix A.
[100] Arkin and Fieldhouse (note 15), p. 28; Harkavy (note 8), p. 244. See also 'Solar flare expected to disrupt systems of communication', *New York Times*, 27 June 1988, p. A10; 'Solar discharge sends storm over earth', *New York Times*, 7 May 1988, p. 36.
[101] Campbell (note 16); Arkin and Fieldhouse (note 15).
[102] Arkin and Fieldhouse (note 15), p. 227.
[103] See, for a general analysis, Kupchan, C., *The Persian Gulf and the West* (Allen & Unwin: Boston, Mass., 1987), ch. 8, under 'The out-of-area problem for NATO'.
[104] See 'U.S. says raid hurt Libya's ability to "direct and control" terror', *New York Times*, 16 Apr. 1986, p. A20.
[105] The role of the Adana base in this operation is discussed in 'Atomic unit in Germany among forces dispatched', *New York Times*, 17 July 1958, p. A1.
[106] See 'U.S. military is no secret in Egypt', *Washington Post*, 20 June 1980, p. 1.
[107] Luttwak, E. and Laqueur, W., 'Kissinger and the Yom Kippur War', *Commentary*, vol. 58, no. 3 (Sep. 1974), pp. 33–40; 'Portugal bargains for U.S. military aid with strategic mid-Atlantic base', *Christian Science Monitor*, 24 Mar. 1981; 'Spain reportedly urges U.S. to quit air base near Madrid', *International Herald Tribune*, 25 Feb. 1975; 'Secret U.S.–Spain airlift accord told', *Washington Post*, 11 Oct. 1976, p. A24. Regarding the FRG's role in 1973, see Laqueur, W., *Confrontation: The Middle East and World Politics* (Quadrangle/The New York

Times Book Co: New York, 1974), pp. 178 and 207, which states that US war *matériel* was shipped by way of Bremerhaven.

[108] El Shazly, S. (Lt. Gen.), *The Crossing of the Suez* (American Mideast Research: San Francisco, Calif., 1980), p. 252; Heikal, M., *The Road to Ramadan* (Collins: London, 1975), pp. 229–51.

[109] 'U.S. role in Zaire grows with Gabon, Senegal airlift', *Washington Post*, 7 June 1978, p. A21.

[110] See 'Use of bases may become thorny issue for U.S., Spain', *Washington Post*, 30 Jan. 1979, regarding Spain's reluctance to allow U.S. F-15s bound for Saudi Arabia to land at Torrejón. See also 'Portugal cool to idea of U.S. sending spare parts to Iran via Azores base', *Christian Science Monitor*, 4 Nov. 1980, p. 11.

[111] See '2 AWACS aircraft sent to bolster Sudan after raid', *New York Times*, Mar. 19, 1984, p. A1, which also reports on other similar uses of AWACS on behalf of Egypt and Saudi Arabia, a new form of gunboat diplomacy.

[112] Watson, B., *Red Navy at Sea: Soviet Naval Operations on the High Seas, 1956–1980* (Westview Press: Boulder, Colo., 1982).

[113] See the map in Arkin and Fieldhouse (note 15), p. 141.

[114] US Department of Defense, *Report of the Secretary of Defense Caspar W. Weinberger to the Congress for Fiscal Year 1988* (annual report) (US Government Printing Office: Washington, DC, 1988), p. 230; 'Marines prepare for duty in Asia', *New York Times*, 10 Apr. 1985, p. A17; 'Marines train for swift raids and hostage rescues', *New York Times*, 6 Aug. 1987, p. 6.

[115] 'U.S. eyes updating of airbase in Azores', *Washington Star*, 25 July 1980, p. 5.

[116] 'Spain reportedly urges U.S. to quit air base near Madrid', *International Herald Tribune*, 25 Feb. 1975.

[117] See for instance Digby, J., Millot, M. D. and Schwabe, W. L., 'How nuclear war might start—scenarios from the early 21st century', RAND note M/2614/NA (RAND Corp.: Santa Monica, Calif., Oct. 1988).

[118] Epstein, J. M., 'Horizontal escalation: sour notes of a recurrent theme', *Naval Strategy and National Security*, eds S. Miller and S. Van Evera (Princeton University Press: Princeton, N.J., 1988), pp. 102–14.

[119] See Blaker, J. R., Tsagronis, S. J. and Walter, K. T., *U.S. Global Basing: U.S. Basing Options*, Report for the U.S. Department of Defense, HI-3916-RR (Hudson Institute: Alexandria, Va., Oct. 1987).

[120] Hayes, J. H. *Alternatives to Overseas Bases* (RAND Corp.: Santa Monica, Calif., Aug. 1975).

[121] For a goal review of evolving changes in the US nuclear strategic posture, see Schneider, B. R., 'Evaluating the strategic triad', *Journal of Defense and Diplomacy*, vol. 7, no. 1 (Jan. 1989), p. 34.

[122] See, among numerous analyses of the implications of Israel's presumed nuclear weapons cache, Harkavy, R., *Spectre of a Middle Eastern Holocaust: The Strategic and Diplomatic Implications of the Israeli Nuclear Weapons Program*, Monograph Series in World Affairs (University of Denver: Denver, Colo., 1977); Feldman, S., *Israeli Nuclear Deterrence: A Strategy for the 1980s* (Columbia University Press: New York, 1982).

About the contributors

Dr Hartmut Bebermeyer (Federal Republic of Germany) is a former director at the FRG Ministry of Economics, and member of the European Study Group on Alternative Security Policy, Bonn. He is the author of *Regieren ohne Management?—Planen als Führungsinstrument moderner Regierungsarbeit* (1974) and *Das Beziehungsfeld politische Planung und strategische Unternehmensplanung* (1985) and of numerous contributions in books and periodicals on economic and defence issues.

Dr Saadet Deger (Turkey) is Leader of the SIPRI world military expenditure project. She was formerly a Research Fellow at Birkbeck College, University of London. She is a member of the UN Working Group of Experts for the UN Regional Centre for Peace and Disarmament in Africa and of the Editorial Board of *Defence Economics*. She is the author of *Military Expenditure in Third World Countries: The Economic Effects* (1986) and co-editor of *Defence, Security and Development* (1987). She is the author of chapters in the SIPRI Yearbooks 1988, 1989 and 1990, and has written extensively for international journals in her field.

Dr Simon Duke (UK) is an Assistant Professor at the Pennsylvania State University and was formerly a post-doctoral Research Fellow at the Mershon Center, Ohio State University. He was a Researcher on the SIPRI project 'Europe after an American withdrawal' in 1986–89, and is the author of the project's first volume, *United States Military Installations and Forces in Europe* (1989). His publications also include *US Defence Bases in the United Kingdom* (1987).

Dr Sami Faltas (Netherlands) is Director of the Eindhoven University Studium Generale, a centre for cultural events and courses on Science and Society. He is the author of *Arms Markets and Armament Policy: The Changing Structure of Naval Industries in Western Europe* (1986) and of articles on military research and development and on weapon sales to South Africa.

General Sir Martin Farndale, K.C.B. (UK) served for 20 years and at every command rank in the British Army of the Rhine. His last appointment was Commander-in-Chief. He is a Graduate of the Royal Military Academy at Sandhurst and the Staff College at Camberley. He was commissioned into the Royal Artillery and has served on operational tours in the Far, Near and Middle East. He has also served in the Policy and Operations Staffs at the British Ministry of Defence.

Paul J. Friedrich (Federal Republic of Germany) is a political scientist and works as an independent analyst of international political and business affairs for corporate and investment clients. He is based in Bonn.

Dr David Greenwood (UK) is Director of the Centre for Defence Studies at the University of Aberdeen, Scotland. He is the author of *Budgeting for Defence* (1972) and of numerous monographs, articles, contributions to symposia and research reports, mainly on the economic aspects of British and European defence.

Dr Robert E. Harkavy (USA) is a Professor of Political Science at Pennsylvania State University, specializing in national security policy, arms control and US foreign policy, and formerly served with the US Atomic Energy Commission and the Arms Control and Disarmament Agency. He is the author of *The Arms Trade and International Systems* (1975), *Spectre of a Middle Eastern Holocaust* (1978), *Great Power Competition for Overseas Bases* (1982) and the SIPRI volume *Bases Abroad: The Global Foreign Military Presence* (1989).

Dr Keith Hartley (UK) is Professor of Economics and Director of the Institute for Research in the Social Sciences, University of York. He has a major research interest in the economics of defence policy, is the founding editor of *Defence Economics*, Secretary-General of the International Defence Economics Association and the author of *NATO Arms Co-operation* (1983).

Joseph R. Higdon (USA) is a Senior Vice President for Capital Strategy Research in Washington, DC, and an observer and analyst of governmental and political impact in business for the past 20 years. He is currently co-ordinator of political research for a US-based global money management firm.

Nicholas Hooper (UK) is a Research Fellow at the Institute of Social and Economic Research, University of York, working on the economics of defence. He has previously worked as an industrial development analyst for Aramco in Saudi Arabia and as a statistician for the National Economic Development Office in London.

Dr Hilmar Linnenkamp (Federal Republic of Germany) is Lecturer on Conflict Research at the Armed Forces Command and General Staff College in Hamburg. He was formerly Counselor at the Planning Division of the Army Staff and, from 1979 to 1987, Head of Section of the Planning Staff of the FRG Ministry of Defence. He has published widely on issues of defence policy, defence management, systems analysis and arms control.

Alice C. Maroni (USA) is a Specialist in National Defense Issues at the Congressional Research Service in Washington, DC. She is a Graduate of Mount

Holyoke College, the Fletcher School of Law and Diplomacy and the US National War College. Her particular area of expertise is the US defence budget. She writes frequently on the congressional defence budget debate, the defence budget process, NATO burden-sharing and budget reform.

Dr Steven E. Miller (USA) leads the SIPRI project on Conventional Stability in Europe. He was previously Assistant Professor in the Department of Political Science and Research Associate in the Defense and Arms Control Studies Program at the Massachusetts Institute of Technology. He is a former Assistant Director of the Center for Science and International Affairs, John F. Kennedy School of Government, Harvard University, where he retains his position as co-editor of *International Security*. He is editor or co-editor of a number of books, including *Conventional Forces and American Defense Policy* (1986) and *Naval Strategy and National Security* (1988).

Dr José Molero (Spain) is Professor of Applied Economics at the School of Economics, University of Madrid (Complutense). His work is con-centrated on the industrial transformation of the Spanish economy within an international framework, with special attention to technological change. His publications include *Tecnología e Industrialización* (1982), *El Estado y el Cambio Tecnológico en la Industrialización Tardía* (in collaboration with J. Braña and M. Buesa) (1984) and *Estructura Industrial de España* (in collaboration with M. Buesa) (1988).

John Pike (USA) is Associate Director for Space Policy of the Federation of American Scientists, Washington, DC. He is the author of the chapter on the military use of outer space in the *SIPRI Yearbooks 1989* and *1990*.

Dr Athanassios Platias (Greece) is Assistant Professor at the Panteios University of Political and Social Sciences in Athens. He was formerly a Social Sciences Research Council/MacArthur Fellow in International Peace and Security at the Center for International Affairs, Harvard University, and at the Program in Science and Technology for International Security at the Massachusetts Institute of Technology.

Jane M. O. Sharp (UK) is Leader of the SIPRI project on Europe after an American withdrawal (1987–90). She is author of the chapters on conventional arms control in the *SIPRI Yearbooks 1988, 1989* and *1990*. She previously held research fellowships at the Center for European Studies and the Center for Science and International Affairs at Harvard University, and directed a three-year project on the Warsaw Treaty Organization at the Peace Studies Program at Cornell University (1981–84), resulting in *The Warsaw Pact: Alliance in Transition* (1984). She was the 1984–85 Peace Fellow at the Bunting Institute, and the 1985–86 George B. Kistiakowsky Fellow of the American Academy of

Arts and Sciences. Her publications focus on East–West arms control and on alliance politics in both halves of Europe.

Dr Jennifer Sims (USA) is currently US Coordinator of the multinational Nuclear History Program at the University of Maryland, a collaborative effort involving researchers in France, the Federal Republic of Germany, the United Kingdom and the United States. She has worked at the US Arms Control and Disarmament Agency (Bureau of Non-Proliferation) and at the United Nations Secretariat (Office of the Under-Secretary General for Political and General Assembly Affairs). She has been a Research Associate at the International Institute for Strategic Studies (London) and at the Instituto per gli Studi di Politica Internazionale (Milan), where she was also Deputy to the Director for Strategic Programs. She is the author of *SDI: Issues for the Alliance* (1988), *The Weapons-Stability Nexus: Origins of the Cambridge Approach to Nuclear Controls* (forthcoming) as well as articles on strategic issues and Italian defence policies for the *Revista Internazionale*.

Dr Paul B. Stares (UK) is a Research Associate in the Foreign Policy Studies programme at the Brookings Institution, Washington, DC. He is a former Research Fellow at the Centre for the Study of Arms Control and International Security at the University of Lancaster and was a Rockefeller Foundation International Relations Fellow and Guest Scholar at Brookings before assuming his present position. He is the author of numerous studies on military space issues, including *The Militarization of Space: US Policy 1945–1984* (1986), *Space and National Security* (1987) and *Command Performance: The Neglected Dimension of European Security* (forthcoming).

Captain George Thibault (US Navy, Ret., USA) is Director of the Strategy Analysis Center, Booz Allen & Hamilton, Inc., an international consulting firm based in Washington, DC. He was formerly Chairman of the Department of Military Strategy, National War College, Washington, DC, and Assistant to the Director of Central Intelligence. He has published widely on national security and intelligence matters. His most recent books include *The Art and Practice of Military Strategy* (1986) and *Dimensions of Military Strategy* (1987).

Christian Thimann (Federal Republic of Germany) is a graduate student of economics at the University of Bonn and since 1989 at the London School of Economics. He is a member of the European Study Group on Alternative Security Policy, Bonn, and the author of 'Drei militärische Modelle auf einen Blick' and 'Das Personalmodell einer alternativen Verteidigung', in *Vertrauensbildende Verteidigung* (1989).

Index

A-129 helicopter 293
AB-212 helicopter 286
Abingdon 135
ABM Treaty 11
ACE HIGH network 172, 220, 423
Aegean 377, 390
AFCENT: US forces in 81, 359, 453
Afghanistan 268
AFNORTH 303, 327, 332, 333, 334, 453
AFSATCOM 468
AFSOUTH 369, 402, 453
Agnano 278
Agusta 280, 295
Aiken, Senator George 15
aircraft carriers 321–22, 324
AirLand Battle 358, 402, 403–11, 413–15, 422
Akrotiri 407, 470
Albania 370
Albatross missile 286
Alcochete 212
Alconbury 135, 142, 409
alliances: economics of 132–33
Allied Command Europe Mobile Force (AMF) 83, 277, 316, 320, 442
Allied Contributions to the Common Defense 90, 229, 292
America see United States of America
American Express Bank 105, 106
AM-X aircraft 288, 292
Andreotti, Giulo 282
Angola 206, 210, 212
Ankara 380
Ansbach 98, 110
Aras River 382
Archangelsk 319
Arctic 314, 468
Army and Air Force Exchange Service (AAFES) 101, 102, 103, 105
Army Contracting Agency Europe 112
Army Tactical Missile System 440
ASAT (anti-satellite activities) 459, 480
Ascension Island 464, 466, 468, 470, 475
Ashdown, Paddy 18
ASW (anti-submarine warfare) 285, 308, 461, 462, 465
Athens 226, 227, 378

Atlantic Fleet (USA) 43, 323, 324, 325, 326
Augsburg 98, 408
Augusta Bay 377
Austria 283, 375, 436
Aviano 276, 277, 373, 377, 380, 461
AWACS aircraft 220, 307, 308, 401, 418, 422, 423
Azores 84, 205, 206, 207, 208, 211, 212, 215, 371, 372, 373, 389, 467

B-1 aircraft 459
B-26 aircraft 206
B-52 aircraft 158, 185, 200, 459, 460, 461, 468
Backfire aircraft 286, 383
Baker, James 47
Baker-Nunn camera 408, 464
BALTAP 303, 304, 326–32, 361
Baltic Fleet (USSR) 330–31
Baltic Sea 326, 327, 329, 330, 442
Barents Sea 162, 171, 305, 310, 412, 465
Beja 212
Belbasi 268
Belgium:
 arms trade and 67
 cruise missiles in 123
 military expenditure 292
 NATO and 120
 US forces in: costs of 8; numbers of 83, 85, 122
Benelux countries:
 arms trade and 129
 civilian employees 123, 127
 option A: US withdrawal under 122–29
 option B: US withdrawal under 122–29
 US forces in: income 126–27; military expenditure 128, 129; NATO and 120; numbers of 123, 124, 126; political setting 36, 119–22; public opinion 120, 121; withdrawal 122–30
 see also Belgium; Luxembourg; Netherlands
Bentwaters 135, 142
Beretta guns 280
Berlin, West 82, 83, 98, 99, 106, 355, 408
Bermuda 413, 458, 459, 464, 465, 467, 470
Big Bird satellites 403

Bitburg 100–1, 106
Blackjack bomber 167
Black Sea 247, 276, 377, 378
Blinder aircraft 286
BMEWS (Ballistic Missile Early Warning System) 165, 406, 407, 462, 463, 480
Bø 172
Bodø 314
Boeing 409, 422
Börfink 414, 418
Bornholm Island 469
Bosporus 378
Brattland 173, 467
Bravo, Gregorio López 186
Brawdy 135, 464, 465
Bremerhaven 111, 362
Brentano, Heinrich von 5
Brezhnev Doctrine 18
Bright Dawn 424
British Aerospace 143
British Army on the Rhine (BAOR) 11
Brize Norton 171–72
Brunswick 413
Brussels 123
Brussels Treaty 4, 13, 16, 36, 119, 120
Brzezinski, Zbigniew 8, 55
Bulgaria 229, 230, 370, 377, 380
Burt, Richard 214
Burtonwood 135, 142
Bush, President George 12, 22, 25, 42, 404

C3I:
 Italy 286
 option A: US withdrawal under 421, 422–23
 option B: US withdrawal under 421–422
 US, contribution of 400–24
 US withdrawal and 363–64, 421–24
C-130 aircraft 286
Caetano, Marcelino 206, 213
Caligaris, Luigi 284
Callaghan, James 42
Camp Darby 277, 377
Camp David agreement 481
Camp Ederle 277, 377
Canada 316, 317, 460, 461, 463
Canary Islands 210–11, 464
Carl Gustav gun 316
Carlucci, Frank 208, 211
Carter, President Jimmy 8, 11, 42, 55
CAST Brigade 442
Castiella, Fernando María 186

CATRIN system 286, 292
CENTAG 356, 444–45, 447, 448:
 US forces 81, 359, 361, 362
Central Europe: US withdrawal:
 mission gaps and 42–43, 349–66
 option A: US withdrawal under 44, 350, 446–48
 option B: US withdrawal under 44, 351, 446–47, 448
CFE (Conventional Armed Forces in Europe) Negotiation 12, 19, 25, 36, 42, 49, 94, 114, 324, 358, 431, 435, 437
CH-47 helicopter 286
Chad 284
Chalet satellites 405, 469
Chandler, Geoffrey 219
Chase National Bank 181
chemical weapons 436, 439
Cheney, Richard 9, 19
Chicksands 135, 408
China 476, 478
Chirac, Jacques 17, 213
Churchill, Winston S. 205
Cigli 380
civilian employees 83, 84, 86, 88, 89; see also under names of countries
Clear, Alaska 165, 407, 463
Coast Guard 65
Cobra Dane 464
COCOM 78, 198
Cohen, Senator William 9
collocated operating bases (COB) 38, 83, 360; see also under names of countries
Comiso 9, 41, 276, 280, 281–82, 283
command 401–2, 453–54
Common Market see EC
communications 401, 418–21, 467–68
Council of Europe 25
Crete 219, 220, 226, 378
crisis management 353, 354, 355, 362
Crotone 42, 277, 375
Croughton 135
cruise missiles 9, 10, 81, 276, 395, 471
CSCE (Conference on Security and Co-operation in Europe) 14, 19, 22–26, 366
 institutionalization of 25, 28, 48
Cyprus 39, 221–23, 232, 255, 390, 407
Czechoslovakia 22, 375

Danish Straits 303, 304, 326, 327, 331, 359
Dardanelles 378
Dardo missile 286

INDEX 493

Darmstadt 98, 100, 357
data fusion systems 413
DECA (Defence and Economic Cooperation Agreement) 224, 225, 226, 248, 255, 257, 260, 266, 271
Decimomannu 377
Defense Meteorological Satellite Program (DMSP) 405–6, 470–71
Defense Support Program 406, 462, 463
Defense Communication System (DCS) 220, 373, 375, 418, 424
defensive defence 451
Delors, Jaques 25–26
Demirel, Süleyman 255
Denmark:
  air bases in 155
  air traffic and 155, 156, 159, 167
  arms trade and 67
  base agreements 157–61
  bases, policy on 82, 158
  Collocated Operating Base programme 159, 165
  economy 155
  employment 165
  forces, size of 331
  geographic importance 327, 329, 330
  FRG and 327
  hardened air shelters in 155, 156
  NATO role 154
  nuclear weapons policy 158, 159
  US forces in: costs 87; impact of 155, 156; numbers of 83, 84, 185; withdrawal of 165–68
deterrence, gaps in 351–55
DEW (Distant Early Warning) 156, 160, 167, 311, 463
DICA (Defence Industrial Cooperation Agreement) 226
Dillon, C. Douglas 5
dollar, decline of 62, 69, 102, 103, 104, 105, 109, 110, 127
Dumas, Roland 26, 48
Duncan, Robert 410
Dunkirk, Treaty of 4, 13

E-3 aircraft 288, 307, 422
EC 13–16, 25–26, 27, 75, 76, 77, 78, 186, 188, 214, 294, 295
Ecevit, Bülent 255
Eden, Anthony 13
Edzell 135, 142, 464
Eekelen, Willem van 17

EF-111 aircraft 364
EFTA 25, 205
Egypt 473, 474, 475, 478
EH-101 helicopter 292, 293, 295
electronic warfare 364
Eppelmann, Rainer 23
Eureka project 214
Eurogroup 15, 16, 213
Europe:
  US forces in: costs of 90; numbers of 355–56, 444
  US withdrawal: costs of replacing 70; out-of-area problems and 473–76
  see also under names of countries
European Bank for Reconstruction and Development (EBRD) 25
European Coal and Steel Community (ECSC) 13–14
European Defence Community 13–14, 36
European defence force 138
European defence industry 148, 149–50
European Fighter Aircraft (EFA) 288, 293, 295
European Long Term Initiative on Defence 214
European Troop Strength (ETS) ceiling 8
Export-Import Bank 181, 182, 183, 184, 186, 187, 206, 207
Eyskens, Gaston 17

F-4 aircraft 308, 358, 364
F-5 aircraft 225, 319
F-14 aircraft 322
F-15 aircraft 307, 308, 358, 361, 459, 462
F-16 aircraft 201, 276, 277, 319, 373, 375, 385, 388, 459, 461
F-18 aircraft 189, 191, 195, 199
F-100 aircraft 253
F-104 aircraft 253, 286, 288
F-111 aircraft 357, 359, 459, 461, 471
Faeroe Islands 156, 310, 311
Fairford 9, 135, 142
Falin, Valentin 22
Falklands/Malvinas War 138, 209, 405, 436
FIAT 295
Flaming Arrow 421
Florennes 123
Flores 212
FLTSATCOM 468
force multipliers 396, 403, 452
Forces Range Accuracy Control System 220

494 INDEX

Ford, President Gerald 8, 211
FOTL (Follow-on to Lance) missile 11, 12
France:
  arms trade and 67
  military expenditure 133, 138, 388
  NATO and 438, 448
  US costs in 87
  US economic interest 74
  US forces in, number of 83, 84
Franco, General Francisco 180, 181, 187
Franconia 98
Frankfurt 98, 362
FROG missile 284
Fulda 110
Fylingdales 135, 166, 407, 408, 463

G-91 aircraft 212
G-222 aircraft 286
Gabon 474
Gaeta 277, 375
Galvin, General John 24
Gama, Jaime 209, 215
Garlstedt 108, 356
Garoufalias, Petros 233
Gardermoen 162
Gaulle, General Charles de 12
GEC 143
Genscher, Hans-Dietrich 12, 24, 26
GEODSS system 464
German Democratic Republic 22, 330, 331
Germany, Federal Republic:
  air defence 95
  arms trade and 67
  Baltic and 331
  Bundesbank 106
  civilian employees in 97, 102, 103, 104–6, 109, 112
  dollar/Mark exchange 101, 102, 115
  GNP/GDP 114
  military expenditure 133, 138, 292, 388
  option A: US withdrawal under 36, 115–16, 117
  option B: US withdrawal under 36, 116–17
  public opinion in 76–77
  R&D 173
  US forces in: as employees 104–6; assets 471; characteristics 100–1; construction work and 107, 108, 110; cost-benefit analysis 112–14; costs 86, 87, 90, 99, 107–9; economic benefits 109–12; economic impact 97–117;
  economic situation of 102–4, 109–10; goods and services 106–7, 109–12; housing in 110, 111, 113; incomes of 103, 104, 105, 106; legal basis 97; maintenance costs 108, 113; man œuvres 108, 113; numbers 83, 84, 85, 98–99, 114, 355–56; real estate and 107, 115, 116, 117; reduction debate 98; self-sufficiency 101–2; social costs 109, 113, 116; withdrawal 36, 114–17
  Wartime Host Nation Support Agreement 90–91, 107–8, 116, 117, 359
Germany, unification of 14, 18, 19, 20, 22, 23, 24, 25, 27, 28, 47, 355, 433
Gibraltar 185, 186, 211, 475
Gibraltar Straits 382, 412
Gießen 100
Giraud, André 11
Giscard d'Estaing, Valéry 14, 25, 42
GIUK gap 170, 305, 307, 311, 412, 465, 480
Gorbachev, President Mikhail 11, 18, 20, 22, 23, 36, 76, 94, 365, 434, 435
GPS satellites 466, 467
Grafenwöhr 110
Graves, Brigadier-General Howard D. 422
Graydon, Michael 327
Great Britain see United Kingdom
Greece:
  aid to 191, 219, 224, 225, 231–38
  arms trade and 232, 233
  civil war 232
  Cyprus and 221, 222, 223, 234
  domestic politics 220–21
  military expenditure 228–31, 234, 235
  military rule 221, 223, 234
  NATO and 39, 225, 229
  nuclear weapons and 220
  out-of-area operations and 389–90
  Turkey and 220, 221–24, 232, 233, 234, 390, 391
  US forces in: assets 472; costs and benefits 218–39; costs in 87, 89, 226; numbers of 84, 85; origins of 219–20; real estate and 227; security costs 220–24; treaties 219; withdrawal 218, 238–39
Greek Thrace 371, 377–78
Greenham Common 9, 135, 142
Greenland:
  base agreements 157–61
  nuclear weapons policy 158

INDEX 495

population 155
US forces in: costs of 87; impact of 155, 156; numbers of 84, 85, 168; withdrawal 166–67
Greenland Sea 311
Green Pine network 468
Grindavík 468
Gromyko, Andrei 21
Grumman Company 409, 411

Hanau 100
HAWK missile 116, 286, 287
Heidelberg 277, 356
Helguvík 168
Helios satellite 288, 293, 296, 422
Hellenikon 9, 219, 226, 227, 228, 378, 407, 470
Heraklion 226
Heseltine, Michael 213, 214
High Wycombe 135
Hispano-American Council 189
Hohenfels 10
Holst, Jørgen 311–12, 320, 322
Holy Loch 82, 135, 142, 309–10, 326, 459, 460, 471, 479
Hong Kong 446
Hungary 22, 380

Iberian Pact 210
Iceland:
 ASW and 307
 base agreements 160–61, 305
 importance of 305, 307
 population 155, 305
 radar stations in 307–8
 US forces in: costs of 87; impact of 156, 160; numbers of 84, 85, 305; withdrawal of 168–70, 431
Iceland Defence Force (IDF) 160, 161, 168, 169, 305
Icelandic Prime Contractors (IPC) 168, 169
International Monetary Fund (IMF) 211
Improved Hawk missile 320–21, 357
Incirlik 266, 267, 375, 380, 461, 471, 472
India 476
INF Treaty 9, 11, 81, 276, 282, 357, 358, 385, 439, 457, 459
intelligence 401, 402–18, 422
International Civil Aviation Organization (ICAO) 159
investment funding 63, 65, 86
Invictus Arrangement 162, 163, 172, 312

Iraklion 219, 378
Iran 466, 473, 474, 476, 478
Iskenderun 380
Israel 205, 209, 473, 389, 474, 476, 478, 480, 481
Italy:
 arms industry 293, 294, 295
 arms trade and 67, 279, 283, 295, 296
 Carabinieri 278, 287
 defence co-operation and 293, 294–96
 economy 290–93
 military expenditure 287–78, 290–93, 296
 option A: US withdrawal under 41, 42
 option B: US withdrawal under 42, 289
 R&D 293, 294, 295, 296
 rapid deployment force 286, 287
 US forces in: assets 4 72; civilian employees 279; costs and benefits 86, 87, 274–97; numbers 83, 84, 85; perimeter security 278; status of 275–76; withdrawal of 274, 283–96, 297, 375–77
Izmir 380

Jackson, Senator Henry 7
Japan 456, 457
Jean, General Carlo 284
Jessheim 171
Johnson, President Lyndon B. 7
JSTARS aircraft 401, 409, 410
Jumpseat satellite 405
Jutland 327, 329, 331, 356, 359

Kampe, Vice Admiral Helmut 330
Kapaun 406, 463, 471
Karasjok 171
KC-135R aircraft 308
Keflavík 156, 160, 161, 168, 169, 170, 174, 175, 305–9, 311, 413, 464, 469
Kennan, George 4, 205
KH satellites 403, 414
Kissinger, Henry 8, 47, 211, 457
Kohl, Chancellor Helmut 11, 14
Kola Peninsula 311, 319–20, 324, 470
Kongsberg 171
Korean War 4, 255
Krauss, Melvyn 8

Lacrosse satellite 404, 414
Lajes 205, 206, 207, 209, 371, 373, 389, 413, 474, 480
Lakenheath 135

La Maddalena 277, 377
Lance missile 392, 440, 461
LANDJUT 361
La Spezia 470
Lebanon 286, 287, 389, 473, 475
Liberia 474, 475
Libya 205, 209, 284, 392, 473, 474, 476, 478
Lisbon 215, 371
London 135, 416
Loran-C system 165, 173, 311, 467
Luxembourg:
  arms trade and 67
  US forces in: costs 87; numbers of 84, 85, 122

M-47 tank 287
Machrihanish 135, 462
McNamara, Robert 10, 15, 245
MAD 244, 245
Madeira 205, 209, 211
Madrid Pacts 180, 183, 185
Magnum satellites 405
Mainz 98
Maizière, Lothar de 23
Malta 287, 293
Mansfield, Senator Mike 7, 8
Marine Amphibious Brigade (MAB) 162, 163–64, 171, 172, 312
Marine Expeditionary Brigade (MEB) 43, 312, 317–18, 320, 321, 332, 333, 442
Marines 81, 317, 333
Marshall Plan 4, 15
Marta, Antonio 212
Martlesham Heath 135
Mattingley, Mack 214
Mediterranean:
  choke-points 386
  SLOCs 276, 285, 369, 383, 389
  US withdrawal and 382–84
Menwith Hill 135, 405, 408, 469
Mestersvig 310
MFO 293
Middle East 389, 440, 475, 480, 481
MiG-23 aircraft 284
MiG-31 aircraft 319
MILAN missile 235
Mildenhall 135, 357, 407, 420, 467, 470
Military Airlift Command (MAC) 378
Military Traffic Management Command 123
Milstar satellite 403, 414

MIRVs 460
missile launches, telemetry 162, 469, 481
Mitterrand, President François 12, 20, 25, 26, 47
Mobile Force 332, 377
Molesworth 135
Mons 356
Montreux Convention 392
Morocco 185, 474, 475, 478, 481
Morón 371, 375
Motorola 409
Mozambique 210, 212
multinational companies 74, 75, 76
Murphy, General D. J. 307, 323

NADGE (NATO Air Defence Ground Environment) 172–73, 286, 321
Naples 276, 277, 278, 375, 390, 474
National City Bank 181
NATO (North Atlantic Treaty Organization):
  air defences 320
  arms production rationalization 129
  arms trade in 66, 67, 69
  borders 246
  burden sharing in 54–56, 80, 133, 141
  command structure 401–2
  Conventional Defense Improvement Program 67
  co-operative defence programmes 56, 66–69, 71, 94–95, 129, 150–51
  Co-operative Research and Development Program 67
  Council 438, 439
  Defence Planning Committee 199
  equipment pre-positioning 59
  equipment standardization 66, 67, 69, 71
  flexible response 94, 133, 244, 245, 354, 432, 438, 450
  Follow-on Forces Attack 329, 439
  forward defence 94, 438, 450
  headquarters 123
  Independent European Programme Group (IEPG) 94, 95, 149–50, 210, 213, 214
  Infrastructure Fund 68, 116, 155, 165, 166, 168, 171, 175, 311
  Infrastructure Programme (NIP) 107–8, 116, 117, 165, 172, 226, 259, 262
  land area 246
  manpower 230, 231, 232
  military expenditure 54–56, 87, 133, 141, 228, 249, 442

INDEX 497

Montebello decision 358
North Atlantic Defence System 307
Nuclear Planning Group 10, 211, 358
nuclear weapons and 358, 393–94, 434, 437, 439, 446, 450
  option A and: US withdrawal under 47, 440, 442–43, 448–50, 454
  option B and: US withdrawal under 47, 440, 442–44, 446–48, 450, 454
  origins of 119
  out-of-area operations and 389–90
  radicalism and 119–20
  reinforcements 327, 359–60, 363
  strategy changes 450–51
  US forces, coping without 431–35
see also Central Europe; Northern Flank; Southern Flank and under names of member countries
Navstar satellites 406
Nea Makri 9, 219, 224, 226, 228, 471
Netherlands:
  air force 82
  arms trade and 67
  cruise missiles in 121, 123
  NATO and 120, 121
  US forces in: costs 87; numbers of 84, 85, 122
NATO Frigate Replacement 293, 295
NH-90 helicopters 293
Nike missile 116, 288
Nike-Hercules missile 280, 461
Nimrod aircraft 462
NORAD (North American Aerospace Defense Command) 167, 463
NORSAR (Norwegian Seismic Array) 172
NORTHAG 81, 359, 361, 362, 423, 445, 446–47, 453
North Atlantic: SLOCs 304, 324, 325
North Atlantic Council 203, 213
North Atlantic Relay System 419–20
North Atlantic Treaty 79, 157
Northern Flank:
  base agreements 156–65
  bases in 304–13
  communications 310
  intelligence and 310–12
  option A: US withdrawal under 303, 332, 441–42
  option B: US withdrawal under 43, 154, 303, 442–43
  R&D and 174
  reinforcement 154, 303, 313

US forces in: withdrawal of 37, 154–75, 303–33, 441–44
see also Denmark; Iceland; Norway
Northern Fleet (USSR) 319, 320, 325, 330
North Warning System 167, 463, 480
Norway:
  arms trade and 67
  base agreements 162–65
  basing policy 82, 162, 163, 305, 311, 312
  Collocated Operating Base agreement 162–63, 171, 172, 312, 332
  Home Guard 315–16
  Host Nation Support agreement 309
  intelligence and 311
  military expenditure 316
  NATO role 154, 162
  naval support 312–13
  northern, defending 314–18
  nuclear weapons policy 163
  pre-positioned equipment in 60
  R&D 154, 173
  reinforcement 312, 316, 333, 442
  sensitivities 154
  sigint in 470
  US forces in: assets 471; costs of 87; impact of 156; numbers of 84, 85, 173; withdrawal of 162, 170–74
Norwegian Sea 163, 303, 305, 307, 308, 310, 313, 321, 323, 333, 465
NPT (Non-Proliferation Treaty) 11, 15
nuclear explosion detection 406, 466, 470
nuclear weapons:
  campaigns against 120–21
  deterrence 354, 360, 366, 459
  see also under names of
Nunn, Senator Sam 7, 8, 9, 24, 44, 56, 293
Nuremberg 98

Oakhanger 135, 464
Offenbach 98
oil 56, 260, 290, 390, 481
Olivetti 295
Oman 473, 475, 481
Omega facility 173
OPEC 474, 475
'open skies' meeting 23, 48, 114
option A xv, 48, 58–59
  see also under NATO, Central Europe; Northern Flank; Southern Flank; NATO and under names of countries
option B xv–xvi, 4, 48, 58–59

see also under NATO, Central Europe;
Northern Flank; Southern Flank; NATO
and under names of countries
Orion aircraft *see* P-3 aircraft
Ørland 163, 165

P-3 aircraft 307, 310, 411, 412, 413, 423, 462, 465, 471, 480
Pakistan 268, 476, 477, 481
Papagos, Alexandros 219
Papandreou, Andreas 238, 472
Papandreou, George 221, 225, 233, 234
Patriot system 108, 116, 280, 357
Pave Paws radar 407, 464
peace movement 120–21
Pershing 2 missile 9–10, 357, 439
Persian Gulf 56, 218, 244, 287, 293, 389, 457, 474, 475, 476, 478, 481
Pinetree line 463
Pires de Mirana, Pedro 209
Pirinclik 268, 408, 472
Plesetsk 162
Pleven, René 13
POL 377, 378
Polaris missile 460
POMCUS 7, 359, 362, 457, 458
Porto Santo 209
Portugal:
  aid to 204, 206–8, 212–14, 215
  arms industry 211–14
  arms trade and 67, 204
  balance of payments 207
  colonies 204, 206, 210
  EC and 21, 212, 215
  economy 211
  France and 212
  Germany and 212, 213
  military expenditure 209
  NATO and 38–39, 204, 209–11
  nuclear weapons and 209
  option A: US withdrawal under 215
  option B: US withdrawal under 215
  out-of-area operations and 389
  Spain and 210–11
  UK and 212–13
  US economic interest 74
  US forces in: assets 472; costs and benefits of 87, 204–15; numbers 84, 85, 215; treaties 205; withdrawal 371–75
Portuguese-American Development Fund (FLAD) 207, 208
Portuguese Guinea 210, 212, 210, 212

port visits 384, 459
Poseidon missile 460
Powers, Gary 407
Preveza 220
PX (Post Exchange) shops 101, 111

radar, missile warning 407–8
Ramstein 102, 106, 276, 414, 417
Rapid Deployment Force 277, 474
Rapid Reinforcement Plan 159, 277, 321
Rapid Reinforcement Programme 359, 360
Rapier system 136
Ravenal, Earl 55
Raytheon 280
RB-47 171
Reagan, President Ronald 8, 11, 17, 77, 121, 208, 225, 250, 457
'Reforger' exercises 355, 359, 441
Regency Net 421
Remotely piloted vehicles 410
reserves 451–52
Reykjavík 160, 161
Rolls Royce 143
Romania 380
Roosevelt, President Franklin D. 205
Rota 200, 371, 373, 413, 416, 460, 464, 469
Roth, Senator William 8
Rupp, Rainer 213
Ruppertsweiler 418
Russia *see* Union of Soviet Socialist Republics
Russett, Bruce 228
Rühe, Volker 11
Ryolite satellites 405, 469

SA missiles 284
St Mawgan 135, 462
Salazar, António 204, 205, 206, 213
Santa Maria 205
San Vito dei Normanni 276, 464, 471
Sardinia 377
satellites 385, 403–6, 421, 422, 423, 424, 464, 466, 469, 471–72, 478, 480; *see also under names of*
Saudi Arabia 475, 481
Schlesinger, James 9
Schleswig-Holstein 329
Schmidt, Helmut 5, 14, 42, 211
Scud missile 284
Sculthorpe 135
SDI 8, 17, 389
Sea Sparrow missile 286

INDEX 499

Sea Watch programme 415
Seckenheim 317, 356
Second fleet (USA) 402
Second World War *see* World War II
Seek Igloo 463
Selenia 279, 295
Sembach 98, 357
Senegal 474
SHAPE 123, 356, 416, 417
Shetland Islands 310, 311
Shevardnadze, Eduard 20, 22, 23
Shultz, George 271
Sicily 276, 377
Sicily, Channel of 285
Siemens 419
signals intelligence 405, 408, 409, 410, 421, 423, 468–70, 481
Sigonella–Catania 277
SIK 167
Silk Purse 420
Sinop 268
Sixth Fleet (USA):
  Greece and 21
  Italy and 274, 276, 285, 288, 289, 297, 375
  size of 83
  Spain and 200
  Turkey and 247
  withdrawal of 385
Skogan, John Kristen 307
Skubiszewski, Krzysztof 25, 48
Skynet satellites 424
Smith, Colonel G. Sidney 410
Soares, Mario 208, 211
Soesterberg 126
Sola 162
Somalia 475
Søndre Strømfjord 155, 160, 166, 168, 310, 311
SOSUS 172, 307, 326, 411, 412, 416, 423, 465, 477, 478
Souda Bay 219, 226, 378
Southern Flank:
  leadership 390–91, 397
  nuclear weapons and 392–93
  out-of-area operations 389–90
  option A: US withdrawal under 449–50
  option B: US withdrawal under 450
  reinforcement 387–89
  technology and 394–95
  theatres 370, 449
  US withdrawal from 369–98, 449–50

space shuttle 404, 405
space-track system 408
Spada missile 288
SPADATS 463–64
Spain:
  aid to 180, 181–82, 183, 184, 186, 187, 189, 190, 191, 199, 200
  air force 82
  arms industry 185, 188, 193–96
  arms trade and 67, 193–96, 200
  base, economic evaluation 191–98
  defence co-operation and 189
  EC and 186, 188
  economy 180
  geographical situation 180
  investments in 183, 184, 186
  NATO and 188, 199, 200, 210, 371
  nuclear weapons and 185, 187, 189, 199, 200
  Technical Aid Programme 185
  technology transfer and 196–98, 199
  US forces in: assets 472; civilian employees 192, 193; costs to Spain 191; costs to USA 86, 87, 192; economic aspects 179–201; economic compensation 181, 187, 198, 199; numbers of 83–85, 191–93; origins of 180; reduction 179, 385; secrecy 179; treaties on 179, 180–91; withdrawal of 371–75
Spangdahlem 461
Sperry 279
Spinola, General Antonio de 204
SPOT satellites 422
SR-71 aircraft 359, 364, 407, 470, 474
Stalin, Joseph 391
START talks 459, 460, 479
Status of Forces Agreement 97, 126, 157
Sterling, Admiral 205
Stevens, Senator Ted 8
Stinger missile 287
Stuttgart 98, 356, 362, 408
Su-27 aircraft 319
Subic Bay 477
submarine depot ships 81
submarines:
  France 384
  Germany 331
  USA 309
  USSR 309, 320, 325, 389, 465
Sudan 475
Suez Canal 474
Suitland 416

500 INDEX

SURTASS 411, 412, 423, 480
Sweden 327

T-62 tank 284
T-72 tank 284
TACAMO aircraft 467
tanks: numbers of 83
technology, emerging 394, 395, 396
Terceira 205, 373
Thatcher, Margaret 11, 12, 26
Thule 160, 165, 166, 168, 174, 175, 310,
   407, 408, 463, 464, 468, 471
Thurso 135
Titan rockets 405, 406, 412
Tornado aircraft 286, 288, 423
Torrejón 188, 192, 200, 201, 277, 371, 373,
   380, 385, 388, 461, 471, 472
TR-1 aircraft 359, 364, 409
training 60, 440, 441
Trapani 286
Trident system 136, 137, 460, 479
Trøndelag 312
Truly, Richard H. 416
Truman Doctrine 4, 218, 219, 255
Tu-22 aircraft 284
Tunisia 284, 392, 474, 478, 481
Turkey:
   aid to 191, 247, 252, 255, 257, 258, 271
   arms trade and 67, 263, 264, 269
   Bulgaria and 243
   CENTO and 257
   Cyprus and 221, 222, 223, 232, 244
   debts 258, 262, 263, 271
   economy 256
   geography 243, 245, 247, 254
   Greece and 220, 221–24, 232, 233, 234,
      390, 391
   intelligence and 243, 265, 266, 268, 269
   lira depreciation 261
   Middle East and 243
   military expenditure 267, 270
   NATO and 255
   option A: US withdrawal under 265–67
   option B: US withdrawal under 266–67
   out-of-area operations 390
   population 253, 260
   security 251–54
   sigint in 469–70
   US forces in: civilian employees 267;
      costs and benefits 87, 243–71;
      numbers 84, 85, 260; withdrawal 244,
      249, 268–70, 378–82

USSR and 243, 244, 245, 246, 247, 255,
   269, 382, 387
Turkish Straits 392, 469
Turkish Thrace 371, 378–80
'two-plus-four, talks 19, 23, 42

U-2 aircraft 407, 409, 470
UNIFIL 293
Union of Soviet Socialist Republics:
   aid given by 247
   defence budget 18
   military doctrine 18, 395
   shipbuilding 327
   threat 325, 330
   troop reductions 365
   WTO, role in 360
United Kingdom:
   air force in 82
   arms trade in 67
   balance of payments 136, 143, 145, 152
   civilian employees 136
   Cyprus and 221, 222
   defence industry 142–44
   economy 138–40, 144, 145, 146, 147,
      148
   GCHQ 405
   military expenditure 133, 134, 137–38,
      139, 140–48, 388
   military manpower 134
   option A: US withdrawal under 140
   option B: US withdrawal under 37
   R&D 173
   unemployment 139, 141, 142, 143, 144,
      146, 151, 152
   US costs in 86, 87
   US forces in 94, 358–59: assets 471;
      construction and 136; costs and
      benefits of 132–52; host nation support
      136; numbers of 83, 84, 85, 134, 135;
      withdrawal 138, 140, 143, 151, 442
United Kingdom/Netherlands Landing
   Force 317, 442
United Nations 119, 206
United States of America:
   arms trade and 90
   balance of payments 63
   Buckley AFB 463
   budget deficit 53, 55, 56, 80
   burden sharing and 53, 54–55, 57, 80,
      249
   Bureau of Economic Analysis 88
   Central Europe, mission 361–63

CIA 162, 402, 405, 419
Congress: Iceland and 169–70; military expenditure and 54, 55, 56, 62, 64; property taxes and 125; troop with drawal and 62
decline of 352
defence industrial 66–69, 90
Eastern Test Range 470
Europe, bases in: economics of 79–95; employment 89, 90; income and 89, 90; scale 83–89; scope 81–83
Europe, investments in 74–78
Europe, troops in: costs of 54, 57, 63, 64, 65, 66, 70; income of 100
Europe, withdrawal from: compensation 93–95; costs and benefits 35, 53–71, 91–93; options 59
forces overseas, costs of 3456–57
Foreign Military Sales Programme 191, 231, 234–38, 257, 258, 265
Fort Hood 356
Fort Meade 469
Fort Ord 308
House Committee on Armed Services 53
Jackson–Nunn Amendment 7, 8
Mansfield amendments 7, 15, 16, 22
Military Assistance Program 129, 231, 234, 258
military expenditure 54, 55, 56–66, 69, 80, 87, 88, 146, 250, 251
Mutual Security Law 182
NATO, role in 360
Navy 61–62, 70, 325, 326
NSA 162, 172
Nunn Amendment 67
Nunn–Roth Amendment 56
option A: US withdrawal under 34–35, 59–60, 61, 457
option B: US withdrawal under 34–35, 59–60, 61, 457
Public Law 480 182, 183, 184
public opinion in 55
Quayle Amendment 67
R&D 68, 69, 173–74
trade balance 68, 69
USA deficit 53, 55, 56, 80
unemployment in 105
Vandenberg AFB 404
for foreign presence see US troops in *under names of countries concerned*
Upper Heyford 135

Vadsø 171
Værnes 163, 165, 172
Vardø 162
Vedrine, Robert 48
Vela satellites 406
Veneto–Friuli Plain 276, 377
Verona 276
Vetan 171
Vicenza 276, 277, 377
Vickers Shipbuilding 143
Viet Nam War 5, 7, 36, 120
Viksjøfjellet 162
Vredeling proposals 150, 213

Warsaw Pact *see* WTO
Washington, George 79
Watkins, Admiral James 412
Weinberger, Casper W. 60, 229
Weizsäcker, Richard von 11
Welford 135
Western European Union (WEU) 13, 16–17, 38, 199, 209, 210, 211, 212, 213, 214
Wethersfield 9, 135
White Cloud satellites 411–12
White Sea 162
Wiesbaden 98, 100
Wildflecken 110
Woensdrecht 123, 126
World War I 245, 255
World War II 4, 13, 18, 21, 22, 160, 255, 305
Worldwide Command and Control System (WWMCCS) 422
Wörner, Dr Manfred 213
WTO (Warsaw Treaty Organization): changes in 114, 366, 388
force reductions 19, 2594
interior lines 434
threat 18–20, 24–25, 433–36
Würzburg 98, 110

Yom Kippur War (1973) 209, 252
Younger, George 214
Ytri-Njardvík 168
Yugoslavia 283, 370, 375, 391, 436
Yumurtalik 380

Zaire 209
Zaragoza 371, 375, 377, 475
Zircon satellite 423
Zweibrücken 9, 114

UA
26
.E9
E93
1990

76.00

UA
26
.E9
E93

1990